Affine Algebraic Geometry

The Russell Festschrift

Volume 54

CRM
PROCEEDINGS &
LECTURE NOTES

Centre de Recherches Mathématiques
Montréal

Affine Algebraic Geometry

The Russell Festschrift

Daniel Daigle
Richard Ganong
Mariusz Koras
Editors

The Centre de Recherches Mathématiques (CRM) of the Université de Montréal was created in 1968 to promote research in pure and applied mathematics and related disciplines. Among its activities are special theme years, summer schools, workshops, postdoctoral programs, and publishing. The CRM is supported by the Université de Montréal, the Province of Québec (FQRNT), and the Natural Sciences and Engineering Research Council of Canada. It is affiliated with the Institut des Sciences Mathématiques (ISM) of Montréal. The CRM may be reached on the Web at www.crm.math.ca.

American Mathematical Society
Providence, Rhode Island USA

The production of this volume was supported in part by the Fonds Québécois de la Recherche sur la Nature et les Technologies (FQRNT) and the Natural Sciences and Engineering Research Council of Canada (NSERC).

2010 *Mathematics Subject Classification.* Primary 14Rxx, 13–XX.

Library of Congress Cataloging-in-Publication Data

Affine algebraic geometry : the Russell festschrift / Daniel Daigle, Richard Ganong, Mariusz Koras, editors.

p. cm. — (CRM proceedings & lecture notes ; 54)

Includes bibliographical references.

ISBN 978-0-8218-7283-3 (alk. paper)

1. Geometry, Affine. I. Russell, P. (Peter), 1939– II. Daigle, Daniel, 1959– III. Ganong, Richard, 1947– IV. Koras, Mariusz, 1951–

QA477.A34 2011
516′.4—dc23

2011030116

∞ The paper used in this book is acid-free and falls within the guidelines
established to ensure permanence and durability.
This volume was submitted to the American Mathematical Society
in camera-ready form by the Centre de Recherches Mathématiques.
Visit the AMS home page at http://www.ams.org/

10 9 8 7 6 5 4 3 2 1   16 15 14 13 12 11

Photo: Karol Palka

Peter Russell, Spring 2011
(Then Director of the CRM)

Contents

Preface ix

Foreword xi

Biography of Peter Russell (through June, 2011) xiii

Bibliography of Peter Russell xv

Newton Trees at Infinity of Algebraic Curves
Pierrette Cassou-Noguès 1

Polynomials $f(X, Y, Z)$ of low LND-degree
Daniel Daigle 21

Corrigendum to Our Paper "Birational Transformations of Weighted Graphs"
Hubert Flenner, Shulim Kaliman, and Mikhail Zaidenberg 35

Bivariate Analogues of Chebyshev Polynomials with Application to Embeddings of Affine Spaces
Gene Freudenburg 39

The Pencil of Translates of a Line in the Plane
Richard Ganong 57

A Geometric Proof of Boutot's Result on Singularities of Quotients
R. V. Gurjar 73

On the Present State of the Andersén – Lempert Theory
Shulim Kaliman and Frank Kutzschebauch 85

Group Actions on Affine Cones
Takashi Kishimoto, Yuri Prokhorov, and Mikhail Zaidenberg 123

$\mathbb{C}^*$ in $\mathbb{C}^2$ is Birationally Equivalent to a Line
Mariusz Koras 165

Initial Algebras and the Jung – van der Kulk Theorem
Shigeru Kuroda 193

Holomorphic Curves on Irregular Varieties of General Type Starting from Surfaces
Steven S.-Y. Lu 205

Locally Nilpotent Derivations of Affine Domains
L. Makar-Limanov 221

Equivariant Derivations and Additive Group Actions
Kayo Masuda 231

Frobenius Sandwiches of Affine Algebraic Surfaces
M. Miyanishi 243

Automorphism Groups of Koras – Russell Threefolds of the First Kind
L. Moser-Jauslin 261

Recent Progress in the Geometry of $\mathbb{Q}$-Acyclic Surfaces
Karol Palka 271

On the Makar-Limanov, Derksen Invariants, and Finite Automorphism Groups of Algebraic Varieties
Vladimir L. Popov 289

Embeddings of Hyperbolas
Avinash Sathaye 313

Groups of Russell Type and Tango Structures
Yoshifumi Takeda 327

Preface

This volume grew out of an international conference which was held the week of 1 – 5 June 2009 at McGill University, in honour of Professor Peter Russell, on the occasion of his 70th birthday and his retirement from McGill. Some 40 mathematicians from ten countries were invited; 25 talks were given. The present volume contains 19 refereed articles, essentially all in the area of affine algebraic geometry, almost all contributed by participants in the conference. The papers largely reflect the themes running through the very substantial body of work done by Peter in this relatively young branch of algebraic geometry, in a career spanning over four decades.

Peter is described in one of the articles here as someone "who asks the right questions". He has certainly been responsible in part for substantial progress over the last 40 years in affine geometry, by both asking many right questions and helping to answer them. Peter's generosity, with both financial support and support in the form of encouragement, ideas, suggestions and inspiring enthusiasm and energy, is well known to researchers in the field whose mathematical paths have intersected his. The contributors to this volume include former students of Peter's, postdoctoral fellows and academic guests of his at McGill, and many of his co-authors, all of whom can testify to his liberality.

The number of articles in his bibliography of which he is a joint author bespeaks Peter's view of mathematics as a cooperative venture. Several of the contributors to this volume plainly acknowledge their intellectual indebtedness to him.

One of the leaders in affine algebraic geometry for several decades, Peter, never retiring and, it seems, not quite retired, remains mathematically vigorous, having very recently completed his term as Director of Montréal's Centre de recherches mathématiques. All who have known him professionally or personally surely join us in extending to Peter both our gratitude and our appreciation for all that he has given us, and our hope that much more is to come.

Daniel Daigle
Richard Ganong
Mariusz Koras
July 2011

Foreword

The editors have kindly invited me to add a few personal reminiscences to these proceedings. I apologize in advance for being somewhat idiosyncratic and selective.

My acquaintance with affine algebraic geometry, loosely speaking the study of affine spaces and algebraic varieties closely resembling them, dates to 1963. My teacher Maxwell Rosenlicht mentioned the Jung – van der Kulk theorem on automorphisms of polynomials in two variables in his course on algebraic geometry. Later I learned about Nagata's famous automorphism in dimension 3, which was a pointer to the enormous difficulties in understanding automorphisms of affine space in higher dimensions. I of course also heard about the Jacobian problem.

It took me time to absorb all this, but my fascination with the subject was born. There is an intriguing interplay of algebra and geometry. Basic questions are easy to ask, the right answers often are much less obvious, and once formulated they may be very difficult to prove. Over the years I was privileged to witness a tremendous development and maturing of this subject and from time to time to make a contribution myself. As highlights in the early 1970s let me mention the epimorphism theorem of Abhyankar, Moh and Suzuki on lines in the plane, and the proof of the Serre conjecture on projective modules over polynomial rings.

For me a very momentous event was the introduction in the mid1970s, by Iitaka, of logarithmic genera, in particular the logarithmic Kodaira dimension, into the study of non-compact varieties, so in particular of affine varieties. It put affine algebraic geometry firmly into the classical context of the geometry of complete algebraic varieties. Right away, for instance, the Kodaira dimension of Ramanujam's fascinating contractible surface was found to be two, and the question whether exotic affine spaces exist could be settled: in contrast to the situation in dimension two, affine three-space as an algebraic variety is not determined by its topology. Another great success came shortly after when Fujita, Miyanishi and Sugie used Kodaira dimension to establish cancellation for the affine plane. With contributions from many authors the logarithmic methods led to a beautiful flowering of a theory of open surfaces that parallels the results and the richness of the classical theory. I was very happy to be involved, in the 1990s, in settling the linearizability of algebraic actions of the multiplicative group on affine three-space. This is an unexpectedly profound result that technically drew on all the developments I have mentioned above. It also needed a brilliant new idea that was invented for the purpose — the introduction of the Makar-Limanov invariant and its use in the study of exotic affine spaces.

Most of my remarks above tacitly refer to developments in characteristic zero, but problems in positive characteristic geometry always have been close to my heart — my thesis was on a monkey field topic. In affine algebraic geometry positive characteristic poses very special challenges. In dimension two, for instance, the automorphism theorem holds in all characteristics; the epimorphism theorem is not true in positive characteristic. Even what should replace it, what the right theorem is about lines in the plane, is an entirely open question. In fact, if one leaves aside the Jacobian problem, then in characteristic zero the geometry of the affine plane can be considered as settled, and the unknown territory begins in dimensions three and four. In positive characteristic very fundamental questions remain in dimension two. This is not for a lack of trying; on many of them a lot of hard work already has been done that was in the end abandoned due to their intractability. It deserves to be taken up again.

My professional journey of forty-plus years was undertaken in the company of numerous students and collaborators, and of course many colleagues who were important for me as partners in discussion, teachers and collaborators in spirit. All of them became lifelong friends. It was a special gift for me that so many of them were able to come to the conference and to contribute to these proceedings. Listening to their talks I could reflect with satisfaction on the great progress our subject had made. With equal satisfaction, though, I could tell myself that the basic task of affine algebraic geometry, that is, to understand affine spaces and polynomial rings, is far from being completed. It still stands as a beautiful and very fundamental mathematical challenge for the next generation.

Peter Russell

Biography of Peter Russell (through June 2011)

Personal

Born 7 September 1939 in Essen, Germany

Education

Vordiplom, Universität Hamburg, 1962

Ph.D., University of California, Berkeley, 1966
Thesis: *The category of commutative algebraic groups*
Supervisor: M. Rosenlicht

Positions Held

University of California, Berkeley
Teaching Fellow, 1962 – 65
Research Fellow, 1965 – 66

Harvard University
Benjamin Peirce Instructor, 1966 – 69

McGill University
Assistant Professor, 1969 – 76
Associate Professor, 1976 – 82
Professor, 1982 – 2009
Professor Emeritus, 2009

Purdue University
Visiting Assistant Professor, 1972 – 73

Bhaskaracharya Pratishthana, Pune, India
Visiting Associate Professor, 1976 – 77

McGill University
Chair, Dept. of Mathematics and Statistics, 1988 – 94

Canadian Mathematical Society
Vice President, Quebec Region, 1991 – 93

Institut des sciences mathématiques, Montréal
Director, 1995 – 96 and 2000 – 04

Centre de recherches mathématiques, Montréal
Director, 2009 – 2011

Other

Prize for Mathematical Research of the Ministry of Education, Poland for the Year 2000 (with M. Koras)

Theses Directed

P. Roberts, Ph.D., 1974
Complexes over local rings

R. Ganong, Ph.D., 1978
On plane curves with one place at infinity

L. Paradis, Ph.D., 1981
Affine contractibility of plane curves
(Supervised jointly with A. Lascu, Université de Montréal)

L. Sangines, Ph.D., 1983
On quadratic planes

J. Malard, M.Sc., 1983
The classification of ruled surfaces and rank 2 *vector bundles over a curve of genus* 1

D. Daigle, Ph.D., 1987
Birational endomorphisms of the affine plane

J. Bell, M.Sc., 1999
The equivariant Grothendieck ring of certain contractible threefolds

T. W. Dong, Ph.D., 1999
The spectral theory of ordinary derivations of $K[X,Y]$ *and weaker forms of the Jacobian conjecture*

Bibliography of Peter Russell

1. *Forms of the affine line and its additive group*, Pacific J. Math. **32** (1970), 527–539.
2. *Field generators in two variables*, J. Math. Kyoto Univ. **15** (1975), no. 3, 555–571.
3. *Continuous derivations of the ring of Witt-vectors*, J. Pure Appl. Algebra **6** (1975), no. 3, 259–263.
4. (with J. Labute) *On K_2 of truncated polynomial rings*, J. Pure Appl. Algebra **6** (1975), no. 3, 239–251.
5. *On Hironaka's additive groups associated with points in projective space*, Math. Ann. **224** (1976), no. 2, 97–109.
6. *Simple birational extensions of two dimensional affine rational domains*, Compositio Math. **33** (1976), no. 2, 197–208.
7. *Good and bad field generators*, J. Math. Kyoto Univ. **17** (1977), no. 2, 319–331.
8. *Simple Galois extensions of two-dimensional affine rational domains*, Compositio Math. **38** (1979), no. 3, 253–276.
9. (with A. Sathaye) *On finding and cancelling variables in $k[X, Y, Z]$*, J. Algebra **57** (1979), no. 1, 151–166.
10. *Hamburger–Noether expansions and approximate roots of polynomials*, Manuscripta Math. **31** (1980), no. 1-3, 25–95.
11. *On affine-ruled rational surfaces*, Math. Ann. **255** (1981), no. 3, 287–302.
12. (with T. Kambayashi) *On linearizing algebraic torus actions*, J. Pure Appl. Algebra **23** (1982), no. 3, 243–250.
13. (with R. Ganong) *Derivations with only divisorial singularities on rational and ruled surfaces*, J. Pure Appl. Algebra **26** (1982), no. 2, 165–182.
14. (with M. Miyanishi) *Purely inseparable coverings of exponent one of the affine plane*, J. Pure Appl. Algebra **28** (1983), no. 3, 279–317.
15. *Factoring the Frobenius morphism of an algebraic surface*, Algebraic Geometry, Bucharest 1982 (Bucharest, 1982) (L. Bădescu and D. Popescu, eds.), Lecture Notes in Math., vol. 1056, Springer, Berlin, 1984, 366–380.
16. (with R. Ganong) *The tangent bundle of a ruled surface*, Math. Ann. **271** (1985), no. 4, 527–548.

17. (with M. Koras) *$\mathbb{G}_m$-actions on $\mathbb{A}^3$*, Proceedings of the 1984 Vancouver Conference in Algebraic Geometry (Vancouver, 1984) (J. Carrell, A. V. Geramita, and P. Russell, eds.), CMS Conf. Proc., vol. 6, Amer. Math. Soc., Providence, RI, 1986, 269–276.

18. (with M. Koras) *On linearizing "good" $\mathbb{C}^*$-actions on $\mathbb{C}^3$*, Group Actions and Invariant Theory (Montréal, QC, 1988) (A. Białynicki-Birula, J. Carrell, P. Russell, and D. Snow, eds.), CMS Conf. Proc., vol. 10, Amer. Math. Soc., Providence, RI, 1989, 93–102.

19. (with M. Koras) *Codimension 2 torus actions on affine n-space*, Group Actions and Invariant Theory (Montréal, QC, 1988) (A. Białynicki-Birula, J. Carrell, P. Russell, and D. Snow, eds.), CMS Conf. Proc., vol. 10, Amer. Math. Soc., Providence, RI, 1989, 103–110.

20. *Gradings of polynomial rings*, Algebraic Geometry and Its Applications (West Lafayette, IN, 1990) (C. L. Bajaj, ed.), Springer, New York, 1994, 365–373.

21. (with S. M. Bhatwadekar) *A note on geometric factoriality*, Canad. Math. Bull. **38** (1995), no. 4, 390–395.

22. *Sufficiently homogeneous closed embeddings of $\mathbb{A}^{n-1}$ into $\mathbb{A}^n$ are linear*, Canad. J. Math. **48** (1996), no. 6, 1286–1295.

23. (with M. Koras) *Contractible threefolds and $\mathbb{C}^*$-actions on $\mathbb{C}^3$*, J. Algebraic Geom. **6** (1997), no. 4, 671–695.

24. (with Sh. Kaliman, M. Koras, and L. Makar-Limanov) *$\mathbb{C}^*$-actions on $\mathbb{C}^3$ are linearizable*, Electron. Res. Announc. Amer. Math. Soc. **3** (1997), 63–71.

25. (with M. Koras) *$\mathbb{C}^*$-actions on $\mathbb{C}^3$: the smooth locus of the quotient is not of hyperbolic type*, J. Algebraic Geom. **8** (1999), no. 4, 603–694.

26. (with D. Daigle) *Affine rulings of normal rational surfaces*, Osaka J. Math. **38** (2001), no. 1, 37–100.

27. (with D. Daigle) *On weighted projective planes and their affine rulings*, Osaka J. Math. **38** (2001), no. 1, 101–150.

28. *Normal crossing systems of plane curves with one place at infinity*, Sitzungsberichte der Berliner Mathematischen Gesellschaft, Berliner Math. Gesellschaft, Berlin, 2001, 163–176.

29. *Some formal aspects of the theorems of Mumford–Ramanujam*, Algebra, Arithmetic and Geometry, Parts I, II (Mumbai, 2000) (R. Parimala, ed.), Tata Inst. Fund. Res. Stud. Math., vol. 16, Narosa Publ. House, New Delhi, 2002, 557–584.

30. *Multiple planes ramified over one-place curves*, Algebra, Arithmetic and Geometry with Applications (West Lafayette, IN, 2000) (C. Christensen, G. Sundaram, A. Sathaye, and C. Bajaj, eds.), Springer, Berlin, 2004, 673–685.

31. (with D. Daigle) *On log $\mathbb{Q}$-homology planes and weighted projective planes*, Canad. J. Math. **56** (2004), no. 6, 1145–1189.

32. (with M. Koras) *Linearization problems*, Algebraic Group Actions and Quotients (Wykno, 2000) (J. A. Wiśniewski, ed.), Hindawi Publ. Corp., Cairo, 2004, 91–107.

33. (with M. Koras) *Contractible affine surfaces with quotient singularities*, Transform. Groups **12** (2007), no. 2, 293–340.

34. *Embedding problems in affine algebraic geometry*, Polynomial Automorphisms and Related Topics (Hanoi, 2006) (H. Bass, N. V. Chau, and S. Maubach, eds.), Publishing House for Science and Technology, Hanoi, 2007, 113–135.

35. (with P. Cassou-Noguès) *Birational morphisms $\mathbb{C}^2 \to \mathbb{C}^2$ and affine ruled surfaces*, Affine Algebraic Geometry (T. Hibi, ed.), Osaka Univ. Press, Osaka, 2007, 57–105.

36. (with R. V. Gurjar, K. Masuda, and M. Miyanishi) *Affine lines on affine surfaces and the Makar-Limanov invariant*, Canad. J. Math. **60** (2008), no. 1, 109–139.

37. (with R. V. Gurjar and M. Koras) *Two dimensional quotients of $\mathbb{C}^n$ by a reductive group*, Electron. Res. Announc. Math. Sci. **15** (2008), 62–64.

38. (with P. Cassou-Noguès and M. Koras) *Closed embeddings of $\mathbb{C}^*$ in $\mathbb{C}^2$*. I, J. Algebra **322** (2009), no. 9, 2950–3002.

39. (with R. V. Gurjar, M. Koras, and M. Miyanishi) *Affine normal surfaces with simply connected smooth locus*, Math. Ann., to appear.

40. (with R. V. Gurjar, M. Koras, and M. Miyanishi) *A homology plane of general type can have at most a cyclic quotient singularity*, J. Algebraic Geom., to appear; available at `arXiv:1012.4120`.

Other publications

41. S. S. Abhyankar, *Algebraic space curves*, Sem. Math. Sup., vol. 43, Presses Univ. Montréal, Montréal, QC, 1971. Notes written by P. Russell and A. Sathaye.

42. *Open problems on open varieties* (P. Russell, ed.), Technical Report 311, Institut Fourier, Grenoble, 1995; available at `arXiv:alg-geom/9506006`.

43. *Open algebraic surfaces*, CMS Notes **33** (2001), no. 8, 18–20. Book review.

44. (with G. Freudenburg) *Open problems in affine algebraic geometry*, Affine Algebraic Geometry (J. Gutierrez, V. Shpilrain, and J.-T. Yu, eds.), Contemp. Math., vol. 369, Amer. Math. Soc., Providence, RI, 2005, 1–30.

45. *Affine algebraic geometry*, Oberwolfach Rep. **4** (2007), no. 1, 5–82. Abstracts from the workshop held January 7–13, 2007, organized by H. Flenner, P. Russell, and M. Zaidenberg.

Centre de Recherches Mathématiques
CRM Proceedings and Lecture Notes
Volume **54**, 2011

Newton Trees at Infinity of Algebraic Curves

Pierrette Cassou-Noguès

Dédié à Peter Russell, avec mes amitiés

ABSTRACT. In this paper, we give the construction of a Newton tree at infinity of an algebraic curve. A byproduct of this construction is the explicit computation of rational maps (not morphisms) that convert a polynomial into an other polynomial. This is very useful to construct new polynomials such that bad field generators, polynomials with no singularities for example, and has been already used in [2–4, 6]. We end up the article showing that if two polynomials satisfy $J(f,g) = 1$ then there is no face parallel to the diagonal on their Newton polygon.

1. Introduction

In this paper, we give a complete description of the construction of Newton trees at infinity for algebraic curves. This construction is based on Newton algorithm and uses all Newton polygons obtained when performing the algorithm. The idea is the following. Let $f \in \mathbb{C}[x,y]$. Consider the Newton polygon at infinity of f, and choose a face S of this Newton polygon. Assume $p\alpha + q\beta = N$ is the equation of the line supporting S, with $N \geq 0$. To this face corresponds a face polynomial F_S, which factorizes in polynomials of degree one in $\mathbb{C}[x,y]$. We choose a root μ of this polynomial and consider a Newton map $\sigma_{(p,q,\mu)}$ such that $\sigma_{(p,q,\mu)}(f)(x_1,y_1) = f(\mu^{q'} x_1^{-p}, x_1^{-q}(y_1 + \mu^{p'})) = f_1(x_1,y_1)$, where (p',q') satisfies $qq' - pp' = 1$. We have $f_1(x_1,y_1) = x_1^{-N} g_1(x_1,y_1)$ with $g_1(x_1,y_1) \in \mathbb{C}[x_1,y_1]$. So the result is not a polynomial but a rational function with denominator a power of x_1. Then we will consider the Newton algorithm in the ring of rational functions with a denominator a power of x_1. The first part of the paper gives this algorithm and the construction of Newton trees in this case. It is the most important part since the Newton trees for algebraic curves, follow from this construction. The Newton algorithm for the ring of rational functions with a denominator a power of x, generalizes the Newton algorithm in $\mathbb{C}[\![x,y]\!]$, done in [5]. However we give a complete description.

2010 *Mathematics Subject Classification.* 14H20, 14H50, 14R15.
Key words and phrases. Newton trees at infinity, Jacobian conjecture.
The author is partially supported by MTM2010-21740 C02-01 and MTM2010-21740 C02-02.
This is the final form of the paper.

We consider the local Newton polygon. To each face of the Newton polygon and each root of the face polynomial, we consider a Newton map. We consider the Newton polygon of the result and if it is not empty we do it again. We stop when the Newton polygon is empty. To each Newton polygon we associate a decorated graph which is equivalent (in the sense that it is constructed from the Newton polygon, but also allows to construct the Newton polygon) and the Newton tree is a decorated tree obtained by gluing the graphs associated to all the Newton polygons. The Newton tree of an algebraic curve $f(x, y) = t$ is obtained by gluing these Newton trees to the graph of the Newton polygon at infinity of $f(x, y) = t$. To each vertex of the Newton tree, we associate a rational parametrization that we call a curvette. There are particular vertices that we call dicritical vertices and which play a particular role. In particular the parameterizations of the curvettes at the dicritical vertices give rational maps, which are not morphisms, that convert the polynomial to another polynomial. These properties are used already in [2–4]. At the end of the article we apply this to show that if two polynomials f and g satisfy $J(f, g) = 1$ where $J(f, g)$ is the Jacobian of f and g, then there is no face parallel to the diagonal on their Newton polygon at infinity.

There are topics which are not discussed in this paper, such that the computation of the intersection number of two algebraic curves from their Newton tree, the computation of the Milnor number, as well as the comparison of the Newton tree with splice diagrams at infinity [7, 8] or with the graph dual of the resolution of the polynomial map.

One other topic which will be discussed in subsequent papers this the relation between the dicritical vertices and some generalized Danielewsky surfaces on which the polynomial map extends.

2. Newton trees

In this section we construct Newton trees at infinity for polynomials.

Let $K = \mathbb{C}[\![x, y]\!]$ and K_x denote the subring of the quotient field of K of rational functions whose denominator is a power of x. To construct Newton trees at infinity for polynomials we first need to construct Newton trees for elements in K_x. As Newton trees in K are already constructed in [5], we consider elements in K_x, which are not in K.

2.1. Newton trees for K_x.

2.1.1. *Newton polygons.* Let $n \in \mathbb{N}$, for any set $E \subset \mathbb{Z}_{\geq -n} \times \mathbb{N}$, denote by $\Delta(E)$ the smallest convex set containing $E + \mathbb{R}^2_+ = \{a + b, a \in E, b \in \mathbb{R}^2_+\}$. A subset $\Delta \subset \mathbb{R}^2$ is a *Newton diagram* if there exists an integer n and a set $E \subset \mathbb{Z}_{\geq -n} \times \mathbb{N}$ such that $\Delta = \Delta(E)$. The smallest set $E_0 \subset \mathbb{Z} \times \mathbb{N}$ such that $\Delta = \Delta(E_0)$ is called the *set of vertices* of Δ. The set of vertices of a Newton diagram is finite. Let $E_0 = \{v_0, \ldots, v_m\}$ with $v_i = (\alpha_i, \beta_i) \in \mathbb{Z} \times \mathbb{N}$ and $\alpha_{i-1} < \alpha_i, \beta_{i-1} > \beta_i$ for $i = 1, \ldots, m$. For $i = 1, \ldots, m$ we denote by $S_i = [v_{i-1}, v_i]$ and by l_{S_i} the line supporting the segment S_i. We call $\mathcal{N}(\Delta) = \{S_i, i = 1 \ldots, m\}$ the *Newton polygon* of Δ. The Newton polygon $\mathcal{N}(\Delta)$ is empty if and only if $\Delta = (\alpha_0, \beta_0) + \mathbb{R}^2_+$. The integer $h(\Delta) = \beta_0 - \beta_m$ is called the *height* of Δ.

Let

$$f(x, y) = \sum_{(\alpha,\beta)\in\mathbb{Z}\times\mathbb{N}} c_{\alpha,\beta} x^\alpha y^\beta \in K_x.$$

We define

$$\operatorname{Supp} f = \{(\alpha,\beta) \in \mathbb{Z}\times\mathbb{N} \mid c_{\alpha,\beta} \neq 0\}.$$

Let $\Delta(f) = \Delta(\operatorname{Supp} f)$ and $\mathcal{N}(f) = \mathcal{N}(\Delta(f))$. For all $S \in \mathcal{N}(f)$ let $\operatorname{in}(f,S) = \sum_{(\alpha,\beta)\in S} c_{\alpha,\beta}x^\alpha y^\beta$. If l_S has equation $p\alpha + q\beta = N$, with $\gcd(p,q) = 1$ then $\operatorname{in}(f,S) = x^{a_S}y^{b_S}F_S(x^q,y^p)$ where $F_S(x,y) = \prod_i (x-\mu_i y)^{\nu_i}$ and $(a_S,b_S) \in \mathbb{Z}\times\mathbb{N}$.

2.1.2. *Newton algorithm.* In the following definition we denote by K_{1,x_1} the ring $\mathbb{C}[\![x_1,y_1]\!]_{(x_1)}$.

Let $(p,q) \in \mathbb{N}^2$, $\gcd(p,q) = 1$. Let $(p',q') \in \mathbb{N}^2$ such that $qq' - pp' = 1$. Let $\mu \in \mathbb{C}^*$. We define

$$\begin{aligned} \sigma_{(p,q,\mu)}: \quad & K_x \to K_{1,x_1} \\ & f(x,y) \mapsto f\big(\mu^{q'}x_1^p, x_1^q(y_1+\mu^{p'})\big). \end{aligned}$$

We say that $\sigma_{(p,q,\mu)}$ is a *Newton map*. It depends on (p',q') but this doesn't affect our results. In fact, if $(p'+iq, q'+ip)$ is another pair, then

$$f\big(\mu^{q'+ip}x_1^p, x_1^q(y_1+\mu^{p'+iq})\big) = f\big(\mu^{q'}(x_1\mu^i)^p, (x_1\mu^i)^q(y_1\mu^{-iq}+\mu^{p'})\big).$$

Lemma 2.1. *Let $f(x,y) \in K_x$ and $f_1(x_1,y_1) = \sigma_{(p,q,\mu)}(f) \in K_{1,x_1}$. If either there doesn't exist $S \in \mathcal{N}(f)$ such that the supporting line l_S has equation $p\alpha+q\beta = k$ for some k, or if there exists $S \in \mathcal{N}(f)$ such that the supporting line l_S has equation $p\alpha + q\beta = k$ for some k, and $F_S(\mu,1) \neq 0$ then $f_1(x_1,y_1)$ is equal to $x_1^{n_1}u(x_1,y_1)$, where $n_1 \in \mathbb{Z}$ and u is a unit in $\mathbb{C}[\![x_1,y_1]\!]$.*

If $f_1(x_1,y_1)$ is equal to $x_1^{n_1}y_1^{m_1}u(x_1,y_1)$, where $(n_1,m_1) \in \mathbb{Z}\times\mathbb{N}$ and u is a unit in $\mathbb{C}[\![x_1,y_1]\!]$, we say for short that f_1 is a monomial times a unit. From this lemma, we see that there are a finite number of (p,q,μ) such that $\sigma_{(p,q,\mu)}(f)$ is eventually not a monomial times a unit in $\mathbb{C}[\![x_1,y_1]\!]$. These triples are given by the equations of the faces of the Newton polygon and the roots of the corresponding face polynomials.

We say that $f \in K_x$ is in *good coordinates* if

(1) The Newton polygon of f has at least two faces, or
(2) $\beta_m \neq 0$ or
(3) if $\beta_m = 0$ and
  (a) either l_{S_m} has equation $p\alpha + q\beta = N$ with $p \neq 1$ or
  (b) if $p = 1$, then F_{S_m} has at least two factors.

Lemma 2.2. *If $f \in K_x$ is not in good coordinates, there exist changes of variables in $\mathbb{C}[\![x,y]\!]$ in which it is in good coordinates.*

Proof. Assume that f is not in good coordinates.

$$\operatorname{in}(f,S_m) = x^k(x^q-\mu y)^\nu$$

with $k \in \mathbb{Z}, (q,\nu) \in \mathbb{N}^2$. Let ρ be the automorphism of $\mathbb{C}[\![x,y]\!]$ defined by $x \mapsto x$, $y \mapsto y + (1/\mu)x^q$. We can notice that it can only decrease the height of the Newton polygon. If f is not in good coordinates, then

$$\operatorname{in}(f,S_{m'}) = x^{k'}(x^{q'}-\mu' y)^{\nu'}$$

with $q' > q$. We apply the new automorphism $x \mapsto x$, $y \mapsto y + (1/\mu')x^{q'}$. The process is not necesseraly finite. For example, $f(x,y) = y + x^2 + xy$ is not in good coordinates. To put it in good coordinates, we have to change y in $y + h(x)$ where $h(x) = x^2(1+x)^{-1}$. □

Let $f \in K_x$, $f \notin K$ in good coordinates. Let $\sigma = \sigma_{(p,q,\mu)}$ be a Newton map. We denote by f_σ the result of $\sigma(f)$ after a change of variables so that f_σ is in good coordinates. Let $\Sigma_n = (\sigma_1, \dots, \sigma_n)$ where σ_i is a Newton map for all i, we define f_{Σ_n} by induction: $f_{\Sigma_1} = f_{\sigma_1}$, $f_{\Sigma_i} = (f_{\Sigma_{i-1}})_{\sigma_i}$.

Theorem 2.3. *For all $f(x,y) \in K_x$, there exists a natural integer n_0 such that for any $\Sigma_n = (\sigma_1, \dots, \sigma_n)$ where σ_i is a Newton map for all i with $n \geq n_0$, f_{Σ_n} is a monomial up to a unit.*

PROOF. From Lemma 2.1, we first observe that the number of Newton maps σ, such that f_σ is not a monomial times a unit is finite, bounded by the sum on all faces S of the number of roots of F_S. What we have to show is that the number of successive Newton maps we have to perform so that f_Σ is a monomial up to a unit, is also finite.

We start with f in good coordinates. In this system of coordinates, we denote by h the height of $\Delta(f)$. We argue by induction on the height of $\Delta(f)$. If $h = 0$, then f is a monomial up to a unit, and $n = 0$. Consider the case where $h > 0$. In that case, $\mathcal{N}(f)$ is not empty . Choose a face of $\mathcal{N}(f)$, S, and a root of F_S with multiplicity ν. Let $\alpha p + \beta q = N$ be the equation of the supporting line of S. Then $f_\sigma(x_1, y_1) = x_1^N g_1(x_1, y_1)$, $N \in \mathbb{Z}, g_1(x_1, y_1) \in \mathbb{C}[\![x_1, y_1]\!]$ and the height of $\Delta(f_1)$ is less or equal to $\nu < h$ since f is in good coordinates. □

If f is in good coordinates, we define the *depth* of f, $d(f)$ by induction. If f is a monomial up to a unit, then $d(f) = 0$. Otherwise $d(f) = \max d(f_\sigma) + 1$ where the maximum is taken over all faces S of the Newton polygon and all roots of F_S. Note that the definition of the depth depends on the choice of good coordinates at each step of the Newton algorithm.

Given $f \in K_x$ in good coordinates, the Newton process consists in applying successive Newton maps attached to successive Newton polygons until the result is a monomial times a unit.

Newton trees are trees that encode the Newton process. They are build by induction. First of all, we need to build a graph associated to a Newton diagram.

2.1.3. *Graph associated to a Newton diagram.* It is a vertical linear graph with vertices, edges connecting vertices and two arrows at the top and bottom of the graph.

Let Δ be a Newton diagram. If $\Delta = (\alpha_0, \beta_0) + \mathbb{R}_+^2$, the graph has no vertices, it has one edge and two arrows. The arrow on the top is decorated by (α_0) and the arrow at the bottom is decorated by (β_0).

Now let $\{v_0, \dots, v_m\}$ be the set of vertices of Δ. Each face $S_i = [v_{i-1}, v_i]$ is represented by a vertex on the graph and two vertices are connected by an edge if the corresponding faces have a common point. We add an edge to the vertices representing S_1 and S_m ended by an arrow decorated by (α_0) at the top and (β_m) at the bottom. The extremities of the edges are decorated the following way: A vertex corresponds to a face S whose supporting line has equation $p\alpha + q\beta = N$. We decorate the extremity of the edge above the vertex by q, and the extremity of the edge under the vertex by p. We decorate the vertex by (N). We can recover the whole Newton polygon from the graph since we can read the equations of the supporting lines of the faces on the graph.

2.1.4. *Newton tree of $f \in K_x$.* We assume that f is in good coordinates. We build the Newton tree of f by induction on the depth. Assume that f has depth 0.

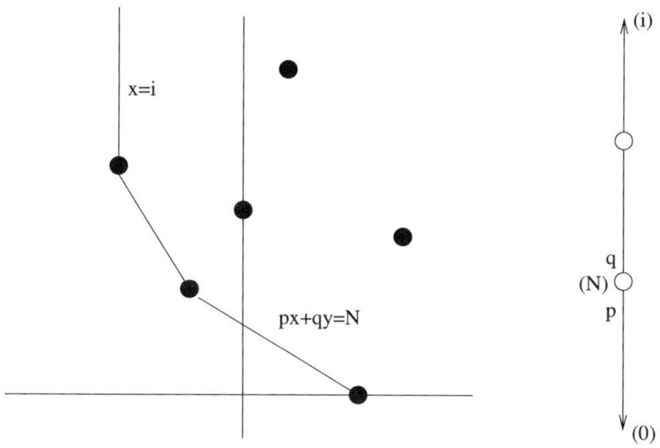

FIGURE 1

Then f is a monomial times a unit and we define its Newton tree as the graph of its Newton diagram.

Assume that we have constructed the Newton tree for all f of depth less or equal to $n-1$. Let $f \in K_x$ in good coordinates of depth n. On one hand we can construct the graph associated to its Newton diagram. On the other hand, for each face of the Newton polygon and each root of the face polynomials the resulting functions are of depth at most $n-1$. Then by hypothesis, we can construct their Newton trees. To obtain the Newton tree of f, we delete the upper arrow of the Newton tree of f_σ and glue the edge to the corresponding vertex on the graph of the Newton diagram, for all σ. In the process, the edges that are glued become *horizontal edges.* All others are *vertical edges.* The vertices on a vertical line correspond to faces on a Newton polygon, the edges on an horizontal line are produced by successive Newton maps. Horizontal edges have no decorations on their left extremities. They have to be considered as decorated with 1.

Let v be a vertex of a Newton tree. If v corresponds to a face of the Newton polygon of f, we say that v has no preceding vertex and we define $\mathcal{S}(v) = \{v\}$. Let v be a vertex on a Newton tree. It is on the Newton polygon of f_Σ. The Newton tree of f_Σ has been glued on a vertex v_1 which is called the *preceding vertex* of v. If v_1 does not correspond to a face of the Newton polygon of f, we can consider its preceding vertex v_2. Then we can define $\mathcal{S}(v) = \{v_i, \dots, v_2, v_1, v\}$, where v_i has no preceding vertex and v_j is the preceding vertex of v_{j-1}, $j = 2, \dots, i$.

The final Newton tree is decorated the following way. The decorations of the arrows and vertices are not changed. The decorations of the edges are changed. Let v be a vertex on the Newton tree. If $\mathcal{S}(v) = \{v\}$ then the decorations near v are not changed. If $\mathcal{S}(v) = \{v_i, \dots, v_2, v_1, v\}$ and if the decoration near v on the Newton tree such that $\mathcal{S}(v) = \{v_{i-1}, \dots, v_2, v_1, v\}$ (that is on the Newton tree which is glued at v_i), are (q,p), after the gluing, the decorations near v are $(q + p_i q_i p_{i-1}^2 \cdots p_1^2 p, p)$.

Example 2.4. Consider

$$f(x,y) = x^{-6}((x+y^2)^2(x+y)^3 - x^3y^3 + x^5y^2).$$

First we have to make a change of coordinates so that f is in good coordinates. The change of coordinates is $x = x, y = y - x$. We get a Newton polygon with two

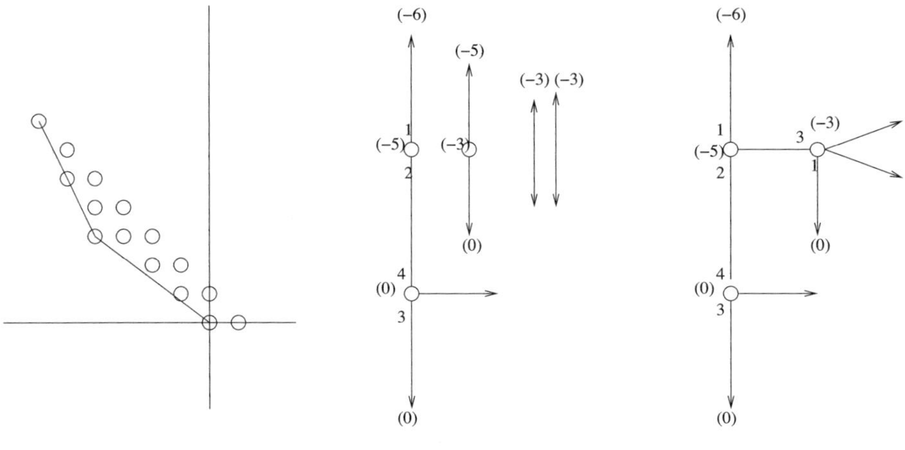

FIGURE 2

faces. First face has equation $2\alpha+\beta=-5$ with face polynomial $(x+y^2)^2$. The corresponding Newton map is $x=x_1^2, y=x_1(y_1+1)$. Then

$$f_\sigma(x_1,y_1)=x_1^{-5}(4y_1^2-8y_1+3+\cdots).$$

We have one face with a face polynomial with two factors of multiplicity one. We get monomial times unit after the next step. The second face has equation $3\alpha+4\beta=0$, with face polynomial (y^3+x^4). We get a monomial times a unit in one step.

2.1.5. *Combinatorial properties of Newton trees.* We recall some properties proved in [5] for $\mathbb{C}[\![x,y]\!]$.

Proposition 2.5. *If v_0 is the preceding vertex of v decorated respectively by (q_0,p_0) and (q,p), we have*

$$q=p_0q_0p+\widetilde{m}.$$

where $(\widetilde{m},p)$ are the decorations of v on the Newton tree where $\mathcal{S}(v)=\{v\}$.

PROOF. We recall the proof given in [5] for $\mathbb{C}[\![x,y]\!]$, but nothing is changed.

We work by induction on the number of elements of $\mathcal{S}(v)$. If $\mathcal{S}(v)=\{v_0,v\}$, $q=p_0q_0p+\widetilde{m}$ is the definition of q. Assume that $\mathcal{S}(v)=\{v_i,v_{i-1},\ldots,v_0,v\}$, and that on the Newton tree where $\mathcal{S}(v)=\{v_{i-1},\ldots,v_0,v\}$, the decorations $(\tilde{q}_0,p_0)$ and $(\tilde{q},p)$ of v satisfy $\tilde{q}=p_0\tilde{q}_0p+\widetilde{m}$.

We have

$$\begin{aligned}
q_0&=p_iq_ip_{i-1}^2\cdots p_1^2p_0+\tilde{q}_0\\
q&=p_iq_ip_{i-1}^2\cdots p_1^2p_0^2p+\tilde{q}\\
q&=p_iq_ip_{i-1}^2\cdots p_1^2p_0^2p+p_0\tilde{q}_0p+\widetilde{m}\\
q&=p_0p(p_iq_ip_{i-1}^2\cdots p_1^2p_0+\tilde{q}_0)+\widetilde{m}.
\end{aligned}$$

□

Definition 2.6. Consider a path on a Newton tree. We say that a number is adjacent to this path if it is a decoration near a vertex v on the path, on an edge, issued from the vertex v, not belonging to the path. If the path contains an arrow, the decorations of the arrow is a number adjacent to the path.

Definition 2.7. Consider an edge on a Newton tree, its edge determinant is the difference between the product of the numbers on the edge and the product of the numbers adjacent to the edge.

Corollary 2.8. (1) *In the process of gluing the determinants of vertical edges remain constant.*
(2) *All edges determinants are strictly positive integers.*

PROOF. This is a consequence of the previous proposition. □

Proposition 2.9. *The decoration* (N) *of a vertex* v *on a Newton tree is the sum over all the arrows* $\mathcal{F}$ *of the tree, of the products of the numbers adjacent to the paths* $[v, \mathcal{F}]$.

PROOF. We recall the proof of [5]

We prove the proposition by induction on the number of elements of $\mathcal{S}_v$. If $\mathcal{S}_v = \{v\}$, v is on the first Newton polygon. Let S be the face of the Newton polygon corresponding to v, and (α, β) its origin. We have

$$\alpha = \alpha_0 + \sum_i |S_i|_1, \beta = |S|_2 + \sum_j |S_j|_2 + \beta_m$$

where $i \in I$, and I is the set of indices of faces before S, and $j \in J$ where J is the set of indices of faces after S and $|S|_1$ is the width of S and $|S|_2$ its height. Since $N = \alpha p + \beta q$ the property is true for $\mathcal{S}_v = \{v\}$.

Now let $\mathcal{S}_v = \{v_{k-2}, \ldots, v_0, v\}$ and $\mathcal{S}_{v_0} = \{v_{k-2}, \ldots, v_0\}$. We can apply the induction hypothesis to $\mathcal{S}_{v_0}$. In particular, we can write

$$N_0 = N_0^1 + \tilde{h}(\Delta) p_0 q_0$$

where Δ is the Newton polygon to which belongs the face S corresponding to v and

$$\tilde{h}(\Delta) = \sum_i |S_i|_2 + |S|_2 + \sum_j |S_j|_2 + \beta'_m.$$

Using the equation of the line supporting S, we have

$$N = \Big(N_0 + \sum_i |S_i|_1\Big) p + \Big(|S|_2 + \sum_j |S_j|_2 + \beta'_{m'}\Big) \widetilde{m}$$

$$N = (N_0^1 + \tilde{h}(\Delta) p_0 q_0) p + \Big(\sum_i |S_i|_1\Big) p + \Big(|S|_2 + \sum_j |S_j|_2 + \beta'_{m'}\Big) \widetilde{m}.$$

Then,

$$N = N_0^1 p + \Big(|S|_2 + \sum_j |S_j|_2 + \beta'_{m'}\Big)(\widetilde{m} + p_0 q_0 q) + p\Big(\sum_i |S_i|_1 / \widetilde{m_i}(\widetilde{m_i} + p_0 q_0 p_i)\Big).$$

Then the proposition is proved. □

Remark 2.10. (1) From this proposition we see that when the tree is constructed, the decorations of the vertices are not needed any more because we can compute them from the decorations of the edges. But, anyway, we have to keep in mind that we know them from the beginning.
(2) Notice that the decorations of the vertices and arrows are not enough to recover the decorations of the edges. Here is an example.

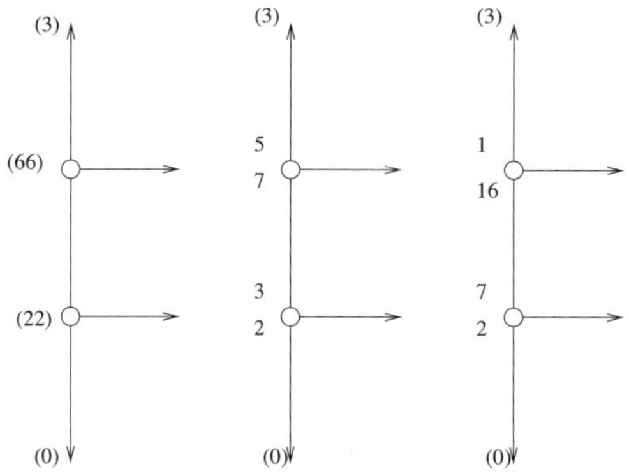

FIGURE 3

On a Newton tree, we can add vertices (of valency 2), and decorate the edge above and under (or left and right) with natural numbers with the condition that all the edge determinants are positive. When we have decorated the edge above and under a vertex, we can compute its decoration using the previous proposition.

The edges of the tree connecting a vertex to an arrow decorated with (0) are called *leaves.*

We give a partial order on the vertices of a Newton tree. The smallest element V_0 is the vertex connected to the arrow decorated with a negative integer. We say that $v < v'$ if v is on the path between V_0 and v'.

Proposition 2.11. *If $v < v'$ two consecutive vertices on a Newton tree and if (N) and (N') are the negative decorations of v and v' and (p, q) and (p', q') the decorations near v and v', we have $N \leq N'/p'$.*

Corollary 2.12. *The set of vertices on a Newton tree decorated with negative (N) is connected.*

PROOF. First we assume that $v < v'$ are two consecutive vertices on the same vertical line. Then v and v' correspond to faces on the same Newton polygon with equation respectively $p\alpha + q\beta = N$ and $p'\alpha + q'\beta = N'$. Then N/p and N'/p' are the intersection of the faces with the x-axis. the convexity gives the inequality $N/p < N'/p'$.

If $v < v'$, and v is the preceding vertex of v'. Then if v is decorated with (N), the vertical face of the Newton polygon to which the face corresponding to v' belongs, cuts the x-axis at $x = N$. Then $N < N'/p'$. □

Let v be a vertex on a Newton tree. Assume $\mathcal{S}_v = \{v_1, \ldots, v_{i-1}, v = v_i\}$, decorated with $(p_1, q_1), (p_2, q_2), \ldots, (p_i, q_i)$ where for $2 \leq j \leq i$, $q_j = p_{j-1}q_{j-1}p_j + \widetilde{m_j}$. There is a sequence of Newton maps and changes of variables associated to $\mathcal{S}_v$.

$$x = x_1^{p_1}\mu_1, \qquad y = x_1^{q_1}(y_1 + \mu_1')$$

$$x_1 = x_1', \qquad y_1 = y_1' + a_{1,1}x_1' + \cdots + a_{1,l_1}x_1'^{l_1}$$

$$\vdots$$

$$x_{i-1} = x_i^{p_i}\mu_i, \qquad y_{i-1} = x_i^{\widetilde{m_i}}y_i$$

$$x_i = x_i', \qquad y_i = y_i' + a_{i,1}x_i' + \cdots + a_{i,l_i}x_i'^{l_i}.$$

The composition of the sequence of Newton maps and changes of variables, gives a parameterization in $\mathbb{C}[y_i'][x_i'^{-1}, x_i']$:

$$x = x_i'^{p_1p_2\cdots p_i}\mu, \qquad y = x_i'^{q_1p_2\cdots p_i}\mu_0' + \cdots + x_i'^{q_1p_2\cdots p_i+k}y_i'.$$

To verify this statement we proceed by induction on the number of steps. Let

$$x = x_1^{p_1}\mu_1, \qquad y = x_1^{q_1}(y_1 + \mu_1')$$
$$x_1 = x_1', \qquad y_1 = y_1' + a_{1,1}x_1' + \cdots + a_{1,l_1}x_1'^{l_1}$$
$$x_1' = x_2^{p_2}\mu_2, \qquad y_1' = x_2^{q_2}y_2.$$

We have $1/l_1 > p_2/q_2$. Then

$$x = x_2^{p_1p_2}\mu_1\mu_2$$
$$y = x_1'^{q_1}(y_1' + a_{1,1}x_1' + \cdots + a_{1,l_1}x_1'^{l_1} + \mu_1')$$
$$y = x_2^{p_2q_1}\mu_2^{q_1}(x_2^{q_2}y_2 + a_{1,1}\mu_2x_2^{p_2} + \cdots + a_{1,l_1}\mu_2^{l_1}x_2^{l_1p_2} + \mu_1').$$

Such a parameterization defines an element of $\mathbb{C}[\![x, y]\!]$ that we call a *curvette* of v. We denote it by $\mathcal{C}_v$. If we consider the Newton tree of $f\mathcal{C}_v$, it is obtained by adding an arrow to the vertex v on the Newton tree of f.

Example 2.13. We come back to Example 2.4. We have the following maps:

$$x = x_1^2 \quad y = x_1(y_1 + 1); \quad x_1 = x_2 \quad y_1 = x_2(y_2 - \tfrac{1}{2}); \quad x_2 = x_3, \quad y_2 = x_3y_3$$
$$x = x_3^2 \quad y = x_3 - 1/2x_3^2 + y_3x_3^3$$

is the parametrization of $\mathcal{C}_v$. (see Figure 4)

Let v be a vertex on a Newton tree. Assume the parametrization of $\mathcal{C}_v$:

$$x = x_i^n, y = x_i^m a_0 + x_i^{m+1}a_1 + \cdots + x_i^{m+k}y_i.$$

By definition, we have

$$f(x_i^n, x_i^m a_0 + x_i^{m+1}a_1 + \cdots + x_i^{m+k}y_i) = x_i^N(P_{0,v}(y_i) + x_iP_{1,v}(y_i) + \dots)$$

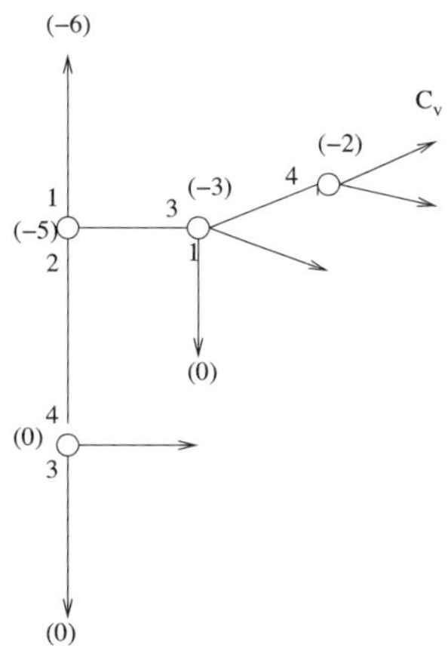

FIGURE 4

where N is the decoration of the vertex v and $P_{0,v}(y_i)$ has degree at least one if v is not a vertex added on a leaf.

2.1.6. *Dicritical vertices.* There are vertices which play a particular role for $f \in K_x$.

Definition 2.14. A dicritical vertex of $f \in K_x$ is a vertex decorated with (0) which is not on a leaf.

A dicritical vertex can be a vertex in the Newton tree or a vertex added. From the definition we see that if $f \in \mathbb{C}[\![x,y]\!]$, it has no dicritical vertex since all the vertices are decorated with positive multiplicities.

Let $f \in K_x$, $f \notin \mathbb{C}[\![x,y]\!]$.

Let v be a dicritical vertex of f, let $t \in \mathbb{C}$.

$$f(x_i^n, x_i^m a_0 + x_i^{m+1} a_1 + \cdots + x_i^{m+k} y_i) + t = t + P_{0,v}(y_i) + x_i P_{1,v}(y_i) + \cdots .$$

Assume that the vertex v is decorated with (p,q). First we notice that $P_{0,v}(y_i)$ is a polynomial in y_i^p. In fact the last map we are considering is $x_{i-1} = x_i^p$, $y_{i-1} = x_i^{\widetilde{m}} y_i$ and the face polynomial is $F_S(x_{i-1}, y_{i-1}) = x_{i-1}^{a_S} y_{i-1}^{b_S} \prod (x_{i-1}^{\widetilde{m}} - \mu_j y_{i-1}^p)^{\nu_j}$. We write $P_{0,v}(y_i) = Q_{0,v}(y_i^p)$. The degree of $Q_{0,v}$ is called the *degree of the dicritical vertex.*

We assume that $Q_{0,v}(0) = 0$. Let I be the set of roots of the discriminant of $Q_{0,v} + t$ with respect to t. If $t \notin I \cup \{0\}$ then $t + Q_{0,v}$ has only simple roots and we are at the last step of the Newton process. Then there are $\deg Q_{0,v}$ edges with arrows attached to v. Then if we consider the Newton tree of $f + t$ for t generic, v is the last vertex on each path between V_0 and the arrows attached to v, in the case where $\deg Q_{0,v} > 1$, or $p > 1$. There is only one dicritical vertex on the Newton tree between V_0 and an arrow. Moreover, on the Newton tree of $f + t$ for t generic, all vertices are decorated with $N \leq 0$.

Now we will show that if we consider an arrow decorated with a positive multiplicity, then there is always one dicritical vertex (eventually added) between V_0 and the arrow.

Consider an arrow and the vertex where the arrow is attached by an edge. If v is not a dicritical vertex we have

$$f(x_i^n, x_i^m a_0 + x_i^{m+1} a_1 + \cdots + x_i^{m+k} y_i) + t = x_i^N (P_{0,v}(y_i) + x_i P_{1,v}(y_i) + \cdots)$$

where $N < 0$ is the decoration of the vertex v. Since there is only one edge between v and the arrow, $P_{0,v}(y_i) = Q_{0,v}(y_i^p) \widetilde{Q}_v(y_i^p)$ with $\deg Q_{0,v} = 1$. We can find $k' \in \mathbb{N}$, $a_k, \ldots, a_{k+k'} \in \mathbb{C}$ such that

$$\begin{aligned} f(x_i^n, x_1^m a_0 + x_i^{m+1} a_1 + \cdots + x_i^{m+k} a_k + \cdots + x_i^{m+k+k'} y_i') + t \\ = t + (R_{0,v}(y_i) + x_i R_{1,v}(y_i) + \cdots) \end{aligned}$$

where $\deg R_{0,v}(y_i) = 1$.

We have proved

Proposition 2.15. *On the Newton tree of $f + t$, for t generic, on each path between V_0 and an arrow decorated with positive multiplicity there is one and only one dicritical vertex.*

2.2. Newton trees for $\mathbb{C}[x,y]$.

2.2.1. *Newton polygon at infinity of an algebraic curve.* Let E be a finite set in $\mathbb{N}^2$. We consider

(1) $\Delta_{0,\infty}(E)$, the smallest convex set containing $E + (\mathbb{R}_+ \times \mathbb{R}_-)$
(2) $\Delta_{\infty,\infty}(E)$, the smallest convex set containing $E + (\mathbb{R}_- \times \mathbb{R}_-)$
(3) $\Delta_{\infty,0}(E)$, the smallest convex set containing $E + (\mathbb{R}_- \times \mathbb{R}_+)$

The *Newton polygon at infinity* of E, denoted $\mathcal{N}_\infty(E)$ is the union of the compact faces of $\Delta_{0,\infty}(E)$, $\Delta_{\infty,\infty}(E)$ and $\Delta_{\infty,0}(E)$. Let $S \in \mathcal{N}_\infty(E)$, we write its equation $p\alpha + q\beta = N$ with $(p,q,N) \in \mathbb{Z}^2 \times \mathbb{N}$.

Let $f \in \mathbb{C}[x,y]$.

$$f(x,y) = \sum_{(\alpha,\beta)\in\mathbb{N}\times\mathbb{N}} c_{\alpha,\beta}x^\alpha y^\beta$$

and

$$\operatorname{Supp} f = \{(\alpha,\beta) \in \mathbb{N}\times\mathbb{N} \mid c_{\alpha,\beta} \neq 0\}.$$

We denote by $\mathcal{N}_\infty(f) = \mathcal{N}_\infty(\operatorname{Supp} f)$. If $S \in \mathcal{N}_\infty(f)$,

(1) $S \in \Delta_{\infty,\infty}(\operatorname{Supp} f)$, then

$$f_S(x,y) = x^{a_S}y^{b_S}\prod(x^q+\mu_i y^p)^{\nu_i}, \quad \text{and} \quad f_S(v^{-p}, v^{-q}w) = v^{-N}w^{b_S}\prod(1+\mu_i w^p)^{\nu_i}.$$

(2) $S \in \Delta_{0,\infty}(\operatorname{Supp} f)$, then

$$f_S(x,y) = x^{a_S}y^{b_S}\prod(x^q y^{-p}+\mu_i)^{\nu_i} \quad \text{and} \quad f_S(v^{-p}w, v^{-q}) = v^{-N}w^{a_S}\prod(w^q+\mu_i)^{\nu_i},$$

(3) if $S \in \Delta_{\infty,0}(\operatorname{Supp} f)$, then

$$f_S(x,y) = x^{a_S}y^{b_S}\prod(x^{-q}y^p+\mu_i)^{\nu_i} \quad \text{and} \quad f_S(v^{-p}, v^{-q}w) = v^{-N}w^{b_S}\prod(w^p+\mu_i)^{\nu_i}.$$

In the case of polynomials, we are considering the following Newton maps.

(1) If $p > 0$, $q > 0$, $(p,q) = 1$. Let $(p',q') \in \mathbb{N}^2$ such that $qq' - pp' = 1$ and define

$$\begin{aligned}\sigma_{(p,q,\mu)} \colon \mathbb{C}[x,y] &\to \mathbb{C}[v^{-1}, v, w] \\ f(x,y) &\mapsto f(\mu^{q'}v^{-p}, v^{-q}(w+\mu^{p'})).\end{aligned}$$

(2) If $p < 0$, $q > 0$, $(p,q) = 1$. Let $(p',q') \in \mathbb{N}^2$ such that $qq' - pp' = 1$ and define

$$\begin{aligned}\sigma_{(p,q,\mu)} \colon \mathbb{C}[x,y] &\to \mathbb{C}[v^{-1}, v, w] \\ f(x,y) &\mapsto f(v^{-p}(w+\mu^{q'}), \mu^{p'}v^{-q}).\end{aligned}$$

(3) If $p > 0$, $q < 0$, $(p,q) = 1$. Let $(p',q') \in \mathbb{N}^2$ such that $p'p - qq' = 1$ and define

$$\begin{aligned}\sigma_{(p,q,\mu)} \colon \mathbb{C}[x,y] &\to \mathbb{C}[v^{-1}, v, w] \\ f(x,y) &\mapsto f(\mu^{q'}v^{-p}, v^{-q}(w+\mu^{p'}).\end{aligned}$$

When we apply the Newton maps, for each face of $\mathcal{N}_\infty(f)$ and each root of the face polynomials, we get elements in $\mathbb{C}[v^{-1}, v, w]$ that we have studied in the previous sections. Now we are able to construct the Newton tree of an algebraic curve.

Let $f(x,y) \in \mathbb{C}[x,y]$. We consider the algebraic curve $f(x,y) = 0$. We consider $\mathcal{N}_\infty(f)$. We first associate a graph to $\mathcal{N}_\infty(f)$. Assume that $f(x,y) = x^a y^b g(x,y)$, with $g \in \mathbb{C}[x,y]$ and not divisible by x and y. To each face of $\mathcal{N}_\infty(f)$, we associate a vertex and two vertices are linked by an edge if and only if the faces have a common point. We go around the Newton polygon clockwise and along the graph from top to bottom. We add an edge with an arrow decorated with (a) to the top and an edge with an arrow decorated with (b) to the bottom. Each vertex is decorated with (N) if the corresponding face has equation $q\alpha + p\beta = N$ and the extremity of the edge above is decorated with q and the extremity of the edge under is decorated with p.

Now to each vertex of the graph corresponds a face of $\mathcal{N}_\infty(f)$, and to each root of the face polynomial corresponds a Newton map and the result is an element f_1 of $\mathbb{C}[x^{-1}, x, y]$. We put it in good coordinates and we construct the Newton tree of f_1. Then we delete the top arrow of the Newton tree of f_1 and we glue the top edge to the vertex we are considering on the graph of $\mathcal{N}_\infty(f)$. To decorate the tree, we keep the decorations of the graph of $\mathcal{N}_\infty(f)$. If v is a vertex on the Newton tree of f_1 glued at v_0. Assume the decorations near v_0 are (p_0, q_0). Assume that $\mathcal{S}(v) = \{v_1, \ldots, v_{i-1}, v\}$, where the decorations near v_j are (q_j, p_j) and the decorations near v are (q, p), then after the gluing the decorations near v are $(p_0 q_0 p_1^2 \cdots p_{i-1}^2 p - q, p)$, and the decoration of v is $(-N)$.

Example 2.16. Consider

$$f(x,y) = x^6y^4 + (4x^5 + 3x^4)y^3 + (6x^4 + 11x^3 + 3x^2)y^2 + (4x^3 + 13x^2 + 2x + 1)y + x^2 + 5x + 5.$$

The Newton polygon has two faces: Clockwise the first one has equation $-x + 2y = 2$ with face polynomial $y(x^2y+1)^3$ and the second one has equation $x - y = 2$ with face polynomial $x^2(xy+1)^4$. We have

$$f(v(w+1), 1/v^2) = v^{-2}(8w^3 - v + \cdots)$$
$$f(1/v, v(w-1)) = v^{-2}(w^4 + 2w^2v + 2w - v + \cdots)$$

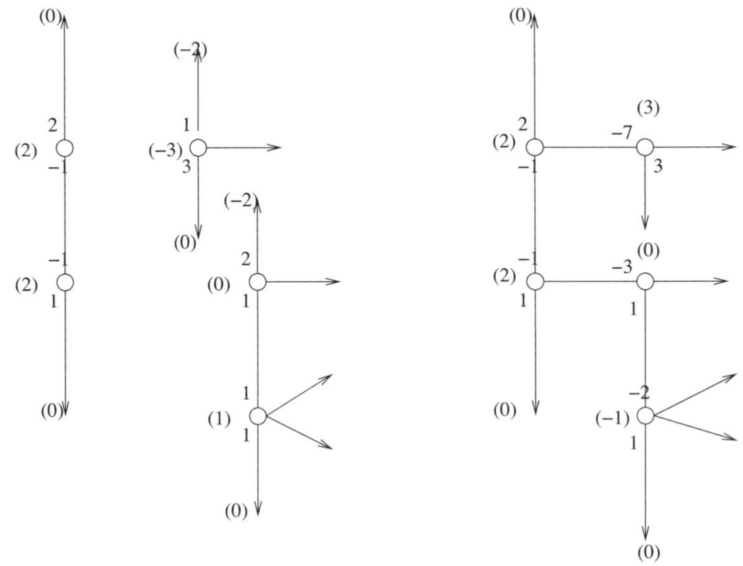

FIGURE 5

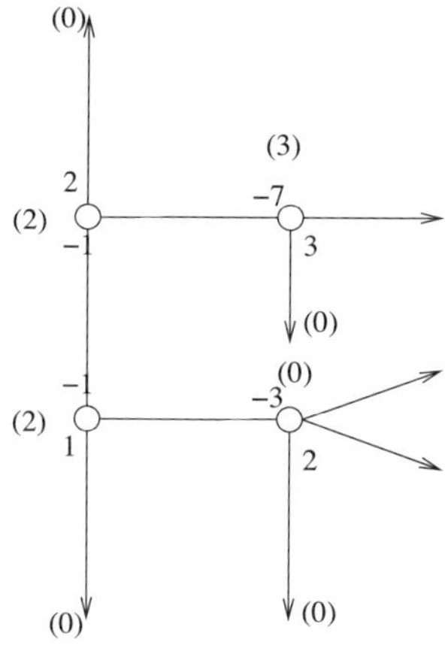

FIGURE 6

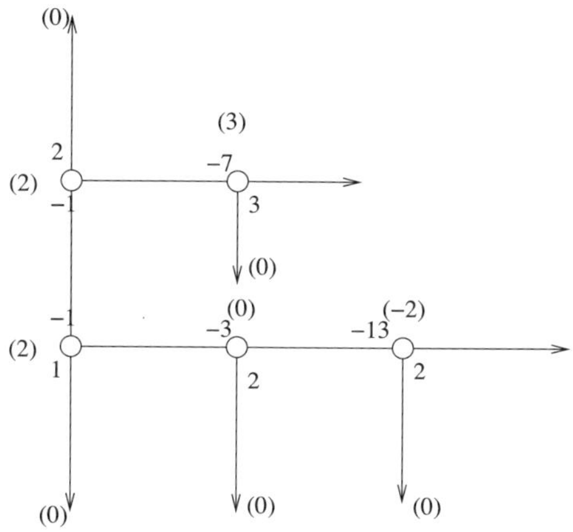

FIGURE 7

We have represented the graph associated to the Newton polygon at infinity of f, then the two Newton trees of $f(v(w+1), 1/v^2)$ and $f(1/v, v(w-1))$, and finally the Newton tree at infinity of the curve $\{f(x,y)=0\}$ in Figure 5.

The Newton tree of $\{f(x,y)-t=0\}$ for $t \neq 0, -1$ is in Figure 6 and the Newton tree of $f(x,y)+1=0$ is in Figure 7.

2.2.2. *Combinatorial properties of Newton trees at infinity.*

Proposition 2.17. *If v_0 is the preceding vertex of v decorated respectively by (p_0, q_0) and (p, q), we have*

$$q = p_0 q_0 p - \widetilde{m}.$$

where $(\widetilde{m}, p)$ are the decorations of v on the Newton tree where $\mathcal{S}(v) = \{v\}$.

PROOF. Assume that $\mathcal{S}(v) = \{v_i, \ldots, v_1, v_0, v\}$. We have glued on the graph of a Newton polygon at infinity a Newton tree of a function in $\mathbb{C}[v^{-1}, v, w]$, on which $\mathcal{S}(v) = \{v_{i-1}, \ldots, v_1, v_0, v\}$. On this Newton tree, by Proposition 2.5, we know that, if the decorations of v_0 and v are respectively $(p_0, \widetilde{q_0})$ and $(p, \tilde{q})$, then

$$\tilde{q} = p_0 \widetilde{q_0} p + \widetilde{m}.$$

But

$$q_0 = p_i q_i p_{i-1}^2 \cdots p_1^2 p_0 - \widetilde{q_0}$$
$$q = p_i q_i p_{i-1}^2 \cdots p_1^2 p - \tilde{q}.$$

Then

$$q = p_0 p q_0 - \widetilde{m}.$$

□

Corollary 2.18. *On a Newton tree at infinity, all edge determinants are negative.*

PROOF. The previous proposition shows that for horizontal edges, the edge determinants are negative. Consider a vertical edge $[v, v']$. Assume that $\mathcal{S}(v) = \{v_i, \ldots, v_1, v_0, v\}$ and $\mathcal{S}(v') = \{v_i, \ldots, v_1, v_0, v'\}$. We have glued on the graph at infinity a Newton tree of a function in $\mathbb{C}[v^{-1}, v, w]$ where $\mathcal{S}(v) = \{v_{i-1}, \ldots, v_1, v_0, v\}$ and $\mathcal{S}(v') = \{v_{i-1}, \ldots, v_1, v_0, v'\}$. We know that on the local tree, if the decorations near v and v' are respectively $(p, \tilde{q})$ and $(p', \widetilde{q'})$, $p'\tilde{q} - p\widetilde{q'} > 0$. Since

$$q = p_i q_i p_{i-1}^2 \cdots p_1^2 p - \tilde{q}$$
$$q' = p_i q_i p_{i-1}^2 \cdots p_1^2 p' - \widetilde{q'}$$

then $p'q - pq' < 0$. □

We have also

Proposition 2.19. *The decoration (N) of a vertex v on a Newton tree is the sum over all the arrows $\mathcal{F}$ of the tree, of the products of the numbers adjacent to the paths $[v, \mathcal{F}]$.*

PROOF. Again, we use induction on the number of elements of $\mathcal{S}_v$. If this number is 1, the proof is the proof of the local case.

With the same notations as in the local case, using the equation of the line supporting S, we have

$$N = -\left(\left(-N_0 + \sum_i |S_i|_1\right)p + \left(|S|_2 + \sum_j |S_j|_2 + \beta'_{m'}\right)\widetilde{m}\right).$$

Then,

$$N = N_0^1 p + \left(|S|_2 + \sum_j |S_j|_2 + \beta'_{m'}\right)(-\widetilde{m} + p_0 q_0 q) + p\left(\sum_i |S_i|_1 / \widetilde{m_i}(-\widetilde{m_i} + p_0 q_0 p_i)\right).$$

Then the proposition is proved. □

On a Newton tree at infinity we can add vertices. The decorations near these vertices should respect the rule that all edge determinant is negative. We can decorate these vertices using proposition 1.15. The vertex (eventually added) on the first vertical line whose nearby decorations are $(1, 1)$, is called the root of the Newton tree at infinity. The decoration of the root is the degree of the polynomial. We define a partial order on the vertices of a Newton tree at infinity, by $v < v'$ if v is on the path between the root and v'. Now we have:

Proposition 2.20. *If $v < v'$ on a Newton tree at infinity and if (N) and (N') are the positive decorations of v and v' and (p, q) and (p', q') the decorations near v and v', we have $N \geq N'/r'$, where r' is the decoration p' or q' which is not on the path between v and v'.*

Corollary 2.21. *The set of vertices on a Newton tree at infinity decorated with positive (N) is connected.*

Let v be a vertex of a Newton tree at infinity. We have $\mathcal{S}_v = \{v_0, v_1, \ldots, v_i = v\}$, where v_0 corresponds to a face of the Newton polygon at infinity and $\{v_1, \ldots, v_i = v\}$ is the sequence of vertices in a local Newton tree. Assume the decoration near v_0 are (p_0, q_0) with $p_0 > 0$.

We have a sequence of Newton maps corresponding to $\{v_1, \ldots, v_i = v\}$, which gives a parametrization

$$x_1 = x_i^{p_1 \cdots p_i}, y_1 = x_i^{q_1 p_2 \cdots p_i} a_0 + \cdots + x_i^{q_1 p_2 \cdots p_i + k} y_i$$

of an element in $C[\![x, y]\!]$. Now to each root of the face polynomial associated to v_0, we have considered the Newton maps

$$x = x_1^{-p_0}, y = x_1^{-q_0}(y_1 + c).$$

Then we get a meromorphic parameterization

$$x = x_i^{-p_0 p_1 \cdots p_i}, y = x_i^{-q_0 p_1 \cdots p_i} b_0 + \cdots + x_i^{-q_0 p_1 \cdots p_i + k'} y_i.$$

This meromorphic parametrization is called a curvette of v.

We have

$$f(x_i^{-p_0 p_1 \cdots p_i}, x_i^{-q_0 p_1 \cdots p_i} b_0 + \cdots + x_i^{-q_0 p_1 \cdots p_i + k'} y_i) = x_i^{-N}(P_{0,v}(y_i) + x_i P_{1,v}(y_i) + \cdots.$$

If the decoration near v_0 are (p_0, q_0) with $p_0 < 0$, we write the first parameterization

$$x_1 = y_i^{p_1 \cdots p_i} a_0 + \cdots + y_i^{p_1 p_2 \cdots p_i + k} x_i, \qquad y_1 = y_i^{q_1 p_2 \cdots p_i}.$$

We are exchanging the role of x_1 and y_1 in the Newton algorithm. The Newton map associated to v_0 and a root of the face polynomial

$$x = y_1^{-p_0}(x_1 + c), \qquad y = x_1^{-q_0}$$

and we get a meromorphic parameterization

$$x = y_i^{-p_0 p_1 \cdots p_i} b_0 + \cdots + y_i^{-p_0 p_1 \cdots p_i + k'} x_i, \qquad y = y_i^{-q_0 p_1 \cdots p_i}$$

and

$$f(y_i^{-p_0 p_1 \cdots p_i} b_0 + \cdots + y_i^{-p_0 p_1 \cdots p_i + k'} x_i, y_i^{-q_0 p_1 \cdots p_i}) = y_i^{-N}(P_{0,v}(x_i) + y_i P_{1,v}(x_i) + \cdots.$$

2.2.3. *Dicriticals of a polynomial.* We consider $f(x,y) \in \mathbb{C}[x,y]$ and consider the algebraic curves $\{f(x,y)-t=0\}$ for all $t \in \mathbb{C}$. There exists a finite set I such that the Newton tree at infinity of $\{f(x,y)-t=0\}$ for all $t \notin I$, doesn't depend on t. We call this tree, the Newton tree at infinity of the polynomial.

Consider a vertex v decorated with (N). We have a meromorphic parameterization such that

$$f(x_i^{-p_0p_1\cdots p_i}, x_i^{-q_0p_1\cdots p_i}b_0 + \cdots + x_i^{-q_0p_1\cdots p_i+k}y_i) \\ = x_i^N(P_{0,v}(y_i) + x_iP_{1,v}(y_i) + \cdots$$

or

$$f(y_i^{-p_0p_1\cdots p_i}b_0 + \cdots + y_i^{-p_0p_1\cdots p_i+k}x_i, y_i^{-q_0p_1\cdots p_i}) \\ = y_i^N(P_{0,v}(x_i) + y_iP_{1,v}(x_i) + \cdots .$$

If the decoration of v is (0), then

$$f(x_i^{-p_0p_1\cdots p_i}, x_i^{-q_0p_1\cdots p_i}b_0 + \cdots + x_i^{-q_0p_1\cdots p_i+k}y_i) + t \\ = t + P_{0,v}(y_i) + x_iP_{1,v}(y_i) + \cdots$$

or

$$f(y_i^{-p_0p_1\cdots p_i}b_0 + \cdots + y_i^{-p_0p_1\cdots p_i+k}x_i, y_i^{-q_0p_1\cdots p_i}) + t \\ = t + P_{0,v}(x_i) + y_iP_{1,v}(x_i) + \cdots .$$

The vertices on this tree (eventually added) decorated with (0), which are not on leaves, i.e., such that $\deg P_{0,v} \geq 1$ are called *dicritical vertices of f.*

If we come back to Example 2.16. The dicritical vertices are shown in black in Figure 8.

The vertex, whose nearby decorations are $(-2,1)$ is not a dicritical vertex. It is decorated with (0), but is on a leaf and

$$f(1/v, -v+v^2w) = t + 6 + v(-1-4w+\cdots .$$

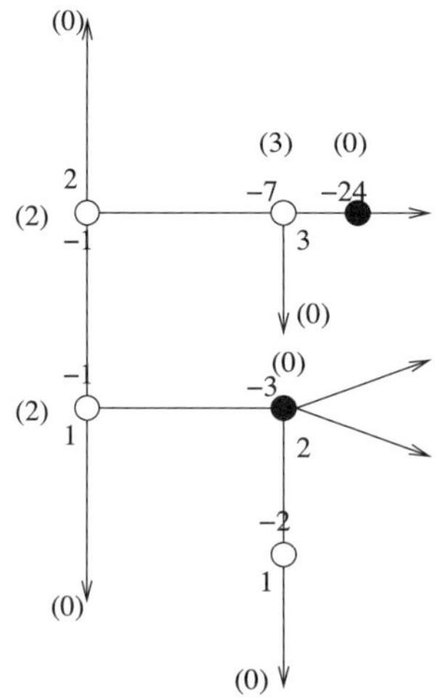

FIGURE 8

Proposition 2.22. *There is one and only one dicritical vertex on each path between the root and the arrows with positive multiplicity.*

Proposition 2.23. *Let S be a face of the Newton polygon of a polynomial $f \in \mathbb{C}[x,y]$, with equation $p\alpha + q\beta = N$, $(p,q,N) \in \mathbb{Z}^2 \times \mathbb{N}$. Then there exist $(n,m) \in \mathbb{N}^*$ and $(a_0, a_1, \cdots a_{m-1}) \in \mathbb{C}^* \times \mathbb{C}^{m-1}$ such that, if $p > 0$*

$$f(v^{-pn}, a_0 v^{-qn} + a_1 v^{-qn+1} + \cdots + v^{-qn+m} w) + t = t + P_0(w) + vP_1(w) + \cdots$$

and if $p < 0$

$$f(a_0 v^{-pn} + a_1 v^{-pn+1} + \cdots + v^{-pn+m} w, v^{-qn}) + t = t + P_0(w) + vP_1(w) + \cdots$$

with $\deg P_0(w) \geq 1$.

3. Application

Let f and g be two polynomials in $\mathbb{C}[x,y]$. We denote by F the map from $\mathbb{C}^2$ to $\mathbb{C}^2$ given by $(x,y) \mapsto (f(x,y), g(x,y))$. The Newton tree of F is the bicolored tree of $(f-t)(g-t')$ for t, t' generic, where arrows corresponding to $f-t$ are coloured in blue and $g-t'$ are coloured in red.

We assume that $J(f,g) = 1$, where $J(f,g)$ is the Jacobian of f and g.

Proposition 3.1. *On each path between the root and a dicritical vertex of f, v, there is no dicritical of g except v if it is a common dicritical.*

PROOF. Assume there is a dicritical vertex of g, v', different from v on the path between the root and v. We assume that $\mathcal{S}_{v'} = \{v_0, \ldots, v'\}$ and the decorations near v_0 are (p_0, q_0). We assume $p_0 > 0$. The proof is the same if $p_0 < 0$. That means that we have $(n,m) \in \mathbb{N}^2$, $a_0 \in \mathbb{C}^*$, $(a_1, \ldots, a_{m-1}) \in \mathbb{C}^{m-1}$ such that

$$\begin{aligned} g(v^{-p_0 n}, a_0 v^{-q_0 n} + a_1 v^{-q_0 n+1} + \cdots + a_{m-1} v^{-q_0 n+m-1} + v^{-q_0 n+m} w) \\ = Q_0(w) + vQ_1(w) + \cdots \end{aligned}$$

with $\deg Q_0(w) \geq 1$.

$$\begin{aligned} f(v^{-p_0 n}, a_0 v^{-q_0 n} + a_1 v^{-q_0 n+1} + \cdots + a_{m-1} v^{-q_0 n+m-1} + v^{-q_0 n+m} w) \\ = P_0(w) v^{-N} + P_1(w) v^{-N+1} + \cdots \end{aligned}$$

with $N > 0$ and $\deg P_0(w) \geq 1$. Computing the Jacobian and we see that $Q_0'(w) P_0(w)$ should be a constant. But it is a polynomial of degree at least one. □

Proposition 3.2. *Let $(p,q) \in \mathbb{Z}^2$ with if $p > 0$, $q \in Z$ and if $p < 0$, $q \in \mathbb{N}$, such that $p + q \leq 0$. Let $(n,m) \in \mathbb{N}^2$. Assume there exist $(a_0, a_1, \ldots, a_{m-1}) \in \mathbb{C}^* \times \mathbb{C}^{m-1}$ such that if $p > 0$*

$$\begin{aligned} f(v^{-pn}, a_0 v^{-qn} + a_1 v^{-qn+1} + \cdots + a_{m-1} v^{-qn+m-1} + v^{-qn+m} w) \\ = P_0(w) + vP_1(w) + \cdots \end{aligned}$$

or if $p < 0$

$$\begin{aligned} f(a_0 v^{-pn} + a_1 v^{-pn+1} + \cdots + a_{m-1} v^{-pn+m-1} + v^{-pn+m} w, v^{-qn}) \\ = P_0(w) + vP_1(w) + \cdots \end{aligned}$$

then, if $p > 0$

$$g(v^{-pn}, a_0 v^{-qn} + a_1 v^{-qn+1} + \cdots + a_{m-1} v^{-qn+m-1} + v^{-qn+m} w) \in \mathbb{C}[v,w]$$

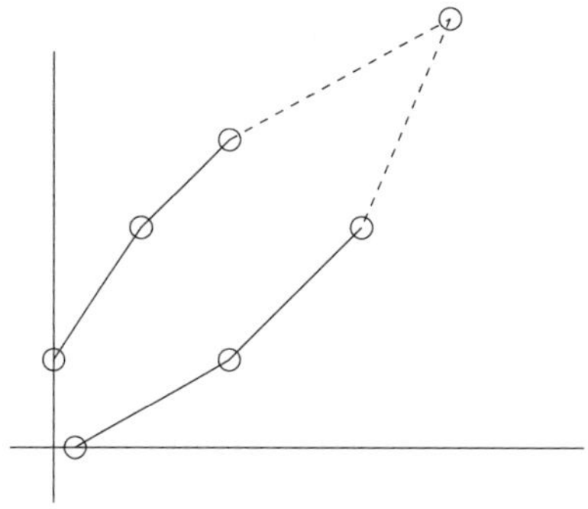

FIGURE 9

or if $p < 0$

$$g(a_0 v^{-pn} + a_1 v^{-pn+1} + \cdots + a_{m-1} v^{-pn+m-1} + v^{-pn+m} w, v^{-qn}) \in \mathbb{C}[v, w].$$

This proposition gives us faces of the Newton polygon where f and g have a common dicritical. These are the faces parallel to the diagonal and the faces in the corner delimited by these faces the axes and the origin (see Figure 9).

PROOF. It is a simple computation of the Jacobian. □

Proposition 3.3. *Let* $(p, q) \in \mathbb{Z}^2$ *with* $p + q \geq 0$. *Let* $n \in \mathbb{N}$ *and* $m \in \mathbb{N}, m > (p+q)n$, $(a_0, a_1, \ldots, a_{m-1}) \in \mathbb{C}^* \times \mathbb{C}^{m-1}$ *such that, if* $p > 0$

$$f(v^{-pn}, a_0 v^{-qn} + a_1 v^{-qn+1} + \cdots + v^{-qn+m} w) = P_0(w) v^N + P_1(w) v^{N+1} + \cdots$$

or if $p < 0$,

$$f(a_0 v^{-pn} + a_1 v^{-pn+1} + \cdots + v^{-pn+m} w, v^{-qn}) = P_0(w) v^N + P_1(w) v^{N+1} + \cdots$$

with $N \geq 0$ *and* $\deg P_0 \geq 1$, *then* $a_{(p+q)n} = 0$.

PROOF. The proof is inspired by David Wright [10] The condition on the Jacobian means that $\mathrm{d}f \wedge \mathrm{d}g = \mathrm{d}x \wedge \mathrm{d}y$. Let $p = f\,\mathrm{d}g - x\,\mathrm{d}y$. Then $p \in \Omega_{\mathbb{C}[x,y]}$, and p is an exact differential. We choose m big enough such that $f\,\mathrm{d}g \in \Omega_{\mathbb{C}[v,w]}$.

$$\begin{aligned} x\,\mathrm{d}y = (&-qna_0 v^{(p+q)n-1} + \cdots + pna_{(p+q)n} v^{-1} + \cdots \\ &+ (-qn + m - 1) a_{m-1} v^{-(p+q)n+m-2} + w v^{(p+q)+m-1})\,\mathrm{d}v + v^{(p+q)n+m}\,\mathrm{d}w. \end{aligned}$$

Then if $a_{(p+q)n} \neq 0$, p is not an exact differential. □

In particular, we have:

Corollary 3.4. *There is no face parallel to the diagonal in the Newton polygon at infinity of* f.

Proposition 3.3 is much stronger than the corollary, but it is not easy to use its strength!

One can notice that A. Bystrikov [1] using also D. Wright ideas has proved a particular case of the corollary. Oka [9] has proven some restrictions on the faces of the Newton polygon for polynomials having a Jacobian mate. Our result is not contained in Oka's.

References

1. A. S. Bystrikov, *On linear chains of blow-ups related to the Jacobian conjecture*, Mat. Zametki **75** (2004), no. 5, 783–786 (Russian); English transl., Math. Notes **75** (2004), no. 5-6, 726–730.
2. P. Cassou-Noguès, *The effect of rational maps on polynomial maps*, Ann. Polon. Math. **76** (2001), no. 1-2, 21–31.
3. ———, *Bad field generators*, Affine Algebraic Geometry (Seville, 2003) (J. Gutierrez, V. Shpilrain, and J.-T. Yu, eds.), Contemp. Math., vol. 369, Amer. Math. Soc., Providence, RI, 2005, pp. 77–83.
4. ———, *A note on a note by Mutsuo Oka*, Acta Math. Vietnam. **32** (2007), no. 2-3, 243–246.
5. P. Cassou-Noguès and A. Libgober, *Multivariable Hodge theorical invariants of germs of plane curves.* II, in preparation.
6. P. Cassou-Noguès and P. Russell, *Birational morphisms* $\mathbb{C}2 \to \mathbb{C}^2$ *and affine ruled surfaces*, Affine Algebraic Geometry (T. Hibi, ed.), Osaka Univ. Press, Osaka, 2007, pp. 57–105.
7. D. Eisenbud and W. Neumann, *Three-dimensional link theory and invariants of plane curve singularities*, Ann. of Math. Stud., vol. 110, Princeton Univ. Press, Princeton, NJ, 1985.
8. W. D. Neumann, *Complex algebraic plane curves via their links at infinity*, Invent. Math. **98** (1989), no. 3, 445–489.
9. M. Oka, *Note on boundary obstruction to Jacobian conjecture of two variables*, Acta Math. Vietnam. **32** (2007), no. 2-3, 123–139.
10. D. Wright, *Affine surfaces fibered by affine lines over the projective line*, Illinois J. Math. **41** (1997), no. 4, 589–605.

Institut de Mathématiques de Bordeaux, Université Bordeaux 1, 351, cours de la Libération, 33405 Talence cedex, France

E-mail address: Pierrette.Cassou-nogues@math.u-bordeaux1.fr

Centre de Recherches Mathématiques
CRM Proceedings and Lecture Notes
Volume **54**, 2011

Polynomials $f(X, Y, Z)$ of low LND-degree

Daniel Daigle

To my teacher, Peter Russell

Abstract. Let B be the polynomial ring in three variables over an algebraically closed field $\mathbf{k}$ of characteristic zero. We give a necessary condition that a polynomial $f \in B$ must satisfy if there exists a nonzero locally nilpotent derivation $D\colon B \to B$ such that $D^2(f) = 0$. In the case $f = X^a + Y^b + Z^c$, we determine the values of a, b, c for which such a derivation exists.

1. Introduction

Let R be an integral domain of characteristic zero. A derivation $D\colon R \to R$ is *locally nilpotent* if for each $x \in R$ there exists $n > 0$ such that $D^n(x) = 0$. We use the notations

$$\begin{aligned} \text{LND}(R) &= \text{set of all locally nilpotent derivations } D\colon R \to R \\ \text{KLND}(R) &= \{\ker(D) \mid D \in \text{LND}(R) \text{ and } D \neq 0\} \end{aligned}$$

where $\ker(D) = \{x \in R \mid D(x) = 0\}$.

We say that R is *rigid* if $\text{KLND}(R) = \varnothing$ (i.e., if $\text{LND}(R) = \{0\}$).

Assume that R is not rigid and consider $D \in \text{LND}(R) \setminus \{0\}$. If $f \in R \setminus \{0\}$, define $\deg_D(f)$ to be the least $i \in \mathbb{N}$ satisfying $D^{i+1}(f) = 0$; also define $\deg_D(0) = -\infty$. Then it is well known that $\deg_D\colon R \to \mathbb{N} \cup \{-\infty\}$ is a degree function on R, i.e., for all $f, g \in R$ there holds: $\deg_D(f) = -\infty \Leftrightarrow f = 0$, $\deg_D(fg) = \deg_D(f) + \deg_D(g)$ and $\deg_D(f + g) \leq \max(\deg_D(f), \deg_D(g))$. Also well known and easy to see is the fact that two elements of $\text{LND}(R) \setminus \{0\}$ determine the same degree function if and only if they have the same kernel. So each $A \in \text{KLND}(R)$ determines a degree function $\deg_A\colon R \to \mathbb{N} \cup \{-\infty\}$. For $f \in R$, we have $\deg_A(f) \leq 0$ if and only if $f \in A$; if $\deg_A(f) = 1$, we call f a *preslice* of A (or a preslice of D, where $D \in \text{LND}(R)$ is such that $\ker D = A$). Discussions between this author and

---

2010 *Mathematics Subject Classification.* Primary 14R10; Secondary: 14R20, 13N15.

Key words and phrases. Locally nilpotent derivation, polynomial ring, affine space, affine variety, graded ring, Pham – Brieskorn variety.

Research supported by NSERC Canada (grant RGPIN104976-2005).

This is the final form of the paper.

Gene Freudenburg led to the idea that one should consider the function

$$\text{LNDEG}\colon R \to \mathbb{N} \cup \{-\infty\}, \quad \text{LNDEG}(f) = \min\{\deg_A(f) \mid A \in \text{KLND}(R)\},$$

and that it would be interesting to characterize the elements $f \in R$ which satisfy

$$\text{LNDEG}(f) \leq 1,$$

or equivalently,

$$D^2(f) = 0 \text{ for some } D \in \text{LND}(R) \setminus \{0\}.$$

We call $\text{LNDEG}(f)$ the "LND-degree" of f, even though LNDEG is not a degree function (the authors of [5] call $\text{LNDEG}(f)$ the "absolute degree" of f, and denote it $|f|_R$).

Consider the polynomial ring $B = \mathbf{k}[X, Y, Z]$, where $\mathbf{k}$ is an algebraically closed field of characteristic zero. Then Theorem 3.11, below, gives a necessary condition that $f \in B$ must satisfy if $\text{LNDEG}(f) \leq 1$. That result is one of the tools used in Section 4 for answering the following question, posed to this author by Freudenburg: *Let* $f = X^a + Y^b + Z^c \in B = \mathbf{k}[X,Y,Z]$; *for which* a, b, c *does there exist* $D \in \text{LND}(B) \setminus \{0\}$ *satisfying* $D^2(f) = 0$? Corollary 4.8 gives the answer. Freudenburg's motivation for asking this question is the fact that its answer, combined with certain results of [5], implies rigidity of the integral domain $\mathbf{k}[X,Y,Z,T]/(X^a+Y^b+Z^c+T^d)$ for certain values of a, b, c, d.

Conventions. All rings are commutative and have a unity. The set of units of a ring R is denoted R^*. If $r \in R$, we denote by R_r the localization $S^{-1}R$ where $S = \{1, r, r^2, \dots\}$. If R is an integral domain, $\operatorname{Frac} R$ is its field of fractions.

If A is a subring of a ring B then the notation $B = A^{[n]}$ means that B is isomorphic as an A-algebra to the polynomial ring in n variables over A. If $K/\mathbf{k}$ is a field extension, we write $K = \mathbf{k}^{(n)}$ to indicate that K is a purely transcendental extension of $\mathbf{k}$, of transcendence degree n.

2. Gradings and homogeneous derivations

In this section we recall some notations, definitions and known facts concerning gradings and homogeneous derivations.

Let R be an integral domain of characteristic zero and let $\mathfrak{g}$ be a $\mathbb{Z}$-grading of R, with notation $R = \bigoplus_{i\in\mathbb{Z}} R_i$. A derivation $D\colon R \to R$ is $\mathfrak{g}$-*homogeneous* if there exists $d \in \mathbb{Z}$ satisfying $D(R_i) \subseteq R_{i+d}$ for all $i \in \mathbb{Z}$. We write

$$\text{LND}(R, \mathfrak{g}) = \{D \in \text{LND}(R) \mid D \text{ is } \mathfrak{g}\text{-homogeneous}\},$$
$$\text{KLND}(R, \mathfrak{g}) = \{\ker(D) \mid D \in \text{LND}(R, \mathfrak{g}) \text{ and } D \neq 0\}.$$

Note that if $A \in \text{KLND}(R, \mathfrak{g})$ then $A = \bigoplus_{i\in\mathbb{Z}} A_i$ where $A_i = R_i \cap A$, i.e., A inherits a grading from $(R, \mathfrak{g})$.

We often write "homogeneous" when we mean $\mathfrak{g}$-homogeneous (i.e., homogeneous with respect to $\mathfrak{g}$), and "$\deg(f)$" when we mean $\deg_{\mathfrak{g}}(f)$. We observe:

2.1. Lemma. *Let R be an integral domain and a finitely generated algebra over a field $\mathbf{k}$ of characteristic zero, and assume that R is not rigid. For any $\mathbb{Z}$-grading $\mathfrak{g}$ of R, the following hold.*

(1) $\text{KLND}(R, \mathfrak{g}) \neq \varnothing$.

(2) *For each* $A \in \text{KLND}(R)$, *there exists* $\mathcal{A} \in \text{KLND}(R, \mathfrak{g})$ *satisfying*

$$\deg_{\mathcal{A}} \bar{f} \leq \deg_A f \quad \text{for all } f \in R,$$

where $\bar{f}$ *stands for the highest* $\mathfrak{g}$*-homogeneous component of* f.

(3) *For any* $\mathfrak{g}$*-homogeneous* $h \in R$,

$$\text{LNDEG}(h) = \min\{\deg_A(h) \mid A \in \text{KLND}(R, \mathfrak{g})\}.$$

In particular, the following conditions are equivalent:
- *there exists* $D \in \text{LND}(R) \setminus \{0\}$ *satisfying* $D^2(h) = 0$
- *there exists* $D \in \text{LND}(R, \mathfrak{g}) \setminus \{0\}$ *satisfying* $D^2(h) = 0$.

PROOF. Assertions (1) and (3) follow from (2), and (2) follows from the well-known fact (cf. for instance [10, Lemmas 2.7 and 2.8]) that, given $D \in \text{LND}(R) \setminus \{0\}$, the homogenization $\mathcal{D}$ of D (in the sense of [10, Definition 2.6]) is a nonzero element of $\text{LND}(R, \mathfrak{g})$ which satisfies $\deg_{\mathcal{D}} \bar{f} \leq \deg_D f$ for all $f \in R$. □

We are particularly interested in $\mathbb{Z}$-gradings of polynomial rings. The following definitions and facts (2.2 and Theorem 2.3) can be found in Chapter 2 of [7].

2.2. Let $\mathbf{k}$ be a field of characteristic zero, $B = \mathbf{k}^{[n]}$ a polynomial ring over $\mathbf{k}$ and $\mathfrak{g}$ a $\mathbb{Z}$-grading of B. A *homogeneous coordinate system* of B is an ordered n-tuple $\gamma = (X_1, \dots, X_n)$ of elements of B such that $B = \mathbf{k}[X_1, \dots, X_n]$ and such that each X_i is homogeneous (with respect to $\mathfrak{g}$). If there exists a homogeneous coordinate system of B, we say that the grading $\mathfrak{g}$ is *coordinatizable*.

Fix a $\mathbb{Z}$-grading $\mathfrak{g}$ of B which is coordinatizable and nontrivial. Then each homogeneous coordinate system $\gamma = (X_1, \dots, X_n)$ of B determines an n-tuple $\alpha(\mathfrak{g}, \gamma) = (\alpha_1, \dots, \alpha_n)$, as follows: first, let $d_i = \deg(X_i)/\gcd\big(\deg(X_1), \dots, \deg(X_n)\big)$, $1 \leq i \leq n$; then,

$$\alpha_i = \gcd\{d_j \mid j \in \{1, \dots, n\} \setminus \{i\}\}.$$

Note that $\alpha_1, \dots, \alpha_n$ are pairwise relatively prime natural numbers. It can be shown that, up to permutation, the n-tuples $\big(\deg(X_1), \dots, \deg(X_n)\big)$ and $\alpha(\mathfrak{g}, \gamma)$ are independent of the choice of γ. So the cardinality of the set $\{i \mid \alpha_i \neq 1\}$ is completely determined by $\mathfrak{g}$; this cardinality is called the *type* of $\mathfrak{g}$. We also define the type of the trivial grading to be zero.[1] So each coordinatizable grading $\mathfrak{g}$ of B has a type, and $\operatorname{type} \mathfrak{g} \in \{0, 1, \dots, n\}$.

Also define $d(A) = \gcd\{\deg(h) \mid h \text{ is a nonzero homogeneous element of } A\}$ for each $A \in \text{KLND}(B, \mathfrak{g})$.

2.3. Theorem (Kolhatkar). *Let* $\mathbf{k}$ *be a field of characteristic zero,* $B = \mathbf{k}^{[n]}$ *a polynomial ring, and* $\mathfrak{g}$ *a* $\mathbb{Z}$*-grading of* B *which is coordinatizable and nontrivial. Let* $\gamma = (X_1, \dots, X_n)$ *be a homogeneous coordinate system of* B*. Define* $\alpha(\mathfrak{g}, \gamma) = (\alpha_1, \dots, \alpha_n)$ *and* $\operatorname{type}(\mathfrak{g})$ *as in 2.2.*

(1) $\{d(A) \mid A \in \text{KLND}(B, \mathfrak{g})\} = \{\alpha_1, \dots, \alpha_n\}$

(2) *Suppose that* $A \in \text{KLND}(B, \mathfrak{g})$ *and* $i \in \{1, \dots, n\}$ *are such that* $d(A) = \alpha_i \neq 1$*. Then the set of homogeneous prime elements* h *of* B *satisfying* $\deg_A(h) = 1$ *is equal to* $\{\lambda X_i \mid \lambda \in \mathbf{k}^*\}$.

(3) *For each* $A \in \text{KLND}(B, \mathfrak{g})$ *we have* $\{X_i \mid i \text{ satisfies } \alpha_i \notin \{1, d(A)\}\} \subset A$*, so in particular* $|\{X_1, \dots, X_n\} \cap A| \geq \operatorname{type}(\mathfrak{g}) - 1$.

---

[1]The trivial grading is of course not the only one of type 0. For nontrivial $\mathfrak{g}$, we have $\operatorname{type}(\mathfrak{g}) = 0$ if and only if $d_1, \dots, d_n$ are $(n-1)$-wise relatively prime.

3. A criterion

The aim of this section is to prove Theorem 3.11 (stated below). We first recall some known facts.

3.1. Let R be a domain containing $\mathbb{Q}$, $D\colon R \to R$ a nonzero locally nilpotent derivation and $A = \ker D$. The following facts are well known, see for instance [4] or [9].

(a) *A is a factorially closed subring of R (that is, the conditions $x, y \in R \setminus \{0\}$ and $xy \in A$ imply $x, y \in A$). Consequently, $A^* = R^*$; if k is any field contained in R then D is a k-derivation; A is algebraically closed in R (i.e., each element of R which is a root of a nonzero polynomial $f(T) \in A[T]$ belongs to A); if R is a UFD then so is A.*

(b) *If $s \in R$ is any preslice of D, and if we write $a = Ds$, then $R_a = A_a[s] = {A_a}^{[1]}$.*

(c) *The transcendence degree of R over A is 1.*

We also need the following consequence of a result of Miyanishi [8]:

3.2. *Let $B = \mathbf{k}^{[3]}$ where $\mathbf{k}$ be a field of characteristic zero.*

(a) *If $D \in \text{LND}(B) \setminus \{0\}$, then $\ker D = \mathbf{k}^{[2]}$.*

(b) *If $D \in \text{LND}(B, \mathfrak{g}) \setminus \{0\}$ where $\mathfrak{g}$ is a $\mathbb{Z}$-grading of B, then $\ker D = \mathbf{k}[f, g]$ for some $\mathfrak{g}$-homogeneous elements $f, g \in B$.*

3.3. Lemma. *Let R be a UFD containing $\mathbb{Q}$ and let $\mathfrak{g}$ be a nontrivial $\mathbb{Z}$-grading on R (say $R = \bigoplus_{i \in \mathbb{Z}} R_i$). There exists $D \in \text{LND}(R, \mathfrak{g})$ such that the grading on $\ker(D)$ is trivial if and only if $R = R_0[v] = R_0^{[1]}$ for some homogeneous element $v \in R$ of nonzero degree.*

PROOF. Assume that $D \in \text{LND}(R, \mathfrak{g})$ is such that the grading on $\ker(D)$ is trivial. Note that $D \neq 0$; so R has transcendence degree 1 over $A = \ker D$. As $A \subseteq R_0 \subseteq R$ and R is not algebraic over R_0, R_0 is algebraic over A; as A is algebraically closed in R,

$$A = R_0.$$

Pick a homogeneous preslice $s' \in R$ of D. Note that $R^* \subset A$, so $s' \notin R^*$; consequently s' is a (nonempty) product of prime elements. As $\deg_D(s') = 1$, some prime factor of s' is a preslice of D. So there exists a preslice $s \in R$ of D which is prime and homogeneous. Let $\alpha = D(s) \in A \setminus \{0\}$, then $R_\alpha = A_\alpha[s] = A_\alpha^{[1]}$. If $r \in R \setminus \{0\}$ is homogeneous then $\alpha^N r = \sum_{i=0}^m a_i s^i$ for some $N, m \geq 0$, $a_i \in A = R_0$. Because $\alpha^N r$ is homogeneous, it follows that $\alpha^N r = a_i s^i$ for some $i \geq 0$. Then $s^i \mid \alpha^N r$ and $\gcd(s^i, \alpha) = 1$ imply that $s^i \mid r$, so $r = as^i$ for some $a \in R_0$. We have shown that $R = R_0[s] = R_0^{[1]}$. The converse is trivial. □

3.4. Definition. Let $R = \mathbf{k}[X_1, \dots, X_n] = \mathbf{k}^{[n]}$, where $n \geq 2$ and $\mathbf{k}$ is a field of characteristic zero. Given $(a_1, \dots, a_n) \in \mathbb{Z}^n$, let $\mathfrak{g}(a_1, \dots, a_n)$ denote the $\mathbb{Z}$-grading of R obtained by stipulating that X_i is homogeneous of degree a_i. Note that $\mathfrak{g}(a_1, \dots, a_n)$ is a coordinatizable grading and that $(X_1, \dots, X_n)$ is a homogeneous coordinate system (cf. 2.2). We say that $\mathfrak{g}(a_1, \dots, a_n)$ is an *admissible* grading if $\{i \mid a_i \neq 0\}$ has cardinality at least 2.

3.5. Lemma. *Let* $\mathfrak{g} = \mathfrak{g}(a_1,\dots,a_n)$ *be an admissible* $\mathbb{Z}$*-grading of* $R = \mathbf{k}[X_1,\dots,X_n] = \mathbf{k}^{[n]}$*, where* $n \geq 2$ *and* $\mathbf{k}$ *is a field of characteristic zero. Then for each* $D \in \operatorname{LND}(R,\mathfrak{g})$*, the grading on* $\ker(D)$ *is nontrivial.*

PROOF. By contradiction, suppose that there exists $D \in \operatorname{LND}(R,\mathfrak{g})$ such that the grading on $\ker(D)$ is trivial. Write $R = \bigoplus_{i\in\mathbb{Z}} R_i$. It follows from Lemma 3.3 that $R = R_0^{[1]}$, so in particular (i) R_0 is factorially closed in R, and (ii) R has transcendence degree 1 over R_0. By (i), all nonzero a_i have the same sign; then, by (ii), only one a_i is nonzero so $\mathfrak{g}$ is not admissible. □

Until the end of this section, we assume that $\mathbf{k}$ is an algebraically closed field of characteristic zero and $B = \mathbf{k}[X,Y,Z] = \mathbf{k}^{[3]}$. The following notion is needed in the proof of Lemma 3.9:

3.6. Definition. Let U be a nonsingular algebraic surface over $\mathbf{k}$. We say that U is *completable by rational curves* if there exists an open immersion $U \hookrightarrow \Omega$ such that Ω is a nonsingular projective surface and $\Omega \setminus U$ is a union of rational curves.

3.7. Remark. Let U be an open subset of a (possibly singular) surface S and assume that U is nonsingular and completable by rational curves. If $C \subset S$ is a curve lying in the complement of U then C is a rational curve.

3.8. Notations. Let R be an $\mathbb{N}$-graded ring and $h \in R$ a homogeneous element. Then we write $V(h)$ for the closed subset $\{\mathfrak{p} \in \operatorname{Spec} R \mid h \in \mathfrak{p}\}$ of $\operatorname{Spec} R$, $V_+(h)$ for the closed subset $\{\mathfrak{p} \in \operatorname{Proj} R \mid h \in \mathfrak{p}\}$ of $\operatorname{Proj} R$ and $D_+(h) = \operatorname{Proj}(R) \setminus V_+(h)$. As is well known, if $\deg(h) > 0$ then $D_+(h) \cong \operatorname{Spec} R_{(h)}$ where $R_{(h)}$ is the component of degree zero of the $\mathbb{Z}$-graded localized ring R_h. If $(a,b,c) \in \mathbb{N}^3 \setminus \{(0,0,0)\}$ and $\mathbf{k}[X,Y,Z]$ is equipped with the $\mathbb{N}$-grading $\mathfrak{g}(a,b,c)$, then $\operatorname{Proj} \mathbf{k}[X,Y,Z]$ is denoted $\mathbb{P}(a,b,c)$.

3.9. Lemma. *Let* $(a,b,c) \in \mathbb{N}^3$ *be such that at most one of* a,b,c *is zero, and consider the* $\mathbb{N}$*-grading* $\mathfrak{g} = \mathfrak{g}(a,b,c)$ *of* $B = \mathbf{k}[X,Y,Z]$*. Let* h *be a nonconstant homogeneous element of* B *and consider the closed set* $V_+(h) \subset \mathbb{P}(a,b,c)$*. If* $D^2(h) = 0$ *for some* $D \in \operatorname{LND}(B,\mathfrak{g}) \setminus \{0\}$*, then each irreducible component of* $V_+(h)$ *is a rational curve.*

PROOF. We first consider the case where $D(h) = 0$. To prove that $V_+(h) \subset \mathbb{P}(a,b,c)$ is a union of rational curves, we may assume that $\gcd(a,b,c) = 1$. Let $A = \ker(D)$, then (cf. 3.2) $A = \mathbf{k}[f,g] = \mathbf{k}^{[2]}$ where f,g are homogeneous elements of B. By Lemma 3.5, the grading on A is nontrivial; so the integer $d = \gcd(\deg f, \deg g)$ is nonzero and hence satisfies $d \geq 1$, which implies (by [1, Corollary 3.10]) that $B_{(fg)} = (A_{(fg)})^{[1]}$. So for some v transcendental over $A_{(fg)}$ we have

$$B_{(fg)} = (A_{(fg)})[v] = (A_{(fg)})^{[1]}. \tag{3.1}$$

Define $p = (\deg f)/d$, $q = (\deg g)/d$ and $\xi = f^q/g^p$; then $A_{(fg)} = \mathbf{k}[\xi,\xi^{-1}]$. As h is a homogeneous element of A, we have $h = \lambda f^r g^s \prod_{k=1}^m (f^q - \lambda_k g^p)$ for some $r,s,m \in \mathbb{N}$ and $\lambda, \lambda_k \in \mathbf{k}^*$, so we may write $fgh = \lambda f^i g^j \prod_{k=1}^m (\xi - \lambda_k)$ (where $i,j \in \mathbb{N}$). Let us record this as

$$fgh = f^i g^j \varphi \quad \text{for some } i,j \in \mathbb{N} \text{ and } \varphi \in \mathbf{k}[\xi] \setminus \{0\} \subseteq A_{(fg)} \setminus \{0\}.$$

This implies that $A_{(fgh)} = A_{(fg)}[1/\varphi]$ and $B_{(fgh)} = B_{(fg)}[1/\varphi]$ so, by (3.1),

$$B_{(fgh)} = B_{(fg)}[1/\varphi] = A_{(fg)}[v, 1/\varphi] = A_{(fgh)}[v].$$

Note that v is transcendental over $A_{(fgh)}$, because it is transcendental over $A_{(fg)}$ and $A_{(fgh)} = A_{(fg)}[1/\varphi]$ is a localization of $A_{(fg)}$. Consequently,

$$B_{(fgh)} = (A_{(fgh)})^{[1]} = \mathbf{k}\big[\xi, \xi^{-1}, \varphi^{-1}\big]^{[1]}.$$

We have $\deg(fgh) > 0$, because the grading on A is not trivial. It follows that the open subset $D_+(fgh)$ of $\mathbb{P}(a,b,c)$ is isomorphic to $\operatorname{Spec} B_{(fgh)}$, which is isomorphic to $C_0 \times \mathbb{A}^1_{\mathbf{k}}$, where $C_0 = \operatorname{Spec} \mathbf{k}[\xi, \xi^{-1}, \varphi(\xi)^{-1}]$ is an affine line minus finitely many points; so $D_+(fgh)$ is isomorphic to $\mathbb{P}^1_{\mathbf{k}} \times \mathbb{P}^1_{\mathbf{k}}$ minus finitely many projective lines; in particular, $D_+(fgh)$ is completable by rational curves. If $C \subset \mathbb{P}(a,b,c)$ is an irreducible component of $V_+(h)$, then C lies in the complement of $D_+(fgh)$ and hence is a rational curve by Remark 3.7. Thus $V_+(h)$ is a union of rational curves.

Next, we consider the case where $D(h) \neq 0$ and $D^2(h) = 0$. Noting that $\deg_D(h) = 1$, we see that $h = h_0 h_1$ where $h_0 \in A$, h_1 is an irreducible element of B, $D(h_1) \neq 0$ and $D^2(h_1) = 0$, and of course h_0, h_1 are homogeneous. The first part of the proof implies that $V_+(h_0) \subset \mathbb{P}(a,b,c)$ is a union of rational curves. It is well known that if s is an irreducible element of B satisfying $\Delta^2(s) = 0$ and $\Delta(s) \neq 0$ for some $\Delta \in \text{LND}(B)$, then $V(s) \subset \operatorname{Spec}(B)$ is a rational surface;[2] thus $V(h_1) \subset \operatorname{Spec}(B)$ is a rational surface. As $V(h_1) \subset \operatorname{Spec}(B)$ is a cone over $V_+(h_1) \subset \mathbb{P}(a,b,c)$, it is birational to $V_+(h_1) \times \mathbb{P}^1$; so $V_+(h_1)$ is a rational curve. So $V_+(h) = V_+(h_0) \cup V_+(h_1)$ is a union of rational curves. □

3.10. Notation. Let $f \in B = \mathbf{k}[X,Y,Z]$ be a nonconstant polynomial. Given $(a,b,c) \in \mathbb{N}^3 \setminus \{(0,0,0)\}$, let $\Gamma(f; a,b,c) = V_+(h) \subset \mathbb{P}(a,b,c)$ where h denotes the highest $\mathfrak{g}(a,b,c)$-homogeneous component of f.

3.11. Theorem. *Let $B = \mathbf{k}[X,Y,Z] = \mathbf{k}^{[3]}$ where $\mathbf{k}$ is an algebraically closed field of characteristic zero, and let $f \in B$ be a nonconstant polynomial such that $D^2(f) = 0$ for some $D \in \text{LND}(B) \setminus \{0\}$. Then, for each choice of $(a,b,c) \in \mathbb{N}^3$ such that at most one of a,b,c is zero, each irreducible component of $\Gamma(f;a,b,c)$ is a rational curve.*

PROOF. Let (a,b,c) be as in the statement, let $\mathfrak{g} = \mathfrak{g}(a,b,c)$ and let h be the highest $\mathfrak{g}$-homogeneous component of f. If $D \in \text{LND}(B) \setminus \{0\}$ satisfies $D^2(f) = 0$, then by Lemma 2.1 there exists $\widetilde{D} \in \text{LND}(B, \mathfrak{g}) \setminus \{0\}$ satisfying $\widetilde{D}^2(h) = 0$, so the result follows from Lemma 3.9. □

4. An application

The aim of this section is to prove that $P(3)$ is true, where $P(n)$ is the following assertion:

4.1. Statement $P(n)$. *Let $\mathbf{k}$ be a field of characteristic zero, $B = \mathbf{k}[X_1, \dots, X_n] = \mathbf{k}^{[n]}$ and $f = X_1^{a_1} + \dots + X_n^{a_n} \in B$ where $(a_1, \dots, a_n) \in \mathbb{N}^n$. If $\min(a_1, \dots, a_n) \geq 2$ and at most one a_i is equal to 2, then no element D of $\text{LND}(B) \setminus \{0\}$ satisfies $D^2(f) = 0$.*

To be precise, "$P(n)$ is true" means that the assertion is true for all fields $\mathbf{k}$ of characteristic zero. However, we observe that it's enough to prove $P(n)$ under the assumption that $\mathbf{k}$ is algebraically closed, so we are free to make this assumption

[2]The composite $V(s) \hookrightarrow \operatorname{Spec} B \to \operatorname{Spec}(\ker \Delta) = \mathbb{A}^2_{\mathbf{k}}$ is a birational morphism.

whenever it is convenient. It is trivial that $P(1)$ is true. We prove $P(2)$ and $P(3)$ below, and also show that if $P(n-1)$ is true then $P(n)$ is at least partially true.

The hypersurfaces defined by $X_1^{a_1}+\cdots+X_n^{a_n}=0$ in affine n-space are known as Pham – Brieskorn varieties.

4.2. Let $\mathbf{k}$ be a field of characteristic zero and $n \ge 2$. Let the notation be as in 4.1 and suppose that $\min(a_1,\dots,a_n) \ge 2$ and that at most one a_i is equal to 2. Let $M=\operatorname{lcm}(a_1,\dots,a_n)$, let $a_i^*=M/a_i$ $(1\le i\le n)$ and note that $\gcd(a_1^*,\dots,a_n^*)=1$. Let $\mathfrak{g}$ be the $\mathbb{N}$-grading of B defined by stipulating that X_i is homogeneous of degree a_i^*, and note that f is $\mathfrak{g}$-homogeneous of degree M. Thanks to Lemma 2.1 we know that, in order to prove $P(n)$,

it's enough to show that no $D \in \text{LND}(B,\mathfrak{g})\setminus\{0\}$ satisfies $D^2(f)=0$.

Observe that $\mathfrak{g}$ is a coordinatizable grading of B and that $\gamma=(X_1,\dots,X_n)$ is a homogeneous coordinate system of B; so the concepts of 2.2 can be used here. In particular, we have the n-tuple $\alpha(\mathfrak{g},\gamma)=(\alpha_1,\dots,\alpha_n)$, defined by

$$\alpha_i=\gcd\{a_j^* \mid j\in\{1,\dots,n\}\setminus\{i\}\},$$

and we have $\operatorname{type}\mathfrak{g}\in\{0,1,\dots,n\}$ defined by $\operatorname{type}(\mathfrak{g})=|\{i \mid \alpha_i\neq 1\}|$. It is convenient to define

$$\operatorname{cotype}(a_1,\dots,a_n)=\operatorname{type}\mathfrak{g}.$$

4.3. Lemma. *Let $n\ge 2$ and assume that $P(n-1)$ is true. Then the following assertions are also true (where the notations and assumptions are as in 4.2):*

(1) *If $D\in \text{LND}(B)\setminus\{0\}$ satisfies $D(X_i)=0$ for some $i\in\{1,\dots,n\}$, then $D^2(f)\neq 0$.*

(2) *$P(n)$ is true for all $(a_1,\dots,a_n)$ satisfying* $\operatorname{cotype}(a_1,\dots,a_n)>1$.

PROOF. Consider $D\in\text{LND}(B)\setminus\{0\}$ satisfying $D(X_i)=0$ for some $i\in\{1,\dots,n\}$, and let us show that $D^2(f)\neq 0$. Without loss of generality, we may assume that $D(X_n)=0$. Let $K=\mathbf{k}(X_n)$, $\mathcal{B}=K[X_1,\dots,X_{n-1}]=K^{[n-1]}$ and $F=X_1^{a_1}+\cdots+X_{n-1}^{a_{n-1}}\in\mathcal{B}$. Since $D(X_n)=0$, D extends to $\mathcal{D}\in\text{LND}(\mathcal{B})\setminus\{0\}$. If $D^2(f)=0$ then $\mathcal{D}^2(f)=0$, so $\mathcal{D}^2(F)=0$, which contradicts the assumption that $P(n-1)$ is true. So $D^2(f)\neq 0$ and assertion (1) is proved.

Suppose that $\operatorname{cotype}(a_1,\dots,a_n)>1$. We have to show that $D^2(f)\neq 0$ for all $D\in\text{LND}(B,\mathfrak{g})\setminus\{0\}$, so let us consider such a D. By Theorem 2.3,

$$|\{X_1,\dots,X_n\}\cap\ker D|\ge \operatorname{type}(\mathfrak{g})-1,$$

and since $\operatorname{type}(\mathfrak{g})-1=\operatorname{cotype}(a_1,\dots,a_n)-1>0$ we have $D(X_i)=0$ for some i, so we are done by the first assertion. □

4.4. Lemma. *$P(2)$ is true.*

PROOF. As we already noted, we may assume that $\mathbf{k}$ is algebraically closed. Let the notations and assumptions be as in 4.2, with $n=2$. In view of Lemma 4.3 (and since $P(1)$ is true), we may assume that $\operatorname{cotype}(a_1,a_2)\le 1$. As $\alpha_1=a_2^*=M/a_2$ and $\alpha_2=a_1^*=M/a_1$, this assumption implies that one of M/a_1, M/a_2 is equal to 1, so one of a_1,a_2 divides the other. Without loss of generality, we may assume that $a_1\mid a_2$. Then for some $m\ge 1$ we have $f=X_1^{a_1}+X_2^{ma_1}=\prod_{i=1}^{a_1}g_i$, where $g_i=X_1+\lambda_iX_2^m$ for all $i=1,\dots,a_1$, and where $\lambda_1,\dots,\lambda_{a_1}$ are distinct elements of $\mathbf{k}^*$.

Now suppose that there exists $D \in \text{LND}(B, \mathfrak{g})$ such that $D^2(f) = 0$, and let $A = \ker D$. It is not possible to have $\deg_D(g_i) = 0$ (i.e., $g_i \in A$) for two values of i, because that would imply $A = \mathbf{k}[X_1, X_2]$, a contradiction. As $1 = \deg_D(f) = \sum_{i=1}^{a_1} \deg_D(g_i)$ and $a_1 \geq 2$, it follows that $a_1 = 2$ and that $\{\deg_D(g_1), \deg_D(g_2)\} = \{0, 1\}$. Then

$$1 = \deg_D(g_1 - g_2) = \deg_D\big((X_1 + \lambda_1 X_2^m) - (X_1 + \lambda_2 X_2^m)\big) = m \deg_D(X_2),$$

so $m = 1$ and hence $(a_1, a_2) = (2, 2)$, a contradiction. So D does not exist. □

4.5. Lemma. *Let S be a nonsingular surface over an algebraically closed field $\mathbf{k}$ of characteristic zero, let $C \subset S$ be a curve and let $P \in S$ be a point. Recall that the completion of the local ring $\mathcal{O}_{S,P}$ is isomorphic to $\mathbf{k}[\![x, y]\!] = \mathbf{k}^{[\![2]\!]}$ and suppose that the local equation of C at P is of the form*

$$x^a + y^b + \sum_{i,j} c_{ij} x^i y^j \in \mathbf{k}[\![x, y]\!],$$

where $a, b \geq 0$ and where $bi + aj > ab$ for all i, j such that $c_{ij} \neq 0$. Then

$$\sum_Q \mu(Q)(\mu(Q) - 1) = ab - a - b + \gcd(a, b), \tag{4.1}$$

where Q (in the left-hand side) runs in the set of all singular points of C infinitely near P, and where $\mu(Q)$ is the multiplicity of the singular point Q of C.

PROOF. We may assume that $a \leq b$. We may also assume that $P \in C$, because otherwise the left-hand side of (4.1) is the empty sum and the right-hand side adds up to zero (because $\min(a, b) = 0$). Let

$$x_0 = q_1 x_1 + x_2, \quad \dots, \quad x_{p-2} = q_{p-1} x_{p-1} + x_p, \quad x_{p-1} = q_p x_p$$

be the Euclidean algorithm of (a, b), where $x_0 = b$, $x_1 = a$, all x_i and q_i are positive integers and $x_1 > \cdots > x_p \geq 1$ (so that $\gcd(a, b) = x_p$). Define n and $r_1, r_2, \dots, r_n$ by

$$(r_1, r_2, \dots, r_n) = (\underbrace{x_1, \dots, x_1}_{q_1 \text{ times}}, \dots, \underbrace{x_{p-1}, \dots, x_{p-1}}_{q_{p-1} \text{ times}}, \underbrace{x_p, \dots, x_p}_{q_p \text{ times}});$$

then it is an exercise in elementary arithmetic to show that

$$\sum_{i=1}^{n} r_i(r_i - 1) = ab - a - b + \gcd(a, b). \tag{4.2}$$

On the other hand, the minimal resolution of the singularity of the germ (C, P) is a sequence of n blowings-up $S = S_0 \xleftarrow{\pi_1} S_1 \xleftarrow{\pi_2} \cdots \xleftarrow{\pi_n} S_n$, where π_i is the blowing-up of S_{i-1} at a point Q_i, $Q_1 = P$, and $\pi_i(Q_{i+1}) = Q_i$ for $1 \leq i < n$, and where the multiplicity of the point $Q_i \in S_{i-1}$ on the strict transform $C_{i-1} \subset S_{i-1}$ of $C_0 = C$ is equal to r_i. Here we note that the local equation of C_{n-1} at Q_n has the form

$$x^{r_n} + y^{r_n} + \text{higher degree terms} = 0,$$

so Q_n is an ordinary singularity of C_{n-1} of multiplicity r_n, and the nth blowing-up resolves the singularity. This implies that the left-hand sides of (4.2) and (4.1) are equal. □

4.6. Lemma. *Let* $\mathbf{k}$ *be an algebraically closed field of characteristic zero and consider* $F = Y^M + X^N + X^n \in \mathbf{k}[X,Y] = \mathbf{k}^{[2]}$, *where* $M \geq 2$ *and* $N > n \geq 0$. *Let* $C \subset \mathbb{A}^2_{\mathbf{k}}$ *be the zero-set of* F. *Then* C *is rational if and only if one of the following conditions is satisfied*:

(a) $F = Y^2 + X^{2k+2} + X^{2k}$, $k \geq 0$;
(b) $F = Y^M + X^{kM+1} + X^{kM}$, $k \geq 0$, $M \geq 2$;
(c) $F = Y^M + X^{kM} + X^{kM-1}$, $k \geq 1$, $M \geq 2$.

PROOF. The closure $\overline{C} \subset \mathbb{P}^2$ of $C \subset \mathbb{A}^2$ is the zero-set of the homogeneization $F^* \in \mathbf{k}[X,Y,Z]$ of F. We compute the geometric genus g of $\overline{C}$.

Consider first the case where $M < N$. Then $\deg(F) = N$ and $F^* = Y^M Z^{N-M} + X^N + X^n Z^{N-n}$, so the singular points of $\overline{C}$ belong to $\{P_0, P_\infty\}$ where $P_0 = (0:0:1)$ and $P_\infty = (0:1:0)$. The local equations f_0 and f_∞ of $\overline{C}$ at P_0 and P_∞ respectively are:

$$f_0 = F = Y^M + X^N + X^n \in \mathbf{k}[\![X,Y]\!], \qquad (a,b) = (M,n),$$
$$f_\infty = F^*(X,1,Z) = Z^{N-M} + X^N + X^n Z^{N-n} \in \mathbf{k}[\![X,Z]\!], \quad (a,b) = (N, N-M),$$

where the pairs (a,b) of Lemma 4.5 are also indicated. Set $d_0 = \gcd(M,n)$ and $d_\infty = \gcd(N, N-M) = \gcd(M,N)$, then by Lemma 4.5 and the genus formula we get

$$\begin{aligned}2g &= (N-1)(N-2) - (Mn - M - n + d_0) - (N(N-M) - N - (N-M) + d_\infty) \\ &= (M-1)(N-n) + 2 - d_0 - d_\infty.\end{aligned}$$

So, in the case where $M < N$, we obtain:

(4.3) $2g = (M-1)(N-n) + 2 - d_0 - d_\infty, \quad d_0 = \gcd(M,n), \quad d_\infty = \gcd(M,N).$

We claim that (4.3) continues to be valid in the case where $M \geq N$. Indeed, in this case we have $\deg(F) = M$ and $F^* = Y^M + X^N Z^{M-N} + X^n Z^{M-n}$, so $\operatorname{Sing}(\overline{C}) \subseteq \{P_0, P_\infty\}$ where $P_0 = (0:0:1)$ and $P_\infty = (1:0:0)$. The local equations f_0 and f_∞ of $\overline{C}$ at P_0 and P_∞ respectively are:

$$f_0 = F = Y^M + X^N + X^n \in \mathbf{k}[\![X,Y]\!], \qquad (a,b) = (M,n),$$
$$f_\infty = F^*(1,Y,Z) = Y^M + Z^{M-N} + Z^{M-n} \in \mathbf{k}[\![Y,Z]\!], \quad (a,b) = (M, M-N),$$

where the pairs (a,b) of Lemma 4.5 are also indicated. Set $d_0 = \gcd(M,n)$ and $d_\infty = \gcd(M, M-N) = \gcd(M,N)$, then by Lemma 4.5 and the genus formula we get

$$\begin{aligned}2g &= (M-1)(M-2) - (Mn - M - n + d_0) - (M(M-N) - M - (M-N) + d_\infty) \\ &= (M-1)(N-n) + 2 - d_0 - d_\infty,\end{aligned}$$

which is identical to (4.3) (and the values of d_0 and d_∞ also agree with (4.3)). So (4.3) is valid in all cases.

Using (4.3), it is easy to check that if one of conditions (a–c) holds (see the statement of the lemma) then $g = 0$, so C is rational.

Conversely, suppose that $g = 0$; using (4.3), we show that one of conditions (a)–(c) must hold. Note that d_0, d_∞ satisfy (in particular) $d_0 \mid M$ and $d_\infty \mid M$, so $\max(d_0, d_\infty) \leq M$.

If $N - n \geq 3$ then $0 = 2g \geq 3(M-1) + 2 - d_0 - d_\infty = (M-1) + (M-d_0) + (M-d_\infty) \geq M-1$, so $M \leq 1$, a contradiction. So we must have $N - n \in \{1,2\}$.

If $N-n=2$ then $0=2g=2(M-1)+2-d_0-d_\infty=(M-d_0)+(M-d_\infty)$ implies that $d_0=M=d_\infty$, which implies that $M\mid n$ and $M\mid N$; then M divides the number $N-n=2$, and since $M\geq 2$ we get $M=2$. As $M=2$ divides n and $N=n+2$, condition (a) holds.

If $N-n=1$ then $0=2g=(M-1)+2-d_0-d_\infty$, so $d_0+d_\infty=M+1$. This implies that $\max(d_0,d_\infty)>M/2$, so $\max(d_0,d_\infty)=M$. As one of d_0,d_∞ is equal to M, we have $M\mid n$ or $M\mid N$; in the former case (b) holds, and in the latter case (c) holds. □

4.7. Proposition. *$P(3)$ is true.*

PROOF. We may assume that $\mathbf{k}$ is algebraically closed. Let $n=3$ and let the notations and assumptions be as in 4.2. The first part of the proof consists in showing that if (a_1,a_2,a_3) satisfies the hypothesis of $P(3)$ and $\operatorname{cotype}(a_1,a_2,a_3)<3$, then one of the following conditions must hold:[3]

(i) the curve $C_{(a_1,a_2,a_3)}=V_+(f)\subset\mathbb{P}(a_1^*,a_2^*,a_3^*)$ is not rational;
(ii) $\operatorname{cotype}(a_1,a_2,a_3)=2$;
(iii) up to permutation, (a_1,a_2,a_3) is (b,a,a) for some $a\geq 3$ and $b\geq 2$ such that $\gcd(a,b)=1$.

Suppose that (a_1,a_2,a_3) satisfies the hypothesis of $P(3)$ and $\operatorname{cotype}(a_1,a_2,a_3)<3$. This condition on the cotype means that $\gcd(a_i^*,a_j^*)=1$ for some $i,j\in\{1,2,3\}$, so, replacing if necessary (a_1,a_2,a_3) by a permutation of it, we may arrange that

$$\gcd(a_2^*,a_3^*)=1. \tag{4.4}$$

Observe that $a_2^*\mid a_3$. Indeed, by definition of a_i^* we have $a_ia_i^*=M$ for all i, where $M=\operatorname{lcm}(a_1,a_2,a_3)$. So $a_2a_2^*=a_3a_3^*$, and we get $a_2^*\mid a_3$ by (4.4). The integer $a_3/a_2^*=a_2/a_3^*$ plays a role in the analysis below, and it is convenient to give it a name. Define

$$\rho=a_3/a_2^*=a_2/a_3^*\in\mathbb{N}\setminus\{0\}.$$

Then we note that

$$(a_1,a_2,a_3)=(a_1,\rho a_3^*,\rho a_2^*)\quad\text{and}\quad M=\rho a_2^*a_3^*. \tag{4.5}$$

By (4.4), we may choose $r,s\in\mathbb{N}\setminus\{0\}$ such that

$$a_2^*r-a_3^*s=-1. \tag{4.6}$$

Consider the submodule W of $\mathbb{Z}^3$ defined by

$$W=\{(i_1,i_2,i_3)\in\mathbb{Z}^3\mid a_1^*i_1+a_2^*i_2+a_3^*i_3=0\}$$

and the elements $w_1=(1,ra_1^*,-sa_1^*)$ and $w_2=(0,a_3^*,-a_2^*)$ of W. We claim that $\{w_1,w_2\}$ is a basis of the $\mathbb{Z}$-module W. Linear independence is clear, so it suffices to check that $\{w_1,w_2\}$ generates W. Let $w=(i_1,i_2,i_3)\in W$. As $\{w_1,w_2\}$ is a basis of the $\mathbb{Q}$-vector space $\mathbb{Q}\otimes_{\mathbb{Z}}W$, there exist $q_1,q_2\in\mathbb{Q}$ such that $w=q_1w_1+q_2w_2$. Then $q_1=i_1\in\mathbb{Z}$, so $q_2w_2=w-q_1w_1\in\mathbb{Z}^3$; it follows that $q_2a_2^*$ and $q_2a_3^*$ are integers, so $q_2\in\mathbb{Z}$ by (4.4). So $\{w_1,w_2\}$ is indeed a basis of W. Now define $u,v\in B_{(X_3)}$ by

$$u=X_1X_2^{ra_1^*}X_3^{-sa_1^*}\quad\text{and}\quad v=X_2^{a_3^*}X_3^{-a_2^*}.$$

Consider a monomial $\mathcal{M}=X_1^{i_1}X_2^{i_2}X_3^{i_3}\in\mathbf{k}[X_1^{\pm1},X_2^{\pm1},X_3^{\pm1}]$ satisfying $(i_1,i_2,i_3)\in W$; then we have $(i_1,i_2,i_3)=aw_1+bw_2$ for some $a,b\in\mathbb{Z}$, so $\mathcal{M}=u^av^b\in$

[3] In fact this implication is also the proof of Corollary 4.10.

$\mathbf{k}[u^{\pm 1}, v^{\pm 1}]$. As each element of $B_{(X_3)}$ is a linear combination (over $\mathbf{k}$) of monomials $X_1^{i_1}X_2^{i_2}X_3^{i_3}$ satisfying (in particular) $(i_1,i_2,i_3) \in W$, we have

$$\mathbf{k}[u,v] \subseteq B_{(X_3)} \subseteq \mathbf{k}\big[u^{\pm 1}, v^{\pm 1}\big]. \tag{4.7}$$

In particular, the elements $X_1^{a_1}/X_3^{a_3}$ and $X_2^{a_2}/X_3^{a_3}$ of $B_{(X_3)}$ can be expressed in terms of u, v as

$$\frac{X_1^{a_1}}{X_3^{a_3}} = \frac{u^{a_1}}{v^{a_3 r}} \quad \text{and} \quad \frac{X_2^{a_2}}{X_3^{a_3}} = v^{\rho}. \tag{4.8}$$

As $\deg(X_3) = a_3^* > 0$, the open subset $D_+(X_3)$ of $\mathbb{P}(a_1^*, a_2^*, a_3^*)$ is isomorphic to $\operatorname{Spec} B_{(X_3)}$ and a birational morphism $\beta\colon D_+(X_3) \to \mathbb{A}^2$ is determined by the inclusion $\mathbf{k}[u,v] \hookrightarrow B_{(X_3)}$. The irreducible curve $C_{(a_1,a_2,a_3)} \subset \mathbb{P}(a_1^*, a_2^*, a_3^*)$ meets the open set $D_+(X_3)$ and we may consider the closure Γ of $\beta\big(C_{(a_1,a_2,a_3)} \cap D_+(X_3)\big)$ in $\mathbb{A}^2$. It follows from (4.7) that $C_{(a_1,a_2,a_3)} \cap D_+(X_3)$ is not contracted by β, so $\Gamma \subset \mathbb{A}^2$ is an irreducible curve which is birational to $C_{(a_1,a_2,a_3)}$. The equation of Γ can be obtained by dividing $X_1^{a_1} + X_2^{a_2} + X_3^{a_3} = 0$ by $X_3^{a_3}$, using (4.8) and clearing denominators; this gives:

(4.9) $\Gamma \subset \mathbb{A}^2$ is the zero-set of the polynomial $u^{a_1} + v^{a_3 r + \rho} + v^{a_3 r} \in \mathbf{k}[u,v]$.

In view of Lemma 4.6, if Γ is rational then one of the following conditions holds:

(a) $a_1 = 2$, $a_3 r$ is even and $\rho = 2$;
(b) $a_1 \mid a_3 r$ and $\rho = 1$;
(c) $a_1 \mid (a_3 r + \rho)$ and $\rho = 1$.

In case (a), we have $a_1 = 2$ and $\rho = 2$, so (4.5) gives $(a_1,a_2,a_3) = (2, 2a_3^*, 2a_2^*)$ and $M = 2a_2^* a_3^*$. Since at most one a_i is equal to 2, it follows that $\min(a_2^*, a_3^*) > 1$. As $a_1^* = M/a_1 = a_2^* a_3^*$, we have $(a_1^*, a_2^*, a_3^*) = (a_2^* a_3^*, a_2^*, a_3^*)$ and consequently $(\alpha_1, \alpha_2, \alpha_3) = (1, a_3^*, a_2^*)$. This implies that $\operatorname{type} \mathfrak{g} = 2$, because $\min(a_2^*, a_3^*) > 1$. So in case (a) we have $\operatorname{cotype}(a_1,a_2,a_3) = 2$, i.e., condition (ii) holds.

If (b) or (c) holds then $\rho = 1$, so (4.5) gives $(a_1,a_2,a_3) = (a_1, a_3^*, a_2^*)$ and $M = a_2^* a_3^*$. Because $\min(a_1,a_2,a_3) > 1$, it follows that $\min(a_2^*, a_3^*) > 1$. Also, "(b) or (c)" implies that a_1 divides one of the following numbers:

$$a_3 r = a_2^* r = a_3^* s - 1 \quad \text{or} \quad a_3 r + \rho = a_3 r + 1 = a_2^* r + 1.$$

In the first (resp. second) case, a_1 is relatively prime to a_3^* (resp. a_2^*). As a_1 divides $M = a_2^* a_3^*$ and is relatively prime to one of a_2^*, a_3^*, it follows that a_1 divides one of a_2^*, a_3^*. Consider i,j such that $\{i,j\} = \{2,3\}$ and $a_1 \mid a_i^*$. Then $a_1^* = M/a_1 = a_2^* a_3^*/a_1 = (a_i^*/a_1) a_j^*$ and consequently

$$\begin{aligned} \alpha_1 &= \gcd(a_2^*, a_3^*) = 1, \\ \alpha_i &= \gcd(a_1^*, a_j^*) = \gcd\big((a_i^*/a_1)a_j^*, a_j^*\big) = a_j^* > 1, \\ \alpha_j &= \gcd(a_1^*, a_i^*) = \gcd\big((a_i^*/a_1)a_j^*, a_i^*\big) = \gcd\big((a_i^*/a_1), a_i^*\big) = a_i^*/a_1. \end{aligned}$$

If $a_1 \neq a_i^*$ then $\alpha_i \neq 1$ and $\alpha_j \neq 1$, so $\operatorname{type} \mathfrak{g} = 2$ and hence condition (ii) holds.

If $a_1 = a_i^*$ then $(a_1,a_2,a_3) = (a_i^*, a_3^*, a_2^*)$, so condition (iii) holds.

We have shown that if (i) does not hold then (ii) or (iii) holds, so the first part of the proof is complete.

To prove the proposition, we have to show that no $D \in \operatorname{LND}(B, \mathfrak{g})$ satisfies $D^2(f) = 0$. This follows from Lemma 3.9 or Theorem 3.11 in case (i), and from Lemmas 4.3 and 4.4 in case (ii). There only remains to show that D doesn't exist in case (iii); for this, we use Theorem 2.3 and some results in the classification of homogeneous locally nilpotent derivations of $\mathbf{k}^{[3]}$.

Assume that $(a_1, a_2, a_3) = (b, a, a)$ for some $a \geq 3$ and $b \geq 2$ such that $\gcd(a, b) = 1$. Then $(a_1^*, a_2^*, a_3^*) = (a, b, b)$, so the grading $\mathfrak{g}$ is defined by $\deg(X_1) = a$ and $\deg(X_2) = b = \deg(X_3)$. Also note that $(\alpha_1, \alpha_2, \alpha_3) = (b, 1, 1)$, so $\operatorname{type} \mathfrak{g} = 1$ and $\alpha_1 = b \neq 1$.

Let $D \in \text{LND}(B, \mathfrak{g}) \setminus \{0\}$ and $A = \ker D$. Let us show that $D^2(f) \neq 0$. By Lemma 4.3, we may assume that $X_1 \notin A$.

By Theorem 2.3(1), we have $d(A) \in \{b, 1\}$. If $d(A) = 1$ then $\alpha_1 = b \notin \{1, d(A)\}$, so $X_1 \in A$ by Theorem 2.3(3), contradicting our assumption. So in fact we have $d(A) = b = \alpha_1 \neq 1$. Then Theorem 2.3(2) implies that the set of homogeneous prime elements h of B satisfying $\deg_D(h) = 1$ is equal to $\{\lambda X_1 \mid \lambda \in \mathbf{k}^*\}$. As $f \neq \lambda X_1$, $\deg_D(f) \neq 1$ and there only remains to show that

$$f \notin A. \tag{4.10}$$

We give two proofs of this. The first one was suggested by the referee and relies on [6, Lemma 4], which states that $R = \mathbb{C}[X, Y, Z]/(X^r + Y^s + Z^t)$ satisfies $\text{LND}(R) = \{0\}$ whenever $\min(r, s, t) \geq 2$ and at most one of r, s, t is equal to 2. Now it is clear that if $f \in A$ then $\text{LND}(B/fB) \neq \{0\}$, which contradicts the cited result if $\mathbf{k} = \mathbb{C}$. This proves (4.10) in the case $\mathbf{k} = \mathbb{C}$.

The second proof of (4.10) is valid for general $\mathbf{k}$; it uses the following facts, which follow respectively from Lemma 3.4.3 and Theorem 3.4.2 of [2]:

(4.11) Let u, v be $\mathfrak{g}$-homogeneous elements of B such that $\mathbf{k}[X_1, u, v] = \mathbf{k}[X_1, X_2, X_3]$. Then $u, v \in \mathbf{k}[X_2, X_3]$ are linearly independent linear forms.

(4.12) $A = \mathbf{k}[u, u^e v + \psi(u, X_1)]$ for some pair (u, v) as in (4.11), some $e \in \mathbb{N}$ and some $\psi(u, X_1) \in \mathbf{k}[u, X_1]$ such that $u^e v + \psi(u, X_1)$ is $\mathfrak{g}$-homogeneous and irreducible.

Let us write $A = \mathbf{k}[u, \theta]$, where $\theta = u^e v + \psi(u, X_1)$. Note that $\deg(u) = b$ and $\deg(\theta) = (e+1)b$, while $\deg(f) = ab$. By contradiction, assume that $f \in A$. As f is an element of A which is both homogeneous and irreducible, one of the following conditions must hold for some $\lambda, \mu \in \mathbf{k}^*$:

(a) $f = \lambda u$;
(b) $f = \lambda \theta$;
(c) $f = \lambda u^{e+1} + \mu \theta$.

Condition (a) cannot hold because $\deg(f) \neq \deg(u)$, so (b) or (c) holds, so $e = a - 1$ and $f = \lambda u^a + \mu \theta$ for some $\lambda, \mu \in \mathbf{k}$. Setting $X_1 = 0$ in this equality of polynomials yields $X_2^a + X_3^a = \lambda u^a + \mu\big(u^{a-1} v + \psi(u, 0)\big)$, where $\psi(u, 0) = \nu u^a$ for some $\nu \in \mathbf{k}$. Then $u^{a-1} \mid X_2^a + X_3^a$; as there can be no multiple factor in the prime factorization of $X_2^a + X_3^a$, we obtain $a \leq 2$, which contradicts one of our assumptions. This contradiction proves (4.10), so the proof is complete. □

It is easy to see that if $\mathbf{k}$ is algebraically closed, $n \geq 3$ and $(a_1, \dots, a_n)$ does not satisfy the hypothesis of $P(n)$ then $\text{LNDEG}(\sum_{i=1}^n X_i^{a_i}) = 0$. So the above result implies:

4.8. Corollary. *Let $f = X_1^{a_1} + X_2^{a_2} + X_3^{a_3} \in B = \mathbf{k}[X_1, X_2, X_3] = \mathbf{k}^{[3]}$, where $\mathbf{k}$ is an algebraically closed field of characteristic zero and $(a_1, a_2, a_3) \in \mathbb{N}^3$. Then the following are equivalent:*

(a) $\text{LNDEG}(f) = 0$;
(b) $\text{LNDEG}(f) \leq 1$;
(c) $\min(a_1, a_2, a_3) \leq 1$ *or* $|\{i \mid a_i = 2\}| > 1$.

4.9. Remark. In Corollary 4.8, one cannot drop the hypothesis that $\mathbf{k}$ is algebraically closed. For instance, $(a_1, a_2, a_3) = (2, 2, 2)$ satisfies 4.8(c) but, if $\mathbf{k} = \mathbb{R}$, does not satisfy 4.8(a). Indeed, if 4.8(a) holds then $A = \mathbb{R}[X_1, X_2, X_3]/(X_1^2 + X_2^2 + X_3^2)$ is not rigid; however, [3, Example 2.2] shows that A is rigid.

4.10. Corollary. *Let* $f = X_1^{a_1} + X_2^{a_2} + X_3^{a_3} \in B = \mathbf{k}[X_1, X_2, X_3] = \mathbf{k}^{[3]}$, *where* $\mathbf{k}$ *is an algebraically closed field of characteristic zero. Suppose*:

(a) $\min(a_1, a_2, a_3) \geq 2$ *and at most one of* a_1, a_2, a_3 *is equal to* 2;
(b) a_1, a_2, a_3 *are distinct or* $\gcd(a_1, a_2, a_3) > 1$;
(c) $\operatorname{cotype}(a_1, a_2, a_3) \leq 1$.

Then the curve $C_{(a_1,a_2,a_3)} = V_+(f) \subset \mathbb{P}(a_1^*, a_2^*, a_3^*)$ *is not rational. Moreover, if* g *is any element of* B *satisfying* $\deg_{\mathfrak{g}}(g) < \deg_{\mathfrak{g}}(f)$ (*where* $\mathfrak{g}$ *is defined in* 4.2), *then no* $D \in \text{LND}(B) \setminus \{0\}$ *satisfies* $D^2(f + g) = 0$.

PROOF. Suppose that (a_1, a_2, a_3) satisfies (a)–(c). Then, by the proof of Proposition 4.7, one of the following conditions holds:

(i) the curve $C_{(a_1,a_2,a_3)}$ is not rational;
(ii) $\operatorname{cotype}(a_1, a_2, a_3) = 2$;
(iii) up to permutation, (a_1, a_2, a_3) is (b, a, a) for some $a \geq 3$ and $b \geq 2$ such that $\gcd(a, b) = 1$.

Condition (ii) does not hold by (c), and condition (iii) does not hold by (b), so $C_{(a_1,a_2,a_3)}$ is not rational. By Theorem 3.11, it follows that no $D \in \text{LND}(B) \setminus \{0\}$ satisfies $D^2(f + g) = 0$. □

Finally, we note that the fact that $P(3)$ is true gives us partial information about $P(4)$, by Lemma 4.3. In particular:

4.11. Corollary. $P(4)$ *is true whenever* $\operatorname{cotype}(a_1, a_2, a_3, a_4) > 1$.

References

1. D. Daigle, *Homogeneous locally nilpotent derivations of* $\mathbf{k}[X, Y, Z]$, J. Pure Appl. Algebra **128** (1998), no. 2, 109–132.
2. ———, *Classification of homogeneous locally nilpotent derivations of* $\mathbf{k}[X, Y, Z]$. I: *Positive gradings of positive type*, Transform. Groups **12** (2007), no. 1, 33–47.
3. ———, *Affine surfaces with trivial Makar-Limanov invariant*, J. Algebra **319** (2008), no. 8, 3100–3111.
4. G. Freudenburg, *Algebraic theory of locally nilpotent derivations*, Encyclopaedia Math. Sci., vol. 136, Springer, Berlin, 2006.
5. G. Freudenburg and L. Moser-Jauslin, *Locally nilpotent derivations of rings with roots adjoined*, Technical Report 612, Institut de Mathématiques de Bourgogne, 2010.
6. S. Kaliman and M. Zaidenberg, *Miyanishi's characterization of the affine* 3*-space does not hold in higher dimensions*, Ann. Inst. Fourier (Grenoble) **50** (2000), no. 6, 1649–1669 (2001).
7. R. Kolhatkar, *Homogeneous locally nilpotent derivations and affine ML-surfaces*, Ph.D. Thesis, University of Ottawa, 2010.
8. M. Miyanishi, *Normal affine subalgebras of a polynomial ring*, Algebraic and Topological Theories (Kinosaki, 1984) (M. Nagata, S. Araki, A. Hattori, N. Iwahori, et al., eds.), Kinokuniya, Tokyo, 1986, pp. 37–51.

9. A. van den Essen, *Polynomial automorphisms and the Jacobian conjecture*, Progr. Math., vol. 190, Birkhäuser, Basel, 2000.
10. Z. Wang, *Homogeneization of locally nilpotent derivations and an application to* $k[X,Y,Z]$, J. Pure Appl. Algebra **196** (2005), no. 2-3, 323–337.

Department of Mathematics and Statistics, University of Ottawa, Ottawa, ON K1N 6N5, Canada

E-mail address: ddaigle@uottawa.ca

Centre de Recherches Mathématiques
CRM Proceedings and Lecture Notes
Volume **54**, 2011

Corrigendum to Our Paper "Birational Transformations of Weighted Graphs"

Hubert Flenner, Shulim Kaliman, and Mikhail Zaidenberg

ABSTRACT. We give a corrected and improved version of Corollary 3.33 in [5], and fix some minor inaccuracies in *loc. cit.*

In our paper [5] on birational transformations of weighted graph, we showed that every minimal connected weighted graph $\Gamma \neq [\![-1]\!]$ is birationally equivalent to a standard one, see Theorem 2.15(b).[1] See §2.1 and §2.4 in *loc. cit.* for the definitions of birational transformations, standard weighted graphs, and linear and circular segments. Concerning the uniqueness of this standard form, we claimed in Corollary 3.33 that *a noncircular standard weighted graph is unique in its birational equivalence class up to reversion of its linear segments.*

However in this form the corollary is only true if there are no segments of type $[\![0_{2k+1}]\!]$ ($k \geq 0$), as was pointed out to us by Karol Palka. The correct general version is as follows.

Corollary 3.33. *Every noncircular standard weighted graph is unique in its birational equivalence class up to reversions of its linear segments* ***and changing the weights of its branching vertices****. Similarly, a circular standard graph is unique in its birational equivalence class up to a cyclic permutation of its nonzero weights and reversion.*

In the subsequent papers [6–8] we applied this result only to linear or circular graphs. The latter graphs have no branching vertices. Hence we used a correct version, and so this does not affect any of our subsequent results.

To demonstrate that this correction is needed, following a suggestion of Karol Palka let us consider the weighted tree

Γ:

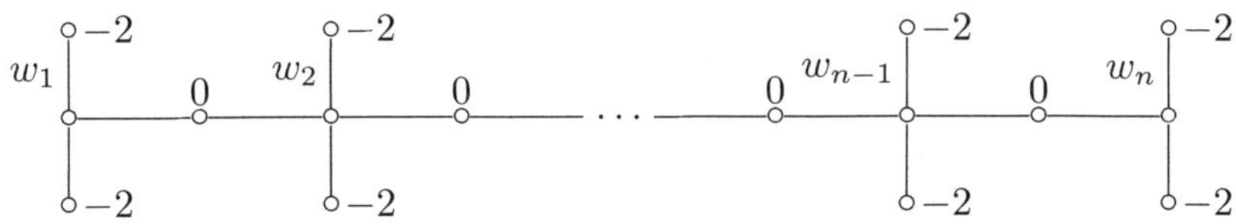

2010 *Mathematics Subject Classification.* Primary 05C99, 14R20, Secondary 14J50.

This is the final form of the paper.

[1]In the formulation of that result in *loc. cit.* the assumption that Γ is connected and $\neq [\![-1]\!]$ is missing and has to be added (we owe this correction to the referee).

Performing inner elementary transformations at zero vertices (see [5, §2.3]), one can change the weights $w_1, \dots, w_n$ of the branching vertices and so replace them by the new weights $w'_1 = \dots = w'_{n-1} = 0$ and $w'_n = \sum_{i=1}^n w_i$, without changing the birational class of Γ. Thus the correction above is indeed necessary.

It is possible to describe completely the possible weights of branching points in the birational equivalence class of a standard graph. For this we let B denote the set of branching points of Γ. Moreover let Γ_0 be the subgraph of Γ, which is the union of B and all linear segments of type $[\![0_{2k+1}]\!]$ of Γ including the edges that join these segments with B. For instance, if Γ has no segments of type $[\![0_{2k+1}]\!]$ then $\Gamma_0 = B$.

With these notations Corollary 3.33 can be replaced by the following more detailed statement.

Corollary 3.33′. *Let Γ be a connected noncircular standard weighted graph. Then Γ is unique in its birational equivalence class up to reversion of its linear segments and the change of weights of branching points described by the following procedure:*

(1) *If a connected component G of Γ_0 contains an end-vertex of Γ*[2] *then the weights of points of $G \cap B$ can be chosen arbitrarily;*

(2) *If G does not contain any end-vertex of Γ then the weights of the points of $G \cap B$ can be chosen arbitrarily modulo preservation of the sum of weights at these branching points.*

In particular, we can change the weights of Γ in such a way that for every connected component G of Γ_0 the new weights of points of $G \cap B$ are all zero in case (1), *and are all zero with one exception in case* (2), *with the new weight in this exceptional position equal to the sum of the weights of the points of $G \cap B$.*

For the proof we need a few preparations. Let us recall first the following results.

(i) Any birational transformation of one standard graph into another one is composed of a sequence of elementary transformations ([5, Theorem 3.1]). In particular, it identifies the branching vertices of both graphs, and transforms linear segments of the first graph into linear segments of the second one.

(ii) A standard linear segment is unique in its birational equivalence class up to reversion ([5, Corollary 3.32]). The reversion of a standard linear segment L of Γ leaves unchanged the weighted graph $\Gamma \ominus L$ ([5, Lemma 2.12]).

Let $\sigma \colon \Gamma \dashrightarrow \Gamma'$ be a birational transformation of weighted graphs and let L be a segment of Γ. We call σ an *L-transformation* if it is composed of blowups and blowdowns of L and their subsequent total transforms. Thus σ induces a bijection $\Gamma \ominus L \to \Gamma' \ominus L'$ that respects the weights of all vertices except possibly the branching points that are adjacent to L and L', respectively. We need the following lemma.

Lemma 1. (a) *Every birational transformation $\sigma \colon \Gamma \dashrightarrow \Gamma'$ of standard graphs can be written as a composition of L-transformations, where L runs through the segments of Γ.*

[2] I.e., a vertex of Γ of degree 1.

(b) *Every birational transformation* $\sigma\colon \Gamma \dashrightarrow \Gamma'$ *of standard graphs admits a domination*

$$\begin{array}{ccc} & \Delta & \\ \swarrow & & \searrow \\ \Gamma & \dashrightarrow & \Gamma' \end{array}$$

such that Δ, Γ *and* Γ' *have the same branching points. In particular, if* L *is a segment of* Γ *and* L' *is the corresponding segment of* Γ' *then the induced birational transformation* $\sigma|L\colon L \dashrightarrow L'$ *can be dominated by a linear graph.*

PROOF. By (i) σ can be written as a composition of elementary transformations $\sigma_1 \cdot \cdots \cdot \sigma_n$. Every such elementary transformation takes place at some segment L of Γ. Obviously elementary transformations taking place at different segments commute. Thus if σ_L is the product of all σ_i taking place at L then σ is the product of all σ_L. This proves (a).

The first part of (b) follows from Lemma 3.8 in [5] while the second is immediate from the first one. □

PROOF OF COROLLARY 3.33′. Let $\sigma\colon \Gamma \dashrightarrow \Gamma'$ be a birational transformation of standard graphs. By Lemma 1(b) they can be dominated by a graph Δ such that Δ, Γ and Γ' have the same branching points. Moreover by Lemma 1(a) σ can be decomposed as a product $\sigma = \prod_L \sigma_L$, where σ_L is an L-transformation.

Let us first describe the effect of σ_L on the weights of the branching points of Γ.

(a) If a branching point is not adjacent to L then its weight remains unchanged under σ_L.

(b) Assume that $L = [\![0_{2k}, w_1, \ldots, w_n]\!]$ with $n \geq 0$ and $w_i \leq -2$ for all i. According to Proposition 3.4 in [5] the induced birational transformation $\sigma|L = \sigma_L|L\colon L \dashrightarrow L'$ is either the reversion or the identity. As observed in (ii), in both cases σ_L does not affect the weights of branching points.

(c) If $L = [\![0_{2k+1}]\!]$ then according to Proposition 3.7 in [5] $\sigma_L|L = \tau^s$ for some $s \in \mathbb{Z}$, where τ denotes the left move (see Definition 3.5 in *loc. cit.*). Comparing with Lemma 2.12(c) in *loc. cit.*, σ_L will increase the weight of one of the branching points adjacent to L by s, while it decreases the weight of the other one (if existent) by s.

In any case, for every connected component G of Γ_0 the sum of the weights of points in $G \cap B$ is invariant under σ_L and hence also under σ.

Finally, applying (c) repeatedly we can move the weight of any vertex v in $G \cap B$ to any other vertex v' in $G \cap B$ along a path connecting them in G. To annihilate all the weights but one, we proceed as follows. We choose a rooted subtree of G containing all the vertices of $G \cap B$ (and the root is one of them), and move recursively the weights to the root starting from the vertices in $G \cap B$ with maximal distance from the root. In the case that G contains an end vertex of Γ, choosing for the root the vertex in $G \cap B$ closest to the end vertex, again by (c) we finish the recursive procedure by annihilating the weight of the root.[3] This proves Corollary 3.33′. □

Remarks. (1) As was pointed out to us by D. Daigle, the classification result Corollary 3.33′ (in its correct form) for trees or forests is implicitly contained

---

[3]This procedure is transparent in the example above.

in [2]. See also [1, 3] for the case of linear chains, where the statements coincide explicitly with ours. His methods can be likely generalized to arbitrary noncircular weighted graphs.

(2) The relations between our results and the earlier classification due to Danilov and Gizatullin [4] are clarified in [5].

We want to take the opportunity to correct two other inaccuracies in our paper [5]. In Lemma 2.17 the case of the zigzag $[\![w_0, \ldots, w_n]\!]$ with all $w_i \leq -2$ is missing and has to be added (this was already mentioned in 2.1.2 of [8]). In Corollary 3.36 we have to add the possibility that a connected component of the divisor D of a completion (X, D) of a normal algebraic surface V is a (-1)-curve. Thus the correct version is as follows.

Corollary 3.36. *Every normal algebraic surface V admits an SNC-completion (X, D) such that every connected component of the dual graph Γ_D is either standard or $[\![-1]\!]$. Moreover, any other such completion can be obtained from (X, D) by a sequence of elementary transformations.*

Acknowledgments. It is pleasure for us to thank Karol Palka, who attracted our attention to the flaw in Corollary 3.33 in [5].

References

1. D. Daigle, *Birational endomorphisms of the affine plane*, J. Math. Kyoto Univ. **31** (1991), no. 2, 329–358.
2. ———, *Classification of weighted graphs up to blowing-up and blowing-down*, available at `arXiv:math.AG/0305029`.
3. ———, *Classification of linear weighted graphs up to blowing-up and blowing-down*, Canad. J. Math. **60** (2008), no. 1, 64–87.
4. M. H. Gizatullin and V. I. Danilov, *Automorphisms of affine surfaces*. I, Izv. Akad. Nauk SSSR Ser. Mat. **39** (1975), no. 3, 523–565, 703 (Russian).
5. H. Flenner, Sh. Kaliman, and M. Zaidenberg, *Birational transformations of weighted graphs*, Affine Algebraic Geometry (T. Hibi, ed.), Osaka Univ. Press, Osaka, 2007, pp. 107–147.
6. ———, *Completions of $\mathbb{C}^*$-surfaces*, Affine Algebraic Geometry (T. Hibi, ed.), Osaka Univ. Press, Osaka, 2007, pp. 149–201.
7. ———, *Uniqueness of $\mathbb{C}^*$- and $\mathbb{C}_+$-actions on Gizatullin surfaces*, Transform. Groups **13** (2008), no. 2, 305–354.
8. ———, *Smooth affine surfaces with non-unique $\mathbb{C}^*$-actions*, J. Algebraic Geom. **20** (2011), no. 2, 329–398.

Fakultät für Mathematik, Ruhr Universität Bochum, Geb. NA 2/72, Universitätsstr. 150, 44780 Bochum, Germany
E-mail address: `Hubert.Flenner@rub.de`

Department of Mathematics, University of Miami, Coral Gables, FL 33124, U.S.A.
E-mail address: `kaliman@math.miami.edu`

Institut Fourier, Université Grenoble I, UMR 5582 CNRS-UJF, BP 74, 38402 St-Martin d'Hères cedex, France
E-mail address: `zaidenbe@ujf-grenoble.fr`

Centre de Recherches Mathématiques
CRM Proceedings and Lecture Notes
Volume **54**, 2011

Bivariate Analogues of Chebyshev Polynomials with Application to Embeddings of Affine Spaces

Gene Freudenburg

To Peter Russell, friend and mentor

ABSTRACT. This paper introduces polynomials $F_n \in \mathbb{Z}[x,y]$ ($n \in \mathbb{Z}$) which, due to their strong algebraic properties, are bivariate analogues of the Chebyshev polynomials $T_n \in \mathbb{Z}[t]$ ($\deg T_n = n$). As an application, we show that these polynomials define a family of plane embeddings $\Phi: \mathbb{A}^2 \to \mathbb{A}^5$ over any field $\mathbf{k}$ of characteristic zero. Each of these embeddings restricts to a line embedding of the form $\phi = (T_i, T_j, T_k): \mathbb{A}^1 \to \mathbb{A}^3$. In general, it is unknown whether these embeddings are rectifiable over $\mathbf{k}$. However, when $\mathbf{k} = \mathbb{R}$, the space curves obtained in this way are generally nontrivial knots in $\mathbb{R}^3$, and are therefore not rectifiable over $\mathbb{R}$.

1. Introduction

Following the appearance of the Epimorphism Theorem in the mid-1970s, which characterizes algebraic embeddings of a line in the plane, Abhyankar considered embeddings of lines in 3-space of the form $\theta(l,m,n) = (t + t^l, t^m, t^n)$ for positive integers l, m, n [1]. Abhyankar conjectured that these embeddings are not rectifiable when none of the three integers belongs to the semigroup generated by the other two. However, Craighero showed that $\theta(5,4,3)$ and $\theta(6,5,4)$ can be rectified [6, 7]; see [9, Section 4.3.1], for another proof. Subsequently, Bhatwadekar and Roy gave families of integers l, m, n for which $\theta(l,m,n)$ can be rectified [4]. The general case remains open, including $\theta(7,6,5)$.

The lines defined by $\theta(l,m,n)$ are topologically trivial over the field of real numbers. Vassiliev [25] observed that any "noncompact" knot can be represented by a polynomial embedding. Shastri [22] gave a proof of this fact and provided specific polynomial embeddings for the trefoil and figure-8 knots. Thus, the Gordon–Lueke Theorem from topology implies that there exist algebraic embeddings of $\mathbb{R}^1$ in $\mathbb{R}^3$ which are *not* rectifiable. It is an important open question whether algebraic embeddings of $\mathbb{C}^1$ in $\mathbb{C}^3$ can always be rectified by algebraic automorphisms; Kaliman [14] showed that they can be rectified by holomorphic automorphisms.

2010 *Mathematics Subject Classification.* Primary 14R10; Secondary 14H50.
Key words and phrases. line embedding, epimorphism of polynomial rings.
This is the final form of the paper.

In the present work, we introduce line embeddings of the form $\phi = (T_i, T_j, T_k)$: $\mathbb{A}^1_{\mathbf{k}} \to \mathbb{A}^3_{\mathbf{k}}$, where $T_n \in \mathbb{Z}[t]$ denotes the Chebyshev polynomial of degree n, and $\mathbf{k}$ denotes a field of characteristic zero. Proposition 5.2 asserts that ϕ is an algebraic line embedding if and only if $\gcd(i,j) = \gcd(i,k) = \gcd(j,k) = 1$. Proposition 5.11 shows that, for a fixed pair i, j of relatively prime positive integers, the embeddings of the form (T_i, T_j, T_k) represent a finite number of algebraic equivalence classes. Over the field of real numbers, these embeddings provide a rich family of nontrivial knots parametrized by polynomials, as illustrated in Table 1.

Given their strong algebraic properties, it is natural to ask: What are the multivariate (algebraic) analogues of the Chebyshev polynomials, and what identities should they satisfy? The univariate Chebyshev polynomials T_n imitate the monomial functions $\mu_n = t^n$ in many important respects, most notably, they commute by composition, and they form a linear basis of $\mathbb{Z}[t]$. In this paper, we use matrix relations to define a family of polynomials $F_n(x,y)$, $n \in \mathbb{Z}$, which are bivariate analogues of the Chebyshev polynomials. Like their univariate counterparts, these satisfy a strong commuting relation, namely:

$$F_n(F_m, \overline{F}_m) = F_m(F_n, \overline{F}_n) = F_{mn} \quad \text{for all } m, n \in \mathbb{Z},$$

where $\overline{F}_n(x,y) = F_n(y,x)$; see Proposition 2.1 below. This property allows us to use the polynomials F_n to define certain plane embeddings $\Phi: \mathbb{A}^2 \to \mathbb{A}^5$ which are equivariant relative to standard $\mathbb{Z}_2$-actions on affine space. By restricting to fixed point sets, we obtain embeddings of $\mathbb{A}^1$ in $\mathbb{A}^3$, and these are precisely the embeddings of the form (T_i, T_j, T_k).

The paper is organized in the following way.

Section 1. Introduction
Section 2. The polynomials F_n
Section 3. Chebyshev polynomials
Section 4. The curves (T_i, T_j)
Section 5. Epimorphisms defined by Chebyshev polynomials
Section 6. Knots defined by Chebyshev polynomials
Section 7. Epimorphisms defined by polynomials F_n

Preliminaries. The ground field $\mathbf{k}$ is any field of characteristic zero, though the field of real numbers is used in discussing knots in Section 5. The polynomial ring in n variables is denoted by $\mathbf{k}^{[n]} = \mathbf{k}[x_1, \dots, x_n]$. The subalgebra generated by $f_1, \dots, f_m \in \mathbf{k}[x_1, \dots, x_n]$ is denoted by $\mathbf{k}[f_1, \dots, f_m]$. Affine n-space over $\mathbf{k}$ is $\mathbb{A}^n$.

For any epimorphism of polynomial rings $\phi^*: \mathbf{k}^{[n]} \to \mathbf{k}^{[m]}$ $(m \le n)$, the corresponding map $\phi: \mathbb{A}^m \to \mathbb{A}^n$ is an embedding; see [24, Corollary B.2.6], or [8, Lemma 4]. Throughout this paper, the term *embedding* will refer to an embedding of one affine space in another induced by an epimorphism of polynomial rings.

$GA_n(k)$ is the group of algebraic automorphisms of $\mathbf{k}[x_1, \dots, x_n]$, or equivalently, of $\mathbb{A}^n$. Two embeddings $\phi, \psi: \mathbb{A}^m \to \mathbb{A}^n$ are said to be *algebraically equivalent* if and only if there exists $\alpha \in GA_n(\mathbf{k})$ with $\alpha\phi = \psi$. An embedding ϕ is *rectifiable* if and only if ϕ is algebraically equivalent to the standard embedding σ defined by $\sigma^* \mathbf{k}[x_1, \dots, x_n] \to \mathbf{k}[y_1, \dots, y_m]$, where $\sigma(x_i) = y_i$ for $1 \le i \le m$ and $\sigma(x_i) = 0$ for $m + 1 \le i \le n$.

The famous Epimorphism Theorem of Abhyankar and Moh [2] and Suzuki [23] asserts that every algebraic embedding of $\mathbb{A}^1$ in $\mathbb{A}^2$ is algebraically equivalent to the standard embedding. We use the following equivalent version.

Assume $\mathbf{k}$ *is a field of characteristic zero. If* $f(t), g(t) \in \mathbf{k}[t]$ *are such that* $t \in \mathbf{k}[f(t), g(t)]$, *then either* $\deg f \mid \deg g$ *or* $\deg g \mid \deg f$.

Many other proofs of this result have subsequently appeared. Of particular note is the paper of Rudolph [20], which gives an elegant proof of the Epimorphism Theorem using knot theory.

Shastri [22] gives the embedding $\phi: \mathbb{R}^1 \to \mathbb{R}^3$ defined by

$$\phi(t) = (t^3 - 3t, t^4 - 4t^2, t^5 - 10t),$$

and shows that $\phi(\mathbb{R}^1)$ is the trefoil knot. In particular, if $F \in k[X,Y,Z]$ is $F = YZ - X^3 - 5XY + 2Z - 7X$, then $\phi^*(F) = t$. While ϕ itself is already quite simple, we see that F can be simplified: $F = (Y+2)(Z-5X) + 3X - X^3$. Thus, if $\alpha \in \mathrm{GA}_3(\mathbb{R})$ is defined by $\alpha = (X, Y-2, Z+5X)$, then

$$\alpha\phi = (t^3 - 3t, t^4 - 4t^2 + 2, t^5 - 5t^3 + 5t) \quad \text{and} \quad \alpha(F) = YZ + 3X - X^3.$$

We recognize that the defining polynomials for the trefoil in this re-parametrization are the (monic) Chebyshev polynomials of degree 3, 4, and 5. (This parametrization was also realized by Koseleff and Pecker, and is given in [15].) It was therefore natural to ask if other combinations of Chebyshev polynomials define embeddings, and if so, what knots they parametrize over the field of real numbers.

Remark. Working over the reals, J. Freudenburg reported Propositions 5.2 and 5.11 in two research reports from October and November 2008 [10, 11], results which describe maps (T_i, T_j, T_k) which define line embeddings, and when two such embeddings are algebraically equivalent. These reports included identification for many of the knots listed in Table 1. Working independently, Koseleff and Pecker subsequently posted the preprint *Chebyshev knots* [15] in December 2008. Their paper studies knots parametrized by Chebyshev polynomials, and several of their results parallel those found in the earlier reports. In some cases, their topological approach enables them to give proofs of results conjectured in [10, 11], most notably, that for $n \geq 1$, $(T_3, T_{3n+1}, T_{3n+2})$ parametrizes the $(2, 2n+1)$ torus knot ([15, Theorem 2]). Other results from their paper are given in Section 6, below.

2. The polynomials F_n

Let R be an integral domain with 1, let $A \in \mathrm{M}_2(R)$, the ring of 2×2 matrices over R, and assume $\det(A) \neq 0$. Then A is invertible over $\mathrm{frac}(R)$, and for each $n \in \mathbb{Z}$, the characteristic equation for A^n is

$$A^{2n} - \mathrm{tr}(A^n)A^n + \det(A^n)I = 0. \tag{1}$$

Multiplying by A^{m-n} $(m \in \mathbb{Z})$ yields:

$$A^{m+n} - \mathrm{tr}(A^n)A^m + \det(A^n)A^{m-n} = 0. \tag{2}$$

As a special case, we obtain:

$$A^{n+1} - \mathrm{tr}(A)A^n + \det(A)A^{n-1} = 0. \tag{3}$$

In particular, let $R = \mathbb{Z}[x,y] = \mathbb{Z}^{[2]}$, and define $A \in \mathrm{SL}_2(R)$ by

$$A = A(x,y) = \begin{pmatrix} x & 1 \\ xy - 1 & y \end{pmatrix}.$$

Given $h \in \mathbb{Z}[x,y]$, define $\bar{h}$ by $\bar{h}(x,y) = h(y,x)$. For each $n \in \mathbb{Z}$, there exist $F_n, G_n \in \mathbb{Z}[x,y]$ of degree n such that $\overline{G} = G$ and

$$A^n = \begin{pmatrix} F_n & G_{n-1} \\ (xy-1)G_{n-1} & \overline{F}_n \end{pmatrix}.$$

For $n \geq 0$, this form follows by induction from equation (3); the general case then follows from the identity

$$A^{-n} = \begin{pmatrix} \overline{F}_n & -G_{n-1} \\ (1-xy)G_{n-1} & F_n \end{pmatrix}.$$

In particular, $F_{-n} = \overline{F}_n$ and $G_{-n} = -G_n$.

The two governing properties of the polynomials F_n are contained in the following proposition.

Proposition 2.1. *For all $m, n \in \mathbb{Z}$:*

(a) $F_{m+n} + F_{m-n} = (F_n + \overline{F}_n)F_m$;
(b) $F_m(F_n, \overline{F}_n) = F_n(F_m, \overline{F}_m) = F_{mn}$.

PROOF. Part (a) follows immediately from equation (2). For part (b), fix $m \geq 0$. We show by induction on n that $F_n(F_m, \overline{F}_m) = F_{mn}$. When $n = 0$, we have $F_0(F_m, \overline{F}_m) = 1 = F_0 = F_{0\cdot m}$.

Assume $F_i(F_m, \overline{F}_m) = F_{mi}$ whenever $0 \leq i \leq n$. By part (a), we have

$$F_{n+1}(x,y) = (x+y)F_n(x,y) - F_{n-1}(x,y). \tag{4}$$

Therefore,

$$\begin{aligned} &F_{n+1}(F_m, \overline{F}_m) \\ &\quad = (F_m + \overline{F}_m)\cdot F_n(F_m, \overline{F}_m) - F_{n-1}(F_m, \overline{F}_m) && \text{by equation (4)} \\ &\quad = (F_m + \overline{F}_m)\cdot F_{mn} - F_{m(n-1)} && \text{by the inductive hypothesis} \\ &\quad = (F_m + \overline{F}_m)\cdot F_{mn} - F_{mn-m} \\ &\quad = F_{mn+m} && \text{by part (a)} \\ &\quad = F_{m(n+1)}. \end{aligned}$$

Therefore, $F_n(F_m, \overline{F}_m) = F_{mn}$ for all m, n. Part (b) now follows by symmetry. □

Corollary 2.2. *For all $m, n \in \mathbb{Z}$:*

(a) $F_{m+n} + \overline{F}_{m-n} = (F_m + \overline{F}_m)F_n$.
(b) $F_{m-n} - \overline{F}_{m-n} = F_m\overline{F}_n - F_n\overline{F}_m$.
(c) *If $\psi_n(x) = F_n(x,0)$, then $F_n = x\psi_{n-1}(x+y) - \psi_{n-2}(x+y)$ for $n \geq 1$.*
(d) $F_n + \overline{F}_n \in \mathbb{Z}[x+y]$.
(e) $F_n(x, x^{-1}) = x^n$ *for all $n \in \mathbb{Z}$.*

PROOF. From equation (2) we have

$$F_{m-n} = (F_n + \overline{F}_n)F_m - F_{m+n}.$$

Interchanging the roles of m and n and using the fact that $F_{n-m} = \overline{F}_{m-n}$ yields

$$\overline{F}_{m-n} = (F_m + \overline{F}_m)F_n - F_{m+n}.$$

This proves (a), and (b) follows by subtraction:

$$F_{m-n} - \overline{F}_{m-n} = [(F_n + \overline{F}_n)F_m - F_{m+n}] - [(F_m + \overline{F}_m)F_n - F_{m+n}] = \overline{F}_nF_m - \overline{F}_mF_n.$$

For part (c), note that $\psi_{-1}(x) = F_{-1}(x,0) = \overline{F}_1(x,0) = 0$ and $\psi_0(x) = 1$. In addition, by setting $y = 0$ in the equation $F_{n+1} = (x+y)F_n - F_{n-1}$, we see that the functions ψ_n satisfy

$$\psi_{n+1} = x\psi_n - \psi_{n-1}.$$

We now proceed to prove (c) by induction on n.

First, $x = F_1 = x\psi_0 - \psi_{-1}$, and the equality holds for $n = 1$. Assume $F_i = x\psi_{i-1}(x+y) - \psi_{i-2}(x+y)$ for $1 \le i \le n$. Then

$$\begin{aligned} F_{n+1} &= (x+y)F_n - F_{n-1} \\ &= (x+y)[x\psi_{n-1}(x+y) - \psi_{n-2}(x+y)] - [x\psi_{n-2}(x+y) - \psi_{n-3}(x+y)] \\ &= x[(x+y)\psi_{n-1}(x+y) - \psi_{n-2}(x+y)] - [(x+y)\psi_{n-2}(x+y) - \psi_{n-3}(x+y)] \\ &= x\psi_n(x+y) - \psi_{n-1}(x+y). \end{aligned}$$

This proves part (c). For (d) note that (c) implies

$$\overline{F}_n = y\psi_{n-1}(x+y) - \psi_{n-2}(x+y).$$

Adding F_n and $\overline{F}_n$ thus gives a polynomial in $x+y$.

Finally, when $y = x^{-1}$, the matrix A^n specializes to

$$\begin{pmatrix} x & 1 \\ 0 & x^{-1} \end{pmatrix}^n = \begin{pmatrix} x^n & * \\ 0 & x^{-n} \end{pmatrix}.$$

Therefore, $F_n(x, x^{-1}) = x^n$, which proves (e). □

3. Chebyshev polynomials

Two versions of Chebyshev polynomials are defined and used in what follows.

3.1. Monic Chebyshev polynomials. For $n \ge 0$, let $P_n \in \mathbb{Z}[t]$ be such that $\operatorname{tr}(A^n) = F_n + \overline{F}_n = P_n(x+y)$. Then P_n is the *monic Chebyshev polynomial* of degree n. The following key algebraic properties of monic Chebyshev polynomials are implied by Proposition 2.1 and Corollary 2.2.

Proposition 3.1. *For every pair $m, n \in \mathbb{Z}$:*

(a) $P_{m+n} + P_{m-n} = P_m P_n$;
(b) $P_m(P_n) = P_n(P_m) = P_{mn}$;
(c) $P_n(x + x^{-1}) = x^n + x^{-n}$.

In particular,

$$P_0 = 2, \qquad P_1 = x, \qquad \text{and} \qquad P_{n+1} = xP_n - P_{n-1} \quad \text{for } n \ge 1.$$

Note that P_n is a monic polynomial only for $n \ge 1$.

3.2. Standard Chebyshev polynomials. The matrix A specializes to

$$B = A(x,x) = \begin{pmatrix} x & 1 \\ x^2 - 1 & x \end{pmatrix}.$$

Define $T_n, U_n \in \mathbb{Z}[x]$ by $T_n(x) = F_n(x,x)$ $(n \ge 0)$ and $U_n(x) = G_n(x,x)$ $(n \ge -1)$. Then

$$B^n = \begin{pmatrix} x & 1 \\ x^2 - 1 & x \end{pmatrix}^n = \begin{pmatrix} T_n & U_{n-1} \\ (x^2-1)U_{n-1} & T_n \end{pmatrix}.$$

The polynomial T_n is the *standard Chebyshev polynomial* of degree n.[1] The two key algebraic properties of standard Chebyshev polynomials follow from Proposition 2.1.

Proposition 3.2. *For every pair $m, n \in \mathbb{Z}$:*
(a) $T_{m+n} + T_{m-n} = 2T_m T_n$;
(b) $T_m(T_n) = T_n(T_m) = T_{mn}$.

In particular,

$$T_0 = 1, \qquad T_1 = x, \qquad \text{and} \qquad T_{n+1} = 2xT_n - T_{n-1} \quad \text{for } n \geq 1.$$

This implies: (1) T_n is an even function for even n, and an odd function for odd n; (2) the leading coefficient of T_n is 2^{n-1}; (3) $T_n(1) = 1$.

It is easily established that $P_n(2x) = 2T_n(x)$. The main advantage of using the nonmonic standard Chebyshev polynomials is that their roots are easier to describe; see Lemma 3.6 below. In addition, we have:

Lemma 3.3. *For all $n \geq 0$, $T_n' = nU_{n-1}$.*

PROOF. Define the matrix

$$C = A(2x, 0) = \begin{pmatrix} 2x & 1 \\ -1 & 0 \end{pmatrix} = \begin{pmatrix} U_1 & U_0 \\ -U_0 & -U_{-1} \end{pmatrix}.$$

Equation (3) implies that the polynomials U_n satisfy $U_{n+1} = 2xU_n - U_{n-1}$. Since powers of C also satisfy $C^{n+1} = 2xC^n - C^{n-1}$, it follows that for all $n \geq 1$,

$$C^n = \begin{pmatrix} U_n & U_{n-1} \\ -U_{n-1} & -U_{n-2} \end{pmatrix}.$$

Therefore,

$$U_n - U_{n-2} = \operatorname{tr}(C^n) = F_n(2x, 0) + \overline{F}_n(2x, 0) = P_n(2x) = 2T_n(x).$$

We now proceed by induction to show $T_n' = nU_{n-1}$, the cases $n = 0, 1$ being clear. Assume $n \geq 1$ and $T_m' = mU_{m-1}$ whenever $0 \leq m \leq n$. Then

$$\begin{aligned} T_{n+1}' &= 2t \cdot T_n' + 2T_n - T_{n-1}' \\ &= 2t \cdot nU_{n-1} + 2T_n - (n-1)U_{n-2} \\ &= n(2tU_{n-1} - U_{n-2}) + (2T_n + U_{n-2}) \\ &= nU_n + U_n \\ &= (n+1)U_n. \end{aligned}$$

This completes the proof of the lemma. □

Remark 3.4. Chebyshev polynomials may be viewed as an alternate set of monomials for the polynomial ring $\mathbb{Z}[x]$, and in some cases provide a distinct computational advantage over the standard monomials $X_n := x^n$. Here are the main formal comparisons of the two families.

(1) $\mathbb{Z}[x] = \mathbb{Z}[X_1] = \mathbb{Z}[P_1]$
(2) $\mathbb{Z}[x] = \sum_{n \geq 0} \mathbb{Z} \cdot X_n = \sum_{n \geq 0} \mathbb{Z} \cdot P_n$
(3) $X_m X_n = X_{m+n}$ and $P_m P_n = P_{m+n} + P_{m-n}$
(4) $X_m(X_n) = X_n(X_m) = X_{mn}$ and $P_m(P_n) = P_n(P_m) = P_{mn}$.

[1] U_n is often referred to as the degree-n Chebyshev polynomial of the second kind.

Rivlin gives the following strong characterization of these two families.

Theorem 3.5 ([19, Theorem 4.4]). *Suppose that $f_n \in \mathbb{Q}[x]$, $n \geq 0$, is a sequence such that (1) $\deg f_n = n$, and (2) $f_m(f_n) = f_n(f_m)$ for every $m, n \in \mathbb{Z}$, $m, n \geq 0$. Then there exists an affine automorphism λ of $\mathbb{Q}[x]$ such that the sequence $\{f_n\}$ is conjugate to either $\{X_n\}$ or $\{P_n\}$ via λ.*

This result can be viewed as a special case of the description of commuting polynomials given by Julia [13] and Ritt [17] in 1922; see also [18, 21].

3.3. Chebyshev polynomials over $\mathbb{R}$.

Lemma 3.6. *Let n be a nonnegative integer.*

(a) *For all $\theta \in \mathbb{R}$,*

$$T_n(\cos\theta) = \cos(n\theta) \quad \textit{and} \quad U_n(\cos\theta) = \frac{\sin((n+1)\theta)}{\sin\theta}.$$

(b) *The roots of T_n are $\cos\big((2k-1)\pi/(2n)\big)$, $k = 1, \ldots, n$, and each of these is a simple root.*

(c) *The roots of U_n are $\cos\big(k\pi/(n+1)\big)$, $k = 1, \ldots, n$, and each of these is a simple root.*

(d) *Given $t \in \mathbb{R}$, $-1 \leq t \leq 1$ if and only if $-1 \leq T_n(t) \leq 1$.*

PROOF. From Proposition 3.1, it follows that

$$2T_n(\cos\theta) = P_n(2\cos\theta) = P_n(\mathrm{e}^{\mathrm{i}\theta} + \mathrm{e}^{-\mathrm{i}\theta}) = \mathrm{e}^{\mathrm{i}n\theta} + \mathrm{e}^{-\mathrm{i}n\theta} = 2\cos(n\theta).$$

Therefore, $T_n(\cos\theta) = \cos(n\theta)$. Combining this identity with the derivative $T_n' = nU_{n-1}$ yields $U_{n-1}(\cos\theta) = \sin(n\theta)/\sin\theta$, which proves (a).

Next, note that

$$\left\{ \cos\left(\frac{2k-1}{2n}\pi\right) \,\middle|\, k = 1, \ldots, n \right\}$$

is a set of n distinct real numbers, and from part (a), each number in this set is a root of $T_n(t)$. Since $\deg T_n = n$, part (b) follows. A similar argument is used to show (c).

For (d), let $t \in \mathbb{R}$ be given, $-1 \leq t \leq 1$, and choose $\theta \in \mathbb{R}$ such that $t = \cos\theta$. Then by part (a), it follows that $-1 \leq T_n(t) \leq 1$. Conversely, note that T_n can have no critical points outside the interval $[-1, 1]$, since $T_n' = nU_{n-1}$ and the roots of U_{n-1} lie in the interval $[-1, 1]$. Therefore, since $U_n(1) = n + 1 > 0$, T_n is increasing for $t > 1$, and since $T_n(1) = 1$, this implies $T_n(t) > 1$ for $t > 1$. Finally, since $|T_n(-t)| = |T_n(t)|$, it follows that $|T_n(t)| > 1$ when $|t| > 1$. □

4. The curves (T_i, T_j)

Let $\mathbf{k}$ be a field of characteristic zero. Given positive integers i and j with $\gcd(i, j) = 1$, define $f \in \mathbf{k}[x, y]$ by $f = T_i(y) - T_j(x)$. Consider the degree function on $\mathbf{k}[x, y]$ defined by $\deg x = i$ and $\deg y = j$. Given $p \in \mathbf{k}[x, y]$, let $\kappa(p)$ denote the highest homogeneous summand of p relative to this grading. Clearly, if $f = gh$ for $g, h \in \mathbf{k}[x, y]$ of positive degree, then $\kappa(f) = \kappa(g)\kappa(h)$, where $\kappa(g)$ and $\kappa(h)$ are of positive degree. However, $\kappa(f) = 2^{i-1}y^i - 2^{j-1}x^j$, and it is well known that this polynomial is irreducible when i and j are coprime. Therefore, f is irreducible.

Let C be the irreducible curve defined by $f = T_i(y) - T_j(x) = 0$. Since $T_i \circ T_j = T_j \circ T_i$, it follows that C is parametrized by $\alpha = (T_i, T_j)\colon \mathbb{A}^1 \to \mathbb{A}^2$. Note that by

Lemma 3.3, $\alpha'(t) = (iU_{i-1}, jU_{j-1})$, so that by Lemma 3.6(c), it follows that $\alpha'(t) \neq 0$ for all $t \in k$. Since C admits a polynomial parametrization, it is a rational curve with one place at infinity.

The curves parametrized by (T_i, T_j) are open versions of the classical Lissajous curves; see [12, 15, 16].

Proposition 4.1. *Assume i and j are positive integers with $\gcd(i,j) = 1$, and let C be the plane curve defined by $T_i(y) - T_j(x) = 0$, and parametrized by $\alpha(t) = (T_i(t), T_j(t))$.*

(1) *The singularities of C consist of $\frac{1}{2}(i-1)(j-1)$ nodes.*
(2) *The nodes of C are precisely points of the set*

$$S = \left\{ \left(\cos\left(\frac{\lambda\pi}{j}\right), \cos\left(\frac{\mu\pi}{i}\right) \right) \,\middle|\, 1 \le \lambda \le j-1, 1 \le \mu \le i-1, \lambda \equiv \mu \pmod 2 \right\}.$$

(3) *Given the node $P = (\cos(\lambda\pi/j), \cos(\mu\pi/i))$,*

$$\alpha^{-1}(P) = \left\{ \cos\left(\frac{k_1\pi}{ij}\right), \cos\left(\frac{k_2\pi}{ij}\right) \right\},$$

where $k_1 = \lambda iu + \mu jv$, $k_2 = \lambda iu - \mu jv$, and $u, v \in \mathbb{Z}$ satisfy $iu + jv = 1$.

(4)

$$\alpha^{-1}(S) = \left\{ \cos\left(\frac{k\pi}{ij}\right) \,\middle|\, 1 \le k < ij, k \notin i\mathbb{Z}, k \notin j\mathbb{Z} \right\}.$$

PROOF. If $f(x,y) = T_i(y) - T_j(x)$, then the singular points of C are defined by the system

$$\begin{cases} f(x,y) = T_i(y) - T_j(x) = 0 \\ f_x(x,y) = -T_j'(x) = 0 \\ f_y(x,y) = T_i'(y) = 0. \end{cases}$$

By Lemma 3.3 and Lemma 3.6(c), the roots of $T_j'(x)$ are $\cos(k\pi/j)$ $(k = 1, \dots, j-1)$, and the roots of $T_i'(x)$ are $\cos(k\pi/i)$, $(k = 1, \dots, i-1)$. It follows that the set of singular points of C is $S \cap C$.

Let $P = (a, b)$ be a point of S, where

$$a = \cos\left(\frac{\lambda\pi}{j}\right) \quad \text{and} \quad b = \cos\left(\frac{\mu\pi}{i}\right)$$

as above. Then

$$\begin{aligned} f(P) &= T_i\left(\cos\left(\frac{\mu\pi}{i}\right)\right) - T_j\left(\cos\left(\frac{\lambda\pi}{j}\right)\right) \\ &= \cos\left(i\frac{\mu\pi}{i}\right) - \cos\left(j\frac{\lambda\pi}{j}\right) \\ &= \cos(\mu\pi) - \cos(\lambda\pi) \\ &= 0, \end{aligned}$$

since $\lambda \equiv \mu \pmod 2$. Therefore, $S \subset C$, and S is precisely the set of singular points of C.

It remains to show that P is an ordinary double point of C. For $X = x - a$ and $Y = y - b$, let $G \in \mathbf{k}^{[2]}$ be such that $G(X,Y) = F(x,y)$, and write $G(X,Y) = \sum_{d\geq 0} G_d(X,Y)$, where G_d is homogeneous of degree d in X and Y. Then:

$$\begin{aligned} G(X,Y) = T_i(y) - T_j(x) &= T_i(Y+b) - T_j(X+a) \\ &= \sum_{m=0}^{i} \frac{T_i^{(m)}(b)}{m!} Y^m - \sum_{n=0}^{j} \frac{T_j^{(n)}(a)}{n!} X^n. \end{aligned}$$

In particular,

$$G_2 = \tfrac{1}{2}\big(T_i''(b)Y^2 - T_j''(a)X^2\big).$$

Note that $f_y(P) = T_i'(b) = 0$ and $f_x(P) = -T_j'(a) = 0$. Since the roots of $T_i'(t) = iU_{i-1}(t)$ and $T_j'(t) = jU_{j-1}(t)$ are simple, it follows that $T_i''(b) \neq 0$ and $T_j''(a) \neq 0$. Therefore, G_2 factors as the product of two distinct linear forms. It follows that P is a node of C. So parts (a) and (b) are proved.

In order to prove part (c), let $\alpha = (T_i, T_j)$, and set $t_m = \cos(k_m\pi/ij)$ for k_m defined above ($m = 1,2$). If $t_1 = t_2$, then

$$\frac{k_1\pi}{ij} = \pm\frac{k_2\pi}{ij} + 2N\pi \text{ for some } N \in \mathbb{Z} \implies \lambda iu + \mu jv = \pm(\lambda iu - \mu jv) + 2ijN.$$

Thus, either $2\mu jv = 2ijN$ (positive case) or $2\lambda iu = 2ijN$ (negative case). In the first case, the fact that $iu + jv = 1$ implies $\mu \equiv 0 \pmod i$, which is impossible since $1 \leq \mu \leq i-1$. Likewise, the second case yields $\lambda \equiv 0 \pmod j$, which is impossible since $1 \leq \lambda \leq j-1$. Therefore, $t_1 \neq t_2$.

On the other hand, $\alpha(t_1) = \alpha(t_2)$, since for $m = 1,2$,

$$\begin{aligned} \alpha(t_m) &= \left(T_i\left(\cos\frac{k_m\pi}{ij}\right), T_j\left(\cos\frac{k_m\pi}{ij}\right)\right) \\ &= \left(\cos\left(i\frac{k_m\pi}{ij}\right), \cos\left(j\frac{k_m\pi}{ij}\right)\right) \\ &= \left(\cos\left(\frac{(\lambda iu \pm \mu jv)\pi}{j}\right), \cos\left(\frac{(\lambda iu \pm \mu jv)\pi}{i}\right)\right) \\ &= \left(\cos\left(\frac{(\lambda(1-jv) \pm \mu jv)\pi}{j}\right), \cos\left(\frac{(\lambda iu \pm \mu(1-iu))\pi}{i}\right)\right) \\ &= \left(\cos\left(\frac{\lambda\pi}{j} + (-\lambda \pm \mu)v\pi\right), \cos\left(\frac{\mu\pi}{i} + (\lambda \pm \mu)u\pi\right)\right) \\ &= \left(\cos\left(\frac{\lambda\pi}{j}\right), \cos\left(\frac{\mu\pi}{i}\right)\right). \end{aligned}$$

This last equality uses the fact that $\lambda \equiv \mu \pmod 2$. So part (c) is proved.

For part (d), reasoning as above gives

$$k_1 \equiv k_2 \equiv \lambda \pmod j, \quad k_1 \equiv \mu \pmod i, \quad \text{and} \quad k_2 \equiv -\mu \pmod i.$$

Since $0 < \lambda < j$, it follows that $k_1, k_2 \notin j\mathbb{Z}$. Similarly, $0 < \mu < i$ implies that $k_1, k_2 \notin i\mathbb{Z}$. Therefore,

$$\alpha^{-1}(S) \subseteq \left\{\cos\left(\frac{k\pi}{ij}\right) \,\middle|\, 1 \leq k < ij, k \notin i\mathbb{Z}, j \notin i\mathbb{Z}\right\}.$$

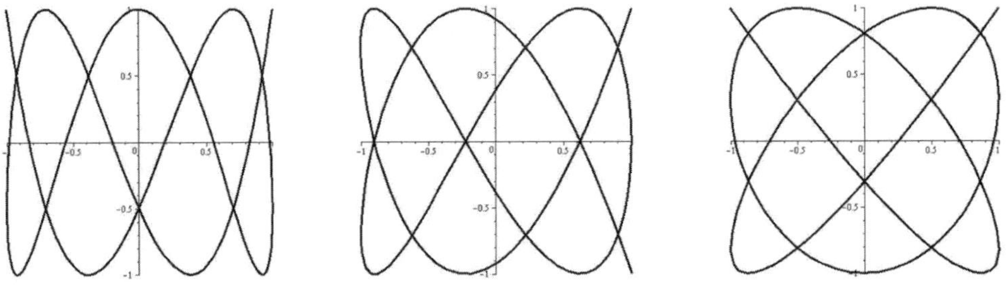

FIGURE 1. The curves (T_3, T_8), (T_4, T_7) and (T_5, T_6) over $\mathbb{R}$

The equality now follows by comparing cardinalities of these two sets. □

Remark 4.2. The much-studied curves $y^n - x^m$ may be viewed as a degeneration of the curves $T_n(y) - T_m(x)$. The Lin–Zaidenberg theorem implies that, when $\gcd(m, n) = 1$, the curve $y^n - x^m$ has a unique embedding in the plane up to algebraic equivalence; see [3]. In the case m and n are distinct prime numbers, Abhyankar and Sathaye [3] generalized this as follows.

Uniqueness Theorem. *If m and n are distinct prime numbers, and if*

$$g = y^n - x^m + \sum_{in+jm<mn} a_{ij} x^i y^j$$

with $a_{ij} \in \mathbf{k}$, then g has only one place at infinity and g has a unique plane embedding, up to algebraic equivalence.

Therefore, the curves $T_n(y) - T_m(x) = 0$ admit only one planar embedding, up to algebraic equivalence, when m and n are distinct primes.

5. Epimorphisms defined by Chebyshev polynomials

Given positive integers i, j, k, define the *pairwise greatest common divisor* of i, j, k by

$$\mathrm{pgcd}(i, j, k) := \max\{\gcd(i, j), \gcd(i, k), \gcd(j, k)\}.$$

Lemma 5.1. *Let i, j, k be positive integers. The following are equivalent.*

(i) $\mathrm{pgcd}(i, j, k) = 1$
(ii) *There exists positive $n \in \mathbb{Z}$ such that i, j, k divide $n, n+1, 2n+1$ in some order.*

PROOF. For any nonzero $n \in \mathbb{Z}$, $\mathrm{pgcd}(n, n+1, 2n+1) = 1$. Therefore, (ii) implies (i).

Conversely, if (i) holds, then at most one of i, j, k is even. So we may assume that k is odd. By the Chinese Remainder Theorem, the system of congruences

$$\begin{aligned} x &\equiv 0 \pmod{i} \\ x + 1 &\equiv 0 \pmod{j} \\ 2x + 1 &\equiv 0 \pmod{k} \end{aligned}$$

has a solution $x \in \mathbb{Z}$, which is unique modulo ijk. In case k is even, we may interchange the roles of k and i. Therefore, (i) implies (ii). □

5.1. Triples which yield embeddings.

Proposition 5.2. *Given integers $i, j, k \geq 2$, define $\phi: \mathbb{Z}[x, y, z] \to \mathbb{Z}[t]$ by $\phi(x) = T_i$, $\phi(y) = T_j$, and $\phi(z) = T_k$. Then ϕ is surjective if and only if* $\text{pgcd}(i, j, k) = 1$.

PROOF. Assume first that ϕ is surjective. Set $d = \gcd(i, j)$, and write $i = ad$ and $j = bd$ for integers a and b. Then

$$T_1 \in \mathbb{Z}[T_i, T_j, T_k] = \mathbb{Z}[T_a(T_d), T_b(T_d), T_k] \subset \mathbb{Z}[T_d, T_k] \subset \mathbb{Q}[T_d, T_k].$$

By the Epimorphism Theorem, it follows that either $k \mid d$ or $d \mid k$. If $k \mid d$, write $d = ck$. Then $T_1 \in \mathbb{Z}[T_c(T_k), T_k] = \mathbb{Z}[T_k]$, which implies $k = 1$, a contradiction. Therefore, $d \mid k$. Write $k = ed$. Then $T_1 \in \mathbb{Z}[T_d, T_e(T_d)] = \mathbb{Z}[T_d]$, which implies $d = 1$. In the same way, $\gcd(i, k) = \gcd(j, k) = 1$. Therefore, $\text{pgcd}(i, j, k) = 1$.

Conversely, assume that $\text{pgcd}(i, j, k) = 1$. By Lemma 5.1, there exists $n \in \mathbb{Z}$ such that i, j, k divide $n, n + 1, 2n + 1$ in some order. Since n depends only on its congruence class modulo ijk, we may assume n is positive.

Suppose $a, b, c \in \mathbb{Z}$ are such that $n = ai$, $n + 1 = bj$, and $2n + 1 = ck$ (so a, b, c are positive). Then the identity $T_1 + T_{2n+1} = 2T_n T_{n+1}$ yields

$$T_1 = 2T_{ai}T_{bj} - T_{ck} = 2T_a(T_i)T_b(T_j) - T_c(T_k) \in \mathbb{Z}[T_i, T_j, T_k].$$

The other cases follow by symmetry. Therefore, ϕ is a surjection. □

Remark 5.3. Given integers i, j, k with $\text{pgcd}(i, j, k) = 1$, the proof above gives an algorithm for finding integers a, b, c such that $T_1 = 2T_a(T_i)T_b(T_j) - T_c(T_k)$, based on the Euclidean algorithm.

Remark 5.4. In [15], the authors point out that Comstock [5] determined the number of self-intersections of the space curves parametrized by

$$\big(\cos(at), \cos(bt), \cos(ct)\big).$$

In particular, Comstock proved that such a curve is nonsingular if and only if $\text{pgcd}(a, b, c) = 1$.

Note that if $1 \in \{i, j, k\}$, then $T_1 \in \mathbb{Z}[T_i, T_j, T_k]$, and (T_i, T_j, T_k) is a surjection. Combining this with Proposition 5.2 gives a complete description of which triples (i, j, k) yield surjections, or equivalently, embeddings $\mathbb{A}^1 \to \mathbb{A}^3$ over a field $\mathbf{k}$ of characteristic zero. We next investigate when two such embeddings are algebraically equivalent.

5.2. Algebraically equivalent embeddings.

Example 5.5. Over the field $\mathbf{k} = \mathbb{R}$, the graph of the embedding $\phi := (T_3, T_4, T_7)$ is given in Figure 2. From this, it is evident that the image of ϕ is a trivial knot. That ϕ is rectifiable is seen as follows. Define $\epsilon \in GA_3(\mathbb{R})$ by $\epsilon = (x, y, 2xy - z): \mathbb{R}^3 \to \mathbb{R}^3$. Since $T_1 = 2T_3T_4 - T_7$, it follows that $\epsilon(T_3, T_4, T_7) = (T_3, T_4, T_1)$. The fact that (T_3, T_4, T_1) is a rectifiable embedding is a consequence of the following lemma. Note that rectifiability is a priori a much stronger property for an embedded real line than that of being topologically trivial.

Lemma 5.6. *If $p(t), q(t), r(t) \in \mathbf{k}[t]$ and $1 \in \{\deg p(t), \deg q(t), \deg r(t)\}$, then $\big(p(t), q(t), r(t)\big)$ is a rectifiable embedding.*

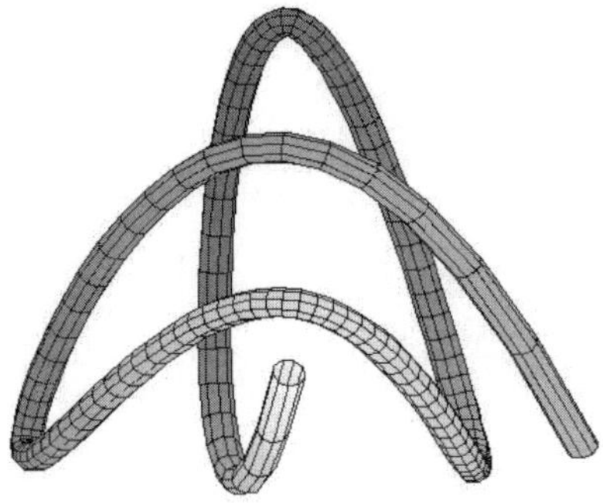

FIGURE 2. The unknot (T_3, T_4, T_7)

PROOF. Suppose $p(t) = at+b$ for $a, b \in \mathbf{k}$ and $a \neq 0$. If $\beta, \gamma \in \mathrm{GA}_3(\mathbf{k})$ are defined by $\beta = (a^{-1}(x-b), y, z)$ and $\gamma = (x, T_j(x) - y, T_k(x) - z)$, then $\gamma\beta\big(p(t), q(t), r(t)\big) = (t, 0, 0)$. □

Definition 5.7. An *elementary involution* $\epsilon\colon \mathbb{A}^3 \to \mathbb{A}^3$ is any polynomial map having one of the following forms:

$$(x,z,y),\quad (y,x,z),\quad (z,y,x)$$
$$(f(y,z)-x,y,z),\quad (x,f(x,z)-y,z),\quad (x,y,f(x,y)-z)$$
$$(g(z)-x,h(z)-y,z),\quad (g(y)-x,y,h(y)-z),\quad (x,g(x)-y,h(x)-z),$$

where $f \in \mathbf{k}^{[2]}$ and $g, h \in \mathbf{k}^{[1]}$.

Definition 5.8. Define an equivalence relation on the set of ordered triples of positive integers by $(i,j,k) \sim (a,b,c)$ if and only if there exists a sequence of elementary involutions $\epsilon_1, \dots, \epsilon_m$ $(m \geq 0)$ such that $(T_i, T_j, T_k) = \epsilon_1\epsilon_2\cdots\epsilon_m(T_a, T_b, T_c)$.

Definition 5.9. Let i, j be positive integers such that $\gcd(i,j) = 1$. The *remnant* of i and j is

$$\langle\!\langle i,j\rangle\!\rangle = \{k \in \mathbb{N} \mid k > \max(i,j), k \notin \langle i,j\rangle, \mathrm{pgcd}(i,j,k) = 1\}.$$

Here, $\langle i,j\rangle$ denotes the semigroup of nonnegative integers generated by i and j.

It is well known that when $\gcd(i,j) = 1$, the complement of $\langle i,j\rangle$ in the set of positive integers is finite, and its largest element is $ij - i - j$ (the Frobenius number of i and j). It follows that $\langle\!\langle 2,j\rangle\!\rangle$ is empty for odd integers j, and that $\max\langle\!\langle i,j\rangle\!\rangle = ij - i - j$ when $i, j > 2$.

Definition 5.10. The ordered triple (i,j,k) of positive integers is a *reduced triple* if and only if either (a) $1 \in \{i,j,k\}$, or (b) $3 \leq i < j$, $\gcd(i,j) = 1$, and $k \in \langle\!\langle i,j\rangle\!\rangle$.

Proposition 5.11. *Let i, j, k be positive integers with* $\mathrm{pgcd}(i,j,k) = 1$. *Then there exists a reduced triple (I,J,K) such that $(i,j,k) \sim (I,J,K)$. In particular, the embeddings (T_i, T_j, T_k) and (T_I, T_J, T_K) are algebraically equivalent.*

PROOF. We may assume $i < j < k$. The proposition obviously holds if (i,j,k) is a reduced triple, so assume (i,j,k) is nonreduced. Then $2 \leq i < j < k$ and $k \in \langle i,j\rangle$. (Note that in case $i = 2$, $k > j$ implies that $k \in \langle i,j\rangle$.) Write $k = ai + bj$ for positive integers a and b. Then

$$T_{|ai-bj|} = 2T_{ai}T_{bj} - T_{ai+bj} = 2T_a(T_i)T_b(T_j) - T_k.$$

TABLE 1. Remnant for small-degree pairs

(i,j)	$\frac{1}{2}(i-1)(j-1)$	$\langle\langle i,j\rangle\rangle$
$(3,4)$	3	5
$(3,5)$	4	7
$(3,7)$	6	8, 11
$(4,5)$	6	7, 11
$(3,8)$	7	13
$(3,10)$	9	11, 17
$(4,7)$	9	9, 13, 17
$(3,11)$	10	13, 16, 19
$(5,6)$	10	7, 13, 19
$(3,13)$	12	14, 17, 20, 23
$(4,9)$	12	11, 19, 23
$(5,7)$	12	8, 9, 11, 13, 16, 18, 23
$(3,14)$	13	19, 25
$(5,8)$	14	9, 11, 17, 19, 27
$(3,16)$	15	17, 23, 29
$(4,11)$	15	13, 17, 21, 25, 29
$(6,7)$	15	11, 17, 23, 29
$(3,17)$	16	19, 22, 25, 28, 31
$(5,9)$	16	11, 13, 16, 17, 22, 26, 31

Define the elementary involution $\epsilon = (x, y, T_a(x)T_b(y) - z)$. Then $\epsilon(T_i, T_j, T_k) = (T_i, T_j, T_{|ai-bj|})$, which implies $(i, j, k) \sim (i, j, |ai - bj|)$. In addition, $\text{pgcd}(i, j, ai + bj) = 1$ implies $\text{pgcd}(i, j, |ai-bj|) = 1$. Observe that the total degree of the triple has been reduced, i.e., $i + j + |ai - bj| < i + j + k$. Since the degree can only be reduced a finite number of times, we eventually arrive at a reduced triple (I, J, K) equivalent to (i, j, k). □

Corollary 5.12. *If (T_2, T_j, T_k) is an embedding, then it is a rectifiable embedding.*

PROOF. By the preceding proposition, $(2, j, k) \sim (I, J, K)$ for a reduced triple (I, J, K). The proof of the proposition shows that $\min\{2, j, k\} \le \min\{I, J, K\}$, which implies $I \le 2$. By definition, 2 cannot appear in a reduced triple, and therefore $I = 1$. The result now follows from Lemma 5.6. □

Corollary 5.13. *If (T_i, T_j, T_{i+j}) is an embedding, then it is a rectifiable embedding.*

PROOF. Assume $i < j$. In the proof of Proposition 5.11, the first reduction replaces the triple $(i, j, i + j)$ with $(i, j, j - i)$, which is equivalent to $(i, j - i, j)$. Note that the triple $(i, j - i, j)$ maintains the form $(a, b, a + b)$. Therefore, this process produces a reduced triple $(I, J, I + J)$ equivalent to $(i, j, i + j)$. It follows that $1 \in \{I, J, I + J\}$. The result now follows by Lemma 5.6. □

The results of this section indicate that every embedding of the form (T_i, T_j, T_k) is algebraically equivalent, via a sequence of elementary involutions, to one for which $k \in \langle\langle i, j\rangle\rangle$. Thus, for a fixed pair of integers i and j with $0 < i < j$ and $\gcd(i, j) = 1$, the set of reduced triples (i, j, k) is finite. Moreover, Lemma 5.6 and Corollary 5.12 show that we need only consider cases in which $i \geq 3$. Table 1 shows the remnant $\langle\langle i, j\rangle\rangle$ for relatively prime pairs i, j such that $3 \leq i < j$ and $\frac{1}{2}(i-1)(j-1) \leq 16$. Recall that $\frac{1}{2}(i-1)(j-1)$ is the number of nodes in the projection of (T_i, T_j, T_k) onto the xy-plane, and that $\max\langle\langle i, j\rangle\rangle = ij - i - j$ if $i, j > 2$

6. Knots defined by Chebyshev polynomials

Throughout this section, we assume that the underlying field is $\mathbf{k} = \mathbb{R}$, and consider the knot type associated with a line embedding in $\mathbb{R}^3$ defined by Chebyshev polynomials.

Table 1 shows that there are 63 reduced triples (i, j, k) with $i \geq 3$ and $\frac{1}{2}(i-1)\times(j-1) \leq 16$. Table 2 gives details for the knots defined by triples up to $\frac{1}{2}(i-1)(j-1) \leq 14$, in addition to a few other identified cases.

For small degree cases, the knot types in Table 2 could be determined empirically, whereas computer software was used to identify the more complicated knots. In a couple of cases, it was possible to determine the crossing number of a knot, even though the particular knot could not be identified; these cases are indicated by $(*)$ in the table. The reader will note the striking variety among the knots appearing in the table, in which very few knot types occur more than once. There are, however, six pairs of matching knot types appearing in Table 2, the smallest pair being $(3, 5, 7)$ and $(5, 7, 8)$. Each parametrizes the 4_1-knot, and it is an open

TABLE 2. Knot types for reduced triples of small degree

	(i, j, k)	xy nodes	knot type		(i, j, k)	xy nodes	knot type
1.	$(3,4,5)$	3	3_1	23.	$(4,9,11)$	12	9_{18}
2.	$(3,5,7)$	4	4_1	24.	$(4,9,19)$	12	$11a_{180}$
3.	$(3,7,8)$	6	5_1	25.	$(4,9,23)$	12	$12a_{541}$
4.	$(3,7,11)$	6	6_3	26.	$(5,7,8)$	12	4_1
5.	$(4,5,7)$	6	5_2	27.	$(5,7,9)$	12	6_3
6.	$(4,5,11)$	6	6_2	28.	$(5,7,11)$	12	$4_1\#4_1$
7.	$(3,8,13)$	7	7_7	29.	$(5,7,13)$	12	$12n_{356}$
8.	$(3,10,11)$	9	7_1	30.	$(5,7,16)$	12	$12n_{798}$
9.	$(3,10,17)$	9	9_{31}	31.	$(5,7,18)$	12	$12n_{321}$
10.	$(4,7,9)$	9	7_5	32.	$(5,7,23)$	12	$12a_{222}$
11.	$(4,7,13)$	9	8_7	33.	$(3,14,19)$	13	$11a_{119}$
12.	$(4,7,17)$	9	9_{20}	34.	$(3,14,25)$	13	$13(*)$
13.	$(3,11,13)$	10	8_3	35.	$(5,8,9)$	14	7_3
14.	$(3,11,16)$	10	9_{17}	36.	$(5,8,11)$	14	7_7
15.	$(3,11,19)$	10	10_{45}	37.	$(5,8,17)$	14	7_4
16.	$(5,6,7)$	10	5_2	38.	$(5,8,19)$	14	$11(*)$
17.	$(5,6,13)$	10	10_{159}	39.	$(5,8,27)$	14	$14(*)$
18.	$(5,6,19)$	10	10_{116}	40.	$(3,16,17)$	15	11_1
19.	$(3,13,14)$	12	9_1	41.	$(6,7,11)$	15	8_{15}
20.	$(3,13,17)$	12	10_{37}	42.	$(5,9,11)$	16	8_{12}
21.	$(3,13,20)$	12	$11a_{175}$	43.	$(7,8,9)$	21	7_5
22.	$(3,13,23)$	12	$12a_{499}$	44.	$(9,10,11)$	36	9_{18}

FIGURE 3. Knots for triples $(3,7,8)$, $(4,5,11)$ and $(5,7,9)$

question whether these two embeddings are algebraically equivalent. Note also that every knot but one in the table is prime, namely, $(5,7,11)$, which is the composite of two 4-crossing knots. Figure 3 illustrates three knots from Table 2.

The following two results, due to Koseleff and Pecker, concern two important families of knots parametrized by Chebyshev polynomials.

Theorem 6.1 ([15, Theorem 1]). *Given coprime positive integers i, j, let K denote the knot parametrized by (T_i, T_j, T_{ij-i-j}). Then the projection of K onto the xy-plane is alternating.*

Theorem 6.2 ([15, Theorem 2]). *Given $n \geq 1$, $(T_3, T_{3n+1}, T_{3n+2})$ parametrizes the $(2, 2n+1)$ torus knot.*

The knot K determined by i and j in Theorem 6.1 is the *alternate harmonic knot* of type (i,j). Note that the crossing number of the alternate harmonic knot of type (i,j) is $\frac{1}{2}(i-1)(j-1)$. Table 3 below lists these knots up to 10 crossings.

The main result in the paper of Koseleff and Pecker is the following remarkable fact.

Theorem 6.3 ([15, Section 4]). *Every knot in $\mathbb{R}^3$ is topologically equivalent to a knot parametrized by*

$$\big(T_i(t), T_j(t), T_k(t+c)\big)$$

for positive integers i, j, k and some constant $c \in \mathbb{R}$, where $\operatorname{pgcd}(i,j,k) = 1$.

TABLE 3. Alternate harmonic knots

(i,j)	knot type
$(3,4)$	3_1
$(3,5)$	4_1
$(3,7)$	6_3
$(4,5)$	6_2
$(3,8)$	7_7
$(3,10)$	9_{31}
$(4,7)$	9_{20}
$(3,11)$	10_{45}
$(5,6)$	10_{116}

In particular, every knot has a topological projection onto a curve of the form $T_i(y) = T_j(x)$.

7. Epimorphisms defined by polynomials F_n

We now return our attention to the polynomials F_n defined earlier.

Let $G = \{1, \mu\}$ denote the two element group, and define actions of G on $\mathbb{Z}^{[5]} = \mathbb{Z}[x_1, x_2, y_1, y_2, z]$ and $\mathbb{Z}^{[2]} = \mathbb{Z}[u, v]$ by

$$\mu(x_1, x_2, y_1, y_2, z) = (x_2, x_1, y_2, y_1, z) \quad \text{and} \quad \mu(u, v) = (v, u).$$

Define G-invariant ideals $I = (x_1 - x_2, y_1 - y_2) \subset \mathbb{Z}^{[5]}$ and $J = (u - v) \subset \mathbb{Z}^{[2]}$. Then $\mathbb{Z}^{[5]}/I = \mathbb{Z}^{[3]}$ and $\mathbb{Z}^{[2]}/J = \mathbb{Z}^{[1]}$. Over a field $\mathbf{k}$ of characteristic zero, I and J define the fixed points for the induced G-actions on $\mathbb{A}^5$ and $\mathbb{A}^2$.

Given a triple of positive integers (i, j, k), define $\Phi: \mathbb{Z}[x_1, x_2, y_1, y_2, z] \to \mathbb{Z}[u, v]$ by

$$\Phi = \big(F_i(u,v), \overline{F}_i(u,v), F_j(u,v), \overline{F}_j(u,v), P_k(u+v)\big).$$

Then Φ is equivariant relative to the given G-actions, and Φ induces a well-defined map $\phi: \mathbb{Z}^{[5]}/I \to \mathbb{Z}^{[2]}/J$. Geometrically, $\Phi^*: \mathbb{A}^2 \to \mathbb{A}^5$ is a G-equivariant morphism of algebraic varieties, and its restriction to the fixed points of the G-actions is $\phi^*: \mathbb{A}^1 \to \mathbb{A}^3$.

Proposition 7.1. *With the notation and hypotheses above*:

(a) $\phi = \big(P_i(u+v), P_j(u+v), P_k(u+v)\big)$;

(b) Φ *is surjective if and only if* $\mathrm{pgcd}(i, j, k) = 1$.

PROOF. Consider the following exact sequences.

$$\begin{array}{ccccccccc}
0 & \longrightarrow & I & \longrightarrow & \mathbb{Z}[x_1, x_2, y_1, y_2, z] & \xrightarrow{\pi} & \mathbb{Z}[x_1 + x_2, y_1 + y_2, z] & \longrightarrow & 0 \\
 & & & & \downarrow \Phi & & \downarrow \phi & & \\
0 & \longrightarrow & J & \longrightarrow & \mathbb{Z}[u, v] & \xrightarrow{\rho} & \mathbb{Z}[u + v] & \longrightarrow & 0.
\end{array}$$

The map π admits a section $\sigma: \mathbb{Z}[x_1 + x_2, y_1 + y_2, z] \hookrightarrow \mathbb{Z}[x_1, x_2, y_1, y_2, z]$, and ϕ factors as $\phi = \rho\Phi\sigma$. Thus,

$$\phi(x_1 + x_2) = \rho\Phi\sigma(x_1 + x_2) = \rho\Phi(x_1 + x_2) = \rho(F_i + \overline{F}_i) = \rho(P_i(u+v)) = P_i(u+v);$$
$$\phi(y_1 + y_2) = \rho\Phi\sigma(y_1 + y_2) = \rho\Phi(y_1 + y_2) = \rho(F_j + \overline{F}_j) = \rho(P_j(u+v)) = P_j(u+v);$$
$$\phi(z) = \rho\Phi\sigma(z) = \rho\Phi(z) = \rho(P_k(u+v)) = P_k(u+v).$$

This proves part (a)

Next, consider the matrices used to define the polynomials $F_n(u, v)$, namely,

$$A = \begin{pmatrix} u & 1 \\ uv - 1 & v \end{pmatrix} \quad \text{and} \quad A^n = \begin{pmatrix} F_n & G_{n-1} \\ (uv-1)G_{n-1} & \overline{F}_n \end{pmatrix}.$$

Multiplying equation (1) by A yields

$$0 = A^{2n+1} - \mathrm{tr}(A^n)A^{n+1} + \det(A^n)A = A^{2n+1} - P_n(u+v)A^{n+1} + A.$$

Therefore, for all $n \in \mathbb{Z}$, we have

$$A = P_n(u+v)A^{n+1} - A^{2n+1}. \tag{5}$$

Assume $\mathrm{pgcd}(i,j,k) = 1$ and i is odd. Then by Lemma 5.1, there exist positive $a, b, c, n \in \mathbb{Z}$ such that $n = ak$, $n+1 = bj$, and $2n+1 = ci$. From equation (5), it follows that

$$\mathrm{diag}(A) = \begin{pmatrix} u \\ v \end{pmatrix} = P_n(u+v)\begin{pmatrix} F_{n+1} \\ \overline{F}_{n+1} \end{pmatrix} - \begin{pmatrix} F_{2n+1} \\ \overline{F}_{2n+1} \end{pmatrix}.$$

Using the fact that $\overline{F}_{n+1} = \overline{F}_b(F_j, \overline{F}_j)$ and $\overline{F}_{2n+1} = \overline{F}_c(F_i, \overline{F}_i)$, this equality becomes

$$\begin{pmatrix} u \\ v \end{pmatrix} = P_a(P_k(u+v))\begin{pmatrix} F_b(F_j, \overline{F}_j) \\ \overline{F}_b(F_j, \overline{F}_j) \end{pmatrix} - \begin{pmatrix} F_c(F_i, \overline{F}_i) \\ \overline{F}_c(F_i, \overline{F}_i) \end{pmatrix}.$$

Therefore, $u, v \in \mathbb{Z}[F_i, \overline{F}_i, F_j, \overline{F}_j, P_k(u+v)]$ when i is odd. If i is even, then j is odd, and the roles of i and j can be interchanged. Therefore, Φ is surjective whenever $\mathrm{pgcd}(i,j,k) = 1$.

Conversely, assume Φ is surjective. Then ϕ is also surjective. From part (a), we have that ϕ is defined by $(P_i(u+v), P_j(u+v), P_k(u+v))$. Since $P_n(2x) = 2T_n(x)$, it follows from Proposition 5.2 that $\mathrm{pgcd}(i,j,k) = 1$. This proves part (b). □

References

1. S. S. Abhyankar, *Lectures on expansion techniques in algebraic geometry*, with by B. Singh, Tata Inst. Fund. Res. Lectures on Math. and Phys., vol. 57, Tata Inst. Fund. Res., Bombay, 1977.
2. S. S. Abhyankar and T. T. Moh, *Embeddings of the line in the plane*, J. Reine Angew. Math. **276** (1975), 148–166.
3. S. S. Abhyankar and A. Sathaye, *Uniqueness of plane embeddings of special curves*, Proc. Amer. Math. Soc. **124** (1996), no. 4, 1061–1069.
4. S. M. Bhatwadekar and A. Roy, *Some results on embedding of a line in 3-space*, J. Algebra **142** (1991), no. 1, 101–109.
5. E. H. Comstock, *The real singularities of harmonic curves of three frequencies*, Wisconsin Acad. Sci. Arts Lett. Trans. **11** (1896), 452–464.
6. P. C. Craighero, *A result on m-flats in* $\mathbf{A}^n_k$, Rend. Sem. Mat. Univ. Padova **75** (1986), 39–46.
7. ———, *A remark on Abhyankar's space lines*, Rend. Sem. Mat. Univ. Padova **80** (1988), 87–93 (1989).
8. A. Durfee and D. O'Shea, *Polynomial knots*, available at `arXiv:math/0612803`.
9. G. Freudenburg, *Algebraic theory of locally nilpotent derivations*, Encyclopaedia Math. Sci., vol. 136, Springer, Berlin, 2006.
10. J. K. Freudenburg, *Parametrizing knots with Chebyshev polynomials*, Siemens Competition in Math, Science, and Technology, 2008.
11. ———, *Parametrizing knots with Chebyshev polynomials*, Intel Science Talent Search, 2008.
12. L. Granát and M. Fiedler, *Rational curves with a maximum number of nodes*, Časopis Pěst. Mat. **79** (1954), 157–161 (Czech).
13. G. Julia, *Mémoire sur la permutabilité des fractions rationnelles*, Ann. Sci. École Norm. Sup. (3) **39** (1922), 131–215.
14. Sh. Kaliman, *Isotopic embeddings of affine algebraic varieties into* $\mathbf{C}^n$, The Madison Symposium on Complex Analysis (Madison, WI, 1991) (A. Nagel and E. L. Stout, eds.), Contemp. Math., vol. 137, Amer. Math. Soc., Providence, RI, 1992, pp. 291–295.
15. P.-V. Koseleff and D. Pecker, *Chebyshev knots*, available at `arXiv:0812.1089`.
16. D. Pecker, *Simple constructions of algebraic curves with nodes*, Compositio Math. **87** (1993), no. 1, 1–4.
17. J. F. Ritt, *Prime and composite polynomials*, Trans. Amer. Math. Soc. **23** (1922), no. 1, 51–66.
18. ———, *Permutable rational functions*, Trans. Amer. Math. Soc. **25** (1923), no. 3, 399–448.
19. T. J. Rivlin, *Chebyshev polynomials: From approximation theory to algebra and number theory*, 2nd ed., Pure Appl. Math. (N. Y.), Wiley, New York, 1990.
20. L. Rudolph, *Embeddings of the line in the plane*, J. Reine Angew. Math. **337** (1982), 113–118.
21. A. Schinzel, *Selected topics on polynomials*, Univ. Michigan Press, Ann Arbor, MI, 1982.

22. A. R. Shastri, *Polynomial representations of knots*, Tohoku Math. J. (2) **44** (1992), no. 1, 11–17.
23. M. Suzuki, *Propriétés topologiques des polynômes de deux variables complexes, et automorphismes algébriques de l'espace* $\mathbf{C}^2$, J. Math. Soc. Japan **26** (1974), 241–257.
24. A. van den Essen, *Polynomial automorphisms and the Jacobian conjecture*, Progr. Math., vol. 190, Birkhäuser, Basel, 2000.
25. V. A. Vassiliev, *Cohomology of knot spaces*, Theory of Singularities and Its Applications (V. I. Arnol′d, ed.), Adv. Soviet Math., vol. 1, Amer. Math. Soc., Providence, RI, 1990, pp. 23–69.

Department of Mathematics, Western Michigan University, Kalamazoo, MI 49008, USA

E-mail address: gene.freudenburg@wmich.edu

Centre de Recherches Mathématiques
CRM Proceedings and Lecture Notes
Volume **54**, 2011

The Pencil of Translates of a Line in the Plane

Richard Ganong

To my teacher, Peter Russell

Abstract. Let R be the polynomial ring in two variables over an algebraically closed field $\mathbf{k}$ of positive characteristic. Let $f \in R$ define a plane curve which is isomorphic to the affine line over $\mathbf{k}$. A long-standing question asks whether all fibres over closed points of the map defined by the inclusion $\mathbf{k}[f] \subset R$ are then lines. One can consider the $\mathbf{k}$[f]-subalgebras of R gotten by iterating "relative Frobenius". It turns out that these are eventually planes if and only if the above question has a positive answer. That answer is in turn equivalent to geometric rationality of the generic curve of the above pencil. A positive answer to the question, viewed as above, would provide an eminently satisfying link with the complete picture one has of embeddings of the line in the plane in zero characteristic, and could be regarded as supplying a "recipe" (terminology from Abhyankar) for all embeddings of the line in the plane.

1. Introduction

This article is in part an exposition of known results, and in part a report of the modest progress made by the author — much of it over three decades ago, much of it in unpublished parts of his doctoral dissertation — toward answering the above question (Question 1.1 below). The question remains open. The perspective afforded by the passage of so much time has affected the presentation of the work, and has made the author markedly more aware of how very large a part Peter Russell played not only in the obtaining of some of the results belatedly presented here but in the author's professional and even personal development. The author acknowledges Russell's generous help, friendship and inspiring example with gratitude and humility. The author would like also to thank sincerely here some other geometers who have helped him greatly at various times over the years, notably Bobby Lascu, Avinash Sathaye, Masayoshi Miyanishi, and Daniel Daigle. His appreciation for their help only deepens with time.

This section recalls some (mostly standard) notation and terminology, and some well-known facts, and recounts some history.

2010 *Mathematics Subject Classification.* Primary 14R10; Secondary 14M20, 14R25, 13N15.
Key words and phrases. line, plane, polynomial ring, affine space, affine variety, Frobenius.
This is the final form of the paper.

$\mathbf{k}$ always denotes a field, with algebraic closure $\overline{\mathbf{k}}$. When the characteristic of $\mathbf{k}$ is positive, it is always denoted by p. The group of units of a ring A is denoted A^*. x, y always denotes a generating pair for a bivariate polynomial algebra R over $\mathbf{k}$. For each positive integer n, $\mathbf{k}^{[n]}$ denotes a ring, having $\mathbf{k}$ as a subring, which is isomorphic as a $\mathbf{k}$-algebra to an n-dimensional polynomial algebra over $\mathbf{k}$. One calls $\mathbf{k}^{[1]}$ and $\mathbf{k}^{[2]}$ a *line* and a *plane*, respectively (over $\mathbf{k}$), and uses the same terms for the associated affine schemes $\mathbb{A}^1_{\mathbf{k}}$, $\mathbb{A}^2_{\mathbf{k}}$. $\subset$ denotes proper containment $0 \notin \mathbb{N}$.

Given a nonzero, proper principal ideal fR of R, the canonical surjection $\pi_{fR}: R \to R/fR$ defines a closed embedding of the $\mathbf{k}$-curve $\mathrm{Spec}(R/fR)$ into the plane. Given a nonconstant $f \in R$ such that this curve is isomorphic to a line, one speaks of "*a line f in the plane R*" (suppressing "R" when the reference is clear). The phrase indicates that the object of interest is the embedding.

Given nonconstant polynomials $f, g \in R$, one says that the canonical embeddings they define are *equivalent* iff there are $\mathbf{k}$-algebra isomorphisms $\alpha: R/fR \to R/gR$ and $\varphi: R \to R$ such that $\alpha \circ \pi_{fR} = \pi_{gR} \circ \varphi$, i.e., iff the corresponding curves are isomorphic via a map induced by an automorphism of the plane. This is the case iff there is such an automorphism φ such that, for some $c \in \mathbf{k}^*$, $\varphi(f) = cg$. One says that f is a *variable of R* if there is such an automorphism φ with g being x; this is tantamount to saying that there is a polynomial $h \in R$ such that $R = \mathbf{k}[f, h]$.

Any variable f of R is a line in R. The converse holds if $\mathbf{k}$ is of characteristic 0, or if a certain condition on the bidegree $(\deg_x f, \deg_y f)$ holds (quaintly named "Epimorphism" Theorem of Moh and Abhyankar, [AM75]). M. Suzuki (1974) also proved the converse for the case $\mathbf{k} = \mathbb{C}$. The bidegree condition here is unsatisfying, by virtue of being by no means necessary — what distinguishes variables f from other lines is the residue field degree of the place at infinity of the $\mathbf{k}(f)$-curve $\mathbf{k}(f) \otimes_{\mathbf{k}[f]} R$ ([Rus75, Theorem 4.5]; see also [Nag71]). That residue field degree divides both entries of the bidegree above, for any choice of generating pair x, y for R. (See, e.g., [Gan79, 1.6.1, 1.6.2, 1.9, 1.15] — work which gives the full force of the "Epimorphism" Theorem but was passed over in the bibliographies of at least two Miyanishi works — work, also, inspired (notably) by Lemma 2.8 of [Miy78], which proves the characteristic 0 version of the same theorem.) The author would also like to call attention quite boldly here to the fact that

the first published example of a line which is not a variable seems to be due to Beniamino Segre.

(See [Seg56, p. 18].[1]) This priority was not acknowledged in [Nag71] or in [AM75]; the authors of the latter paper explicitly call attention to this Segre article in theirs. Many authors since 1975 have wrongly accorded priority to Nagata (e.g., [Moh88, p. 268; Miy11, remark above Example 4.3]) or to Moh or Abhyankar.

Lines in the plane which are not variables have been called "bad" lines in the literature. We feel that "exotic" would be a much more apt label. The focus of this article, these lines are significantly more interesting than variables.

Given a nonconstant polynomial $f \in R$, the curves $R/(f - \lambda)R$ $(\lambda \in \mathbf{k})$ — and their canonical embeddings in the plane R — are the *translates* in R of the curve R/fR. It is evident that this notion is well defined (it does not depend on the generator chosen for fR). One speaks, more briefly, of "the translates of f";

[1]In the final sentence of the paper, letting $q > 1$ be an integer prime to p, Segre exhibits "la curva piana razionale intera... avendo l'equazione non parametrica $y^{p^2} = x^{pq} + x$".

whether this refers to (plane embeddings of) curves $R/(f-\lambda)R$ or to polynomials $f-\lambda$ will always be clear from the context. The spectra of the translates of f are precisely the fibres over $\mathbf{k}$-rational points of the morphism of affine schemes defined by the inclusion $\mathbf{k}[f] \subset R$.

It is evident that a translate of a variable of R is a variable of R.

This article is concerned with the following question, in which one might as well assume that $\mathbf{k}$ has positive characteristic:

1.1. Question. Let f be a line in R. Are all translates of f then lines in R?

To this writer's knowledge, the earliest appearance of this question in the literature is Q(1.2) in [Sat76]. Avinash Sathaye informed the author (email, August 2010) that Question 1.1 was conceived in an automobile:

> *Abhyankar reports that he developed this question and others about lines during a trip from Lafayette to Lexington in a car with Bill Heinzer. So, this was probably a 5-hour seminar* (:-)). *This was* 1968, *and there was a conference at Kentucky. The Abhyankar – Eakin – Heinzer paper was an offshoot of the same discussion.*

2. Plane curves with one place at infinity, basic facts about parametrizations of lines

We call an irreducible polynomial $f \in R$, or the $\mathbf{k}$-curve $A = R/fR$, a *plane curve with one place at infinity*, if there is exactly one valuation ring V over $\mathbf{k}$ of the field of fractions $\operatorname{Frac} A$ of A such that $V \not\supseteq A$. If f has one place V at infinity, one calls the place *rational* if the residue field $\kappa(V)$ is $\mathbf{k}$, and *purely inseparable* if $\kappa(V)/\mathbf{k}$ is a purely inseparable extension.

A question which, around 1976, seemed highly apposite to Question 1.1 (since a line in the plane is of course a plane curve with one rational place at infinity) was

2.1. Question. Let $\mathbf{k}$ be algebraically closed, and $f \in R$ have one place at infinity. Do then all translates of f have one place at infinity? (See [Abh77, (11.18)].)

The question was answered negatively in [Gan79]. The whole story is summarized below. (It is convenient also to choose an indeterminate t over $\operatorname{Frac} R$ and note that the generic member of the pencil referred to below is isomorphic as $\mathbf{k}(f)$-algebra to the $\mathbf{k}(t)$-algebra $\mathbf{k}(t)[x,y]/(f-t)$. This allows one to think of the generic member as being embedded in the affine plane $\mathbf{k}(t)[x,y]$.) The field $\mathbf{k}$ mentioned in the theorem is arbitrary.

2.2. Theorem. *Let $f \in R$ have one rational place at infinity. Then:*

(1) *The generic member $\mathbf{k}(f) \otimes_{\mathbf{k}[f]} R \cong \mathbf{k}(f)[x,y]$ of the pencil $\Lambda = \Lambda(f)$ of translates of f in R has one purely inseparable place at infinity.*

(2) *Extend the scalars to $\overline{\mathbf{k}}$ and consider the pencil $\overline{\Lambda}$ of translates of f in $R \otimes_{\mathbf{k}} \overline{\mathbf{k}} \cong \overline{\mathbf{k}}^{[2]}$. Then the general member of $\overline{\Lambda}$ has one place at infinity.*

(3) *Let x, y be any generating pair for R, and embed $\operatorname{Spec}(R \otimes_{\mathbf{k}} \overline{\mathbf{k}(t)})$ into $\mathbb{P}^2_{\overline{\mathbf{k}(t)}}$ in the corresponding natural way. One can then speak of the tree of infinitely near points at infinity of the projective closures of the curves defined by $f-\lambda$ ($\lambda \in \overline{\mathbf{k}}$) and by $f-t$, and one has:*

The multiplicity sequences at infinity of $f-t$ over $\overline{\mathbf{k}(t)}$ and of those $f-\lambda$ which have one place at infinity are the same, for all but finitely many λ.

(4) *All results above are best possible if* $\operatorname{char}\mathbf{k} > 0$. *One can replace "the general" in* (2) *by "every", and omit "but finitely many" in* (3), *iff* $\operatorname{char}\mathbf{k} = 0$ (*see* [Gan79, 1.4, 2.1, 2.13]).

(5) *One even has examples* (ibid., 2.14) *where the pencil* Λ *has "no moving singularity at finite distance"* (*indeed, where all translates of* f *are nonsingular on* $\operatorname{Spec} R$) *and some member has more than one place at infinity.*

In characteristic 0, none of the above pathologies involving the pencil of translates of a plane curve f with one place at infinity occur. One then has "once one place, always one place", and one has the exhaustive Abhyankar – Moh result that the value semigroups of the translates of f are all the same ([Abh77, Section 8], and Section 7 of the 1973 Crelle forerunner of [AM75], by the same authors). Thus, the above pathologies might be taken as hints that the answer to our main question (Question 1.1) is "No". We conjecture otherwise later in this article.

One complicating fact of life in positive-characteristic algebraic geometry is the fact that linear systems can have moving singularities. Lemma 2.4 below is another result in the spirt of Theorem 2.2 (2), and may be viewed as encouraging, if one hopes for a positive answer to Question 1.1. We first give an example to put the lemma in perspective.

2.3. Example. Let m be an integer greater than 1 and prime to p, and let $f(x,y) = x^p - (xy-1)^m$. Then $xy-1$ divides both f_x and f_y, whence it follows that the pencil $\Lambda(f)$ has a moving singularity on $\operatorname{Spec} R$. But the curve $f = 0$ is a special member of the pencil that happens to be an irreducible rational curve which is nonsingular "at finite distance". (One way to see easily that it has more than one place at infinity is to note that the image of x in R/fR is a nontrivial unit of that ring.)

2.4. Lemma. *Let* $\mathbf{k}$ *be a field of positive characteristic,* $f \in R$ *a nonsingular curve with one rational place at infinity. Then the pencil* $\Lambda(f)$ *has no moving singularity. Otherwise said, the surfaces* $R^{p^n}[f]$ *are normal, for all* $n > 0$.

PROOF. Since f remains a plane curve with one place at infinity if we pass to the algebraic closure of $\mathbf{k}$, we may, and do, assume that $\mathbf{k}$ is algebraically closed. We remark in passing that all members of the pencil $\Lambda(f)$ are irreducible, because f has one place at infinity. We prove the lemma by contradiction. Suppose that h is a nonconstant common factor of f_x and f_y. Then, nonsingularity of f implies that the image of h in R/fR is a unit in that ring. Since R/fR has trivial units, there is a $c \in \mathbf{k}^*$ such that f divides the nonzero polynomial $h - c$. Since $f \notin R^p$, f_x or f_y is nonzero. We may (and do) assume that f_x is nonzero. Letting "deg" denote total degree with respect to x, y, we then have: $\deg f \le \deg h \le \deg f_x \le \deg f - 1$, an absurdity.

Let n be a positive integer. Given an element g of a polynomial ring over $\mathbf{k}$, expressed in terms of a fixed set of variables for that ring, we use the standard notation $g^{(p^n)}$ to denote the polynomial whose coefficients are the p^nth powers of the corresponding coefficients of g. One has the $\mathbf{k}$-algebra isomorphism $\mathbf{k}[X,Y,Z]/(Z^{p^n} - f^{(p^n)}(X,Y)) \xrightarrow{\cong} R^{p^n}[f]$, induced by the map which carries X, Y, Z to x^{p^n}, y^{p^n}, f, respectively. So this last ring is a hypersurface, which by the Jacobian criterion has only isolated singularities iff $\Lambda(f)$ lacks moving singularities. (h divides f_x and f_y iff $h^{(p^n)}$ divides both $(f^{(p^n)})_x = (f_x)^{(p^n)}$ and $(f^{(p^n)})_y = (f_y)^{(p^n)}$.) By Serre's criterion, we are done. □

In closing this section, we prove two well-known facts about parametrizations of lines (see also COROLLARY (2.6) of [AM75] for (2) below):

2.5. Lemma. *Let $f \in R = \mathbf{k}[x,y]$ be a line, and suppose that $\alpha: R \to \mathbf{k}[t] \cong \mathbf{k}^{[1]}$ is a surjection, with kernel fR, which carries x to $u(t)$ and y to $v(t)$. Then:*

(1) $\deg_x f = \deg_t v$ *and* $\deg_y f = \deg_t u$, *and*

(2) *there is a* $C \in \mathbf{k}^*$ *such that* $\alpha(f_x) = Cv'(t)$ *and* $\alpha(f_y) = -Cu'(t)$.

PROOF. Both statements being clear when $f \in \mathbf{k}[x]$ or $f \in \mathbf{k}[y]$, we may, and do, assume that $\deg_x f$ and $\deg_y f$ are both positive, from which it readily follows that $\deg_t u$ and $\deg_t v$ are positive. Since f has one place at infinity, f is "monic" in x and y, i.e., the terms in f of highest degree in x [resp. y] are in $\mathbf{k}[x]$ [resp. $\mathbf{k}[y]$].

We prove just the first equality in (1). Let Z be an indeterminate over $\mathbf{k}(t)$, and put $f_1(Z) = f(Z, v(t))$. f_1 is monic in Z and irreducible in $\mathbf{k}[v(t)][Z]$; hence, by the Gauss lemma, irreducible in $\mathbf{k}(v(t))[Z]$, and $f_1(u(t)) = 0$. In view of all of the above, we have that $\deg_t v = [\mathbf{k}(t) : \mathbf{k}(v)] = [\mathbf{k}(v)(u) : \mathbf{k}(v)] = \deg_Z f_1(Z) = \deg_x f$.

Let $g \in R$ be a preimage under α of t. By two applications of the chain rule, we have that $\alpha(g_x)u'(t) + \alpha(g_y)v'(t) = 1$ and $\alpha(f_x)u'(t) + \alpha(f_y)v'(t) = 0$. So $u'(t)$ and $v'(t)$ are coprime in $\mathbf{k}[t]$, and (2) will follow from the second equation here and coprimeness of $\alpha(f_x)$ and $\alpha(f_y)$. We prove the coprimeness by contradiction.

Suppose that $d \in \mathbf{k}[t]$ is a nonconstant common factor of $\alpha(f_x)$ and $\alpha(f_y)$. Let $c \in \overline{\mathbf{k}}$ be a root of d. Put $a := u(c)$, $b := v(c)$. Then f, f_x, f_y all vanish at (a,b). But f is a line in the affine plane over $\overline{\mathbf{k}}$, so this contradicts the Jacobian criterion. □

2.6. Remark. Given a parametrization α of a line f as above, let τ be the automorphism of $\mathbf{k}[t]$ which maps t to Ct, and let $\overline{\alpha} := \tau \circ \alpha$. Putting $\overline{u} := \overline{\alpha}(x)$ and $\overline{v} := \overline{\alpha}(y)$, one checks easily that $\overline{\alpha}(f_x) = \overline{v}'(t)$ and $\overline{\alpha}(f_y) = -\overline{u}'(t)$. We call $\overline{\alpha}$ the "normalized" parametrization associated with α.

3. Examples of lines

In this section, $\mathbf{k}$ is assumed to be algebraically closed.

3.1. Lemma. *Let $f \in R \setminus R^p$.*

(1) *Suppose that $R^p[f]$ is a plane. Then some variable x of R is in $R^p[f]$. Hence, for any y such that $R = \mathbf{k}[x,y]$, $R^p[f] = \mathbf{k}[x, y^p]$. If, furthermore, f is a line in R then, for some such x, $f - x \in R^p$.*

(2) *If f is a line in R then, for all $n \in \mathbb{N}$, f is a line in the surface $R_n := R^{p^n}[f]$. If f is a line in R_n for some n, then f is a line in R.*

(3) *Let $R = \mathbf{k}[x,y]$, and suppose that f is a line in $\mathbf{k}[x,y^p] =: A$, and is a nonsingular curve in R. Then f is a line in R.*

PROOF. (1) The main result of [Gan82] is that if $A \cong \mathbf{k}^{[2]}$ and $R \supset A \supset R^p$ then some variable x of R is in A. For any y as in (1), $R^p[f]$ is a birational extension of $\mathbf{k}[x, y^p]$ and integral over it.

We prove the last sentence of (1): Let f be a line in R. Then, since $f_y = 0$, by Lemma 2.4 we have that, for some nonzero $c \in \mathbf{k}$, $f - cx \in R^p$. Replacing x by the variable cx completes this proof of (1).

(2) Suppose that f is a line in R. Let $n \in \mathbb{N}$. We have a surjection $\alpha: R \to \mathbf{k}[t] \cong \mathbf{k}^{[1]}$ of $\mathbf{k}$-algebras, with kernel fR. This induces in a natural way a surjective $\mathbf{k}$-algebra homomorphism $\alpha_n: R_n \to \mathbf{k}[t^{p^n}] \cong k^{[1]}$ with kernel $fR \cap R_n$. A direct

calculation, using the fact that R_n is a free R^{p^n}-module of rank p^n with basis $\{f^i \mid 0 \le i \le p^n - 1\}$, shows that $fR \cap R_n = fR_n$. (Alternatively, invoke Lemma 2.4, from which this equality of ideals follows since R is integral over R_n.)

The last sentence of (2) drops out of a straightforward calculation: Let f be a line in R_n for some n, and put $q = p^n$. We have a surjection of $\mathbf{k}$-algebras $\alpha: \mathbf{k}[f, x^q, y^q] \to \mathbf{k}[t^q] \cong \mathbf{k}^{[1]}$ with kernel fR_n. Let $\alpha(x^q) = u^q$ and $\alpha(y^q) = v^q$. Let $\beta: R \to \mathbf{k}[t]$ be the unique $\mathbf{k}$-algebra map taking x, y to u, v respectively. Since α is surjective and $\alpha(f) = 0$, β is clearly surjective. Since $f^q = f^{(q)}(x^q, y^q)$, $0 = 0^q = \alpha(f^q) = f^{(q)}(u^q, v^q) = \beta(f)^q$, so $f \in \ker \beta$. If f factored nontrivially in R then f^q would as well in R_n; then the fact that f is a prime element of R_n would yield a contradiction.

(3) Since R/fR is a purely inseparable integral extension of A/fA, the former is a nonsingular curve with one place at infinity. So by Lemma 2.4 the partials of f are coprime. But $f_y = 0$. So, upon replacing x by a constant multiple of x, we have that $f = h^p - x$ for some $h \in R$.

Let t be transcendental over $\mathbf{k}$, and suppose that f generates the kernel of the surjection $\alpha: A \to \mathbf{k}[t^p]$. There are $u, v \in \mathbf{k}[t]$ such that $\alpha(x) = u^p$ and $\alpha(y^p) = v^p$ (recall that $\mathbf{k}$ is perfect). Then α is the restriction to A of the map $\beta: R \to \mathbf{k}[t]$ which takes x to u^p and y to v. Since $\beta(f) = 0$, $u^p \in \mathbf{k}[u^p, v]^p$, so $u \in \mathbf{k}[u^p, v]$, so this last ring is $\mathbf{k}[u, v]$, so β is surjective. Let $\ker \beta = gR$; then g divides f. But an easy exercise with Lemma 2.5 (1) shows that f and g have the same x-degree and the same y-degree. Another way to see that $fR = gR$: By Lemma 2.5 (2), $\beta(g_y) = 0$, so g divides g_y, which must therefore be 0. So f divides g in A, hence in R. □

3.2. Example.

Generalized Segre lines (See [Seg56, Nag71, AM75]). Let d, n be integers greater than 1, with d coprime to p, $h \in \mathbf{k}[x]$ a nonzero polynomial of degree d, and

$$(1) \qquad f := y^{p^n} - x - h^{(p)}(x^p) = (y^{p^{n-1}} - h)^p - x.$$

(If $h = 0$ or either of d, n is 1, then f is a variable in R, so we have avoided those cases. By Van der Kulk's theorem on $\mathrm{Aut}_k R$, to avoid variables here we need only require that h be nonzero and that neither of p^n, dp divide the other—that the (y, x)-bidegree of f be "nonprincipal".)

Again let R_i, for $i \in \mathbb{N}$, denote $R^{p^i}[f]$. Clearly $x \in R_1$, so $x^p \in R_1^p \subset R_2$, so $x = y^{p^n} - h^{(p)}(x^p) - f \in R_2$. Induction on i gives that, for all nonnegative i less than or equal to n, $R_i = \mathbf{k}[x, y^{p^i}]$ is a plane, and $R_{n-1} = \mathbf{k}[f, y^{p^{n-1}} - h]$—$f$ is a *variable* in that plane. By Lemma 3.1 (2), all translates of f are lines in R (this also follows directly by writing $y^{p^n} - \lambda$ as $(y - \mu)^{p^n}$). Noting that the (y^{p^i}, x)-bidegree of f is nonprincipal for $i < n - 1$ by the coprimeness of d to p, we have that f is not a variable in the plane R_i for $i < n - 1$. One sees easily that this implies, in the language of *forms of the line*, that the generic member $\mathrm{Spec}(\mathbf{k}(f) \otimes_{\mathbf{k}[f]} R)$ of the pencil $\Lambda(f)$ is a form of the affine line, of height $n-1$. (Cf. [Miy11, Theorem 4.4 (3)]. Note also that neither condition (5) of Lemma 4.1 of [Miy11] nor condition (4) of Theorem 4.2 of [Miy11] is satisfied by these generalized Segre lines if $n > 2$.)

3.3. Notation. Let G denote the group of $\mathbf{k}$-algebra automorphisms of R. Given a nonconstant $f \in R$, let G_f denote the subgroup of G which carries the pencil $\Lambda(f)$ of ideals of R into itself.

If f is a *variable* and $R = \mathbf{k}[f,g]$, then G_f contains all those elements of G which carry f to a linear polynomial in f and carry g to $wg - z$, where $w \in \mathbf{k}^*$ and $z \in \mathbf{k}[f]$. The infinite group G_f clearly acts transitively on $\Lambda(f)$.

If f is a generalized Segre line as in equation (1), and $\lambda \in \mathbf{k}$, then the automorphism which fixes x and sends y to $y - \mu$, where $\mu^{p^n} = \lambda$, carries f to $f - \lambda$ (hence, $\Lambda(f)$ into itself). So G_f again acts transitively on $\Lambda(f)$. Clearly Question 1.1 has an affirmative answer for lines f for which the action of G_f on $\Lambda(f)$ is transitive. The next proposition gives an example of a line f for which the action of G_f is highly nontransitive.

3.4. Proposition. *Let a, b be integers greater than 1 and not divisible by p. Let*

$$f := (y^p - x^b)^{ap} - x.$$

Then:

(i) *f is a variable in $R_1 := R^p[f] = \mathbf{k}[x, y^p]$, hence a line in R.*

(ii) *The automorphisms φ of R which carry fR to a translate of fR fix fR. Such a map φ takes x to cx and y to $d^{1/p}y$, where $c^{(abp-1)} = 1$ and $d = c^b$.*

(iii) *G_f is isomorphic to the multiplicative group of νth roots of unity in $\mathbf{k}$, where $\nu := abp - 1$. The stabilizer of fR is G_f, and that of $(f - \lambda)R$ is trivial, if $\lambda \neq 0$.*

Proof. The proof of (i) is the same as that for the corresponding statements about generalized Segre lines, above, with the additional observation that $R_1 = \mathbf{k}[f, u]$, where $u := y^p - x^b$.

Let $\varphi \in G$, $\lambda \in \mathbf{k}$, $c \in \mathbf{k}^*$, and $\varphi(f) = c(f - \lambda)$. Since $\varphi(R^p) = R^p \subset R_1$ and $\varphi(f) \in R_1$, φ maps R_1 into itself. Similarly, $\varphi^{-1}(R_1) \subseteq R_1$. So φ restricts to an automorphism of R_1. Moreover, $\varphi(u)$ generates R_1 over $\mathbf{k}[f]$. So $\varphi(u) = du + h$ for some $d \in \mathbf{k}^*, h \in \mathbf{k}[f]$. So

$$(2) \quad \varphi(x) = (du + h)^{ap} - c(f - \lambda) = (d^{ap} - c)u^{ap} + \sum_{i=0}^{a-1} \binom{a}{i} (du)^{pi} h^{p(a-i)} + cx + c\lambda.$$

Suppose h is nonzero and $\deg_f h =: m \geq 1$. Then $\deg_x \varphi(x) = \deg_x h^{ap} = ba^2mp^2$ and $\deg_y \varphi(x) = pa^2mp^2$. But $p \nmid b$ and $b \nmid p$, so this contradicts the Automorphism Theorem (for R). So $h \in \mathbf{k}$. Considering equation (2) again, noting that $a > 1$, $p \nmid a$ and $d \neq 0$, and repeating the above argument twice (for the u^{pa} and $u^{p(a-1)}$ terms), we conclude that $(*)$ $d^{ap} = c$ and $h = 0$. So $\varphi(x) = c(x + \lambda)$, and

$$d(y^p - x^b) = du = \varphi(u) = \varphi(y)^p - (c(x + \lambda))^b.$$

Consideration of the terms of degree 1 in x forces λ to be 0, hence $(**)$ $d = c^b$ and, finally, $\varphi(y) = d^{1/p}y$. $(*)$, $(**)$ yield that c is a νth root of unity as claimed.

An easy calculation using part (ii) gives, more generally, that if $\varphi \in G_f$ maps some $f - \lambda$ to a nonzero constant multiple of some $f - \mu$, then φ is as in (ii). The same calculation shows that, for all λ, the orbit under G_f of the translate $(f - \lambda)R$ is the set $\{(f - \lambda/c)R \mid c^\nu = 1\}$, hence has order ν if $\lambda \neq 0$. The remaining statements in (iii) are clear. □

Proposition 3.4 graphically makes the point that one cannot successfully attack Question 1.1 using automorphisms in this way.

4. Frobenius sandwiches, partial results

We continue to assume in this section (unless we state to the contrary) that the field $\mathbf{k}$ is algebraically closed and $R \cong \mathbf{k}^{[2]}$, and we denote $\operatorname{Frac} R$ by K.

The rings $R_n = R^{p^n}[f]$, which have cropped up several times earlier in this article, are instances of surfaces which have been called *Frobenius sandwiches* ("of exponent n") in the literature — affine algebraic surfaces $\operatorname{Spec} A$ "factoring the n-fold Frobenius of R", by which is meant, here, that $R \supset A \supset R^{p^n}$, where each extension is of degree p^n. The references [Gan82, MR83, Lan83, Dai90, Dai94, Dai96a, Dai96b] should be mentioned here ([Lan83] contains dozens of errors). The main result (Theorem, p. 52) of [Dai96b] is especially closely related to Theorem 4.12 below (Daigle refers to Theorem 4.12 here — or rather the corresponding theorem in this author's 1978 McGill dissertation — as "the nonsingular case" of his result).

4.1. Notation. Let $f, h \in R$, and choose variables x, y generating R. The *Jacobian* of f and h *with respect to* the ordered pair (x, y) is $J_{(x,y)}(f, h) := f_x h_y - f_y h_x$. We define a derivation $D_f : R \to R$, up to a nonzero constant factor in $\mathbf{k}$, by $D_f(h) := J_{(x,y)}(f, h)$, for all $h \in R$. D_f is called *the Jacobian derivation* associated to f.

Suppose that (x', y') is another generating pair for R, and D'_f is defined by $D'_f(h) := J_{(x',y')}(f, h)$, for all $h \in R$. Then $D'_f = cD_f$, where $c := J_{(x,y)}(x', y') \in \mathbf{k}^*$. So D_f does depend on the choice of coordinate system for R, but not heavily. We denote also by D_f the extension of this derivation to $\operatorname{Frac} R$.

Let $A_f := \{h \in R \mid D_f(h) = 0\}$ be the ring of D_f-constants, $K_f := \{h \in K \mid D_f(h) = 0\}$ the field of D_f-constants. A_f is normal, for any $f \in R$. One way to see this is to note that $A_f = R \cap \operatorname{Frac} A_f$. (If B is a normal domain, D is a derivation of it, and B' is the ring of D-constants, then B' is normal, since it is $B \cap \operatorname{Frac} B'$.)

Now suppose that $f \notin R^p$. Then $K^p \subset K_f \subset K$, so $K_f = K^p(f)$ and $[K : K_f] = p$.

4.2. Lemma. *Let $f \in R$ be nonconstant. Then $A_f = R^p[f]$ iff the pencil $\Lambda(f)$ has no moving singularity (which is the case if, e.g., $f \in R$ is nonsingular with one place at infinity).*

PROOF. A_f is a birational, integral extension of $R^p[f]$; the equivalence follows, by the proof of Lemma 2.4. □

It is well known that if D is a derivation of a field L of characteristic p then D^p is again a derivation of the field.

4.3. Definition. Let L be a field of characteristic p. A nonzero derivation D of L into itself is *p-closed* if there is an $a \in L$ such that $D^p = aD$.

Evidently Nathan Jacobson first noticed that, for $D \in \operatorname{Der}(L)$, D is p-closed if and only if $[L : L^D] = p$. For $f \in R \setminus R^p$, then, since $K^p(f) \subseteq K_f \neq K$, by Jacobson's observation we have that D_f is p-closed. Moreover, if $D_f^p = aD_f$, then $a \in A_f$, hence $a \in R$ ([Sam64, pp. 63–64] — the result holds since R is Noetherian normal, hence a Krull domain). By Hochschild's 1955 formula, if $D \in \operatorname{Der}(L)$ is p-closed, then so is bD, for all nonzero $b \in L$. (This also follows immediately from the above Jacobson observation.)

One even has an explicit formula for the "a" above, if $\Lambda(f)$ has no moving singularity. Given such an f, choose a generating pair (x, y) for R and define D_f

accordingly. Put $\nabla := \partial^{2(p-1)}/(\partial x^{p-1}\partial y^{p-1})$ and note that the differential operator ∇ maps R onto R^p (Wilson's theorem makes the computation of explicit preimages simple). The result below appears on p. 395 of [Lan83] as a corollary of a more general result, and should be known as Lang's Theorem:

4.4. Theorem (J. Lang). *Let $f \in R = \mathbf{k}[x,y]$ and suppose that $\Lambda(f)$ has no moving singularity. Then $D_f^p = aD_f$, where*

$$a = \sum_{i=0}^{p-2} f^i \nabla(f^{p-1-i}) \in R^p[f]. \tag{3}$$

4.5. Remark. The author conjectured formula (3) above in 1977, without the restriction on $\Lambda(f)$, and shared this conjecture with Jeffrey Lang at Purdue in 1979. Lang subsequently proved the above version of it, and generously labelled the result "Ganong's Theorem" in several publications.

4.6. Conjecture (Ganong, 1977, 2010). The above theorem holds without the restriction on $\Lambda(f)$.

Now let $f \in R$ be a line, and denote D_f by D. Pick a surjection $\alpha: R \to \mathbf{k}[t]$ with kernel fR, with α normalized, in the sense of Remark 2.6. One then checks easily that

$$\begin{array}{ccc} R & \xrightarrow{\alpha} & \mathbf{k}[t] \\ {\scriptstyle D}\big\downarrow & & \big\downarrow{\scriptstyle \mathrm{d}/\mathrm{d}t} \\ R & \xrightarrow[\alpha]{} & \mathbf{k}[t] \end{array} \tag{4}$$

commutes. We then have this (to the author, somewhat arcane) consequence:

4.7. Corollary. *Let $f \in R$ be a line. Then $\nabla(f^{p-1}) = 0$.*

Proof. Induction and the above commutative diagram give that $\alpha \circ D^p = (\mathrm{d}/\mathrm{d}t)^p \circ \alpha = 0$, so f divides $D^p(h) = aD(h)$ for all $h \in R$. Letting h be x and y gives that $f \mid af_x$ and $f \mid af_y$. By Lemma 2.4, f fails to divide one of its partials; since f is irreducible, $f \mid a$. By Lemma 2.4 and Lang's Theorem, $f \mid \nabla(f^{p-1})$.

Now, $\nabla(f^{p-1}) = w^p$ for some $w \in R$. If $w = 0$ we are done. Suppose w is nonzero, and consider total x,y-degrees: $p \deg w \le (p-1)\deg f - 2(p-1) = (p-1)(\deg f - 2)$, hence $\deg w < \deg f$. But $f \mid w$, so we have a contradiction. □

4.8. Corollary. *Let $f \in R$ be a line, and $p = 2$. Then $D_f^2 = 0$. I.e., for any choice of x,y generating R, f has no terms $x^i y^j$ with i,j both odd.*

At various points in this article it is of interest to determine whether, for a line $f \in R$, the ring R_1 is a plane. From this viewpoint, the next theorem should certainly merit attention.

4.9. Theorem. *If $f \in R$ is a line, then $R_1 := R^p[f]$ is factorial.*

Proof. Let $\mathbf{j}$ be the canonical map from the divisor class group of A_f (which is $R^p[f]$, by Lemma 4.2) to that of R, and

$$\mathcal{L} := \left\{ \text{logarithmic derivatives } \frac{D(q)}{q} \;\middle|\; \frac{D(q)}{q} \in R, 0 \neq q \in \operatorname{Frac} R \right\}.$$

In the notation of [Sam64, p. 62], $\mathcal{L}' := \{D(u)/u \mid u \text{ is a unit in } R\} = \{0\}$, and, by [Sam64, Theorem 2.1 (a)], we have a canonical injection of $\ker \mathbf{j}$ into $\mathcal{L}/\mathcal{L}' = \mathcal{L}$. Let $0 \neq q = b/c \in \operatorname{Frac} R$, with $b, c \in R$ relatively prime, and suppose that $(cD(b) - bD(c))/bc = D(q)/q \in R$. Then $b \mid D(b)$ and $c \mid D(c)$ in R.

Write $b = f^m b'$, with $m \geq 0$ and $f \nmid b'$. Then $D(b) = f^m D(b')$, so $b' \mid D(b')$ in R. Referring to diagram (4), we then have that $\alpha(b') \mid \alpha(D(b')) = d/dt\,(\alpha(b'))$. So $\alpha(D(b')) = 0$, and $f \mid D(b')$. Since f, b' are coprime, $fb' \mid D(b')$. But $D(b')$, if nonzero, has degree (relative to any generating pair x, y for R) at most $\deg b' + \deg f - 2$. So $0 = D(b') = D(b)$. Similarly, $D(c) = 0$. Hence $\mathcal{L} = \{0\}$. But the divisor class group of R is trivial. □

We recall without detailed proof three facts which for decades have been well known to affine algebraic geometers; (1) and (3) will be needed in the proofs of Proposition 4.11 and Theorem 4.12.

4.10. Lemma. *As usual, let $R \cong \mathbf{k}^{[2]}$.*

(1) *Let $f \in R$ have one place at infinity. Then for each $n \gg 0$, the $\mathbf{k}(f)$-curve $\mathbf{k}(f) \otimes_{\mathbf{k}[f^{p^n}]} R^{p^n} = \mathbf{k}(f)[R^{p^n}]$ has one rational place at infinity.*
(2) *Let $f \in R$ be nonconstant and suppose that the pencil $\Lambda(f)$ has no moving singularity. Then the curves in (1) are geometrically regular for all $n \in \mathbb{N}$.*
(3) *Let $\mathbf{k}$ be any field, and let $f \in S \cong \mathbf{k}^{[2]}$ be nonconstant. Then f is a variable in S iff the $\mathbf{k}(f)$-curve $\mathbf{k}(f)[S] \cong \mathbf{k}(f) \otimes_{\mathbf{k}[f]} S$ is a $\mathbf{k}(f)$-line.*

Sketches of proofs. (1) follows from Theorem 2.2 (1) and the existence of an obvious isomorphism $\mathbf{k}(f^{p^{-n}})[R] \xrightarrow{\cong} \mathbf{k}(f)[R^{p^n}]$ extending the isomorphism of the ground fields given by the nth power of Frobenius, together with classical facts about valuation rings of function fields of curves. (2) follows from the fact that the regular singular points of $\mathbf{k}(f)[R] = (\mathbf{k}[f] \smallsetminus \{0\})^{-1} R$ are in bijection with the ideals $hR \subset R$, with $h \in R$ irreducible, which contain the "Jacobian ideal" of f, which we have assumed is R.

Finally, suppose that $\mathbf{k}(f)[S] \cong \mathbf{k}(f)^{[1]}$. Then f is a "field generator" in Russell's terminology ([Rus75]). Theorem 4.5 of [Rus75] asserts that there are variables x, y for S such that either f is *linear* in them (when we are done) or the x, y-degree form of f has two linear factors. If we had the latter case, then, with t transcendental over $\operatorname{Frac} S$, the same would hold for $f - t \in \mathbf{k}(t)[x, y]$, so this plane curve would have at least two places at infinity. But, as was mentioned just before Theorem 2.2, $\mathbf{k}(t)[x, y]/(f - t) \cong \mathbf{k}(f)[S]$, so this line would have at least two places at infinity. □

Hypothesis (Russell, < 1978). *If $f \in R$ is a line, then $R^p[f]$ is a plane.*

The author first heard this statement from Peter Russell, in 1977.[2] The hypothesis is not given here as a "Conjecture" because, to the author's knowledge, Russell has never called it that. (See below for connections with conjectures of Moh.)

4.11. Proposition. *The Russell Hypothesis implies an affirmative answer to Question* 1.1.

[2]Paternity of a mathematical idea does not preclude the existence of other, perhaps even undeclared, unacknowledged and/or prior, parents, of course.

PROOF. Assume the Russell Hypothesis, and let $f \in R$ be a line. Keep the notation of Lemma 3.1 (2). By that lemma, for all $n \in \mathbb{N}$, f is a line in R_n. It is clear that, for all $n \in \mathbb{N}$, $R_{n+1} = R_n^p[f]$. So, by induction and Russell, R_n is a plane for all $n \in \mathbb{N}$. By Lemma 4.10 (1), there is an $m \in \mathbb{N}$ such that $\mathbf{k}(f)[R_m] = \mathbf{k}(f)[R^{p^m}]$ has one rational place at infinity. By the well-known fact that then f and the generic member of the pencil $\Lambda(f)$ on the plane R_m have the same "divisor of singularities" at infinity, this curve has genus 0, hence is rational. So, being regular, it is a line. By Lemma 4.10 (3), $f \in R_m$ is a variable. So $f - \lambda \in R_m$ is a line for all $\lambda \in \mathbf{k}$. So, by Lemma 3.1 (2), all $f - \lambda$ are lines in R. □

4.12. Theorem. *Let $f \in R$ be a line. Then (1)–(4) below are equivalent.*

(1) *For all $n \gg 0$, $R_n := R^{p^n}[f]$ is a plane.*
(2) *$f - \lambda \in R$ is a line for all $\lambda \in \mathbf{k}$.*
(3) *$f - \lambda \in R$ is a line for infinitely many $\lambda \in \mathbf{k}$.*
(4) *For all $n \gg 0$, R_n is a plane and f is a variable in it.*

This theorem, from the author's 1978 doctoral thesis, may be compared with Theorem 4.2 of [Miy11].

PROOF. (2) implies (3) since $\mathbf{k}$ is algebraically closed.

Assume (1) above. (1) and Lemma 4.10 (1) and Lemma 3.1 (2) together assert the existence of an m such that R_m is a plane and f is a line in it, and $\mathbf{k}(f)[R_m]$ has a rational place at infinity. It is then well known (as in the above proof of Proposition 4.11) that f is a variable in R_m. Continuing as in the proof of Proposition 4.11, we have (2) above.

Assume (3). By Lemma 4.10 (1) and Theorem 2.2 (1) and (3), $\mathbf{k}(f)[R_n] \cong \mathbf{k}(f)^{[1]}$ for all n large enough. Now, for any such n, Theorem 2.3.1 of [RS79], tailor-made for just such a situation as ours, takes over: Replace the S, k, K, A there by our $\mathbf{k}[f] \setminus \{0\}$, $\mathbf{k}[f]$, $\mathbf{k}[f]$, R_n, respectively. We have just verified condition (i) of the theorem. Another condition is that, for each $\lambda \in \mathbf{k}$, $h := f - \lambda$ generates a prime ideal of R_n. But this follows from the fact that $hR \cap R_n = hR_n$, which follows from the facts that R is integral over R_n, which ring is normal, and that h is irreducible in R, as has been remarked before. The remaining conditions are all verified easily, and we conclude (4). But we note that what actually gives this last implication is what Russell and Sathaye call a "method of *modification* of a generic choice of generator" ([RS79, 2.2], used earlier in well-known papers of Eakin and Heinzer (1973) and Kambayashi and Miyanishi (1977), and it may be illuminating to trace the argument:

Put $q := p^n$, and let $\mathbf{k}(f)[R_n] = \mathbf{k}(f)[x^q, y^q] = \mathbf{k}(f)[t]$ with $t \in R_n$, as above. The goal is to replace t by a "modification of t", i.e., by an element t' of R_n of the form $(t-a)/b$, with $a \in \mathbf{k}[f]$ and $b \in \mathbf{k}[f] \setminus 0$, such that $x^q, y^q \in \mathbf{k}[f][t']$. Note that then $\mathbf{k}[f][t] \subseteq \mathbf{k}[f][t']$, and that a modification of a modification is a modification. These remarks and what follows make clear that we need only demonstrate the procedure with x^q.

We have that

$$p_1 x^q = g(t) = \sum a_j t^j \quad \text{for some } p_1 \in \mathbf{k}[f] \setminus 0 \text{ and some } a_j \in \mathbf{k}[f], \tag{5}$$

and we may, and do, assume that p_1 shares no irreducible factor $f - \lambda$ with all a_j. If $p_1 \in \mathbf{k}$, we are done. Assume that p_1 is nonconstant. We proceed to modify t to get a relation similar to (5) with the number of irreducible factors of the new p_1

cut down. Let $w = f - \lambda$ be an irreducible factor of p_1. Note that w fails to divide at least one a_j with $j > 0$. Let "$\overline{}$" denote image in the domain R_n/wR_n. We then have a nontrivial relation

$$\sum \overline{a_j}\bar{t}^j = 0$$

for $\bar{t} \in R_n/wR_n$ over $\overline{\mathbf{k}[f]} \cong \mathbf{k}$. Since $\mathbf{k}$ is algebraically closed, there are an $a \in \mathbf{k}[f]$ and a $t' \in R_n$ such that $t - a = wt'$. Now,

$$p_1 x^q = g(wt' + a) = \sum a'_j t'^j \quad \text{for some } a'_j \in \mathbf{k}[f]. \tag{6}$$

Clearly $a'_j \in w\mathbf{k}[f]$ for all $j > 0$. Since $w \mid p_1$ in $\mathbf{k}[f]$, we have that $w \mid a'_0$ as well. So we may cancel w from both sides of (6). □

4.13. Conjecture. The answer to Question 1.1 is "Yes".

4.14. Remarks. (1) The author regards much of the content of this article as somewhat cosmetic. Progress toward proving Conjecture 4.13 will probably come only via the sort of serious work done in, e.g., [Moh88] and other articles of Moh, using his approximate roots, meromorphic series, and "expansion techniques", or in [Rus80] or [Gan82], using power series and Hamburger – Noether expansions. We also predict that some new ideas will be required. It is not unlikely that work of Daniel Daigle (some of which was mentioned above) on what might be called plane Frobenius sandwiches of higher exponent is closely related to conjectures about lines.

(2) An attractive project is to extend, if possible, Theorem 4.9 to the case of R_n, with $n > 1$, using suitable derivations of the algebras R_n and the Samuel theory.

(3) We observe that Conjecture 1 (part (1)) of [Moh88] is, in view of our Lemmata 2.5 (2) and 3.1 (1), equivalent to the Russell Hypothesis, which, in turn, is part (1) of Conjecture 3, ibid. Parts (2) of Moh's Conjectures 1 and 3 concern bounding bidegrees of requisite automorphisms of R relative to the (x, y)-bidegree of the line f.

An old conjecture and an old new result. The author, as early as 1979, while a visitor at Bhaskaracharya Pratishthana, Pune, independently developed the following conjecture, conveyed to Russell at McGill, and to Moh at Purdue, in late 1979. The author's (perhaps faulty) recollection is that Moh at the time claimed to have held it (probably for some years) already:

4.15. Conjecture (Moh, < 1979, 1988,. . . ; Ganong, 1979, 2010). Let $f \in R = \mathbf{k}[x, y]$ be a line, and have (x, y)-bidegree (mp, np), with $1 < m < n$ and $\gcd(m, n) = 1$. Suppose that $f \notin \mathbf{k}[x^p, y]$. Then there is a de Jonquières automorphism $\sigma : R \to R$ fixing y, such that $\sigma(f) \in \mathbf{k}[x, y^p]$.

The bidegree in the conjecture is the simplest possible for an exotic line — the simplest "two characteristic pairs" case. A careful look at [Moh88] would be required to ascertain how closely Conjecture 4.15 corresponds to the work there, but it seems essentially to be present in the proof of [Moh88, Theorem 1].

Let $f \in R = \mathbf{k}[x, y]$ have one place at infinity, assume that the place at infinity of the $\mathbf{k}(f)$-curve $\mathbf{k}(f)[x, y]$ is of degree p over $\mathbf{k}(f)$, and keep the assumptions of Conjecture 4.15 about the (x, y)-bidegree of f. Define $g \in R$ to be the polynomial whose pth power is precisely the sum of the monomials appearing in f which are pth powers. Let

$$f = g^p + \sum \alpha_i g^{p-i}$$

be the "g-adic expansion of f", i.e., $\deg_y \alpha_i < n$ for all i such that $\alpha_i \neq 0$.

4.16. Proposition (Ganong, unpublished, 1977). *With the above assumptions, $\alpha_1 = 0$. That is, g is an approximate pth root of f.*

(Moh calls g a pseudo-approximate root of f, but we grant g full status.) In [Moh88], this conclusion is that of Theorem 2, where the assumption is that f is a line. Note that since then f is not a variable, Moh's assumption implies the assumptions in Proposition 4.16. We do not need the assumption that our (mp, np)-curve is a line (necessarily exotic); nonrationality of the place at infinity of the generic member of the pencil $\Lambda(f)$ is enough. Indeed, the truth of Proposition 4.16 and perhaps its proof show clearly that the existence of a pth approximate root of such a "one-place" f is much more a reflection of the behaviour of the pencil $\Lambda(f)$ than of maximality of the order of the divisor of singularities at infinity of any one member of that pencil. For instance, all one-place curves offered as counterexamples in [Gan79] have approximate pth roots, but their pencils of translates have no line as a special member.

We outline a proof of Proposition 4.16 which uses no machinery of either the Moh sort or the Hamburger-Noether sort; rather, Newton diagrams, well-known facts about irreducible power series, and high school algebra suffice.

Sketch of proof of Proposition 4.16. Embed $\operatorname{Spec} R$ into the projective plane as in Theorem 2.2 (3), given the generators x, y for R. Let Λ be the pencil whose members are the curves C_λ in Theorem 2.2 (3) together with the np-fold line at infinity. Pass to the point P at infinity (with parameters $1/x, y/x$) of R in the projective plane, common to all members of Λ. The Euclidean algorithm on the base-point orders $(n-m)p, np$ relative to those parameters determines uniquely a sequence of blowings-up leading to a base point Q "infinitely near" to P. One has natural parameters u, v at Q, relative to which the proper transforms of the C_λ have defining polynomials

$$f_\lambda := (u - cv)^p + \sum c_{ij} u^i v^j - \lambda u^{qmp} v^{rnp},$$

where

> $c \in \mathbf{k}^*$, $c_{ij} \in \mathbf{k}$ are independent of λ, the Newton polygons of the f_λ are bounded by the triangle with vertices $(p, 0), (0, p)$ and (qmp, rnp), and $q, r \in \mathbb{Z}^+$ satisfy $qmp + rnp = mnp + p$.

The integers q, r are determined by the Euclidean algorithm (see [Gan79], lines 5–6 of the proof of Proposition 2.13). The assumptions that f has one place at infinity and the last base point of Λ is p-fold now imply the existence of a polynomial $\varphi(v) \in \mathbf{k}[v]$ with lowest degree nonzero term cv such that

$$(*) \qquad \text{the only term } u^i v^j \text{ of } f_0\big(u + \varphi(v)\big) \text{ with nonzero coefficient and } (mn+1)i + j \leq (mn+1)p \text{ is } u^p.$$

Let

$$h := f_0(uv, v)/v^p = (u - c)^p + \sum_{j=1}^{mnp} h_j(u) v^j.$$

It is easily deducible from 2.3–2.9 of [Gan79] that the terms in h of v-degree t correspond to the monomials $x^i y^j$ in $f \in R$ with "weight" $in + jm = mnp - t$, and that pth power monomials in f and h correspond. So our claim amounts to saying that $h_j(u) \in \mathbf{k}[u^p]$ for all j such that $1 < j \leq mn$, and $h_j(u) = 0$ for such j not divisible by p. Given the structure of h and φ and the irreducibility of the power series f_0, one sees fairly easily that existence of a φ satisfying $(*)$ forces these conditions:

$(**)$ For all positive $j \le mnp$ not divisible by p, $u^{p+1-\lceil j/mn \rceil} \mid h_j(u+c)$, and, for all such j divisible by p, $u^{p+1-\lceil j/mn \rceil} \mid h_j(u+c) - h_j(c)$.

Note that for $1 < j \le mn$, the power of u in $(**)$ is u^p. In view of the shape of the Newton polygons of the f_λ, then, Proposition 4.16 will follow from this

High school lemma. *Let* $\mathbf{k}$ *be a field of characteristic* $p > 0$, $a \in \mathbb{Z}^+$, $c \in \mathbf{k}^*$, $h(u) = u^a \sum_{i=0}^{s} c_i u^i \in \mathbf{k}[u] \cong \mathbf{k}^{[1]}$, $0 \le s < p$, *and* $u^p \mid h(u+c) - h(c)$. *Then* $h'(u) = 0$, *and if* $h(c) = 0$ *then* $h = 0$.

Proof of lemma. If the sum is 0 we are done, so we may, and do, assume that it is nonzero, and, maximizing a, that $c_0 c_s \ne 0$. Now, $h(u+c) = (u+c)^a \gamma(u)$, for some γ of degree s. If c is a root of h, it follows that $u^p \mid \gamma(u)$, a contradiction, proving the last part of the lemma.

Since $(\mathrm{d}/\mathrm{d}u)u^p = 0$, u^p divides $h'(u+c)$. But $h'(u+c) = (u+c)^{a-1}\delta(u)$, where $\delta(u) = 0$ or $\deg(\delta) \le s$. Either way, u^p divides $\delta(u)$, so $\delta(u) = 0$, so $h'(u+c) = 0$, so $h'(u) = 0$. □

□

References

[Abh77] S. S. Abhyankar, *Lectures on expansion techniques in algebraic geometry*, with by B. Singh, Tata Inst. Fund. Res. Lectures on Math. and Phys., vol. 57, Tata Inst. Fund. Res., Bombay, 1977.

[AM75] S. S. Abhyankar and T.-T. Moh, *Embeddings of the line in the plane*, J. Reine Angew. Math. **276** (1975), 148–166.

[Dai90] D. Daigle, *A property of polynomial curves over a field of positive characteristic*, Proc. Amer. Math. Soc. **109** (1990), no. 4, 887–894.

[Dai94] ———, *Purely inseparable extensions of* $\mathbf{k}[X, Y]$, Proc. Amer. Math. Soc. **121** (1994), no. 1, 1–12.

[Dai96a] ———, *On purely inseparable extensions* $K[X, Y]/K[X', Y']$ *and their generators*, Proc. Amer. Math. Soc. **124** (1996), no. 5, 1337–1345.

[Dai96b] ———, *On pencils of polynomial curves*, J. Pure Appl. Algebra **111** (1996), no. 1-3, 51–57.

[Gan79] R. Ganong, *On plane curves with one place at infinity*, J. Reine Angew. Math. **307/308** (1979), 173–193.

[Gan82] ———, *Plane Frobenius sandwiches*, Proc. Amer. Math. Soc. **84** (1982), no. 4, 474–478.

[Lan83] J. Lang, *The divisor class group of the surface* $Z^{p^n} = G(X, Y)$ *over fields of characteristic* $p > 0$, J. Algebra **84** (1983), no. 2, 391–410.

[Miy78] M. Miyanishi, *Analytic irreducibility of certain curves on a nonsingular affine rational surface*, Proceedings of the International Symposium on Algebraic Geometry (Kyoto, 1977) (M. Nagata, ed.), Kinokuniya, Tokyo, 1978, pp. 575–587.

[Miy11] ———, *Frobenius sandwiches of affine algebraic surfaces*, in this volume.

[MR83] M. Miyanishi and P. Russell, *Purely inseparable coverings of exponent one of the affine plane*, J. Pure Appl. Algebra **28** (1983), no. 3, 279–317.

[Moh88] T.-T. Moh, *On the classification problem of embedded lines in characteristic p*, Algebraic Geometry and Commutative Algebra, Vol. I (H. Hijikata, H. Hironaka, M. Maruyama, H. Matsumura, M. Miyanishi, T. Oda, and K. Ueno, eds.), Kinokuniya, Tokyo, 1988, pp. 267–279.

[Nag71] M. Nagata, *A theorem of Gutwirth*, J. Math. Kyoto Univ. **11** (1971), 149–154.

[Rus75] P. Russell, *Field generators in two variables*, J. Math. Kyoto Univ. **15** (1975), no. 3, 555–571.

[Rus80] ———, *Hamburger – Noether expansions and approximate roots of polynomials*, Manuscripta Math. **31** (1980), no. 1-3, 25–95.

[RS79] P. Russell and A. Sathaye, *On finding and cancelling variables in $k[X, Y, Z]$*, J. Algebra **57** (1979), no. 1, 151–166.

[Sam64] P. Samuel, *Lectures on unique factorization domains*, Tata Inst. Fundam. Res. Lect. Math., vol. 30, Tata Inst. Fund. Res., Mumbai, 1964.

[Sat76] A. Sathaye, *On linear planes*, Proc. Amer. Math. Soc. **56** (1976), 1–7.

[Seg56] B. Segre, *Corrispondenze di Möbius e trasformazioni cremoniane intere*, Atti Accad. Sci. Torino. Cl. Sci. Fis. Mat. Nat. **91** (1956/1957), 3–19.

Department of Mathematics and Statistics, York University, Toronto, ON M3J 1P3, Canada

E-mail address: ganong@yorku.ca

Centre de Recherches Mathématiques
CRM Proceedings and Lecture Notes
Volume **54**, 2011

A Geometric Proof of Boutot's Result on Singularities of Quotients

R. V. Gurjar

To Peter Russell on the occasion of his 70th birthday

ABSTRACT. We will give a new geometric proof of J.-F. Boutot's result on rationality of singularities of a quotient of $\mathbf{C}^n$ modulo a reductive algebraic group G. The proof uses an induction on $\dim G$ and flattening of the quotient morphism.

1. Introduction

We will give a new proof of the following well-known result of J.-F. Boutot ([2]).

Theorem 1. *Let G be a reductive algebraic group over an algebraically closed field of characteristic 0 acting algebraically on a smooth affine variety V. Then $V/\!\!/G$ has rational singularities.*

In fact, our proof also gives the following more general result.

Theorem 2. *Let Y be an affine irreducible variety with rational singularities. Let an algebraic torus T act on Y and $W := Y/\!\!/T$. Then W has rational singularities.*

Boutot's result is more general but in this paper we are interested in the "classical" case. In what follows we will assume that the ground field is the field of complex numbers $\mathbf{C}$. Theorem 1 was also partly proved by the author in ([11]). There the author used the Hochster – Roberts theorem that $V/\!\!/G$ is Cohen – Macaulay ([14]). In particular, the arguments in ([11]) prove Boutot's result when G is semisimple, assuming Hochster – Roberts result. Boutot's proof is cohomological. Here we give a different proof of Boutot's result which appears to be more geometric. In the case when the connected component of identity of G is semisimple we give a very short proof of Boutot's result by making use of a result of R. Elkik – H. Flenner (Proposition 2). A proof for this special case was also given by G. Kempf ([17]) using the Hochster – Roberts result. Our proof of Boutot's result is inspired by a

2010 *Mathematics Subject Classification.* 13A50, 14L24, 14L30.
This is the final form of the paper.

result of M. Miyanishi which says that an affine normal surface over $\mathbf{C}$ which has an $\mathbf{A}^1$-fibration has at most cyclic quotient singularities ([19, Chapter I, Section 6]). In fact, using the arguments given here we can give an easy conceptual proof of Miyanishi's result.

In our proof an analysis of singularity of a suitable normal affine surface with a $\mathbf{C}^*$ action plays an important role (Lemma 1).

We use flattening of the quotient map using results of Hironaka, induction on $\dim G$ and study of $\mathbf{C}^*$-actions such that the quotient is smooth and the quotient map is flat. This is the novel part of our proof. Flattening is used only to make the fibers of the quotient map equidimensional. Our proof can be considered as an "upside" down proof. First we reduce to the case when the reductive group is a 1-dimensional torus $\mathbf{C}^*$ acting on an affine variety with rational singularities with the same quotient. By flattening and resolution of singularities we make the quotient smooth and prove that the (normalized) fiber product has rational singularities. The proof of this part is purely algebraic geometric. Since the proof for semisimple case is essentially due to Elkik – Flenner, our main contribution in this paper is to use induction on $\dim G$ and the proof when G is a torus.

We hope that our proof of Boutot's result throws a different light on rationality of singularities of quotients.

The author feels that the Proposition 1 should be useful in other situations. For example, using this result it was shown in ([11]) that the local fundamental group at any point in $V/\!\!/G$ is finite. This proved a stronger result than a conjecture of C. T. C. Wall.

Acknowledgements. Part of this work was done when the author was visiting the Max Planck Institut für Mathematik in Bonn, Germany. He is thankful to MPI for the warm hospitality.

The author is also very thankful to the referee for critically reading the proof, pointing out mistakes in the earlier versions and suggesting improvements for a better presentation.

2. Preliminaries

Let $\mathbf{A}^n$ denote the affine space of dimension n over $\mathbf{C}$. A morphism $\varphi\colon X \to C$ from a normal algebraic surface X to a smooth curve C is called a $\mathbf{P}^1$-fibration if a general fiber of φ is isomorphic to $\mathbf{P}^1$. Similarly, an $\mathbf{A}^1$-fibration and a $\mathbf{C}^*$-fibration are defined.

If an algebraic variety $X/\mathbf{C}$ is Cohen – Macaulay we say that X is C.M.

A germ of a normal complex space (W, w) is called a *quotient singularity* if it is isomorphic to a germ $(\mathbf{C}^n, \mathbf{O})/\Gamma$, where Γ is a finite group of automorphims of the germ $(\mathbf{C}^n, \mathbf{O})$.

For a linear action of a reductive algebraic group G on $\mathbf{A}^n$ the ring of invariants $\Gamma(\mathbf{A}^n, \mathcal{O})^G$ is a finitely generated, positively graded domain. The point of $\mathbf{A}^n/\!\!/G$ corresponding to the irrelevent maximal ideal of this ring is called the *vertex* of $\mathbf{A}^n/\!\!/G$. The natural quotient morphism $\mathbf{A}^n \to \mathbf{A}^n/\!\!/G$ has been extensively studied.

Now let V be as in Boutot's theorem (Introduction).

It follows from Kempf's results in ([17]) that the rationality of singularities of $W := V/\!\!/G$ is equivalent to the following two results together.

(i) Letting $d = \dim W$, any rational d-form on W which is regular on the smooth locus of W extends to a regular form on a resolution of singularities of W.

(ii) W is Cohen – Macaulay.

In ([11]) a very general result was proved which easily implies that (i) always holds for $V /\!\!/ G$.

Using the étale slice theorem of Luna we can assume that $V = \mathbf{A}^n$ and G acts linearly on $\mathbf{A}^n$. Let $\pi\colon V \to W$ be the quotient map. Let $w_0 \in W$ be the vertex. Then the main result proved in ([11]) is as follows.

Proposition 1. *There is a closed reductive subgroup $H \subset G$ with a suitable linear action of H on some $\mathbf{A}^m$ such that if $Y := \mathbf{A}^m /\!\!/ H$ then the inverse image of the singular locus of Y in $\mathbf{A}^m$ has codimension ≥ 2 in $\mathbf{A}^m$. Further, the analytic local ring of W at w_0 is isomorphic to the analytic local ring of Y at its vertex.*

The proof of this result uses repeated application of Luna's étale slice theorem. From Proposition 1 it is easy to deduce (i) (cf. [17]).

We will use the following result of Elkik – Flenner which gives a criterion for a singularity to be rational ([5; 7, Satz 1.3]).

Proposition 2. *Let k be an algebraically closed field of char. 0 and let X be a normal algebraic variety/k of dimension d. If the canonical divisor K_X of X is Cartier and any regular d-form on the smooth locus of X extends to a regular form on a resolution of singularities of X then X has rational singularities.*

In the terminology of M. Reid ([21]), the Elkik – Flenner result just says that an index-1 canonical singularity is rational.

We will implicitly use the following result about rational singularities.

Let $f\colon U \to W$ be a finite morphism of normal algebraic varieties.

(i) *If U is C.M. then so is W.*

(ii) *If U has rational singularities then so does W.*

(i) follows from the following observations. We can assume that U, W are affine. First, using the trace map we can see that $\Gamma(W, \mathcal{O})$ is a direct summand of $\Gamma(U, \mathcal{O})$ as a $\Gamma(W, \mathcal{O})$-module. Next use a Noether normalization $W \to A^m$, where $m = \dim W$. Since U is C.M., the ring $\Gamma(U, \mathcal{O})$ is a projective $\Gamma(\mathbf{A}^m, \mathcal{O})$-module by the Auslander – Buchsbaum formula ([4, Theorem 19.9]). Being a direct summand of a projective module, $\Gamma(W, \mathcal{O})$ is a projective $\Gamma(\mathbf{A}^m, \mathcal{O})$-module. Thus $\Gamma(W, \mathcal{O})$ is C.M.

(ii) Since U has rational singularities, by Kempf's criterion U is C.M. By (i) above W is C.M. Let ω be a regular m-form on $W - \operatorname{Sing} W$. Since $U \to W$ is a finite morphism ω lifts to a regular m-form on $U - \operatorname{Sing} U$. Let $\widetilde{W} \to W$ be a resolution of singularities, and $\widetilde{U} \to U$ a resolution of singularities dominating $\widetilde{W}$. Since U has rational singularities ω extends to a regular form on $\widetilde{U}$. It is easy to show that this implies ω extends to a regular m-form on $\widetilde{W}$ ([17]).

Finally, we will use a result by V. I. Danilov about the relation between discreteness of the divisor class group and depth of complete local rings in char 0. This last result is undoubtably rather deep, but it shows the connection between Boutot's result and Danilov's results. This is not surprising, in retrospect, because Boutot's theory of local Picard scheme is quite closely related to Danilov's work ([1]).

3. Proof of Theorem 1

We will prove Theorem 1 by induction on $\dim G$, which will also prove Theorem 2 at the same time.

It is easy to reduce to the case when G is connected (see, Proof for the semisimple case below). Hence we will assume that G is connected.

If G is semisimple then the proof is given later. Assume that G is not semisimple. Then we can find normal subgroups H, T where H is semisimple and T is a torus contained in the center of G such that $G = H \cdot T$, $H \cap T$ a finite (normal) subgroup of G. Let T' be a subtorus of T such that $T/T' = \mathbf{C}^*$ and let $G_1 = H \cdot T'$. Let $Y := V/\!/G_1$. By induction hypothesis Y has rational singularities. Clearly $\mathbf{C}^* = T/T'$ acts on Y such that $W = Y/\!/\mathbf{C}^*$. Denote by ψ the quotient map $Y \to W$. We can find regular functions $y_1, y_2, \ldots, y_m$ on Y which generate the coordinate ring of Y and the action of $\mathbf{C}^*$ is given by $\rho_\lambda(y_1, \ldots, y_m) = (\lambda^{a_1}y_1, \ldots, \lambda^{a_r}y_r, y_{r+1}, \ldots, y_m)$. Here $r \leq m$ and $a_i \neq 0$ for $1 \leq i \leq r$. We can assume that no y_i is identically 0 on Y.

Case 1. Suppose that either $a_i > 0$ for all $i \leq r$ or $a_i < 0$ for all $i \leq r$. We can assume that $a_i > 0$ for $i \leq r$. The functions $y_{r+1}, \ldots, y_m$ generate the coordinate ring $\Gamma(W, \mathcal{O})$ of W. The coordinate ring $\Gamma(Y, \mathcal{O})$ is a positively graded ring over $\Gamma(W, \mathcal{O})$ such that $y_1, y_2, \ldots, y_r$ are of weight > 0 and generate a prime ideal P such that $\Gamma(Y, \mathcal{O})/P \cong \Gamma(W, \mathcal{O})$. Denote the affine subvariety of Y with coordinate ring $\Gamma(Y, \mathcal{O})/P \cong \Gamma(W, \mathcal{O})$ as W_0. This just means that $W_0 \subset Y$ is a retract with the retraction map being a morphism. Let $\widetilde{W_0} \to W_0$, $\widetilde{W} \to W$, $\widetilde{Y} \to Y$ be resolutions of singularities. We can assume by further blowing ups that we have morphisms $\widetilde{W_0} \to \widetilde{Y} \to \widetilde{W}$. Since the composite $\widetilde{W_0} \to \widetilde{W}$ is a proper, birational map between smooth varieties (both being resolutions of singularities of the same space W) the induced map on cohomology groups $H^i(\widetilde{W}, \mathcal{O}) \to H^i(\widetilde{W_0}, \mathcal{O})$ are isomorphims for all i. It follows that the maps $H^i(\widetilde{W}, \mathcal{O}) \to H^i(\widetilde{Y}, \mathcal{O})$ are injective. Since Y has rational singularties the latter groups are trivial for $i > 0$. This shows that W has also rational singularities.

Case 2. Suppose that not all $a_i > 0$ or < 0.

In this case a general orbit is closed and hence a general fiber of ψ is $\mathbf{C}^*$. The morphism ψ may not be flat. Our first step is to reduce to a situation of a flat map. Let $\widetilde{W} \to W$ be a resolution of singularities and let $Y \times_W \widetilde{W}$ be the fiber product. Clearly the action of $\mathbf{C}^*$ extends to the fiber product. Using a result on flattening of morphisms by Hironaka ([13]) we can assume (by making suitable further blowing ups of $\widetilde{W}$ with smooth centers, if necessary) the following properties:

(1) The fiber product has a closed subvariety $\widetilde{Y}$ which is $\mathbf{C}^*$-stable, such that the natural map, say $\widetilde{\psi}\colon \widetilde{Y} \to \widetilde{W}$, is surjective and flat.

(2) The set of points p in $\widetilde{W}$, say D, such that the isotropy subgroup of some point in $\widetilde{Y}$ lying over p is non-trivial is a simple normal crossing divisor.

A general fiber of $\widetilde{\psi}$ is an orbit (hence isomorphic to $\mathbf{C}^*$) and every fiber is 1-dimensional.

From now on flatness of $\widetilde{\psi}$ will be used only for equidimensionality of fibers of this map. In order to avoid too many notations, we will denote the normalization of $\widetilde{Y}$ in its function field again by $\widetilde{Y}$ and the natural map $\widetilde{Y} \to \widetilde{W}$ again by $\widetilde{\psi}$. This new map $\widetilde{\psi}$ is not necessarily flat but all its fibers have the same dimension 1.

The map $\widetilde{\psi}$ is affine and any affine open subset of $\widetilde{W}$ is a categorial quotient of its inverse image in $\widetilde{Y}$.

Now we prove the following result which is analogous to Miyanishi's result mentioned in the introduction. In ([8, Theorem 1.12]) an analogous result is proved in arbitrary dimension but the assumptions in ([8]) (C.M. and isolated singularity) are rather strong and so we cannot use this result here.

Proposition 3. *$\widetilde{Y}$ has rational singularities.*

PROOF. Our strategy is to prove that $\widetilde{Y}$ is dominated by a C.M. variety via a finite morphism.

Let D_i be an irreducible component of D and let $D_{i1}, D_{i2}, \dots$ be the irreducible components of the inverse image of D_i in $\widetilde{Y}$. Let D_i be defined by an irreducible element h_i at a general point p_i on it. Choose general points $p_{i1}, p_{i2} \dots$ in $D_{i1}, D_{i2}, \dots$ respectively lying over p_i. Near p_{ij} we can write $h_i = u_{ij} h_{ij}^{e_{ij}}$, where u_{ij} are units near p_{ij} and h_{ij} is a prime element in the local ring of $\widetilde{Y}$ at p_{ij}. Define $e_i = \prod_j e_{ij}$. By Kawamata's Covering Lemma ([16, Theorem 17]), we can find a finite Galois morphism $\tau\colon Z \to \widetilde{W}$ such that Z is smooth and $\tau^*(D_i) = e_i \Delta_i$ for some divisors Δ_i. The union of all the Δ_i has simple normal crossings. Let T be the normalization of the fiber product $\widetilde{Y} \times_{\widetilde{W}} Z$. There is a natural action of $\mathbf{C}^*$ on T such that every fiber of the natural morphism $\varphi\colon T \to Z$ is 1-dimensional, a general fiber is an orbit, the subvariety of points, F, in T with non-trivial isotropy subgroups has codimension ≥ 2 in T and outside F every fiber of φ is reduced. The subvariety F is a union of two subvarieties F_1, F_2, where F_1 consists of fixed points of the $\mathbf{C}^*$ action and F_2 is the closure of the set of points with non-trivial finite isotropy subgroups.

We will first reduce to the case when the codimension of F_1 in T in 2. Suppose that F_1 has codimension ≥ 3. The codimension of F_2 in T is ≥ 2. The image $B = \varphi(F)$ has codimension ≥ 2 in Z. Let $q \in F_1$ and let N be a small affine neighborhood of $p = \varphi(q)$ in Z. Now $B \cap N$ has codimension ≥ 2 in N. The morphism $\varphi^{-1}(N - B) \to N - B$ is a principal $\mathbf{C}^*$-bundle. Since N is smooth this bundle extends to a principal $\mathbf{C}^*$-bundle over N. To see this, we indicate the proof for the corresponding algebraic line bundle L. First we extend the invertible sheaf corresponding to L on $N - B$ to a coherent algebraic sheaf L' on N by Grothendieck's result. Next, we take the double dual L'^{**}. Since N is smooth, L'^{**} is again an invertible sheaf on N (cf. [12]). Now we can assume (by shrinking N) that this bundle is trivial (this is analogous to the statement that an algebraic line bundle over an algebraic variety is locally trivial in the Zariski topology). Since $\varphi^{-1}(N)$ is affine we conclude that $\varphi^{-1}(N)$ is isomorphic to the total space of this extended principal bundle, therefore smooth. This shows that T is smooth and hence $\widetilde{Y}$ has quotient singularities.

Now we can assume that F_1 has codimension 2 in T and its image in Z has codimension 1. The codimension of F_2 in T is ≥ 2 and its image in Z has codimension ≥ 2. The codimension of $F_1 \cap F_2$ in T is ≥ 3 and its image in Z has codimension ≥ 2. Then $B := \varphi(F) = B_1 \cup B_2$ where $B_i = \varphi(F_i)$. Here B_i has codimension $\geq i$ for $i = 1, 2$ in Z. Since the morphism $T \to \widetilde{Y}$ is finite, if we prove that T is C.M. then it will follow that $\widetilde{Y}$ is also C.M. □

Proposition 4. *T is C.M.*

PROOF. For proving this we will take an affine open subset of Z and its inverse image in T, which is $\mathbf{C}^*$-stable. By changing the notation we will now assume that T, Z are affine varieties such that $Z = T/\!\!/C^*$.

We will prove the result by suitably stratifying T.

For any point $q \in T - F$, by Sumihiro's result ([23]) we can find an affine $\mathbf{C}^*$-stable neighborhood U_q of q in T. By shrinking U_q we can assume that every point in U_q has a trivial isotropy subgroup. Further, since every fiber is 1-dimensional, the inverse image of any point in Z contains only finitely many orbits in $T-F$. So we can assume that U_q is so small (affine) neighborhood of q that U_q does not contain any other orbit which maps to the same point $\varphi(q)$. Every orbit contained in U_q is closed in U_q. Since a general orbit in T is closed in T it is clear that U_q contains a general closed orbit in T. By Luna's result ([18, Section III, Corollary 1]) the morphism $U_q \to U_q/\!\!/\mathbf{C}^*$ is a principal $\mathbf{C}^*$-bundle. The natural morphism $\sigma\colon U_q/\!\!/\mathbf{C}^* \to Z$ is birational since U_q contains a general orbit. By the choice of U_q, the point $\varphi(q)$ is isolated in the inverse image of $\sigma(\varphi(q))$. Therefore by Zariski's Main Theorem the morphism $U_q/\!\!/\mathbf{C}^* \to Z$ is an open embedding in a small neighborhood of $\varphi(q)$. From these observations we infer that U_q is smooth is a neighborhood of q, and hence $T - F$ is smooth.

Next, let $q \in F_2 - F_1$. We can choose a $\mathbf{C}^*$-stable affine neighborhood U_q as above such that every point in U_q has a finite isotropy subgroup. As above, by taking U_q small enough we see that $U_q/\!\!/\mathbf{C}^*$ is isomorphic to an open subset of Z, at least locally near the image of $\varphi(q)$, and hence smooth. The image of $F_2 \cap U_q$ in $U_q/\!\!/\mathbf{C}^*$ has codimension ≥ 2. As seen before, this implies that $U_q \to U_q/\!\!/\mathbf{C}^*$ is a principal $\mathbf{C}^*$ fiber bundle. This implies that U_q is smooth and the isotropy subgroup of q is trivial.

We have thus shown that T has no point with a non-trivial finite isotropy subgroup. It follows that the morphism $T - F_1 \to Z$ is smooth and hence T is smooth at all the points in $T - F_1$. □

We will give the details of the proof when $\dim T = 3$ and indicate the necessary modifications for the general case.

So now let $\dim T = 3$. Hence F_1 is a curve and its image in Z is also a curve.

Since T is normal, by a result of Seidenberg a general hyperplane section $H \cap T$ of T is again normal ([22]). It follows that T is C.M. at any point in $H \cap T$. Now consider a general hyperplane section C of Z. By Bertini's theorem on variable singular points the inverse image $S := \varphi^{-1}(C)$ is smooth outside $S \cap F_1$. Since the intersection $S \cap F_1$ is finite and since T is C.M. at a general point in F_1 we know that S is C.M. This implies that S is normal. Now we know that there exists a Zariski open subset U_0 of F_1 such that $F_1 - U_0$ is finite and T has depth 3 at any point in U_0. Assume now that $q \in U_0$. As seen above, the isotropy subgroup of any point of the fiber through q is trivial except for q. Let $p = \varphi(q)$. The fiber of $\varphi|_S$ is reduced outside q.

Lemma 1 (**cf.** [6, Section 4]). *S has at worst an A_m-singularity at q for some m.*

The proof of Lemma 1 will be given in the appendix.

Remarks. We mention the following consequences from the proof of Lemma 1 for future use.

(1) If $\varphi^{-1}(p) = C_0$ is irreducible then the argument shows that q is a smooth point of S and hence also of T.

(2) Since rational double points have no continuous moduli we can see that T is locally analytically, outside a finite set of points in F_1, a product of the germ of an A_m-singularity and a 1-dimensional disc in $\mathbf{C}$. In particular, the integer m is constant outside a finite subset of F_1. We will use this fact below.

CONTINUATION OF THE PROOF WHEN $\dim T = 3$. S is a Cartier divisor at q and has multiplicity at most 2 at q It is easy to see that any normal singular point of multiplicity 2 is locally analytically a hypersurface. Since T is locally analytically C.M. (it may be smooth) $\widetilde{Y}$ is C.M. at the image of q.

By the construction of Z, the subvariety D of singular values for φ is a simple normal crossing divisor and the map $F_1 \to D$ is a an isomorphism.

Finally, let $q \in F_1$ be a point where T has possibly depth ≤ 2 and let $p = \varphi(q)$.

First we consider the case when F_1 is smooth at q. We will reduce the general case to this one by a suitable "trick."

Again, let C be a curve obtained by intersecting Z with a general hyperplane section passing through p and $S_q = \varphi^{-1}(C)$. There is a prime element x_1 in the local ring of Z at p which defines C. Now S_q may be singular at most at q. Let S'_q be the normalization of S in its function field. There is a unique point $q' \in S'_q$ lying over q. The $\mathbf{C}^*$ action extends to S'. This implies that the map $S'_q \to S_q$ is a homeomorphism. By the proof of Lemma 1, S'_q has at worst cyclic quotient singularity at q' and it is also a rational double point.

We will crucially use the following observation.

Claim. *The analytic local ring of T at q, say $\mathcal{O}$ (or its completion $\widehat{\mathcal{O}}$), has finite divisor class group.*

Let the smooth curve F_1 be defined at q by $\{x_2 = 0 = x_3 = \cdots = x_n\}$, where $x_1, x_2, \ldots, x_n$ are local analytic coordinates at q in the ambient affine space in which T is (locally at q) embedded as a closed complex subspace. The curve C above is defined by the prime element x_1. For $|c|$ small and non-zero the section of T near q defined by $\{x_1 = c\}$ defines either a smooth point or a rational double point. We will give the proof when this section is singular.

As remarked above, we can assume that $|c|$ is so small and non-zero that the integer m is constant for all the sections $\{x_1 = c\}$. Hence all the germs of singularities of $\{x_1 = c\}$ are mutually isomorphic. Using the existence of a versal deformation of the A_m-singularity we see that the holomorphic map from $0 < |x_1| < \epsilon$ to the base of the versal deformation is constant, so that $N - \{x_1 = 0\}$ is isomorphic to the product of the germ of an A_m-singularity and the disc $0 < |c| < \epsilon$ (cf. [15]).

Hence the divisor class group of $\mathcal{O}[1/x_1]$ is finite.

Since x_1 is a prime element this implies, by Nagata's Lemma ([4, p. 487]), that $\mathcal{O}$ has finite divisor class group.

In the language of Danilov ([3]), $\mathcal{O}$ has D.G.C. (Discrete Divisor Class Group). Since $\dim T = 3$, this implies that T has depth 3 ([3], Theorem 2, Section 5), and hence T is C.M. at q. Since $T \to \widetilde{Y}$ is a finite morphisms, $\widetilde{Y}$ is C.M. at the image of q.

Next consider the case when F_1 is a union of two smooth curves $C_1 \cup C_2$ at q and $F_1 = \operatorname{Sing} T$ (this uses the fact that D is a simple normal crossing divisor in Z and the map $F_1 \to D$ is an isomorphism). There are regular functions x_1, x_2

on Z whose zeros define curves which are images of C_2, C_1 respectively. Now we argue as above by considering the complement $T - \{x_1 = 0\}$ and showing that the completion of T at q has finite divisor class group, etc.

This completes the proofs of Proposition 4. □

Let ω be a rational 3-form on $\widetilde{Y}$ which is regular on $\widetilde{Y} \setminus \operatorname{Sing} \widetilde{Y}$. To complete the proof of Proposition 3 it suffices to show that ω remains regular on a resolution of singularities of $\widetilde{Y}$. Since $\operatorname{Sing} \widetilde{Y}$ has codimension ≥ 2 in $\widetilde{Y}$ its image in Y cannot contain any divisorial component. It follows that ω is regular on $Y \setminus \operatorname{Sing} Y$. Since Y has rational singularities ω remains regular on any resolution of singularities of Y. By Kempf's criterion of rationality, $\widetilde{Y}$ has rational singularities.

Now consider the case $\dim \widetilde{Y} = 4$.

Now $\dim Z = 3$. Let $D_1, D_2, \ldots$ be the irreducible components of D. The closed subset $F_1 \subset T$ has dimension 2 and maps isomorphically onto D. Using Seidenberg's result repeatedly we see that T is C.M. except possibly at points on a curve Δ and finitely many isolated points in F_1. Using the previous case when $\dim T = 3$, we can in fact see that T is C.M. at all but finitely many points in F_1. For, suppose that $\Delta \neq \varnothing$. Let H be a general hyperplane section of Z and let $T_1 := \varphi^{-1}(H)$. Since $T_1 \cap \Delta$ is a finite set of points such that T_1 is C.M. at except this set of points and $T_1 \cap F_1$ is a curve, by the proof of the previous case we see that T_1 is C.M. at all its points. Then T is also C.M. at all the points in $T_1 \cap \Delta$, showing that $\Delta = \varnothing$.

Let q be a possibly bad point where T is not C.M. and $p := \varphi(q)$. Let H be a general hyperplane section of Z passing through p and let $T_1 = \varphi^{-1}(H)$. If at most two D_i pass through p then the above argument proves that T_1 is C.M. at q. Suppose D_1, D_2, D_3 pass through p. Let F_{1i} be the irreducible component of F_1 lying over $D_i \cap H$ for $i = 1, 2, 3$. With the notation above, we consider a small neighborhood N of q. Let D_1, D_2, D_3 be defined by the vanishing of the prime elements $x_1, x_2, x_1 - x_2$ respectively. Consider the holomorphic map $h\colon N - \{x_1.x_2 = 0\} \to \Delta^*$, where h defines a transverse section to F_{1i}, which is a germ of an A_m-singularity. By a similar argument as above we show that the analyic (or complete) local ring $\mathcal{O}$ of T_1 at q has finite divisor class group, hence it is C.M. Then T is C.M. at q.

We leave the details of the proof of this and $\dim \widetilde{Y} > 4$ to the reader.

Remark. Danilov's result used above only shows that depth of $\mathcal{O}$ is ≥ 3. Therefore we have to reduce to the case $\dim \widetilde{Y} = 3$ by taking hyperplane sections of Z.

Continuation of the proof. For any affine open subset U of $\widetilde{W}$ its inverse image $\widetilde{U}$ in $\widetilde{Y}$ is affine and G-stable such that U is the quotient of $\widetilde{U}$ for the G-action. To prove rationality of singularities of W we must prove that $H^i(\widetilde{W}, \mathcal{O}) = (0)$ for $i > 0$.

The morphism $\widetilde{Y} \to Y$ is proper and birational. Since Y has rational singularities, by an application of Leray Spectral sequence we see that $H^i(\widetilde{Y}, \mathcal{O}) = (0)$ for $i > 0$.

Next we have the following.

Claim. *The structure sheaf of $\widetilde{W}$ is a direct summand of $\widetilde{\psi}_* \mathcal{O}_{\widetilde{Y}}$.*

Recall that for any f.g. ring R over k and a rational action of a reductive group H on R, there is a Reynold's operator $R \to R^H$, which is R^H-linear. This operator is functorial in the sense that for any rational H modules M, N with an H-equivariant homomorphism $\varphi\colon M \to N$ the Reynold's operator for M, N commutes with φ ([9, p. 156]). We apply this to the G-variety $\widetilde{Y}$. For any affine open subset $\widetilde{U}$ of $\widetilde{Y}$ which is G-stable, the Reynold's operator $\widetilde{R} \to \widetilde{R}^G$ for the coordinate ring $\widetilde{R}$ of $\widetilde{U}$ is natural. This shows that $\widetilde{R}^G$ is a direct summand of $\widetilde{R}$ and these direct summand structures patch together to prove the claim.

Using the claim we immediately see that $H^i(\widetilde{W}, \mathcal{O}) = (0)$ for $i > 0$.

The proof of Theorem 2 is now complete. □

PROOF OF THEOREM 1 FOR THE SEMISIMPLE CASE. Assume that the connected component of identity G_0 of G is semisimple. Suppose we have proved that $\mathbf{A}^n /\!/ G_0$ has rational singularities. The morphism $\mathbf{A}^n /\!/ G_0 \to \mathbf{A}^n /\!/ G$ is a finite (proper) morphism. This implies that $\mathbf{A}^n /\!/ G$ also has rational singularities. For, as ramarked in the beginning, since $\mathbf{A}^n /\!/ G_0$ is C.M. so is $\mathbf{A}^n /\!/ G$. We use Proposition 2 to conclude that $\mathbf{A}^n /\!/ G$ has rational singularities.

Now we will assume that G is connected and semisimple. Let R be the coordinate ring of $\mathbf{A}^n$, which is a polynomial ring in n variables over k. It is a well-known fact that since G has no non-trivial characters the ring of invariants R^G is a UFD ([17]). This implies that the canonical divisor of $\mathbf{A}^n /\!/ G$ is Cartier. It is also known that the inverse image of any codimension ≥ 2 subvariety of $\mathbf{A}^n /\!/ G$ has codimension ≥ 2 in $\mathbf{A}^n$ ([17]). This implies, as remarked in the introduction, that any regular d-form on $\mathbf{A}^n /\!/ G$ extends to a regular form on a resolution of singularities of $\mathbf{A}^n /\!/ G$. Thus $\mathbf{A}^n /\!/ G$ has index 1 canonical singularities. By Proposition 2, $\mathbf{A}^n /\!/ G$ has rational singularities.

This completes the proof of Theorem 1. □

Appendix A. Proof of Lemma 1

PROOF. For proving this we use the fact that S has a $\mathbf{C}^*$ action such that $S /\!/ \mathbf{C}^*$ is smooth and such that near p every fiber of the map $S - \{q\} \to C$ is reduced.

The proof uses some well-known properties of singular fibers of a $\mathbf{P}^1$-fibration on a smooth projective surface. We can find an open embedding $S \subset \overline{S}$ such that $\overline{S}$ is smooth outside S and the $\mathbf{C}^*$-fibration on S extends to a $\mathbf{P}^1$-fibration $\overline{\varphi}$ over $\overline{S}$. Since φ is trivial on a dense open subset of C we see that $\overline{S} - S$ contains two irreducible components D_0, D_0' which are cross-sections of $\overline{\varphi}$. In the terminology of ([20]), φ is an *untwisted* $\mathbf{C}^*$-fibration. We use ([20, Lemma 2.9]). This lemma describes all possible singular fibers of a $\mathbf{C}^*$-fibration on a normal affine surface. Since the group $\mathbf{C}^*$ is reductive every fiber of the map $\varphi\colon S \to C$ contains a unique closed orbit. Hence q is in the closure of every orbit lying over p. Hence using the result of Miyanishi–Sugie we infer easily that either $\varphi^{-1}(p) = C_0$ or $\varphi^{-1}(p) = C_0 \cup C_0'$, where C_0, C_0' are both isomorphic to $\mathbf{A}^1$ and meet transversally at q. We will consider only the second case (see, Remark below). Let $\widetilde{S}$ be a minimal resolution of singularities of $\overline{S}$ and Φ be the induced $\mathbf{P}^1$-fibration on $\widetilde{S}$. Let G be the fiber $\Phi^{-1}(p)$. The closures of the proper transforms of C_0, C_0', say $\widetilde{C}_0, \widetilde{C}_0'$, occur with multiplicity 1 in G. By well-known properties of singular fibers of a $\mathbf{P}^1$ fibration on a smooth projective surface ([19, Chapter I, Section 4.4.1]) we can assume that both $\widetilde{C}_0, \widetilde{C}_0'$ are (-1)-curves and G contains no other (-1)-curve.

Claim 1. *$G = \widetilde{C}_0 \cup \widetilde{C}'_0 \cup E$, where E is the (connected) exceptional curve lying over q.*

Clearly $\widetilde{C}_0, \widetilde{C}'_0$ both meet E. If G contains any other irreducible component, say C_1, then since G is connected we can assume that $\widetilde{C}_0 \cap C_1 \neq \varnothing$. Since $\widetilde{C}_0$ is a (-1)-curve the multiplicity of $\widetilde{C}_0$ in G is > 1, a contradiction.

Hence without loss of generality we can assume that $\widetilde{C}_0$ meets D_0, $\widetilde{C}'_0$ meets D'_0.

Claim 2. *E is a linear tree of smooth projective rational curves of self-intersection (-2) and $\widetilde{C}_0, \widetilde{C}'_0$ meet the two end components of E transversally.*

For, after a succession of contractions of (-1)-curves starting with $\widetilde{C}_0$, which all have multiplicity 1 in G, we reach a branch point which has to be a (-2) curve in G with multiplicity 1 in G. Since it meets at least two other irreducible components of the fiber just before its contraction it has multiplicity > 1 in G. This contradiction proves the claim. □

References

1. J.-F. Boutot, *Schéma de Picard local*, C. R. Acad. Sci. Paris Sér. A-B **277** (1973), A691–A694.
2. ———, *Singularités rationnelles et quotients par les groupes réductifs*, Invent. Math. **88** (1987), no. 1, 65–68.
3. V. I. Danilov, *Rings with a discrete group of divisor classes*, Mat. Sb. (N.S.) **83 (125)** (1970), 372–389 (Russian); English transl., Math. USSR-Sb. **12** (1970), 368–386.
4. D. Eisenbud, *Commutative algebra: With a view toward algebraic geometry*, Grad. Texts in Math., vol. 150, Springer, New York, 1995.
5. R. Elkik, *Rationalité des singularités canoniques*, Invent. Math. **64** (1981), no. 1, 1–6.
6. K.-H. Fieseler and L. Kaup, *On the geometry of affine algebraic* $\mathbf{C}^*$*-surfaces*, Problems in the Theory of Surfaces and Their Classification (Cortona, 1988) (F. Catanese, C. Ciliberto, and M. Cornalba, eds.), Sympos. Math., vol. 32, Academic Press, London, 1991, pp. 111–140.
7. H. Flenner, *Rationale quasihomogene Singularitäten*, Arch. Math. (Basel) **36** (1981), no. 1, 35–44.
8. H. Flenner and M. Zaidenberg, *Rational curves and rational singularities*, Math. Z. **244** (2003), no. 3, 549–575.
9. J. Fogarty, *Invariant theory*, W. A. Benjamin, Inc., New York – Amsterdam, 1969.
10. R. V. Gurjar, *Affine varieties dominated by* $\mathbf{C}^2$, Comment. Math. Helv. **55** (1980), no. 3, 378–389.
11. ———, *On a conjecture of C. T. C. Wall*, J. Math. Kyoto Univ. **31** (1991), no. 4, 1121–1124.
12. R. Hartshorne, *Stable reflexive sheaves*, Math. Ann. **254** (1980), no. 2, 121–176.
13. H. Hironaka, *Flattening of analytic maps*, Manifolds — Tokyo 1973 (Proc. Internat. Conf., Tokyo, 1973) (A. Hattori, ed.), Univ. Tokyo Press, Tokyo, 1975, pp. 313–321.
14. M. Hochster and J. L. Roberts, *Rings of invariants of reductive groups acting on regular rings are Cohen – Macaulay*, Advances in Math. **13** (1974), 115–175.
15. A. Kas and M. Schlessinger, *On the versal deformation of a complex space with an isolated singularity*, Math. Ann. **196** (1972), 23–29.
16. Y. Kawamata, *Characterization of abelian varieties*, Compositio Math. **43** (1981), no. 2, 253–276.
17. G. R. Kempf, *Some quotient varieties have rational singularities*, Michigan Math. J. **24** (1977), no. 3, 347–352.
18. D. Luna, *Slices étalés*, Sur les groupes algébriques, Bull. Soc. Math. France Suppl. Mém., vol. 33, Soc. Math. France, Paris, 1973, pp. 81–105.
19. M. Miyanishi, *Noncomplete algebraic surfaces*, Lecture Notes in Math., vol. 857, Springer, Berlin, 1981.
20. M. Miyanishi and T. Sugie, *Homology planes with quotient singularities*, J. Math. Kyoto Univ. **31** (1991), no. 3, 755–788.

21. M. Reid, *Canonical 3-folds*, Journées de Géometrie Algébrique d'Angers, Juillet 1979/Algebraic Geometry, Angers, 1979 (A. Beauville, ed.), Sijthoff & Noordhoff, Alphen aan den Rijn, 1980, pp. 273–310.
22. A. Seidenberg, *The hyperplane sections of normal varieties*, Trans. Amer. Math. Soc. **69** (1950), 357–386.
23. H. Sumihiro, *Equivariant completion*, J. Math. Kyoto Univ. **14** (1974), 1–28.

School of Mathematics, Tata Institute of Fundamental Research, Homi Bhabha road, Navy Nagar, Mumbai 400005, India
E-mail address: gurjar@math.tifr.res.in

Centre de Recherches Mathématiques
CRM Proceedings and Lecture Notes
Volume **54**, 2011

On the Present State of the Andersén – Lempert Theory

Shulim Kaliman and Frank Kutzschebauch

Dedicated to Professor Peter Russell on the occasion of his seventieth birthday

Abstract. In this survey of the Andersén – Lempert theory we present the state of the art in the study of the density property (which means that the Lie algebra generated by completely integrable holomorphic vector fields on a given Stein manifold is dense in the space of all holomorphic vector fields). There are also two new results in the paper one of which is the theorem stating that the product of Stein manifolds with the volume density property possesses such a property as well. The second one is a meaningful example of an algebraic surface without the algebraic density property. The proof of the last fact requires Brunella's technique.

1. Introduction

In this paper we discuss recent developments in the Andersén – Lempert theory. An earlier account of this theory can be found in a nice survey of Rosay [71]. The theory describes complex manifolds such that among other things the local phase flows on their holomorphically convex compact subsets can be approximated by global holomorphic automorphisms which leads to construction of holomorphic automorphisms with prescribed local properties. Needless to say that this implies remarkable consequences for such manifolds some of which are described below in Section 2. The original work of Andersén and Lempert ([2,4]) established that complex Euclidean spaces of dimension at least 2 belong to this class. Their results were extended by Forstnerič and Rosay [38, 71] who discovered new approximation theorems for such spaces. Perhaps, they understood already that a complex manifold has such approximations if it possesses the following density property introduced later by Varolin.

1.1. Definition. A complex manifold X has the density property if in the compact-open topology the Lie algebra $\mathrm{Lie}_{\mathrm{hol}}(X)$ generated by completely integrable holomorphic vector fields on X is dense in the Lie algebra $\mathrm{VF}_{\mathrm{hol}}(X)$ of all holomorphic vector fields on X. An affine algebraic manifold X has the algebraic

2010 *Mathematics Subject Classification.* Primary 32M05, 14R20; Secondary 14R10, 32M25.

The research of the first author was partially supported by NSA Grant no. H982301010185 and the research of the second author was supported by Schweizerischer Nationalfonds grant no. 2000020-124668/1.

density property if the Lie algebra $\mathrm{Lie}_{\mathrm{alg}}(X)$ generated by completely integrable algebraic vector fields on it coincides with the Lie algebra $\mathrm{VF}_{\mathrm{alg}}(X)$ of all algebraic vector fields on it.

Usually the easiest way to establish the density property for an affine algebraic manifold X is to prove the algebraic density property for it since some convenient tools from affine algebraic geometry are available in this case.[1] In the sequence of papers ([24, 55, 56, 82, 83, 86]) the algebraic density property was established for a wide variety of affine algebraic manifolds, including all homogeneous spaces that coincide with quotients of linear algebraic groups with respect to their reductive subgroups provided that the connected components of these quotients are different from $\mathbb{C}$ or complex tori. The last two papers were based on a very effective criterion of the algebraic density property introduced by the authors which will be presented in Section 3.

Andersén, Lempert, Forstnerič, Rosay, Tóth, Varolin, and the authors considered also another property which has similar consequences for automorphisms preserving a volume form.

1.2. Definition. Let a complex manifold X be equipped with a holomorphic volume form ω (i.e., ω is nowhere vanishing section of the canonical bundle). We say that X has the volume density property with respect to ω if in the compact-open topology the Lie algebra $\mathrm{Lie}^{\omega}_{\mathrm{hol}}(X)$ generated by completely integrable holomorphic vector fields ν such that $\nu(\omega) = 0$, is dense in the Lie algebra $\mathrm{VF}^{\omega}_{\mathrm{hol}}(X)$ of all holomorphic vector fields that annihilate ω (note that condition $\nu(\omega) = 0$ is equivalent to the fact that ν is of ω-divergence zero). If X is affine algebraic we say that X has the algebraic volume density property with respect to an algebraic volume form ω if the Lie algebra $\mathrm{Lie}^{\omega}_{\mathrm{alg}}(X)$ generated by completely integrable algebraic vector fields ν such that $\nu(\omega) = 0$, coincides with the Lie algebra $\mathrm{VF}^{\omega}_{\mathrm{alg}}(X)$ of all algebraic vector fields that annihilate ω.

It is much more difficult to establish the algebraic volume density property than the algebraic density property since the criterion mentioned before is not applicable in the volume case. However Andersén [2] established the algebraic volume density property for Euclidean spaces even before the algebraic density property. Some extra manifolds with the algebraic volume density were found by Varolin and eventually the authors [56] proved it for all linear algebraic groups (with respect to the left or right invariant volume forms). Furthermore, they established some features that are straightforward for the algebraic density property and not at all clear in the volume-preserving case. For instance, they prove that the algebraic volume density property for an affine algebraic manifold X implies the volume density property for such an X and that the product of two manifolds with algebraic volume density property has again the algebraic volume density property. Some facts on this subject are contained in Section 4 together with the following new result.

Theorem 1. *Let X and Y be Stein manifolds equipped with (holomorphic) volume forms ω_X and ω_Y respectively. Suppose that X and Y have the volume density property with respect to these forms. Then so does $X \times Y$ with respect to the form $\omega_X \times \omega_Y$.*

[1]Furthermore, the authors do not know examples of such an X with the density property but without the algebraic density property.

Unlike in the algebraic case the proof of this theorem required some nontrivial facts from functional analysis.

In Sections 5 – 8 we consider a new meaningful example of an affine surface without the algebraic density property. The algebraic density property for an affine algebraic manifold implies that this manifold has an infinitely transitive group of *holomorphic* automorphisms (i.e., any natural m every m-tuple of distinct points in the manifold can be transformed into any other such tuple by an automorphism). Thus it makes sense to consider such objects only. The question about examples of infinitely transitive affine algebraic manifolds without the algebraic density property were posed by P. Russell and D. Akhiezer to the first author who was not aware at the time about the third paper of Andersén [3] showing that tori belong to this class.[2] However Andersén's proof is based heavily on a Borel theorem from the Nevanlinna theory which is quite specific for tori. The question whether there are examples besides tori was still open.

We shall show that the surface S given by $x + y + xyz = 1$ in $\mathbb{C}^3$ does not have the algebraic density property while the action of its group of holomorphic automorphisms is infinitely transitive. The proof will be based on remarkable Brunella's technique [12, 14, 15], that allowed him to classify completely integrable algebraic vector fields on the plane (which is based in turn on the seminal preprint of McQuillan [68] and earlier work of Suzuki [80, 81]). In fact, his technique works for other rational affine surfaces (including the tori $(\mathbb{C}^*)^2$) and, therefore, is more productive than the Andersén's approach in the two dimensional case. We present crucial ingredients of his method in Section 5 – 7 and apply them to S in Section 8.

Acknowledgements. It is a pleasure to express our gratitude to Nahum Zobin for explaining us important facts from functional analysis and to the referee for very useful comments that helped to improve the paper.

2. Applications of the Andersén – Lempert theory

Constructions of holomorphic automorphisms of Stein manifolds with prescribed behavior on compact subsets are based on the next central theorem of the Andersén – Lempert theory which was proven in the papers [2] and [4] of Andersén and Lempert for Euclidean spaces. We give a stronger version of this theorem which is due to Forstnerič and Rosay [38]. They considered it also in the case Euclidean spaces only; however it was essentially their contribution that made the original result an important tool as it is now.

Theorem 2. *Let X be a Stein manifold with the density (resp. volume density) property and let Ω be an open subset of X (resp. with $H^{n-1}(\Omega, \mathbb{C}) = 0$). Suppose that $\Phi\colon [0,1] \times \Omega \to X$ is a C^1-smooth map such that*

(1) *$\Phi_t\colon \Omega \to X$ is holomorphic and injective (and resp. volume preserving) for every $t \in [0,1]$,*

(2) *$\Phi_0\colon \Omega \to X$ is the natural embedding of Ω into X, and*

[2] Actually Andersén did not formulate this fact. However he gave a classification of all completely integrable algebraic vector fields on a torus $T = (\mathbb{C}^*)^n$ from which it follows that all of them have divergence zero with respect to the invariant volume form $\omega = \prod_{i=1}^n dz_i/z_i$. In particular, any algebraic vector field with a nonzero divergence cannot belong to $\mathrm{Lie}_{\mathrm{alg}}^{\omega}(T)$. To establish infinite transitivity one needs to consider completely integrable algebraic vector fields of form $\sigma_i = z_i \partial/\partial z_i$ and apply, for instance, Proposition 8.9 below with $f_i = \sum_{j \neq i} z_j$.

(3) $\Phi_t(\Omega)$ *is a Runge subset*[3] *of* X *for every* $t \in [0,1]$.

Then for each $\epsilon > 0$ *and every compact subset* $K \subset \Omega$ *there is a continuous family,* $\alpha\colon [0,1] \to \mathrm{Aut}_{\mathrm{hol}}(X)$ *of holomorphic (and resp. volume preserving) automorphisms of* X *such that*

$$\alpha_0 = \Phi_0 \quad \text{and} \quad |\alpha_t - \Phi_t|_K < \epsilon$$

for every $t \in [0,1]$.

Furthermore, approximations on Stein manifolds with the density property can be chosen with some specific features as in the next result [85].

Theorem 3. *Let* X *be a Stein manifold of dimension* $n \geq 2$ *with the density (resp. volume density) property,* K *be a compact in* X*, and* $x, y \in X$ *be two points outside the holomorphic convex hull* $\widehat{K}$ *of* K*. Suppose that* $x_1, \ldots, x_m \in K$.

Then there exists a (resp. volume-preserving) holomorphic automorphism Ψ *of* K *such that* $\Psi(x_i) = x_i$ *for every* $i = 1, \ldots, m$, $\Psi|_K\colon K \to X$ *is as close to the natural embedding as we wish, and* $\Psi(y) = x$.

Note that $\widehat{K}$ can be always chosen as a disjoint union of arbitrarily small neighborhoods of the points $x_1, \ldots, x_m$. In particular, x and y are automatically outside $\widehat{K}$. Thus Theorem 3 implies that we can always move one point from a given $(m+1)$-tuple into a desired position while keeping the other m points fixed. This yields the following.

2.1. Corollary. *If* X *has the density property (or the volume density property) and the dimension* n *of* X *is at least* 2*. Then the holomorphic automorphisms group* $\mathrm{Aut}_{\mathrm{hol}}(X)$ *acts infinitely transitively on* X.

2.2. Remark. Theorem 3 was proven in [85] for the density property only but a slightly modified argument works in the volume-preserving case. More precisely, let γ be a piece-wise analytic path between y and x that does not meet the holomorphic hull $\widehat{K}$ of K, K_1 be the union of $\widehat{K}$ and a small ball U around y, and $\gamma_1 = U \cap \gamma$. The main step in construction of Ψ, where the adjustment is needed, is a global approximation on X of a holomorphic vector field ν on K_1 which is identically zero on $\widehat{K}$ and which is tangent to γ_1. In the volume-preserving case not only ν but also its global approximation must be of divergence zero which prevents the direct use of the Runge property as in [85]. However, it turns out that the existence of such an approximation is equivalent to approximation of some closed holomorphic $(n-1)$-form α on K_1 that is identically zero on $\widehat{K}$ by a global closed holomorphic $(n-1)$-form on X (see the proof of Claim in Lemma 4.1 for more accurate details). Note that α is exact (indeed it is zero on $\widehat{K}$ and $H^{n-1}(U, \mathbb{C}) = 0$ which enables us to apply de Rham's theorem), i.e., $\alpha = \mathrm{d}\beta$ where β is a holomorphic $(n-2)$-form on K_1. Since K_1 is still holomorphically convex, the Runge property implies that one can approximate β by a global holomorphic $(n-2)$-form β' which yields an approximation of α by the global closed $(n-1)$-form $\mathrm{d}\beta'$. We want to emphasize that this reasoning uses the assumption that $n \geq 2$ which is essential for the volume-preserving case. For $n = 1$, consider, for example, $X = \mathbb{C}^*$ which has the volume

[3]Recall that an open subset U of X is Runge if any holomorphic function on U can be approximated by global holomorphic functions on X in the compact-open topology. Actually, by Cartan's Theorem A this definition implies more: for any coherent sheaf on X its section over U can be approximated by global sections.

density property with respect to the volume form dz/z where z is the coordinate on $\mathbb{C}^*$. However it does not satisfy this analogue of Theorem 3. By the same reason it has no properties 2.7 – 2.9 and 2.13 described below.

Treating X as a Stein n-dimensional manifold in the rest of this section, let us describe some general features of Stein manifolds with density (resp. volume density) property which follow either directly from the density property or from subtle applications of Theorems 2 and 3.

2.3. If X has the density (resp. volume density) property, then there are finitely many completely integrable holomorphic vector fields (resp. of divergence zero) that span the tangent space at each point (see [55] and Lemma 4.1 below). Therefore X admits a spray,[4] i.e., it is elliptic in the sense of Gromov which implies, in particular, the Oka – Grauert – Gromov principle for submersions over Stein spaces with fibers isomorphic to X.

To be more precise let us give the relevant definitions and results.

2.4. Definition. (a) A (dominating) spray on a complex manifold X is a holomorphic vector bundle $\rho\colon E \to X$, together with a holomorphic map $s\colon E \to X$, such that s is identical on the zero section $X \hookrightarrow E$, and for each $x \in X$ the induced differential map sends the fibre $E_x = \rho^{-1}(x)$, viewed as a linear subspace of $T_x E$, surjectively onto $T_x X$.

(b) A fiber-dominating spray for a surjective submersion $h\colon Z \to W$ of complex spaces is a vector bundle $\rho\colon E \to Z$ together with a map $s\colon E \to Z$ identical on the zero section $Z \hookrightarrow E$ and such that $h \circ s = h \circ p$ and for every $z \in Z$ the induced differential map sends $E_z = \rho^{-1}(z)$, viewed as a linear subspace of $T_z E$, surjectively onto the subspace of $T_z Z$ tangent to the fiber $h^{-1}(h(z))$.

(c) Let $h\colon Z \to W$ be a holomorphic submersion of Stein spaces, and $\mathrm{Cont}(W, Z)$ (resp. $\mathrm{Holo}(W, Z)$) be the set of continuous (resp. holomorphic) sections of h. Then h satisfies the Oka – Grauert – Gromov principle if $\mathrm{Holo}(W, Z) \hookrightarrow \mathrm{Cont}(W, Z)$ is a weak homotopy equivalence. In particular,

(i) each continuous section $f^0\colon W \to Z$ of h can be deformed to a holomorphic section $f^1\colon W \to Z$, and

(ii) any two homotopic holomorphic sections are also homotopic through holomorphic sections.

Theorem 4 (Oka – Grauert – Gromov-principle for elliptic submersions [33,43]). *Suppose that $h\colon Z \to W$ is a holomorphic submersion of a complex space Z onto a Stein space W for which every $x \in W$ has a neighborhood $U \subset W$ such that $h^{-1}(U) \to U$ admits a fiber-dominating spray. Then $h\colon Z \to W$ satisfies the Oka – Grauert – Gromov principle.*

To illustrate this principle recall the following result of the authors [55].

Theorem 5. *The density (resp. algebraic density) property holds for smooth analytic (resp. algebraic) hypersurfaces in $\mathbb{C}^{n+2}_{u,v,\bar{x}}$ given by equations of form $uv = p(\bar{x})$.*

[4]There is a small inaccuracy in [55] where the authors proved the existence of a spray. Namely, the metric on the space of holomorphic automorphisms of X should be defined not as suggested in that paper but by formula (4.1) below.

Hence the Oka – Grauert – Gromov principle for submersions implies the following.

2.5. Corollary. *Let $h\colon Z \to W$ be a surjective submersion of complex manifolds such that W is Stein and for every $w_0 \in W$ there is a neighborhood U for which $h^{-1}(U)$ is naturally isomorphic to a hypersurface in $\mathbb{C}^{n+2}_{\bar{x},u,v} \times U$ given by $uv = p(\bar{x}, w)$ where p is a holomorphic function on $\mathbb{C}^n_{\bar{x}} \times U$ (independent of u and v).*

Suppose, furthermore, that $p^*(0) \cap (\mathbb{C}^n_{\bar{x}} \times w)$ is a smooth reduced proper (may be empty) submanifold of $\mathbb{C}^n_{\bar{x}} \times w$ for every point $w \in U$.

Then h satisfies the Oka – Grauert – Gromov principle.

2.6. It is worth mentioning that the group of holomorphic automorphisms acts on a hypersurface from Theorem 5 infinitely transitively by Corollary 2.1. Actually a stronger statement was proven earlier by Zaidenberg and the first author [58] without use of the density property: the group of automorphisms of such a hypersurface generated by elements of algebraic $\mathbb{C}_+$-actions is also infinitely transitive.

2.7. If X has the density property (or the volume density property) and the dimension $n \geq 2$ then for each point $x \in X$ there is an injective but not surjective holomorphic map $f\colon X \to X$ with $f(x) = x$. The images of such maps are called Fatou – Bieberbach domains of the second kind. This was also observed by Varolin in [85] for the density property, for the volume density property this is an equally simple application of the kick-out method of Dixon and Esterle [23].

In the algebraic setting similar morphisms do not exists by virtue of the Ax's theorem [7] that states that any injective morphism of a scheme of a finite type into itself it automatically surjective. Actually any morphism of an affine algebraic variety into itself, which is an embedding outside of codimension (at least) 2 subvariety, is automatically an automorphism [52].

2.8. If X has the density property, then for each point $x \in X$ there is an injective nonsurjective equidimensional holomorphic map $f\colon \mathbb{C}^n \to X$ with $f(0) = x$. This observation is due to Varolin [85]. In particular, all Eisenman measures[5] on X vanish identically. Such maps are called Fatou – Bieberbach maps of the first kind and their images are Fatou – Bieberbach domains.

2.9. Here is a sketch of the proof of the last fact. Take a holomorphically convex neighborhood Ω of x together with a vector field θ on Ω for which x is an attractive point. Approximating the flow of this field by automorphisms of X as in Theorem 2 we obtain an automorphism whose restriction to a neighborhood of x is a contraction (in some metric) to a point near x. The basin of attraction of x for that automorphism will be biholomorphic to $\mathbb{C}^n$. Clearly θ cannot be volume preserving when contracting to a point. Therefore the volume density property does not guarantee the existence of Fatou-Bieberbach domains. Say, it is known that $(\mathbb{C}^*)^n$ has the volume density property but for $n \geq 2$ it is an open question whether it contains a Fatou – Bieberbach domain or not. In particular, it is unknown whether it has the density property as well.

[5]These measures are called sometimes Kobayashi – Eisenman measures. Also later works of Eisenman were published under the name Pelles.

2.10. All Fatou – Bieberbach domains arising as basins of attraction (as indicated in 2.9) or more generally as domains of convergence of sequences of automorphisms of $\mathbb{C}^n$ are always Runge domains. Thus it is natural to ask whether all Fatou – Bieberbach domains in $\mathbb{C}^n$ have to be Runge. This problem was solved by Wold who constructed a Fatou – Bieberbach domain in $\mathbb{C} \times \mathbb{C}^*$ which is not Runge in $\mathbb{C}^2$ (but Runge in $\mathbb{C} \times \mathbb{C}^*$) using the density property of $\mathbb{C} \times \mathbb{C}^*$ [92].

2.11. One of the questions coming from complex dynamical systems is description of the boundaries of Fatou – Bieberbach domains. Say, a surprising result of Stensønes [79]) provides such a domain in $\mathbb{C}^2$ with a smooth boundary which has, therefore, Hausdorff dimension $d = 3$. Furthermore, it was established by methods of complex dynamical systems that such a dimension can take any value $3 \le d < 4$. However the question about a Fatou – Bieberbach domain in $\mathbb{C}^2$ with a boundary of Hausdorff dimension $d = 4$ remained open until Peters and Wold [69] managed to construct it using the Andersén – Lempert theory.

2.12. A question posed by Siu asks whether there exists always a Fatou – Bieberbach domain contained in the complement to a closed algebraic subvariety Z of $\mathbb{C}^n$ such that $\dim Z \le n-2$. The affirmative answer was obtained by Buzzard and Hubbard [19] who used some concrete construction. Another proof of this fact was given by the authors who used a version of the density property for such complements (see Theorem 11 below). More precisely for any point $x \in \mathbb{C}^n \setminus Z$ there is a Fatou – Bieberbach (i.e., holomorphic injective) map $f\colon \mathbb{C}^n \to \mathbb{C}^n \setminus Z$ with $f(0) = x$ (actually, the crucial fact that guarantees such maps is existence of sprays on $\mathbb{C}^m \setminus Z$ which can be extracted from the earlier papers of Gromov [43] and Winkelmann [87]).

In particular all Eisenman measures on $\mathbb{C}^n \setminus Z$ are trivial. It is worth mentioning that closed analytic subsets of $\mathbb{C}^n$ of codimension k may have k-Eisenman hyperbolic complements. More precisely, it was shown in [10] that if a complex manifold Y admits a proper holomorphic embedding into $\mathbb{C}^n$ then it has also another proper holomorphic embedding with $(n - \dim Y)$-Eisenman hyperbolic complement to the image (the proof is based on the Andersén – Lempert theory and a generalized idea from [18]).

2.13. If X has the density (resp. volume density) property and Y is any manifold which admits a proper holomorphic embedding $\varphi\colon Y \hookrightarrow X$ then the following is true. For any given discrete subset $E = \{x_1, x_2, \dots, x_m, \dots\}$ in X there is another proper holomorphic embedding ψ of Y into X whose image contains E. The original proof of this fact was obtained by Globevnik, Forstnerič and Rosay [34] in the case of $X = \mathbb{C}^n$, $n \ge 2$ but it works also in the general case. This proof is based on Theorem 3. Namely, if the image of a proper holomorphic embedding $\varphi_m\colon Y \hookrightarrow X$ contains already $x_1, \dots, x_m$ but not x_{m+1}, one has to chose a compact K_m containing $x_1, \dots, x_m$ so that x_{m+1} is outside the holomorphic hull of K_m. Setting $K = K_m$ and taking y outside the convex hull we construct an automorphism $\Psi =: \Psi_m$ as in Theorem 3 and replace φ_m by $\varphi_{m+1} = \Psi_m \circ \varphi_m$. Then the image of φ_{m+1} contains already $x_1, \dots, x_{m+1}$. It turns out that compacts K_m and automorphisms Ψ_m can be chosen so that the limit $\psi = \lim_{m \to \infty} \varphi_m$ is also a proper holomorphic embedding which implies the desired conclusion.

It is also worth mentioning that by a result of Winkelmann [88] (generalizing the earlier results of Rosay and Rudin [72] and the first author [51]), there is a discrete subset E of X whose complement is n-Eisenman hyperbolic. Hence there exists a proper embedding $\psi\colon Y \hookrightarrow X$ with n-Eisenman hyperbolic complement $X \setminus \psi(Y)$.

2.14. Another interesting problem is to find embeddings of Stein manifolds into $\mathbb{C}^m$ with prescribed interpolation condition on discrete subsets. This is much more difficult than just letting the image contain a given discrete subset (not caring about the preimage points). Beside Theorem 2 some properties of $\mathbb{C}^m$ were used by Forstnerič, Ivarsson, Prezelj and the second author in [35] to prove the result below (for general targets with the density property the problem is still widely open).

More precisely, let X be a Stein manifold of dimension $n > 1$, $\{a_j\}_{j\in\mathbb{N}}$ (resp. $\{b_j\}_{j\in\mathbb{N}}$) be a sequence of distinct points in X (resp. $\mathbb{C}^m$). If $m \geq N = [3n/2] + 1$[6] then there exists a proper holomorphic embedding $f\colon X \hookrightarrow \mathbb{C}^m$ satisfying

$$f(a_j) = b_j \quad \text{for} = 1, 2, \ldots .$$

A more general result is that if X admits a proper holomorphic embedding into $\mathbb{C}^m$ and $\{a_j\}_{j\in\mathbb{N}}$ and $\{b_j\}_{j\in\mathbb{N}}$ are sequences as before such that the image of $\{a_j\}_{j\in\mathbb{N}}$ form a so-called tame subset in $\mathbb{C}^m$ (by definition there is a holomorphic automorphism of $\mathbb{C}^m$ mapping this sequence onto the set of the integer points in a coordinate axis), then the conclusion remains true. Other results in this direction can be found in [63].

2.15. Property 2.13 yields a counterexample to the analytic version of the Abhyankar – Moh – Suzuki theorem which states that every polynomial embedding of $\mathbb{C}$ into $\mathbb{C}^2$ is rectifiable, i.e., the image can be sent to a coordinate line by a polynomial automorphism of $\mathbb{C}^2$. However, there exists a proper nonrectifiable holomorphic embedding $\psi\colon \mathbb{C} \hookrightarrow \mathbb{C}^2$ [34]. Indeed, according to 2.13 we can make $\mathbb{C}^2 \setminus \psi(\mathbb{C})$ 2-Eisenman hyperbolic while the complement $\mathbb{C}^* \times \mathbb{C}$ to the coordinate line is not. In fact, according to a more sophisticated argument from [10, 18] the complement $\mathbb{C}^2 \setminus \psi(\mathbb{C})$ can be made even Kobayashi hyperbolic.

Furthermore, there exist uncountably many nonrectifiable embeddings in a reasonable sense. Let us be more precise.

2.16. Definition. Two proper holomorphic embeddings $\Phi, \Psi\colon X \hookrightarrow \mathbb{C}^n$ are *equivalent* if there exist holomorphic automorphisms $\varphi \in \mathrm{Aut}_{\mathrm{hol}}(\mathbb{C}^n)$ and $\psi \in \mathrm{Aut}_{\mathrm{hol}}(X)$ such that $\varphi \circ \Phi = \Psi \circ \psi$.

2.17. Remark. We would like to emphasize that there is another (weaker) definition of equivalence. It is so-called $\mathrm{Aut}(\mathbb{C}^n)$-equivalence which was used by several authors (e.g., Buzzard, Forstnerič, Globevnik and Varolin) who proved uncountability of certain equivalence classes of embeddings in this weaker sense. In our definition the map $\Psi^{-1} \circ \varphi \circ \Phi$ is well-defined and it is an automorphism of X while for the weaker notion one has to demand that $\Psi^{-1} \circ \varphi \circ \Phi$ is the identity on X.

2.18. The best known results are the following.

In [10] Borell and the second author showed that if

(i) X is a Stein space such that its group of holomorphic automorphisms is a Lie group (with possibly countably many components) and

[6] Such N is chosen because it is the optimal embedding dimension, see 2.26 below.

(ii) there exists a proper holomorphic embedding of X into $\mathbb{C}^m$ where $0 < \dim X = n < m$, then for any $k \geq 0$ there are uncountably many nonequivalent proper holomorphic embeddings $\Psi\colon X \times \mathbb{C}^k \hookrightarrow \mathbb{C}^m \times \mathbb{C}^k$.

Actually the result can be made stronger: in the case of codimension at least 2 these nonequivalent embeddings appear in holomorphic families. More precisely, under the same assumptions on X as in [10] with $k = m - n - 1$ and any natural l there exists a family of holomorphic embeddings of $X \times \mathbb{C}^l$ into $\mathbb{C}^m \times \mathbb{C}^l$ parameterized by $\mathbb{C}^k$, such that for different parameters $w_1 \neq w_2 \in \mathbb{C}^k$ the embeddings $\psi_{w_1}, \psi_{w_2}\colon X \times \mathbb{C}^l \hookrightarrow \mathbb{C}^{n+l}$ are nonequivalent. This result is due to Lodin and the second author [64] (an important ingredient of their proof is a parametric version of Theorem 2 from [62]).

2.19. It is worth mentioning that the last two results include embeddings of $\mathbb{C}^n$ into $\mathbb{C}^m$ for any $n < m$ (respectively $n < m - 1$ for the families), by choosing $X = \mathbb{C}$ in 2.18. Thus the holomorphic analogue of the Abhyankar and Sathaye problem has a negative answer.

An important application of the nonrectifiable embeddings is the construction of nonlinearizable holomorphic actions of reductive Lie groups on affine spaces by Derksen and the second author [20, 21].

2.20. Definition. A holomorphic (resp. algebraic) action of a reductive group G on $\mathbb{C}^n$ is said to be linearizable if there exists a holomorphic automorphism $\alpha \in \mathrm{Aut}_{\mathrm{hol}}(\mathbb{C}^n)$ (resp. an algebraic automorphism $\alpha \in \mathrm{Aut}_{\mathrm{alg}}(\mathbb{C}^n)$), such that $\alpha \circ g \circ \alpha^{-1} \in \mathrm{GL}_n(\mathbb{C})$ for every $g \in G$.

2.21. It was known since 1989 that there exist nonlinearizable algebraic actions of reductive groups. The first example of such an action of $\mathrm{O}(2)$ on $\mathbb{C}^4$ was constructed by Schwarz [77] (see also [61]). Furthermore, by Knop's result [59] for every connected reductive group different from a torus there exists a nonlinearizable algebraic action on some $\mathbb{C}^n$ where n depends on the group. However, Schwarz's and Knop's examples are linearizable in the holomorphic category [46] the case of which was settled later in the framework of the Andersén – Lempert theory. Namely, for any nontrivial complex reductive Lie group G there is a natural N such that for all $n \geq N$ there is a nonlinearizable holomorphic G-action on $\mathbb{C}^n$ [20]. The optimal dimension (minimal N) is not known for any G including $G = \mathbb{C}^*$. All holomorphic actions on $\mathbb{C}^2$ are linearizable by a result of Suzuki [80] but starting with dimension $n = 4$ there are nonlinearizable actions on $\mathbb{C}^n$. The problem of linearization of holomorphic $\mathbb{C}^*$-actions on $\mathbb{C}^3$ is still open while all algebraic $\mathbb{C}^*$-actions on $\mathbb{C}^3$ are known to be linearizable [54].

2.22. Sketch of a construction of a nonlinearizable holomorphic $\mathbb{C}^*$-action on $\mathbb{C}^4$. Suppose $\varphi\colon \mathbb{C} \to \mathbb{C}^2$ is a proper holomorphic embedding. Consider a pseudo-affine modification[7] X of $\mathbb{C}^3_{x,y,u}$ along the divisor $D := \{u = 0\} = \mathbb{C}^2 \times \{0\}$ with center $\varphi(\mathbb{C}) \subset D$. That is, if $f \in \mathrm{Hol}(D)$ is a holomorphic function generating the principal ideal of functions vanishing on $\varphi(\mathbb{C}) \subset D$, then X is biholomorphic to the submanifold of $\mathbb{C}^4_{x,y,u,v}$ given by the equation $f(x,y) = uv$. One of crucial facts

[7]In terminology of [58] and [55] a Stein manifold Z is a basic pseudo-affine modification of a Stein manifold X if Z is obtained from X by blowing up a closed submanifold $C \subset X$ of codimension at least 2 and removing the proper transform of a smooth closed analytic divisor $D \subset X$ containing C.

observed by Asanuma [6] is that $X \times \mathbb{C}$ is biholomorphic to $\mathbb{C}^4$, because $X \times \mathbb{C}$ is the basic pseudo-affine modification of $\mathbb{C}^4$ along the divisor $\mathbb{C}^3 \times \{0\}$ with center $\varphi(\mathbb{C}) \times \{0\} \subset \mathbb{C}^3$ and any proper holomorphic embedding of $\mathbb{C}$ into $\mathbb{C}^n$ with the image contained in a hyperplane is rectifiable (see also [53]). Consider the $\mathbb{C}^*$-action on $X \times \mathbb{C}_w$ given by $\lambda.(x, y, u, v, w) = (x, y, \lambda^2 u, \lambda^{-2} v, \lambda w)$ where $\lambda \in \mathbb{C}^*$. The categorical quotient $(X \times \mathbb{C})//\mathbb{C}^*$ of this action is equipped with a natural so-called Luna stratification for which one of the strata isomorphic to $\mathbb{C}$ is contained in a higher-dimensional stratum isomorphic to $\mathbb{C}^2$ exactly in the same manner as $\mathbb{C} \simeq \varphi(\mathbb{C})$ is contained in $D \simeq \mathbb{C}^2$. In the case of a linearizable action this first stratum must be rectifiable in the second one. However by 2.15 we can suppose that $\varphi(\mathbb{C})$ is not rectifiable in D which yields a nonlinearizable holomorphic $\mathbb{C}^*$-action.

2.23. The linearization problem for $\mathbb{C}^*$-actions on $\mathbb{C}^3$ is related to the question whether the pseudo-affine modification X of $\mathbb{C}^3$ as before is biholomorphic to $\mathbb{C}^3$. If the answer is positive we have a nonlinearizable action on $\mathbb{C}^3$, otherwise it is a counterexample to the holomorphic analogue of Zariski's cancellation problem (i.e., the question whether a complex manifold Y is biholomorphic to $\mathbb{C}^n$ provided that $Y \times \mathbb{C}^k$ is biholomorphic to $\mathbb{C}^{k+n}$).[8] Returning to X we note that it is diffeomorphic to $\mathbb{R}^6$ as a smooth real manifold ([55, Appendix]) and it has the density property by Theorem 5. By a conjecture of Varolin and Tóth [82] such an X must be biholomorphic to $\mathbb{C}^3$, i.e., in the case of a negative answer we can disprove their hypothesis. More potential counterexamples to the conjecture of Varolin and Tóth can be found in [55]. One of the most interesting among them is a modification whose center is the Russell cubic (more precisely, this modification is isomorphic to the algebraic hypersurface in $\mathbb{C}^6$ given by the equation $x + x^2 y + s^2 + t^3 = uv$). It is again diffeomorphic to $\mathbb{R}^{10}$ and has the density property but it is even unknown whether it is different from $\mathbb{C}^5$ as an algebraic variety.

2.24. Similar reasoning as before leads from families of holomorphic embeddings like in 2.18 to families of pairwise nonequivalent $\mathbb{C}^*$-actions. For example, there is a family $\mathbb{C}_w \times \mathbb{C}^* \to \mathrm{Aut}_{\mathrm{hol}}(\mathbb{C}^5)$ of holomorphic $\mathbb{C}^*$-actions on $\mathbb{C}^5$ parametrized by $w \in \mathbb{C}$ such that for different parameters the actions are nonequivalent (i.e., they are not conjugated by an automorphism). Moreover there is a family such that the value of parameter $w = 0$ represents a linear action. It follows also from 2.18 that on $\mathbb{C}^4$ there are uncountably many nonequivalent holomorphic $\mathbb{C}^*$-actions.

2.25. The question whether two embeddings of an affine algebraic (resp. complex) manifold into $\mathbb{C}^n$ are equivalent can be reformulated as a question about extension of an isomorphism of two subvarieties of $\mathbb{C}^n$ to an algebraic (resp. holomorphic) automorphism of the ambient Euclidean space. If Z is a smooth closed algebraic subvariety of $\mathbb{C}^n$ such that $n > 2 \dim Z + 1$ then it is known [50] (see also

[8]Due to abundance of potential counterexamples the authors believe that the answer to the Zariski problem should be negative. Slightly stronger conjectures are usually proven wrong. Say, in their recent preprint Dubouloz, Moser-Jauslin, and Poloni [26] constructed pairwise nonbiholomorphic contractible algebraic manifolds whose direct products with $\mathbb{C}$ are biholomorphic. They gave also examples of contractible affine algebraic manifolds that are biholomorphic but not isomorphic.

[78])[9] that any automorphism of Z extends to an automorphism of $\mathbb{C}^n$. In the holomorphic category the situation is completely different and exploiting Theorem 2, Derksen, Winkelmann, and the second author showed in [22] that there is a proper holomorphic embedding of $\varphi\colon \mathbb{C} \hookrightarrow \mathbb{C}^n$ with any $n \geq 2$ (thus the codimension is arbitrarily big) such that the only automorphism of $\mathbb{C}^n$ mapping the image onto itself is the identical one.

2.26. The classical theorem of Remmert [70] states that any Stein manifold Y of dimension n admits a proper holomorphic embedding into a Euclidean space $\mathbb{C}^N$ of sufficiently high dimension N. The optimal N was found in the papers of Gromov, Eliashberg [29] and Schürmann [74]; in the case of $n \geq 2$ they proved that Y can be embedded into $\mathbb{C}^{[3n/2]+1}$. This result is sharp by virtue of examples of O. Forster [31], who conjectured that the optimal N was $[3n/2]+1$. The case of $n = 1$ is still open, i.e., it is unknown whether any open Riemann surface admits a proper holomorphic embedding into $\mathbb{C}^2$. However, there was a recent breakthrough in this direction — Wold proved the Forster conjecture for all finitely connected domains in $\mathbb{C}$ and for all elliptic curves with finitely many holes (that are not punctures) using the Andersén–Lempert theory in a very clever way [89–91]. Furthermore, Løw, Wold, and the second author [65] showed that one can require additionally an interpolation condition on discrete subsets as in 2.14. Some additional technical development concerning the Forster conjecture can be found in [39, 67].

2.27. A beautiful combination of differential-topological methods with hard analysis (solutions of $\bar{\partial}$-equations with exact estimates) and the Andersén–Lempert theory is required for understanding of how many totally real differentiable embeddings of a real manifold M into $\mathbb{C}^n$ can exist.

If $f_0, f_1\colon M \to \mathbf{C}^n$ are two totally real, polynomially convex real-analytic embeddings of a compact manifold M into $\mathbb{C}^n$, we say that f_0 and f_1 are $\mathrm{Aut}_{\mathrm{hol}}(\mathbb{C}^n)$-equivalent[10] if $f_1 = F \circ f_0$, where $F\colon U \to F(U) \subset \mathbf{C}^n$ is a biholomorphism defined in a neighbourhood U of $f_0(M)$ such that F is the uniform limit in U of a sequence of elements of $\mathrm{Aut}_{\mathrm{hol}}(\mathbb{C}^n)$. Conditions for $\mathrm{Aut}_{\mathrm{hol}}(\mathbb{C}^n)$-equivalence were found in [38], using volume-preserving automorphisms (and an approach using automorphisms preserving the holomorphic symplectic form was considered in [32]).

In the smooth case let $\mathcal{E}^r(M, \mathbf{C}^n)$ be the set of all totally real polynomially convex C^r-embeddings of M into $\mathbb{C}^n$ (for $2 \leq r \leq \infty$). It is proved by Forstnerič and Løw that two embeddings $f_0, f_1 \in \mathcal{E}^\infty(M, \mathbb{C}^n)$ belong to the same connected component (in the space of C^r-embeddings of M into $\mathbb{C}^n$ equipped with the usual topology of uniform convergence of all derivative up to order r) if and only if there exists a sequence $\{\Phi_j\} \subset \mathrm{Aut}_{\mathrm{hol}}(\mathbb{C}^n)$ such that $\Phi_j \circ f_0 \to f_1$ and $\Phi_j^{-1} \circ f_1 \to f_0$ in $C^\infty(M)$ as $j \to \infty$. Precise results in the case $r < \infty$ were obtained in [37].

2.28. Recall that a "long $\mathbb{C}^n$" is a complex manifold X that can be exhausted by open subsets Ω_i which are all biholomorphic to $\mathbb{C}^n$, i.e., $X = \bigcup_{i=1}^\infty \Omega_i$, $\Omega_i \subset \Omega_{i+1}$, and $\Omega_i \cong \mathbb{C}^n$ for all $i \in \mathbb{N}$. (Here, of course, $\Omega_i \subset \Omega_{i+1}$ is not necessarily a Runge

[9]Actually, the theorem proven in [50] and [78] is more general and it includes the singular case. According to one of referees of the first author the smooth case above was also known to Nori. It is worth also mentioning the earlier paper of Jelonek [49] where the similar result was obtained for $n > 4 \dim Z + 1$.

[10]It is unfortunate that in the literature the term "$\mathrm{Aut}_{\mathrm{hol}}(\mathbb{C}^n)$-equivalence" is used in different meanings - another one was mentioned in Remark 2.17.

pair.) The fist examples of such manifolds not biholomorphic to $\mathbb{C}^n$ for $n \geq 3$ were constructed by Fornæss [30] in 1976. The case of $n = 2$ had been resistent until recently when, developing the ideas from 2.10 further, Wold [93] constructed a "long $\mathbb{C}^2$" which is not biholomorphic to $\mathbb{C}^2$. Every known "long $\mathbb{C}^n$" which is not biholomorphic to $\mathbb{C}^n$ (including the case of $n = 2$) is non-Stein.

2.29. The classical approximation theorem of Carleman states that for each continuous function $\lambda\colon \mathbb{R} \to \mathbb{C}$ and a positive continuous function $\epsilon\colon \mathbb{R} \to (0, \infty)$ there exists an entire function f on $\mathbb{C}$ such that $|f(t) - \lambda(t)| < \epsilon(t)$ for every $t \in \mathbb{R}$.

Using the Andersén – Lempert theory together with some explicit shears (which are special automorphisms of $\mathbb{C}^n$ which appear in that theory) Buzzard and Forstnerič [17] were able to prove a similar result for holomorphic embeddings into $\mathbb{C}^n$. Namely, for any proper embedding $\lambda\colon \mathbb{R} \to \mathbb{C}^n$ of class C^r (where $n \geq 2$ and $r \geq 0$) and a positive continuous function $\epsilon\colon \mathbb{R} \to (0, \infty)$ there exists a proper holomorphic embedding $f\colon \mathbb{C} \to \mathbb{C}^n$ such that

$$|f^{(s)}(t) - \lambda^{(s)}(t)| < \epsilon(t) \quad \forall\, t \in \mathbb{R}, 0 \leq s \leq r.$$

Actually this fact remains valid under the additional requirement that the embedding satisfies the interpolation property as in 2.14.

3. Criterion for the algebraic density property

Tóth and Varolin established the algebraic density property for some manifolds including semisimple complex Lie groups [82, 83]. Their proof follows to a great extend the original ideas of Andersén and Lempert and is quite complicated. A new approach suggested by the authors [56] lead to the following.

Theorem 6. *Each linear algebraic group whose connected component is different from $\mathbb{C}_+$ or $(\mathbb{C}^*)^k$, $k \geq 1$ has the algebraic density property.*

The proof was based on the following simple fact.

Theorem 7. *Let X be an affine algebraic manifold with a transitive group $\operatorname{Aut} X$ of algebraic automorphisms and let $\mathbb{C}[X]$ be its algebra of regular functions. Suppose that there is a submodule L of the $\mathbb{C}[X]$-module T of all algebraic vector fields on X such that $L \subset \mathrm{Lie}_{\mathrm{alg}}(X)$ and the fiber of L at some point $x_0 \in X$ contains a generating subset*[11] *of $T_{x_0}X$. Then X has the algebraic density property.*

Proof. The action of $\alpha \in \operatorname{Aut} X$ maps L onto another $\mathbb{C}[X]$-module L_α. The sum of such modules with α running over a subset of $\operatorname{Aut} X$ is again a $\mathbb{C}[X]$-submodule N of T. Let $\mathcal{M}_x \subset \mathbb{C}[X]$ be the maximal ideal that consists of functions vanishing at $x \in X$. By assumption N can be chosen so that $N/\mathcal{M}_{x_0}T$ coincides with $T_{x_0}X = T/\mathcal{M}_{x_0}T$. Furthermore, since X is homogeneous with respect to $\operatorname{Aut} X$ we can suppose that the same is true for every point in X. That is, for the $\mathbb{C}[X]$-module $Q = T/N$ and every $x \in X$ we have $Q/\mathcal{M}_x Q = 0$. Thus $Q = 0$ and $N = T$ (e.g., see [45, Exercise II.5.8]). Since composition with automorphisms preserves complete integrability, all element of N are in $\mathrm{Lie}_{\mathrm{alg}}(X)$ which implies the desired conclusion. □

[11] Our notion of a generating subset is milder than usual. A finite subset F of $T_{x_0}X$ is called a generating subset if the span of the orbit of F under the action of the isotropy group $(\operatorname{Aut} X)_{x_0}$ coincides with $T_{x_0}X$ (say, if X is a simple Lie group then every nonzero vector is a generating set).

Thus the idea of the proof of Theorem 6 is to catch a nontrivial $\mathbb{C}[X]$-module in the $\mathrm{Lie_{alg}}(X)$. In order to demonstrate how to do it we prove the main observation of the Andersén-Lempert theory.

3.1. Proposition. *For $n \geq 2$ the space $\mathbb{C}^n$ has the algebraic density property.*

PROOF. Let $x_1, \dots, x_n$ be a coordinate system on $X = \mathbb{C}^n$ and $\delta_i = \partial/\partial x_i$, i.e., $\operatorname{Ker}\delta_i = \mathbb{C}[x_1, \dots, \hat{x}_i, \dots, x_n]$ and therefore

$$\mathbb{C}^{[n]} = \operatorname{Span}\operatorname{Ker}\delta_1 \cdot \operatorname{Ker}\delta_2.$$

Note also that for $f_i \in \operatorname{Ker}\delta_i$ the algebraic vector fields $f_i\delta_i$ and $x_i f_i \delta_i$ are completely integrable on $\mathbb{C}^n$. This implies that the field

$$[f_1\delta_1, x_1 f_2 \delta_2] - [x_1 f_1 \delta_1, f_2\delta_2] = f_1 f_2 \delta_2$$

belongs to $\mathrm{Lie_{alg}}(X)$. Thus $\mathrm{Lie_{alg}}(X)$ contains all algebraic fields proportional to δ_2 which in combination with Theorem 7 implies the desired conclusion. □

To transfer this argument to other affine algebraic manifolds we have to use locally nilpotent and semisimple derivations instead of partial derivatives. Recall that an algebraic vector field σ on X is locally nilpotent (LND) if its flow is an algebraic $\mathbb{C}_+$-action. Equivalently, σ is a LND if for every $a \in \mathbb{C}[X]$ there exists natural n for which $\sigma^n(a) = 0$. The last algebraic definition enables us to introduce the degree of any regular element $a \in \mathbb{C}[X]$ with respect to σ as $\deg_\sigma(a) = \min\{n - 1 \mid \sigma^n(a) = 0\}$.

An algebraic vector field on X is semisimple if its flow is an algebraic $\mathbb{C}^*$-action.

3.2. Definition. Let σ be a LND on X and δ be either a LND or semisimple. Then pair (σ, δ) is called semicompatible if the the span of $\operatorname{Ker}\sigma \cdot \operatorname{Ker}\delta$ contains a nonzero ideal of $\mathbb{C}[X]$.

A semicompatible pair (σ, δ) is called compatible if one of the following conditions holds:

(1) there exists $a \in \mathbb{C}[X]$ such that $a \in \operatorname{Ker}\delta$ and $\sigma(a) \in \operatorname{Ker}\sigma \setminus 0$, i.e., $\deg_\sigma(a) = 1$.
(2) both σ and δ are LNDs and there exists $a \in \mathbb{C}[X]$ such that $\deg_\sigma(a) = 1 = \deg_\delta(a)$.

Repeating the argument from Proposition 3.1 with σ and δ instead of δ_1 and δ_2 we get the following.

3.3. Proposition. *The existence of a compatible pair yields the existence of a nontrivial $\mathbb{C}[X]$-module in* $\mathrm{Lie_{alg}}(X)$.

Now we can formulate our criterion as the following.

Theorem 8. *Let X be a smooth affine algebraic variety with a transitive automorphism group* $\operatorname{Aut} X$. *Suppose that there are finitely many pairs of compatible vector fields $\{\sigma_i, \delta_i\}$ such that at some point $x_0 \in X$ vectors $\{\delta_i(x_0)\}$ form a generating subset of $T_{x_0}X$. Then X has the algebraic density property.*

3.4. Example. (1) As an obvious application of this theorem we see that the manifold $X = \mathbb{C}^k \times (\mathbb{C}^*)^l$ with $k \geq 1$ and $k + l \geq 2$ has algebraic density property.

(2) A more interesting case is when $X = \mathrm{SL}_2$ (or PSL_2). Denote by

$$A = \begin{pmatrix} a_1 & a_2 \\ b_1 & b_2 \end{pmatrix}$$

an element of X. Consider the following LNDs on $\mathbb{C}[X]$:

$$\delta_1 = a_1 \frac{\partial}{\partial b_1} + a_2 \frac{\partial}{\partial b_2}$$

$$\delta_2 = b_1 \frac{\partial}{\partial a_1} + b_2 \frac{\partial}{\partial a_2}.$$

Then $\operatorname{Ker} \delta_1 = \mathbb{C}[a_1, a_2]$ and $\operatorname{Ker} \delta_2 = \mathbb{C}[b_1, b_2]$, i.e., $\operatorname{Span} \operatorname{Ker} \delta_1 \cdot \operatorname{Ker} \delta_2 = \mathbb{C}[X]$ and the pair (δ_1, δ_2) is semicompatible.

Furthermore, $\deg_{\delta_1}(a) = \deg_{\delta_2}(a) = 1$ for $a = a_1 b_2$ and therefore the pair (δ_1, δ_2) is compatible. Hence SL_2 and PSL_2 have the algebraic density property. It can be shown in the same manner that SL_n and PSL_n have also algebraic density property.

3.5. Definition. Let us choose an identification of elements of SL_2 with (2×2)-matrices with determinant 1. Suppose that $H_1 \simeq \mathbb{C}_+$ (resp. $H_2 \simeq \mathbb{C}_+$) is the unipotent upper (resp. lower) triangular subgroup of SL_2. Then any SL_2-action on X generates H_i-action on X and, therefore, a locally nilpotent vector field δ_i. The pair (δ_1, δ_2) will be called an associated pair of LNDs of the SL_2-action.

Theorem 9. *Let X be a smooth complex affine algebraic variety whose group of algebraic automorphisms is transitive.*

(1) *Suppose that X is equipped with a nondegenerate*[12] *fixed point free SL_2-action. Then the associated pair (δ_1, δ_2) is compatible.*
(2) *Suppose that X is equipped with N nondegenerate fixed point free SL_2-actions. Let $\{\delta_1^k, \delta_2^k\}_{k=1}^N$ be the corresponding pairs of associated locally nilpotent vector fields.*

If $\{\delta_2^k(x_0)\}_{k=1}^N \subset T_{x_0} X$ is a generating set at some point $x_0 \in X$ then X has the algebraic density property.

The second statement is, of course, a consequence of (1) and Theorem 8. The idea of the proof of the first statement can be described as follows. By Luna's slice theorem (e.g., see [25]) any closed SL_2-orbit O possesses an SL_2-invariant neighborhood U' for which there exists a surjective étale SL_2-equivariant morphism $U'' \to U'$ such that U'' is naturally isomorphic to $\mathrm{SL}_2 \times_{I_x} V$[13] where $I_x \subset SL_2$ is the isotropy group of some point $x \in O$ and V is an I_x-invariant subvariety of X (called a slice). Then the associated locally nilpotent derivations δ_1 and δ_2 generate naturally similar derivations δ_1' and δ_2' (resp. δ_1'' and δ_2'') on U' (resp. U''). The straightforward application of Nullstellensatz shows that δ_1 and δ_2 are compatible provided that for every closed orbit O the pair (δ_1', δ_2') described before is compatible. It was shown in [56] that compatibility of δ_1' and δ_2' follows from compatibility of δ_1'' and δ_2''. The proof of compatibility of the associated locally nilpotent derivations for the natural SL_2-action on $\mathrm{SL}_2 \times_{I_x} V$ (especially in the

---

[12]That is, the dimension of general orbits is 3.

[13]That is, elements of U'' are equivalence classes in $\mathrm{SL}_2 \times V$ given by the relation $(s, v) \sim (sg^{-1}, g \cdot v)$ for $g \in I_x$.

case when I_x contains $\mathbb{C}^*$) is the most difficult part which does not work for SL_2-actions with fixed points. Actually, if a nondegenerate SL_2-action has a fixed point the pair (δ_1', δ_2') is not compatible.

As a (nonstraightforward) application of Theorem 9 we have.

Theorem 10. *Let G be a linear algebraic group and R be its proper reductive subgroup.*

Suppose that the connected components of $X = G/R$ are different from $\mathbb{C}_+$ or $(\mathbb{C}^*)^k$.

Then X has the algebraic density property.

The difficulty in the proof of the last result lies in checking the assumption of Theorem 9. That is, one needs to find an SL_2-subgroup Γ of G such that its natural action on X is fixed point free and nondegenerate. This is equivalent to the next fact.

3.6. Proposition. *Let G and R be as before. Then there exists an SL_2-subgroup Γ of G such that*

(1) *$g\Gamma g^{-1}$ is not contained in R for any $g \in G$ and*
(2) *$g_0\Gamma g_0^{-1}$ meets R at a finite set for some $g_0 \in G$.*

Furthermore Γ can be chosen as a principal or subregular SL_2-subgroup of G.

The proof of Proposition 3.6 is surprisingly nontrivial. First using the Jackobson – Morozov theorem (e.g., see [11]) we can switch from studying the conjugate classes of SL_2-subgroups in G to the study of orbits of nilpotent elements in the Lie algebra of G under the adjoint action. The orbit of the largest dimension is called principal and the second largest dimension subregular (they are both unique). To show that an SL_2-subgroup corresponding to one of these orbits satisfy conditions (1) and (2) one needs to use the Dynkin classification of nilpotent orbits (e.g., see [8]) in combination with some difficult facts about the Freudental square [66].

Besides Theorem 9(1) we have another useful criterion for compatibility [56, Proposition 3.9].

3.7. Proposition. *Suppose that either $\Gamma = \mathrm{SL}_2$ and H_1, H_2 are from Definition 3.5 or $\Gamma = H_1 \times H_2$ where $H_1 \simeq \mathbb{C}_+$ and H_2 is one of the groups $\mathbb{C}_+$ or $\mathbb{C}^*$. Let X be a smooth affine algebraic Γ-variety and Y be a normal affine algebraic variety equipped with a trivial Γ-action. Let $r\colon X \to Y$ be a surjective Γ-equivariant morphism and δ_1, δ_2 be the algebraic fields on X generated by the action of H_1 and H_2. Suppose that for every point $y \in Y$ there exists an étale neighborhood $W \to Y$ such that the pair of vector fields induced by δ_1 and δ_2 on the fibered product $X \times_Y W$ is compatible. Then the pair (δ_1, δ_2) is compatible.*

3.8. Non-Stein case: Complement to a subvariety of codimension at least 2. Let Y be an algebraic subvariety of an affine algebraic manifold X such that $\mathrm{codim}_X Y \geq 2$. Then according to Forstnerič $X \setminus Y$ has the algebraic density property if the Lie algebra $\mathrm{Lie}_{\mathrm{alg}}(X, Y)$ generated by completely integrable algebraic vector fields on X that vanishes on Y coincides with the Lie algebra $\mathrm{VF}_{\mathrm{alg}}(X, Y)$ of all algebraic vector fields on X that vanishes on Y.

3.9. Remark. By Hartogs' theorem every algebraic vector field on $X \setminus Y$ can be extended to X. In the case of completely integrable fields this extension must be tangent to Y. In particular, $\mathrm{Lie}_{\mathrm{alg}}(X, Y)$ cannot contain algebraic fields with extension nontangent to Y. Furthermore, in the case of Y without the algebraic density property (say, when Y is singular) $\mathrm{Lie}_{\mathrm{alg}}(X, Y)$ cannot contain all vector fields tangent to Y. Thus the Forstnerič definition is the only reasonable.

Theorem 11 ([56]). *Let Y be a subvariety of $\mathbb{C}^n$ of codimension ≥ 2. Then*

(1) *$\mathbb{C}^n \setminus Y$ has the algebraic density property if $T_y Y \leq n-1$ for every $y \in Y$;*
(2) *in any case there exists natural k such that $\mathrm{Lie}_{\mathrm{alg}}(\mathbb{C}^n, Y)$ contains all algebraic vector fields that vanish on Y with multiplicity at least k.*

It can be shown that even under condition (2) the approximation theorems we had in the affine case are valid for $\mathbb{C}^n \setminus Y$.

4. Volume density property

As we mentioned before, the algebraic volume density property (see Definition 1.2) is a much more delicate thing to prove than the algebraic density property. It is hopeless to look for a $\mathbb{C}[X]$-module inside the Lie algebra $\mathrm{Lie}^{\omega}_{\mathrm{alg}}(X)$ generated by completely integrable algebraic vector fields of divergence zero with respect to an algebraic volume form ω on X. Indeed, let ν be a vector field on X of ω-divergence zero and $f \in \mathbb{C}[X]$. Then $\mathrm{div}_{\omega}(f\nu) = \nu(f)$ is nonzero for general f. However, some important facts remain valid in this volume-preserving case [57].

Theorem 12. *Let X be a smooth hypersurface in $\mathbb{C}^{n+2}_{u,v,\bar{x}}$ given by an equation $uv = p(\bar{x})$. Suppose that $Z \subset \mathbb{C}^n_{\bar{x}}$ is given by $p(\bar{x}) = 0$ and $H^{n-2}(Z, \mathbb{C}) = 0$. Then X has the algebraic volume density property.*

Note that unlike in Theorem 5 we have an extra assumption on the zero locus Z; namely $H^{n-2}(Z, \mathbb{C}) = 0$. It is unknown whether this assumption is essential. However this is enough to show that SL_2 (given by equation $uv = xy + 1$ in $\mathbb{C}^4$) and PSL_2 have the algebraic volume density property with respect to the invariant volume form which is important in the proof of the next result [57].

Theorem 13. *Every linear algebraic group has the algebraic volume density property with respect to the left-invariant (or right-invariant) volume form.*

The proof uses also the second statement of the following theorem which is not at all obvious in the volume-preserving case.

Theorem 14. (1) *For an affine algebraic manifold X equipped with an algebraic volume form ω the algebraic volume density property implies the volume density property (in the holomorphic sense).*

(2) *If affine algebraic manifolds X and Y have the algebraic volume density property with respect to volume forms ω_X and ω_Y, then so does $X \times Y$ with respect to the volume form $\omega_X \times \omega_Y$.*

Sketch of the proof of (1). Let μ be a holomorphic vector field such that $\mu(\omega) = 0$. Our aim is to find an algebraic vector field ν with $\nu(\omega) = 0$ that approximates μ.

In the first step one can show that there is a natural duality between the holomorphic (resp. algebraic) vector fields of ω-divergence zero and the closed $(n-1)$-forms on X given by $\mu \to \iota_{\mu}(\omega)$ where ι_{μ} is the inner product (see [57, Lemmas 3.5]).

Next one needs to use the remarkable Grothendieck's theorem [44] that states that the computation of de Rham cohomology on a smooth affine algebraic variety can be achieved via the complex of algebraic forms. Thus there exists an algebraic $(n-1)$-form τ_{n-1} such that $\iota_\mu(\omega) - \tau_{n-1}$ is exact and equal to $\mathrm{d}\tau_{n-2}$ where τ_{n-2} is a holomorphic $(n-2)$-form. Approximate τ_{n-2} by an algebraic form τ'_{n-2} and set ν equal to the dual of the form $\tau_{n-1} + \mathrm{d}\tau'_{n-2}$. □

Instead of proving (2) we present a more complicated proof of its analogue in the holomorphic case mentioned in the introduction.

Theorem 15. *If Stein manifolds X and Y have the volume density property with respect to forms ω_X and ω_Y, so does $X \times Y$ with respect to the form $\omega_X \times \omega_Y$.*

In preparation for the proof of Theorem 15 we must establish a number of facts. Recall that for a complex manifold X and a holomorphic volume form ω on it the space of holomorphic vectors fields of ω-divergence zero is denoted by $\mathrm{VF}^\omega_{\mathrm{hol}}(X)$ and the Lie algebra generated by the set $\mathrm{IVF}^\omega_{\mathrm{hol}}(X)$ of completely integrable holomorphic vector fields from $\mathrm{VF}^\omega_{\mathrm{hol}}(X)$ is denoted by $\mathrm{Lie}^\omega_{\mathrm{hol}}(X)$. First we shall prove the holomorphic analogue of Lemma 4.2 from [57].

4.1. Lemma. *Let ω be a holomorphic form on a Stein manifold X such that X has the volume density property with respect to ω. Then there exist finitely many vector fields $\delta_1, \delta_2, \ldots, \delta_N \in \mathrm{IVF}^\omega_{\mathrm{hol}}(X)$ such that* $\mathrm{Span}\{\delta_i(x) \mid i = 1, \ldots, N\} = T_xX$ *for every point $x \in X$.*

PROOF. We start with the following.

Claim. *The set $\mathrm{VF}^\omega_{\mathrm{hol}}(X)$ generate T_xX for every $x \in X$.*

Lets us assume first that the dimension of X is bigger than one. Let $x \in X$ and U be a Runge neighborhood of x such that $H^{n-1}(U, \mathbb{C}) = 0$ where $n = \dim X$ (take, for instance, a small sublevel set U of a strictly plurisubharmonic exhaustion function on X with minimum at x). Shrinking U we can assume that in some holomorphic coordinate system $z_1, \ldots, z_n$ on U the form $\omega|_U$ is the standard volume $\mathrm{d}z_1 \wedge \cdots \wedge \mathrm{d}z_n$. Thus the holomorphic vector fields $\partial/\partial z_i$ on U are of divergence zero and they span the tangent space at x. We need to approximate them by global holomorphic fields of divergence zero on X which would yield our claim. As in Theorem 14 consider the inner product $\iota_\nu(\omega) =: \alpha$ for $\nu \in \mathrm{VF}^\omega_{\mathrm{hol}}(U)$. By [57, Lemma 3.5(1)] α is a closed $(n-1)$-form on U and since $H^{n-1}(U, \mathbb{C}) = 0$ we can find an $(n-2)$-form β on U with $\mathrm{d}\beta = \alpha$. Since U is Runge in X we can also approximate β by a global holomorphic $(n-2)$-form $\tilde\beta$ (uniformly on compacts in U). Then the closed holomorphic $(n-1)$-form $\mathrm{d}\tilde\beta$ approximates α and the unique holomorphic vector field θ defined by $\iota_\theta(\omega) = \mathrm{d}\tilde\beta$ approximates ν. Since $\mathrm{d}\tilde\beta$ is closed, the field θ is of divergence zero which concludes the proof of the claim in the case when the dimension of X is bigger than one.

Now assume the dimension of X is equal to one, i.e., X is a smooth Stein curve with the volume density property. In particular, it admits a nontrivial completely integrable holomorphic vector field and, therefore, a nonconstant holomorphic map $\mathbb{C} \to X$. This implies that X is isomorphic to either $\mathbb{C}_+$ or $\mathbb{C}^*$. In each of these cases the volume form ω for which X has the volume density property is the unique (up to a constant factor) invariant volume form on the corresponding group and there exists a nowhere vanishing completely integrable vector field of ω-divergence zero which concludes the proof of the claim.

It follows from the claim and the volume density property that vector fields from $\mathrm{Lie}^\omega_{\mathrm{hol}}(X)$ span the tangent space T_xX at any given point $x \in X$. Observe that every Lie bracket $[\nu, \mu]$ of completely integrable holomorphic vector fields of divergence zero can be approximated by a linear combination of such fields which follows immediately from the equality $[\nu, \mu] = \lim_{t\to 0}(\phi_t^*(\nu) - \nu)/t$ where ϕ_t is the flow generated by μ. Thus the set $\mathrm{IVF}_{\mathrm{hol}}^{\omega}(X)$ generates T_xX at any $x \in X$.

To prove that there are finitely many fields from $\mathrm{IVF}_{\mathrm{hol}}^{\omega}(X)$ that span each tangent space let us start with n fields $\theta_1, \dots, \theta_n$ which span the tangent space at some point x_0 and thus outside a proper analytic subset A. The set A may have countably many irreducible components $A_1, A_2, A_3, \dots$.

It suffices now to find a volume preserving holomorphic automorphism $\Phi \in \mathrm{Aut}_{\mathrm{hol}}^{\omega}(X)$ such that $\Phi(X \setminus A) \cap A_i \neq \varnothing$ for every $i = 1, 2, 3, \dots$. Indeed, for such an automorphism Φ the completely integrable holomorphic vector fields $\Phi_*(\theta_1), \dots,$ $\Phi_*(\theta_n)$ have divergence zero and span the tangent space at a general point in each A_i, i.e., together with the fields $\theta_1, \dots, \theta_n$ they span the tangent space at each point outside an analytic subset B of a smaller dimension than A. Then the induction by dimension implies the desired conclusion.

In order to construct Φ consider a monotonically increasing sequence of compacts $K_1 \subset K_2 \subset \cdots$ in X such that $\bigcup_i K_i = X$ and a closed imbedding $\iota\colon X \hookrightarrow \mathbb{C}^m$. For every continuous map $\varphi\colon X \to \mathbb{C}^m$ denote by $\|\varphi\|_i$ the standard norm of the restriction of φ to K_i. Let d be the metric on the space $\mathrm{Aut}_{\mathrm{hol}}(X)$ of holomorphic automorphisms of X given by the formula

$$d(\Phi, \Psi) = \sum_{i=1}^{\infty} 2^{-i}(\min(\|\Phi - \Psi\|_i, 1) + \min(\|\Phi^{-1} - \Psi^{-1}\|_i, 1) \tag{4.1}$$

where automorphisms $\Phi^{\pm 1}, \Psi^{\pm 1} \in \mathrm{Aut}_{\mathrm{hol}}(X)$ are viewed as continuous maps from X to $\mathbb{C}^m$. This metric makes $\mathrm{Aut}_{\mathrm{hol}}(X)$ a complete metric space. Its subset $\mathrm{Aut}_{\mathrm{hol}}^{\omega}(X)$ of volume-preserving automorphisms is closed and, therefore, it is a complete metric space as well.

Set $Z_i = \{\Psi \in \mathrm{Aut}_{\mathrm{hol}}^{\omega}(X)\colon \Psi(A_i) \cap (X \setminus A) \neq \varnothing\}$. Note that Z_i is open in $\mathrm{Aut}_{\mathrm{hol}}^{\omega}(X)$ and let us show that it is also everywhere dense. Indeed, the flow of any $\theta \in \mathrm{IVF}_{\mathrm{hol}}^{\omega}(X)$ preserves ω because for any vector fields ν its divergence with respect to ω is defined by the formula $\mathrm{div}_{\omega}(\nu)\omega = L_{\nu}(\omega)$ where L_{ν} is the Lie derivative. Since $\mathrm{IVF}_{\mathrm{hol}}^{\omega}(X)$ generates the tangent space at each point of X we can choose θ nontangent to A_i. Then for every $\Psi \in \mathrm{Aut}_{\mathrm{hol}}^{\omega}(X)$ its composition with general elements of the flow induced by θ is in Z_i. That is, a perturbation of Ψ belongs to Z_i which proves that Z_i is everywhere dense in $\mathrm{Aut}_{\mathrm{hol}}^{\omega}(X)$. By the Baire category theorem the set $\bigcap_{i=1}^{\infty} Z_i$ is not empty which yields the existence of the desired automorphism. □

4.2. Lemma. *For a holomorphic field ν on X denote by ω_{ν} the inner product $\iota_{\nu}(\omega)$ of ν with the form ω. If ν is of ω-divergence zero then for every holomorphic function f on X we have* $\mathrm{d}(\omega_{f\nu}) = \nu(f)\omega$.

PROOF. Using the well-known formula $L_{\nu} = \mathrm{d} \circ \iota_{\nu} + \iota_{\nu} \circ \mathrm{d}$ (e.g., see [60, Proposition 3.10]) that relates the outer differentiation d with the Lie derivative L_{ν} and inner product ι_{ν} we get

$$\begin{aligned}\mathrm{d}(\omega_{f\nu}) = L_{f\nu}(\omega) - \iota_{f\nu}\,\mathrm{d}(\omega) = L_{f\nu}(\omega) &= \mathrm{div}_{\omega}(f\nu)\omega \\ &= \big(f\,\mathrm{div}_{\omega}(\nu) + \nu(f)\big)\omega = \nu(f)\omega. \quad \square\end{aligned}$$

4.3. Notation. For a complex manifold X with a volume form ω we define the following linear subspace F_X in the space of holomorphic functions $\mathrm{Hol}(X) = H^0(X, \mathcal{O}(X))$ on X:

$$F_X = \mathrm{Span}\{\theta(\mathrm{Hol}(X)) : \theta \in \mathrm{VF}^{\omega}_{\mathrm{hol}}(X)\}.$$

4.4. Lemma. *Let Ω^i be the sheaf of holomorphic i-forms on a complex manifold X and let $\mathcal{B}^i \subset H^0(X, \Omega^i)$ be the space of exact i-forms. For $n = \dim X$ consider the isomorphism $\epsilon\colon \mathrm{Hol}(X) \to H^0(X, \Omega^n)$ into the space $H^0(X, \Omega^n)$ of holomorphic n-forms given by $\epsilon(f) = f\omega$ and suppose that*

(i) *the $\mathrm{Hol}(X)$-module $\mathrm{VF}_{\mathrm{hol}}(X)$ of holomorphic vector fields on X is generated by $\mathrm{VF}^{\omega}_{\mathrm{hol}}(X)$.*

Then $F_X = \epsilon^{-1}(\mathcal{B}^n)$.
Moreover, suppose that X is Stein and instead of condition (i) *we have*

(ii) *finitely many vector fields $\delta_1, \delta_2, \dots, \delta_N \in \mathrm{VF}^{\omega}_{\mathrm{hol}}(X)$ generating T_xX at each $x \in X$.*

Then $F_X = \mathrm{Span}\{\delta_i(\mathrm{Hol}(X)) \mid i = 1, \dots, N\}$.

Proof. Since ω is nowhere vanishing we have an isomorphism between the spaces of holomorphic vector fields and holomorphic $(n-1)$-forms given by $\nu \to \iota_\nu(\omega)$, i.e., any $(n-1)$-form can be presented as ω_ν for some $\nu \in \mathrm{VF}_{\mathrm{hol}}(X)$. Furthermore, condition (i) implies that the vector space $H^0(X, \Omega^{n-1})$ is generated by elements of type $\omega_{f\delta}$ where $\delta \in \mathrm{VF}^{\omega}_{\mathrm{hol}}(X)$ and $f \in \mathrm{Hol}(X)$. By Lemma 4.2 we have

$$F_X = \mathrm{Span}\{\epsilon^{-1}(\mathrm{d}(\omega_{f\delta})) : f \in \mathrm{Hol}(X), \delta \in \mathrm{VF}^{\omega}_{\mathrm{hol}}(X)\}$$

which in combination with the above description of generators of $H^{n-1}(X, \Omega^{n-1})$ yields the first statement.

To prove the second assertion note that the $\mathcal{O}(X)$-module homomorphism of sheafs

$$\mathcal{O}(X)^N \to TX, \quad (f_1, f_2, \dots, f_N) \mapsto \sum_{i=1}^{N} f_i \delta_i$$

and, therefore (because of the isomorphism $\nu \to \iota_\nu(\omega)$) the homomorphism

$$\mathcal{O}(X)^N \to \Omega^{n-1}, \quad (f_1, f_2, \dots, f_N) \mapsto \omega_{\sum_{i=1}^{N} f_i \delta_i}$$

are surjective. Since X is Stein, the Cartan Theorem B implies the surjectivity on the level of global sections. Together with Lemma 4.2 this implies the second assertion. □

In combination with Lemma 4.1 this implies the following.

4.5. Corollary. *For a Stein manifold X with the volume density property we have finitely many vector fields $\delta_1, \delta_2, \dots, \delta_N \in \mathrm{IVF}^{\omega}_{\mathrm{hol}}(X)$ for which*

$$F_X = \mathrm{Span}\{\delta_i(\mathrm{Hol}(X)) \mid i = 1, \dots, N\}.$$

4.6. Lemma. *If X is Stein, then F_X is a closed linear subspace of $\mathrm{Hol}(X)$ and thus a Fréchet space. Moreover, $\mathrm{Hol}(X)/F_X$ is isomorphic to $H^n(X, \mathbb{C})$.*

PROOF. Since X is Stein the inclusion of the holomorphic de Rham complex into the complex-valued de Rham C^∞-complex

$$\begin{array}{ccccccc}
\cdots \xrightarrow{\mathrm{d}} & H^0(X, \mathcal{E}^{n-1}) & \xrightarrow{\mathrm{d}} & H^0(X, \mathcal{E}^n) & \xrightarrow{\mathrm{d}} 0 \\
& \cup & & \cup & \\
\cdots \xrightarrow{\mathrm{d}} & H^0(X, \Omega^{n-1}) & \xrightarrow{\mathrm{d}} & H^0(X, \Omega^n) & \xrightarrow{\mathrm{d}} 0
\end{array}$$

is a quasi-isomorphism, i.e., the complex cohomology of X can be computed via the lower complex. By de Rham's theorem the boundaries $\mathrm{d}(H^0(X, \mathcal{E}^{n-1}))$ are exactly the n-forms whose integral over any real n-dimensional cycle is zero (e.g., see [42]) and, therefore, $\mathrm{d}(H^0(X, \Omega^{n-1}))$ consists exactly of the holomorphic n-forms which give zero when integrated over any real n-dimensional cycle. This condition is closed in the compact-open topology, which proves the first assertion. The second one follows from

$$\begin{aligned} H^n(X, \mathbb{C}) &= H^0(X, \mathcal{E}^n)/\,\mathrm{d}\big(H^0(X, \mathcal{E}^{n-1})\big) \\ &\cong H^0(X, \Omega^n)/\,\mathrm{d}\big(H^0(X, \Omega^{n-1})\big) \cong \mathrm{Hol}(X)/F_X, \end{aligned}$$

where the last isomorphism is induced by ϵ from Lemma 4.4. □

4.7. Grothendieck's tensor products. For convenience of readers we remind some facts from Grothendieck's theory of nuclear vector spaces which can be found in [75] or [84]. Denote by E, E_i, F, F_i, and H locally convex Hausdorff topological vector spaces. Grothendieck introduced two different completions of the algebraic tensor product $E_1 \otimes E_2$, denoted by $E_1 \widehat{\otimes}_\pi E_2$ and $E_1 \widehat{\otimes}_\epsilon E_2$. The crucial properties of these completions we are going to exploit are the following. For linear continuous isomorphisms into (i.e., embeddings) $E_1 \to F_1$ and $E_2 \to E_2$ the induced linear map $E_1 \widehat{\otimes}_\epsilon E_2 \to F_1 \widehat{\otimes}_\epsilon F_2$ is also an isomorphism into [84, Proposition 43.7]. In the case of metrizable E_1 and E_2 and linear continuous surjective maps $E_1 \to F_1$ and $E_2 \to E_2$ the induced linear map $E_1 \widehat{\otimes}_\pi E_2 \to F_1 \widehat{\otimes}_\pi F_2$ is also a surjection [84, Proposition 43.9].

A space E is called nuclear if for any other locally convex topological vector space F the completions $E \widehat{\otimes}_\epsilon F$ and $E \widehat{\otimes}_\pi F$ coincide which allows us to omit indices π and ϵ in tensor products when appropriate. Nuclear spaces possess nice properties: their subspaces and quotients with respect to closed subspaces are again nuclear [84, Proposition 50.1]. Furthermore, the above claims about surjections and embeddings in combination with [84, Exercise 43.2][14] imply that the Grothendieck's tensor product preserves (short) exact sequences of metrizable nuclear spaces. That is, $(E/H) \widehat{\otimes} F = (E \widehat{\otimes} F)/(H \widehat{\otimes} F)$ where H is a closed subspace of the nuclear space E.

4.8. Example. Let X (resp. Y) be a closed Stein submanifold of $\mathbb{C}^n$ (resp. $\mathbb{C}^m$), i.e., we have the natural surjections $\varphi_X\colon \mathrm{Hol}(\mathbb{C}^n) \to \mathrm{Hol}(X)$ and $\varphi_Y\colon \mathrm{Hol}(\mathbb{C}^m) \to \mathrm{Hol}(Y)$. The spaces of holomorphic functions on Euclidean spaces are known to be nuclear [84, Corollary, p. 530] and, therefore, their quotients $\mathrm{Hol}(X)$ and $\mathrm{Hol}(Y)$ are also nuclear. Furthermore $\mathrm{Hol}(\mathbb{C}^n)\widehat{\otimes}\mathrm{Hol}(\mathbb{C}^m)$ is naturally isomorphic to $\mathrm{Hol}(\mathbb{C}^{n+m})$ [84, Theorem 51.6] and the induced surjective linear map $\varphi_X \widehat{\otimes} \varphi_Y\colon \mathrm{Hol}(\mathbb{C}^{n+m}) \to \mathrm{Hol}(X)\widehat{\otimes}\mathrm{Hol}(Y)$ has the kernel generated by functions that vanish either on $X \times \mathbb{C}^m$

[14]Using this exercise one can show that for surjective maps $\varphi_1\colon E_1 \to F_1$ and $\varphi_2\colon E_2 \to F_2$ the induced linear map $\varphi_1 \widehat{\otimes} \varphi_2\colon E_1 \widehat{\otimes} E_2 \to F_1 \widehat{\otimes} F_2$ has $\operatorname{Ker} \varphi_1 \widehat{\otimes} \varphi_2 = \operatorname{Ker} \varphi_1 \widehat{\otimes} E_2 + E_1 \widehat{\otimes} \operatorname{Ker} \varphi_2$.

or on $\mathbb{C}^n \times Y$. Thus this kernel coincides with the defining ideal of $X \times Y$ in $\mathbb{C}^{n+m}$ and $\mathrm{Hol}(X) \widehat{\otimes} \mathrm{Hol}(Y)$ is naturally isomorphic to $\mathrm{Hol}(X \times Y)$.

We present the proof of the next simple fact because of the lack of references.

4.9. Lemma. *Let F_i be a closed subspace of a complete metrizable nuclear space E_i for $i = 1, 2$. Then $I := (F_1 \widehat{\otimes} E_2) \cap (E_1 \widehat{\otimes} F_2)$ coincides with $F_1 \widehat{\otimes} F_2$.*

PROOF. Set $Q_i = E_i/F_i$. Then the kernel of the natural linear map $E_1 \widehat{\otimes} E_2 \to Q_1 \widehat{\otimes} Q_2$ coincides with $F_1 \widehat{\otimes} E_2 + E_1 \widehat{\otimes} F_2$. In particular, $Q_1 \widehat{\otimes} Q_2$ is naturally isomorphic to the quotient of $(E_1 \widehat{\otimes} E_2)/(E_1 \widehat{\otimes} F_2) = E_1 \widehat{\otimes} Q_2$ with respect to $M := (F_1 \widehat{\otimes} E_2)/I$ which implies that M coincides with the subspace $F_1 \widehat{\otimes} Q_2$ of $E_1 \widehat{\otimes} Q_2$ since $(E_1 \widehat{\otimes} Q_2)/(F_1 \widehat{\otimes} Q_2) = Q_1 \widehat{\otimes} Q_2$. Note also that $F_1 \widehat{\otimes} Q_2 \simeq (F_1 \widehat{\otimes} E_2)/(F_1 \widehat{\otimes} F_2)$. Since $I \supset F_1 \widehat{\otimes} F_2$ and $M := (F_1 \widehat{\otimes} E_2)/I$ we must have $I = F_1 \widehat{\otimes} F_2$. □

We need an extra product $E \epsilon F$ of locally convex topological vector spaces introduced by Schwartz [76]. In general $E \widehat{\otimes}_\epsilon F$ is a subspace of $E \epsilon F$. However we have equality $E \widehat{\otimes}_\epsilon F = E \epsilon F$ provided that both E and F are complete and have the approximation property (see, [76, Corollary 1, p. 47]). The latter means, say for E, that the identical operator on E belongs to the closure of operators $E \to E$ of finite rank in the topology $\mathcal{L}_c(E, E)$ of uniform convergence on convex compact subsets of E (see, [76, Definition, p. 5]). For any nuclear space E the identical operator can be always approximated by operators of finite rank. Thus the result of Bungart [16, Proposition 9.2] about ϵ-product can be reformulated in the special case of nuclear spaces as the following.

4.10. Proposition. *Let X be a Stein manifold and E be a complete nuclear space. Then* $\mathrm{Hol}(X) \widehat{\otimes}_\epsilon E$ *coincides with the space of weakly holomorphic E-valued functions on X (i.e., E-valued functions f on X such that for any continuous linear functional T on E the function $T \circ f$ is holomorphic on X).*

4.11. Convention. For Stein manifolds X and Y we consider holomorphic vector fields on $X \times Y$ tangent to the fibers of the natural projection $X \times Y \to X$ (resp. $X \times Y \to Y$) and we call them vertical (resp. horizontal) fields. Every holomorphic vector field δ on X (resp. μ on Y) generates the natural horizontal (resp. vertical) vector field on $X \times Y$ which by abuse of notation will be denoted by the same symbol.

4.12. Lemma. *Let Notation* 4.3 *and Convention* 4.11 *hold, and let $\delta_1, \delta_2, \ldots, \delta_N \in \mathrm{IVF}^{\omega}_{\mathrm{hol}}(X)$ be as in Corollary* 4.5. *Then for every $f \in F_X \widehat{\otimes} \mathrm{Hol}(Y)$ there exist $\bar{b} = (b_1, \ldots, b_N) \in \mathrm{Hol}(X \times Y)^N$ such that $f = \delta_1(b_1) + \delta_2(b_2) + \cdots + \delta_N(b_N)$.*[15]

PROOF. By Lemma 4.6 the map $\Theta\colon \mathrm{Hol}(X)^N \to F_X$ given by $(a_1, a_2, \ldots, a_N) \mapsto \sum_{i=1}^N \delta_i(a_i)$ is a linear surjection of Fréchet spaces. By Grothendieck's theorem the linear map

$$\Theta \widehat{\otimes} \mathrm{id}_Y \colon (\mathrm{Hol}(X))^N \widehat{\otimes} \mathrm{Hol}(Y) \cong \big(\mathrm{Hol}(X \times Y)\big)^N \to F_X \widehat{\otimes} \mathrm{Hol}(Y)$$

is surjective which implies the desired conclusion □

4.13. Remark. (1) The proof of existence of $\bar{b}$ from Lemma 4.12 can be also extracted from another nontrivial fact — the theorem of Michael (e.g., see [1,

[15] In this formula each δ_j is already a horizontal field on $X \times Y$.

Corollary 17.67; 47, p. 183–186]) that states that for every continuous surjective linear map of Fréchet spaces there exists a homogeneous continuous section.

(2) The same argument implies that for every $f \in F_X \widehat{\otimes} F_Y$ there is $\bar{b} = (b_1, \ldots, b_N) \in (\mathrm{Hol}(X) \widehat{\otimes} F_Y)^N$ for which $f = \delta_1(b_1) + \cdots + \delta_N(b_N)$.

4.14. Proposition. *For every vector field $\mu \in \mathrm{IVF}^{\omega}_{\mathrm{hol}}(Y)$ and each $f \in F_X \widehat{\otimes} \mathrm{Hol}(Y)$ there is a vector field α from the closure of $\mathrm{Lie}^{\omega_X \times \omega_Y}_{\mathrm{hol}}(X \times Y)$ (in the compact-open topology) such that the field $f\mu - \alpha$ is a horizontal vector field on $X \times Y$.*

PROOF. By Lemma 4.12 $f = \delta_1(b_1) + \delta_2(b_2) + \cdots + \delta_N(b_N)$. Fix a closed embedding $\iota \colon Y \to \mathbb{C}^m_{y_1,\ldots,y_m}$ and denote a monomial $y_1^{k_1} \cdots y_m^{k_m}$ by y^k, i.e., k is the multi-index $(k_1, \ldots, k_m)$ with norm $|k| = k_1 + \cdots + k_m$. Then b_i can be presented as $b_i = \sum_k b_{i,k} y^k$ where $b_{i,k} \in \mathrm{Hol}(X)$ and the sum converges uniformly on compacts of $X \times Y$. Let us show that the desired limit can be given by the formula

$$\alpha = \lim_{M \to \infty} \sum_{|k|<M} \sum_{i=1}^{N} [y^k \delta_i, b_{i,k}\mu].$$

Indeed, since y^k (resp. $b_{i,k}$) is in the kernel of δ_i (resp. μ) the involved vector fields are completely integrable and of $(\omega_X \times \omega_Y)$-divergence zero. Moreover

$$[y^k \delta_i, b_{i,k}\mu] = y^k \delta_i(b_{i,k})\mu - b_{i,k}\mu(y^k)\delta_i.$$

Thus

$$\sum_{|k|<M} \sum_{i=1}^{N} [y^k \delta_i, b_{i,k}\mu] = \Big(\sum_{|k|<M} \sum_{i=1}^{N} y^k \delta_i(b_{i,k}) \Big) \mu - \sum_{i=1}^{N} \Big(\sum_{|k|<M} b_{i,k}\mu(y^k) \Big) \delta_i$$
$$= \Big(\sum_{i=1}^{N} \delta_i \Big(\sum_{|k|<M} b_{i,k} y^k \Big) \Big) \mu - \sum_{i=1}^{N} \mu \Big(\sum_{|k|<M} b_{i,k} y^k \Big) \delta_i.$$

By the Weierstrass theorem about differentiation of convergent power series we can send $M \to \infty$ and obtain

$$\alpha = f\mu - \sum_{i=1}^{N} \mu(b^i)\delta_i$$

which yields the desired conclusion. □

4.15. Proof of Theorem 15. Let completely integrable holomorphic vector fields $\mu_1, \ldots, \mu_M$ on Y play the same role as $\delta_1, \ldots, \delta_N$ from Corollary 4.5 play for X. In particular, $\mu_1, \ldots, \mu_M$ generate the tangent vector space at any point of Y. Hence a simple application of the Cartan Theorem B implies that every vector field on $X \times Y$ is of form

$$\gamma = \sum_{i=1}^{M} f_i \mu_i + \sum_{j=1}^{N} g_j \delta_j$$

where f_i and g_j are holomorphic functions on $X \times Y$. We suppose further that γ is of ω-divergence zero where $\omega = \omega_X \times \omega_Y$. Recall that for every holomorphic vector field ν of ω-divergence zero we have $\mathrm{div}_\omega f\nu = \nu(f)$ for every holomorphic functions f. Since μ_i and δ_j have ω-divergence zero, the divergence of γ coincides with

$$\operatorname{div}_\omega \gamma = \sum_{i=1}^{M} \mu_i(f_i) + \sum_{j=1}^{N} \delta_j(g_j) = 0.$$

Note that the first (resp. second) summand can be viewed as a weakly holomorphic function on X (resp. Y) with values in F_Y (resp. F_X).

Thus by Bungart's theorem (Proposition 4.10) we have

$$f := \sum_{i=1}^{M} \mu_i(f_i) \in \operatorname{Hol}(X) \mathbin{\widehat{\otimes}} F_Y \quad \text{and} \quad \sum_{j=1}^{N} \delta_j(g_j) \in F_X \mathbin{\widehat{\otimes}} \operatorname{Hol}(Y).$$

By Lemma 4.9 $\sum_{i=1}^{M} \mu_i(f_i) \in F_X \mathbin{\widehat{\otimes}} F_Y$. Furthermore, by Remark 4.13(2) there exist $h_1, \dots, h_M \in F_X \mathbin{\widehat{\otimes}} \operatorname{Hol}(Y)$ for which

$$f = \mu_1(h_1) + \cdots + \mu_M(h_M).$$

By Proposition 4.14 there exists a vector field $\nu \in \operatorname{Lie}^{\omega}_{\text{hol}}(X \times Y)$ of form

$$\nu = \sum_{i=1}^{M} h_i \mu_i + \sum_{j=1}^{N} e_j \delta_j.$$

Subtracting ν from γ we can suppose from the beginning that

$$\operatorname{div}_\omega \sum_{i=1}^{M} f_i \mu_i = \sum_{i=1}^{M} \mu_i(f_i) = 0$$

which implies that

$$\operatorname{div}_\omega \sum_{i=1}^{N} g_i \delta_i = \sum_{i=1}^{N} \delta_i(g_i) = 0$$

because $\operatorname{div}_\omega \gamma = 0$. Since each μ_i is a vertical vector field on $X \times Y$ this means that the restriction of $\sum_{i=1}^{M} f_i \mu_i$ to any fiber $x \times Y$ is of ω_Y-divergence zero and therefore it belongs to $\operatorname{Lie}^{\omega_Y}_{\text{hol}}(x \times Y)$ by the assumption of the theorem. That is, $\sum_{i=1}^{M} f_i \mu_i$ can be viewed as a weakly holomorphic function on X with values in $\operatorname{Lie}^{\omega_Y}_{\text{hol}}(Y)$. By Bungart's theorem (Proposition 4.10) it belongs to $\operatorname{Hol}(X) \mathbin{\widehat{\otimes}} \operatorname{Lie}^{\omega_Y}_{\text{hol}}(Y)$ which is a subspace of $\operatorname{Lie}^{\omega}_{\text{hol}}(X \times Y)$. The same argument implies that $\sum_{i=1}^{N} g_i \delta_i \in \operatorname{Lie}^{\omega}_{\text{hol}}(X \times Y)$ and thus $\gamma \in \operatorname{Lie}^{\omega}_{\text{hol}}(X \times Y)$ which is the desired conclusion. □

5. Preliminary facts about foliation

A more detailed exposition of the results from this section can be found in [13–15].

5.1. Definition. (1) A foliation $\mathcal{F}$ on a smooth complex surface $\overline{X}$ is given by an open covering $\{U_j\}$ of $\overline{X}$ and holomorphic vector fields $\nu_j \in H^0(U_j, T\overline{X})$ with isolated zeros such that

$$\nu_i = g_{ij} \nu_j \quad \text{on } U_i \cap U_j$$

for invertible holomorphic functions $g_{ij} \in H^0(U_i \cap U_j, \mathcal{O}^*_X)$ where $\mathcal{O}^*_X$ is the sheaf of invertible functions. Gluing orbits of $\{\nu_j\}$ one gets leaves of the foliation $\mathcal{F}$. The singular set $\operatorname{Sing}(\mathcal{F})$ is the discrete subset of $\overline{X}$ whose intersection with each U_j coincides with zeros of ν_j. The cocycle $\{g_{ij}\}$ define a holomorphic line bundle $K_{\mathcal{F}}$ which is called the canonical bundle of the foliation $\mathcal{F}$.

(2) This definition can be extended to the case of $\overline{X}$ with quotient singularities only where $\mathcal{F}$ defined as a foliation on $\overline{X} \setminus \mathrm{Sing}(\overline{X})$. We require additionally that

$$\mathrm{Sing}(\overline{X}) \cap \mathrm{Sing}(\mathcal{F}) = \varnothing.$$

That is, a singular point p of $\overline{X}$ is locally of form $\mathbb{B}^2/\mathbb{Z}_k$ where $\mathbb{B}^2$ is a ball in $\mathbb{C}^2$ equipped with a linear $\mathbb{Z}_k$-action. In particular, the foliation can be lifted to $\mathbb{B}^2 \setminus \{(0,0)\}$ and the requirement is that it can be extended to a foliation on $\mathbb{B}^2$ with a nonvanishing associated vector field ν (and this must be true for any singular point of $\overline{X}$).

Then $K_{\mathcal{F}}$ on $\overline{X}$ is the direct image of the canonical bundle on $\overline{X} \setminus \mathrm{Sing}(\overline{X})$ under the inclusion morphism $\overline{X} \setminus \mathrm{Sing}(\overline{X}) \hookrightarrow \overline{X}$ (in this situation $K_{\mathcal{F}}$ is not a bundle but only a $\mathbb{Q}$-bundle).

(3) Foliation $\mathcal{F}$ is called nef if $K_{\mathcal{F}}$ is nef.

(4) A singularity $p \in \mathrm{Sing}(\mathcal{F})$ is reduced if the linear part of the corresponding vector field at p has eigenvalues λ_1, λ_2 such that either they are nonzero and $\lambda_1/\lambda_2 \notin \mathbb{Q}_+$ or $\lambda_1 \neq 0 = \lambda_2$. The foliation $\mathcal{F}$ is called reduced if all of its singularities are reduced.

(5) The Kodaira dimension $\mathrm{kod}(\mathcal{F})$ of a reduced foliation $\mathcal{F}$ on a projective surface $\overline{X}$ is the Kodaira–Iitaka dimension of its canonical bundle $K_{\mathcal{F}} \in \mathrm{Pic}(\overline{X}) \otimes \mathbb{Q}$. That is,

$$\mathrm{kod}(\mathcal{F}) = \limsup_{n \to +\infty} \frac{\log \dim H^0(\overline{X}, K_{\mathcal{F}}^{\otimes n})}{\log n}.$$

We shall study foliations associated with completely integrable holomorphic vector fields on a Stein surface X and the following result of Suzuki [80, 81] is very important.

Theorem 16. *Let $\mathcal{F}$ be a foliation on a normal Stein surface X.*

(1) *If all leaves of $\mathcal{F}$ are properly embedded in $X \setminus \mathrm{Sing}(\mathcal{F})$ then there is a nonconstant meromorphic first integral of $\mathcal{F}$ on X.*

(2) *Furthermore, if the general leaf of $\mathcal{F}$ is isomorphic to $\mathbb{C}^*$ (we shall call below such foliations of $\mathbb{C}^*$-type) then every leaf is closed in $X \setminus \mathrm{Sing}(\mathcal{F})$ and therefore there is a meromorphic first integral as in* (1).

For such a normal affine algebraic surface X the study of the foliation $\mathcal{F}$ would be much simpler when this first integral were a rational function. We need to consider the situation when this first integral is not rational.

First note that a foliation associated with an algebraic vector fields on X can be extended to a completion $\overline{X}$ of X and we can deal with a projective normal surface. Then we have the following result of Seidenberg (e.g., see [13]).

Theorem 17. *Performing a sequence of blow-ups $\widetilde{X} \to \overline{X}$ we obtain a lifted foliation $\widetilde{\mathcal{F}}$ on $\widetilde{X}$ which is reduced.*

Thus from now on we shall work with reduced foliations on projective surfaces. Miyaoka and Shepherd-Barron (e.g., see [15]) established the following.

Theorem 18. *If $\mathcal{F}$ is a reduced foliation on a projective surface $\overline{X}$ with at most cyclic quotient singularities then $K_{\mathcal{F}}$ is pseudoeffective if and only if $\mathcal{F}$ is not a rational foliation* (*i.e., its general fiber is a rational curve $\mathbb{CP}^1$*).

The Zariski – Fujita decomposition implies that every pseudoeffective $\mathbb{Q}$-bundle is the sum of a nef $\mathbb{Q}$-bundle and a negative part for which the associated divisor can be contracted. This is a basis for the next fact (McQuillan's contraction) (e.g., see [13] or [15]).

Theorem 19. *Let $\mathcal{F}$ be a nonrational reduced foliation on a projective surface $\overline{X}$ with at most cyclic quotient singularities. Then there exists a birational morphism $(\overline{X}, \mathcal{F}) \to (\overline{X}', \mathcal{F}')$ such that $\overline{X}'$ is still projective with at most cyclic quotient singularities, $\mathcal{F}'$ is still reduced, and $K_{\mathcal{F}'}$ is nef.*

5.2. Remark ([14, Section 3; 15, p. 10]). Contraction of $\overline{X}$ to $\overline{X}'$ is a sequence of blowing down of rational curves such that each of them is invariant with respect to the consequent induced foliation and the restriction of the canonical bundle of the foliation to the curve is negative. Every of these curves F contains exactly one singularity p of the foliation which is automatically a regular point of the surface. Furthermore, F is contracted to a point which is a regular point of the induced foliation on the resulting surface (but not in general a regular point of the surface).

The reduced foliation generated by a completely integrable algebraic vector field on X admits a lot of entire tangent curves $\mathbb{C} \to \overline{X}$. If $\mathcal{F}$ has no rational first integral, then such general curve is Zariski dense in $\overline{X}$ (Darboux's theorem) and another result of McQuillan says the following (see [68, Sections IV and V] or [13, Chapter 9, Theorems 1 and 4, Corollary 1].

Theorem 20. *Let $\mathcal{F}$ be a reduced foliation on a smooth projective surface $\overline{X}$ such that $\mathcal{F}$ possesses a tangent nonconstant entire curve that is Zariksi dense in $\overline{X}$. Then the Kodaira dimension* $\mathrm{kod}(\mathcal{F})$ *is either* 0 *or* 1.

(1) *Furthermore, if* $\mathrm{kod}(\mathcal{F}) = 1$ *then either*
  (a) *$\mathcal{F}$ is a Riccati foliation, i.e., there exists a fibration $f\colon \overline{X} \to B$ whose general fiber is a rational curve transverse to $\mathcal{F}$ or*
  (b) *$\mathcal{F}$ is a Turbulent foliation, i.e., there exists a fibration $f\colon \overline{X} \to B$ whose general fiber is an elliptic curve transverse to $\mathcal{F}$.*

(2) *If* $\mathrm{kod}(\mathcal{F}) = 0$ *and $\overline{X} \to \overline{X}'$ is the McQuillan's contraction to a nef reduced foliation $\mathcal{F}'$ on $\overline{X}'$ then there exists a finite covering $r\colon Y \to \overline{X}'$ such that*
  (a) *Y is smooth and r is ramified only over the quotient singularities of $\overline{X}'$.*
  (b) *The canonical bundle $K_{\mathcal{G}}$ of the lifted foliation $\mathcal{G} = r^*(\mathcal{F}')$ is trivial, i.e., $K_{\mathcal{G}} = \mathcal{O}_Y$, and so $\mathcal{G}$ is generated by a global holomorphic vector field with isolated zeros only.*

6. Brunella's construction: the case of $\mathrm{kod}(\mathcal{F}) = 1$

In fact Brunella proved more ([14, Lemmas 1 and 2]).

6.1. Proposition. (1) *Case* (1b) *in Theorem* 20 *cannot hold, i.e., when* $\mathrm{kod}(\mathcal{F}) = 1$ *for a reduced foliation without a rational first integral then it is a Riccati foliation*

(2) *Contracting curves in the fibers of $f\colon \overline{X} \to B$ for the Riccati foliation on a smooth surface $\overline{X}$ one can suppose that each fiber of f belongs to one of five standard types and there is always at least one singular fiber of f of so-called types* (e) *or* (d) (*see* [14, Lemma 2]) *that consists of a union of leaves of the foliation $\mathcal{F}$ (it will be denoted below by F_∞).*

Next we have the following ([14, Lemmas 3–5]).

6.2. Proposition. *Let ν be a completely integrable algebraic vector field on a rational affine algebraic surface X and $\pi\colon \widehat{X} \to X$ be the resolution of the singularities of ν (i.e., the lift $\hat{\nu}$ of ν is a completely integrable vector field on $\widehat{X}$). Suppose that the foliation $\mathcal{F}$ generated by $\hat{\nu}$ on the completion $\overline{X}$ of $\widehat{X}$ is reduced and has no rational first integral. Let $f\colon \overline{X} \to B$ be the corresponding (rational) Riccati fibration (note that $B \simeq \mathbb{CP}^1$ since $\overline{X}$ is rational). Then*

(1) *the vector field $\hat{\nu}$ preserves $f|_{\widehat{X}}$ (i.e., its flow sends fibers into fibers) or in other words $\hat{\nu}$ is a lift of a completely integrable vector field ν_0 on $f(\widehat{X}) \subset B$;*

(2) *the fiber F_∞ is contained in the divisor $D = \overline{X} \setminus \widehat{X}$ and in particular the exceptional divisor E of π is disjoint from F_∞;*

(3) *contracting E we get from f a regular function P on X whose general fibers are either $\mathbb{C}$ or $\mathbb{C}^*$ (i.e., again the flow of ν maps each fiber of P onto a fiber of P).*

In fact for foliations of $\mathbb{C}^*$-type the same conclusion can be made even without assumption that $\operatorname{kod}(\mathcal{F}) = 1$. In order to demonstrate it we need some facts from another Brunella's paper [12] where he used a slightly different terminology. Let L be a leaf of the foliation $\mathcal{F}$ on X and L_0 be a Riemann surface isomorphic to $\{z \in \mathbb{C} \mid 0 \le r < |z| \le 1\}$ which is properly embedded into L. Then L_0 is called a planar isolated end of L. This end is called transcendental if the set $\bar{L}_0 \setminus L_0$ consists of more than one point where $\bar{L}_0$ is the closure of L_0 in $\overline{X}$. Fibration $\mathcal{F}$ is called (in [12]) P-complete for some regular function P on X if there exists a finite set Q such that for every $t \in \mathbb{C} \setminus Q$ the fiber $P^{-1}(t)$ is transversal to $\mathcal{F}$ and there is a neighborhood $U_t \subset \mathbb{C}$ of t for which $P|_{P^{-1}(U_t)}$ is a fibration and $\mathcal{F}|_{P^{-1}(U_t)}$ define a local trivialization of it.

The next fact is the main result in [12].

6.3. Proposition. *Let X be a smooth rational affine surface with an SNC-completion $\overline{X}$ and $\overline{D} = \overline{X} \setminus X$. Suppose that for any other SNC-completion $\overline{\overline{X}}$ of X that dominates $\overline{X}$ (i.e., there is a morphism $\pi\colon \overline{\overline{X}} \to \overline{X}$ identical on X) the following is true.*

(i) *There is a Kähler metric on $\overline{\overline{X}}$ such that the restriction of the associated 2-form to X is exact.*

(ii) *Let Γ be the dual graph of the divisor $\overline{D}$ and let Γ have no linear (-1)-vertices. Suppose also that for any branch point b of Γ such that the subgraph $\Gamma \ominus b$ contains two connected components (among others), each of which is contractible to a (-2)-vertex, the weight of b is at most -2 (note that this condition will hold automatically for the dual graph of $\overline{\overline{D}} = \overline{\overline{X}} \setminus X$).*

Then for any foliation $\mathcal{F}$ generated by a regular vector field on X, which possesses also a transcendental planar isolated end, there exists a regular function $P \in \mathbb{C}[X]$ with general fibers isomorphic either to $\mathbb{C}$ or to $\mathbb{C}^*$ such that $\mathcal{F}$ is P-complete.

6.4. Remark. (1) This regular function P yields, of course, a Riccati foliation which is mentioned by Brunella [12, p. 1241]. He mentions also in [14] that the construction from [12, pp. 1241–1243] implies that one of singular fibers of P is of type (d).

(2) Suppose that $\mathcal{F}$ is a foliation of $\mathbb{C}^*$-type that has no rational first integral. Then general leaves are isomorphic to $\mathbb{C}^*$, they are Zariski dense in $\overline{X}$ and properly embedded into the complement to singularities by Theorem 16. This implies the existence of a transcendental planar isolated end, i.e., Proposition 6.3 is applicable to foliations of $\mathbb{C}^*$-type generated by regular vector fields.

(3) In [12] this Proposition 6.3 is proven only for $X = \mathbb{C}^2$. However the analysis of the proof shows that it is valid for any smooth rational affine surface X satisfying condition (i) and (ii). Condition (i) (which is obviously true for $\mathbb{C}^2$ since the second cohomology of $\mathbb{C}^2$ is trivial) is used in Lemma 3 in [12]. Condition (ii) is also true for $\mathbb{C}^2$ and used in the proof of Lemma 8 in [12]. The next (certainly well-known) fact says more about condition (i).

6.5. Proposition. *Condition* (i) *from Proposition* 6.3 *is automatic for every affine algebraic manifold* X.

PROOF. We can choose an SNC-completion $\overline{X}$ of X so that $\overline{X} \setminus X$ is a support of an ample divisor (for a two-dimensional X the facts that X is affine and that $\overline{X} \setminus X$ is the support of an ample divisor are equivalent). Then this divisor generates an embedding $\overline{X} \hookrightarrow \mathbb{P}^n =: \overline{Y}$ such that for a hyperplane $H \subset \overline{Y}$ the Euclidean space $\overline{Y} \setminus H =: Y \simeq \mathbb{C}^n$ contains X as a closed submanifold. The birational morphism $\pi\colon \overline{\overline{X}} \to \overline{X}$ may be viewed as the blow-up of an ideal sheaf $\mathcal{I}$ [45, Theorem II.7.17]. Consider the ideal sheaf on $\overline{Y}$ generated by $\mathcal{I}$ and the defining equations of $\overline{X}$. The blow-up with respect to this new sheaf yields a birational morphism $\tau\colon \overline{\overline{Y}} \to \overline{Y}$ which is identical over Y, i.e., $\overline{\overline{Y}}$ is another completion of Y. Furthermore, π is the restriction of τ to the proper transform of $\overline{X}$ [45, Corollary II.7.15]. Since $\overline{\overline{Y}}$ is projective it is a subvariety of $\mathbb{P}^N$ which is equipped with a closed 2-form ω as a Kähler manifold. The restriction of ω to Y is an exact form by de Rham's theorem because the second cohomology of Y is trivial. Thus its restriction to X is also exact which is the desired conclusion. □

6.6. Proposition (cf. [14, Proposition 3]**).** *Let ν be a completely integrable algebraic vector field of type $\mathbb{C}^*$ on a smooth affine rational surface X satisfying condition* (ii) *from Proposition* 6.3*. Then either ν possesses a rational first integral or for some regular function P on X with general fibers $\mathbb{C}$ or $\mathbb{C}^*$ the flow of ν transforms fibers of P into fibers of P.*

PROOF. By Theorem 16 the leaves of the associated foliation $\mathcal{F}$ are properly embedded into $X \setminus \operatorname{Sing}(\mathcal{F})$ and therefore there is a meromorphic first integral. Assume that it is not rational. Then there is at least one transcendental end in X (otherwise all leaves are algebraic and the first integral is rational by the Darboux's theorem). Now Proposition 6.3 and Remark 6.4 imply the existence of a Riccati fibration P with general fibers $\mathbb{C}$ or $\mathbb{C}^*$ such that the fiber over ∞ is of type (d) which is the only thing needed for validity of Proposition 6.2 whose proof in [14] does not use other properties of foliations with $\operatorname{kod}(\mathcal{F}) = 1$. □

7. Brunella's construction: the case of $\operatorname{kod}(\mathcal{F}) = 0$

Thus we can consider only completely integrable algebraic fields of type $\mathbb{C}$ for which most of orbits are isomorphic to $\mathbb{C}$. For a smooth affine surface X with a completely integrable algebraic vector field ν we consider its smooth completion $\widetilde{X}$ and the resolution of singularities $\pi\colon \overline{X} \to \widetilde{X}$ of the foliation induced by the vector field so that the resulting foliation $\mathcal{F}$ on $\overline{X}$ is reduced. Then $D = \pi^{-1}(\widetilde{X} \setminus X)$ is the divisor at infinity and E is the exceptional divisor of π over X, i.e., $\pi(E) = \Gamma$ is a finite subset of X. Moreover we have the following [14, p. 442].

7.1. Lemma. *Each irreducible component of D and each irreducible component of the exceptional divisor E of $\pi\colon \overline{X} \to \widetilde{X}$ are $\mathcal{F}$-invariant.*

Following Brunella, consider now the McQuillan's contraction $\tau\colon \overline{X} \to \overline{X}'$ of a reduced foliation $\mathcal{F}$ on $\overline{X}$ to a nef reduced foliation $\mathcal{F}'$ on $\overline{X}'$ as a sequence of blowing downs

$$\overline{X} = \overline{X}_0 \to \overline{X}_1 \to \cdots \to \overline{X}_{n-1} \to \overline{X}_n = \overline{X}'$$

of rational curves $F_j \subset \overline{X}_j$ as in Remark 5.2. Denote by $\mathcal{F}_j$ the foliation induced by $\mathcal{F}$ on $\overline{X}_j$, by D_j and E_j the images of D and E respectively. We call F_j external if it is not contained in $D_j \cup E_j$.

7.2. Lemma (cf. [14, Lemma 6]). *Let F_j be external and let $\widetilde{F}_j$ be the proper transform of F_j in $\widetilde{X}$. Then $\widetilde{F}_j \cap X$ is an algebraic curve in $X \setminus \Gamma$ isomorphic to $\mathbb{C}$. Furthermore, for different external curves F_j and F_i the curves $\widetilde{F}_j \cap X$ and $\widetilde{F}_i \cap X$ are disjoint.*

PROOF. Note that F_j must meet D_j since otherwise its strict transform in $\overline{X}$ will be contained in $\pi^{-1}(X)$ but not in E which is impossible for a complete curve. Then $D_j \cap F_j$ must be $\mathcal{F}_j$-invariant since both D_j and F_j are. Therefore, it is the only singular point of the foliation on F_j (see Remark 5.2). By the same reason F_j cannot meet E_j which implies the first statement. To see the second one we note that F_j and F_i cannot meet outside D since otherwise contrary to Remark 5.2 F_j has another singular point of the foliation besides $D_j \cap F_j$. Thus $\widetilde{F}_j \cap X$ and $\widetilde{F}_i \cap X$ are disjoint. ☐

7.3. Remark. Let L be the union $\bigcup \widetilde{F}_j$ where F_j runs over the set of external curves. By construction L and therefore $X^* = X \setminus (L \cup \Gamma)$ are invariant with respect to the original foliation, i.e., the restriction of ν to X^* is completely integrable.

Then Brunella considers morphism $r\colon Y \to \overline{X}'$ as in Theorem 20 and the commutative diagram generated by it

$$\begin{array}{ccc} Z & \xrightarrow{h} & \overline{X} \\ \Big\downarrow{\scriptstyle g} & & \Big\downarrow{\scriptstyle \tau} \\ Y & \xrightarrow{r} & \overline{X}' \end{array}$$

where $g\colon Z \to Y$ is a birational morphism and $h\colon Z \to \overline{X}$ is a ramified covering such that $r \circ g = \tau \circ h$. Set $R = D \cup E$, $R' = \tau^{-1}(\tau(R))$, and $W = Z \setminus h^{-1}(R')$. Suppose that μ_0 is the holomorphic vector field on projective variety Y that generates fibration $\mathcal{G}$ from Theorem 20. Let μ be its lift to Z via g.

7.4. Lemma (cf. [14, Lemma 7]). *Let μ, W, X^*, L be as before. Then*

(1) *the restriction of h makes W an unramified cover of X^*;*
(2) *the restriction $g|_W\colon W \to Y$ is an embedding;*
(3) *elements of the flow associated with the restriction of $\mu|_W$ are algebraic automorphisms of W.*

PROOF. By Theorem 20 morphism r can be ramified only over singularities of $\overline{X}'$. By construction these singularities are contained in $\tau(R)$. Thus h can be ramified only over R'. For (1) it remains to note that $R' = D \cup E \cup L$.

Since Z is the fibered product of Y and $\overline{X}$ over $\overline{X}'$ we see that W is the fibered product of Y and $\overline{X}\backslash R'$ over $\overline{X}'$. Hence for (2) it suffices to note that the restriction of τ to $\overline{X} \setminus R'$ is an embedding.

We have also (3) since μ arises from a holomorphic vector field μ_0 on the projective variety Y and elements of its flow are algebraic automorphisms by Chow's theorem. □

7.5. Corollary. *Suppose that for any finite collection of disjoint lines in $X \setminus \Gamma$ and every finite unramified covering $W \to X \setminus (\Gamma \cup L)$ where L is the union of these lines the manifold W has a finite number of algebraic automorphisms. Then the case of* $\operatorname{kod}(\mathcal{F}) = 0$ *does not hold.*

8. Geometry of the surface $x + y + xyz = 1$

Consider the hypersurface $S \subset \mathbb{C}^3_{x,y,z}$ given by $x+y+xyz = 1$. It can be viewed as the affine modification of $\pi\colon S \to \mathbb{C}^2$ of $\mathbb{C}^2_{x,y}$ along the divisor $xy = 0$ with center at $xy = 1 - x - y = 0$ (i.e at points $(1,0)$ and $(0,1)$) since $z = (1 - x - y)/(xy)$. Note that S contains four lines L_1, L_2, L_3, and L_4 such that $\pi(L_1) = (0,1)$, $\pi(L_2) = \{y = 1\}$, $\pi(L_3) = (1,0)$, and $\pi(L_4) = \{x = 1\}$.

8.1. Lemma. *There is only one polynomial curve L_5 in S different from L_1, L_2, L_3, L_4 and it is the proper transform of the line $x + y = 1$ in $\mathbb{C}^2$.*

PROOF. Consider a nonconstant morphism $\phi\colon \mathbb{C} \to S$ whose image is not one of the four lines. Its composition with π yields a morphism $\mathbb{C} \to \mathbb{C}^2$ given by $x = p(t)$ and $y = q(t)$ (where p and q are nonconstant polynomials) whose image C may meet the x-axis and the y-axis only at points $(1,0)$ and $(0,1)$ respectively. Furthermore, local branches of C cannot be tangent to any of these axes since otherwise the proper transform of C in S (which is $\phi(\mathbb{C})$) is a Riemann surfaces with at least two punctures contrary to the assumption that it is a polynomial curve. This implies that $1 - p(t)$ is divisible by $q(t)$ while $1 - q(t)$ is divisible by $p(t)$. In particular, $\deg p = \deg q$ and therefore, $p(t) = 1 - q(t)$. This yields the desired conclusion. □

8.2. Remark. One can check that the equations of L_1, L_2, L_3, L_4, and L_5 in S are $x = 0$, $yz + 1 = 0$, $y = 0$, $xz + 1 = 0$, and $z = 0$ respectively.

8.3. Lemma. *Let $\rho\colon S' \to S$ be a finite morphism of smooth surfaces unramified over the complement to the five lines in S. Then S' has a finite number of algebraic automorphisms.*

PROOF. First note that for every polynomial curve C in S' its image $\rho(C)$ is a polynomial curve in S. Thus the number of such curves in S' is finite. Any automorphism α of S' generates their permutation. Without loss of generality we can assume that this permutation is identical. Then the restriction of α yields an automorphism of the surface $H' \subset S'$ that is an unramified finite covering of the Kobayashi hyperbolic surface $H = S\backslash\bigcup_{i=1}^5 L_i$ (and therefore H' is itself hyperbolic). There are at most finite number of such automorphisms and we have the desired conclusion. □

8.4. Proposition. *Let ν be a completely integrable algebraic vector field on S. Then*

(1) *either ν has a first integral which is a regular function with general fibers isomorphic to $\mathbb{C}^*$ or*

(2) *there is a regular function P with general fibers isomorphic to $\mathbb{C}^*$ that is a Riccati fibration for ν, i.e., the flow associated with ν transforms fibers of P into fibers of P.*

PROOF. We need to check conditions (i) and (ii) in Proposition 6.3. The first of them is always true because of Proposition 6.5. As for the second, the surface S possesses an SNC-completion whose dual graph is a cycle consisting of three vertices of weights $1, 0$, and 0 respectively. Thus Proposition 6.3 is applicable.

Assume now that f is a rational first integral of ν. Then general orbits of ν are contained in the fibers of f on the complement S^* to the set of indeterminacy points of f in S. Since S contains only five polynomial curves these fibers must be isomorphic to $\mathbb{C}^*$. Furthermore, since their closures must meet at indeterminacy points we see that such points cannot exist because otherwise the closures produce infinite number of polynomial curves. Thus $S^* = S$ and f is regular which is (1).

In the absence of a rational first integral for ν let $\mathcal{F}$ be the associated foliation on an SNC-completion of S. By Corollary 7.5 and Lemma 8.3 $\operatorname{kod}(\mathcal{F})$ cannot vanish, i.e., it is equal to 1 by Theorem 20. Now Proposition 6.2 implies (2). □

8.5. Lemma. *The surface S is factorial and its Euler characteristics is 2.*

PROOF. Note that $S = T \cup L_1 \cup L_3$ where $T = S \setminus (L_1 \cup L_3)$ is a torus. Hence $\chi(S) = \chi(T) + \chi(L_1) + \chi(L_3) = 2$ by [27]. Then modification $\pi\colon S \to \mathbb{C}^2$ may be viewed as composition of two modifications $\pi'\colon S \to S'$ and $\pi''\colon S' \to \mathbb{C}^2$ where S' is the affine modification of $\mathbb{C}^2$ along $x = 0$ at center $(0, 1)$ (i.e., S' is again isomorphic to $\mathbb{C}^2$ and therefore factorial) and S is obtained as the affine modification along the proper transform of $y = 0$ in S' (which is a $\mathbb{C}^*$-curve) with center at point $(1, 0)$. That is, S is obtained from S' by replacing an irreducible curve by a line L_3. The Nagata lemma (e.g., see [28, Lemma 19.20]) implies that S remains factorial. □

8.6. Proposition. *Let f be a regular function on S whose general fibers are isomorphic to $\mathbb{C}^*$. Then up to a linear transformation f coincides with one of the functions $x^k y^l$ (where k and l are relatively prime nonnegative integers with $k + l \geq 1$), $xz + 1, yz + 1$, and z.*

PROOF. Since the general fibers are isomorphic to $\mathbb{C}^*$ each connected component of singular fibers is either $\mathbb{C}^*$ or a curve of Euler characteristics one (more precisely either a line or a cross) by [94, Theorem 3]. Furthermore, there are exactly two such components because the Euler characteristics of S is 2 by Lemma 8.5. They are disjoint and therefore the description of polynomials curves in S implies that the union of these components contains one of the following pairs of lines: $(L_1, L_3), (L_1, L_4), (L_2, L_3), (L_2, L_5)$, and (L_4, L_5). Consider the case of the first three pairs and tori $T_{13} = S \setminus (L_1 \cup L_3)$, $T_{14} = S \setminus (L_1 \cup L_4)$, and $T_{23} = S \setminus (L_2 \cup L_3)$. The general fiber $F \simeq \mathbb{C}^*_\xi$ is contained in one of these tori, say T_{13} that is isomorphic to $\mathbb{C}^*_x \times \mathbb{C}^*_y$. Hence up to constant factors $x = \xi^k$ and $y = \xi^l$ where k and l are coprime integers since the functions x and y separate points of F. Let m and n be integers such that $mk + nl = 1$. Then $\xi = x^m y^n$ together with $f = x^k y^{-l}$ form a new coordinate system on T such that the restriction of f to F is constant. Note that this f is regular on S only when the powers are nonnegative which yields the first functions mentioned in the formulation.

Note that T_{14} is isomorphic to the torus $\mathbb{C}_x^* \times \mathbb{C}_\zeta^*$ where $\zeta = y/(x-1)$. Thus for T_{14} this function f must be of form $x^k(x-1)^l/y^l$. However for a nonzero l this function is not regular on S and we have to disregard it. Similarly in the case of T_{23} we have to disregard function of the form $y^k(x-1)^l/x^l$.

Now consider the case when the zero fiber of f contains L_4 and L_5 as components. Let us show that $f^{-1}(0) = L_4 \cup L_5$. Indeed, assume to the contrary that this fiber contains another irreducible component L. Since S is factorial by Lemma 8.5 L_4 is given by zeros of $xz+1$, L_5 by zeros of z (see Remark 8.2), and L by zeros of another function h. All three functions are invertible on $T_{45} = S \setminus f^{-1}(0)$ and thus the group of invertible functions has at least three generators. This group is isomorphic to the group of integer first cohomology (e.g., see [40]) and therefore $H^1(T_{45}, \mathbb{Q})$ is at least of rank 3. On the other hand nonzero fibers of f do not contain connected components of Euler characteristics one and therefore T_{45} is a $\mathbb{C}^*$-fibration over $\mathbb{C}^*$. By the technical Lemma 8.7 below its rational cohomology is at most of rank 2. This contradiction shows that L cannot exist.

Since $f^{-1}(0) = L_4 \cup L_5$ up to a constant factor we have $f = z^k(xz+1)^l$ where k and l are natural and, furthermore, relatively prime (since otherwise the general fiber of f is not connected). Then taking into consideration the equation of S we see that the equation of the projection C of the fiber $f^{-1}(c)$ to $\mathbb{C}^2_{x,y}$ is given by

$$(1-x-y)^k(1-x)^l = cx^ky^{k+l}. \tag{8.1}$$

Hence over a neighborhood of $x=0$ (resp. $x=\infty$) up to constant factor the restriction of y to C behaves as $x^{-k/(k+l)}$ (resp. $x^{l/(k+l)}$). This implies that the projection of C to the x-axis has ramification points of order $k+l$ over $x=0$ and $x=\infty$. Differentiating (8.1) with respect to y we get

$$-k(1-x-y)^{k-1}(1-x)^l = c(k+l)x^ky^{k+l-1}.$$

Dividing (8.1) by this equality we get $1-x = \alpha y$ where $\alpha = l/(k+l)$. Plugging αy instead of $1-x$ into (8.1) we obtain a polynomial in y of degree $2k+l$ which yields additional ramification points. Since the projection of C to the x-axis is $(k+l)$-sheeted the Riemann–Hurwitz formula implies that the genus of C is positive which contradicts to the fact that C is isomorphic to $\mathbb{C}^*$.

Now assume that L_4 and L_5 are contained in different fibers of f. Using again Lemma 8.7 below we can see that at least one of these two fibers is irreducible. That is, up to a linear transformation f must coincide with either z or $xz+1$.

By considering the pair (L_2, L_5) instead of (L_4, L_5) we get also $f = yz+1$ as a possibility which concludes the proof. □

To make the proof of Proposition 8.6 complete we need the following.

8.7. Lemma. *Let $f\colon Y \to B$ be a $\mathbb{C}^*$-fibration (i.e., the general fibers of f are isomorphic to $\mathbb{C}^*$) of a factorial affine algebraic surface over the base B isomorphic to a line with k deleted points. Then the dimension of $H_1(Y, \mathbb{Q})$ is at most $k+1$.*

PROOF. Since Y is factorial the removal of an irreducible curve from Y increases the number of generator of the group of regular invertible functions by 1. Therefore, this procedure increases the dimension of the first rational homology by 1 (e.g., see [40]). Thus it suffices to prove the statement of lemma in the case of

f being locally trivial since we can remove the singular fibers. If Y is the direct product $B \times \mathbb{C}^*$ then $\dim H_1(Y, \mathbb{Q}) = k + 1$ by the Kunneth formula. In the general case the computation goes via the Leray spectral sequence but the dimension of homology can only decrease compared with the case of the direct product. This concludes the proof. □

Theorem 21. *Let ν be a completely integrable algebraic vector field on S. Then either*

(1) *ν coincides with a vector field of form*

$$q(z)\left((1+xz)\frac{\partial}{\partial x} - (1+yz)\frac{\partial}{\partial y}\right)$$

where q is a polynomial or

(2) *the restriction of ν to $T = S \setminus (L_1 \cup L_3) \subset \mathbb{C}^2_{x,y}$ is a vector fields of form*

$$p(x^k y^l)\left(lx\frac{\partial}{\partial x} - ky\frac{\partial}{\partial y}\right)$$

where p is a polynomial with $p(0) = 0$ and k and l are relatively prime nonnegative integers.

Proof. By Proposition 8.4 we have a regular function P on S with general fibers isomorphic to $\mathbb{C}^*$ such that P is invariant with respect to the flow of ν. All such possible P's are listed in Proposition 8.6. Suppose that $P = z$. Then this regular function has two singular fibers. Hence each individual fiber of P is preserved by the flow, i.e., P is the first integral. It remains to note that every field that has general fibers of P as orbits is of form (1).

Now let $P = x^k y^l$ with natural relatively prime k and l,

$$\nu_1 = x\frac{\partial}{\partial x}, \quad \text{and} \quad \nu_2 = lx\frac{\partial}{\partial x} - ky\frac{\partial}{\partial y}.$$

Repeating the argument from [14, Proposition 2(2)] we trivialize a neighborhood of a regular fiber of P, isomorphic to $\Delta \times \mathbb{C}^*_\zeta$, in such a way that ν_1 is sent to $\partial/\partial\xi$ (where ξ is a coordinate on the unit disc Δ) and ν_2 is sent to $\zeta\partial/\partial\zeta$. Extending the flow of ν to $\Delta \times \mathbb{CP}^1$ we find that ν is of form $\beta(\xi)\partial/\partial\xi + \alpha(\xi)\zeta\partial/\partial\zeta$ (because this extension must be tangent to the curves $\zeta = 0$ and $\zeta = \infty$). Hence ν is of the form

$$\nu = q(x^k y^l)\nu_1 + p(x^k y^l)\nu_2$$

where both functions p and q are regular on $\mathbb{C}^*_\xi$. However, if we want this rational field on $\mathbb{C}^2$ to be lifted regularly on S we need to require that it vanishes at points $(1, 0)$ and $(0, 1) \in \mathbb{C}^2$. This implies that p and consequently q are polynomials on $\mathbb{C}_\xi$ and $p(0) = q(0) = 0$. Furthermore, since ν is completely integrable q must be constant and we have the desired form.

In the case of P equal to $xz + 1$ or $yz + 1$ the same argument produces a field as in (2) but with the pair (k, l) equal to $(1, 0)$ or $(0, 1)$ which concludes the proof. □

8.8. Corollary. *Every completely integrable algebraic vector field on S has divergence zero with respect to the algebraic volume form $\omega = \mathrm{d}x/x \wedge \mathrm{d}y/y$. In particular, S has no algebraic density property.*

PROOF. First note that the restriction of vector fields of type (2) from Theorem 21 to the torus $T = S\backslash(L_1 \cup L_3)$ is completely integrable. Thus their divergence with respect to ω is zero [3]. It can be checked also by the direct computation which we do for the fields on type (1). We need the following two formulas (e.g., see [60])

$$\operatorname{div}_\omega(f\nu) = f \operatorname{div}_\omega(\nu) + \nu(f) \quad \text{and} \quad \operatorname{div}_{f\omega}(\nu) = \operatorname{div}_\omega(\nu) + L_\nu(f)/f$$

where ν is a vector field, f is a regular function, and L_ν is the Lie derivative. In particular, $\operatorname{div}_\omega(f\nu) = 0$ provided $\operatorname{div}_\omega(\nu) = 0$ and $\nu(f) = 0$. Thus for the fields of type (1) it suffices to consider $\nu = (1+xz)\partial/\partial x - (1+yz)\partial/\partial y$ because z belongs to the kernel of ν. Since $1 + xz = (1-x)/y$ and $1 + yz = (1-y)/x$ the restriction of ν to T can be written in the from

$$\frac{1-x}{y}\frac{\partial}{\partial x} - \frac{1-y}{x}\frac{\partial}{\partial y}.$$

The divergence of the first summand $\big((1-x)/y\big)\partial/\partial x$ with respect to the standard form $\omega_0 = \mathrm{d}x \wedge \mathrm{d}y$ is $-1/y$. Since $\omega = \frac{1}{xy}\omega_0$ its divergence with respect to ω is

$$\frac{-1}{y} + L_\nu\left(\frac{1}{xy}\right)\Big/\left(\frac{1}{xy}\right) = \frac{-1}{y} - \frac{1-x}{xy} = \frac{-1}{xy}.$$

Similarly, the divergence of the second summand $\big((1-y)/x\big)\partial/\partial y$ is $-1/(xy)$ and $\operatorname{div}_\omega(\nu) = 0$ which implies the first statement.

Since $\operatorname{div}_\omega([\nu_1,\nu_2]) = L_{\nu_1}(\operatorname{div}_\omega(\nu_2)) - L_{\nu_2}(\operatorname{div}_\omega(\nu_1))$ (e.g., see [60]) the Lie algebra generated by completely integrable algebraic vector fields contains only fields of divergence zero with respect to ω. Thus algebraic vector fields with nonzero divergence cannot be contained in this algebra which implies the absence of the algebraic density property. □

It remains to show that the group of holomorphic automorphisms of S generated by completely integrable algebraic vector fields is m-transitive for any natural m. For this consider the fields

$$(1+xz)\frac{\partial}{\partial x} - (1+yz)\frac{\partial}{\partial y}, \quad xy\frac{\partial}{\partial x} - (1+yz)\frac{\partial}{\partial z}, \quad \text{and} \quad xy\frac{\partial}{\partial x} - (1+xz)\frac{\partial}{\partial z}$$

on S. They are completely integrable fields from Theorem 21 (the first one is of type (1) and the last two are of type (2) with the pair (k,l) equal to $(0,1)$ and $(1,0)$ respectively), they generate the tangent space of S at each point, and their kernels in $\mathbb{C}[S]$ are $\mathbb{C}[z], \mathbb{C}[y]$, and $\mathbb{C}[x]$ respectively. Therefore, the desired conclusion follows from the next fact.

8.9. Proposition. *Let X be a smooth affine algebraic variety and $\sigma_1,\dots,\sigma_k$ be completely integrable nonzero algebraic vector fields on X. Suppose that the kernel of σ_j in $\mathbb{C}[X]$ is denoted by* $\operatorname{Ker}\sigma_j$ *and the following conditions hold*

(i) *$\sigma_1,\dots,\sigma_k$ generate tangent space at any point of X;*

(ii) *there exist regular functions $f_1 \in \operatorname{Ker}\sigma_1,\dots,f_k \in \operatorname{Ker}\sigma_k$ such that each f_j is not contained in $\bigcup_{l\neq j}\operatorname{Ker}\sigma_l$ and the map $(f_1,\dots,f_k)\colon X \to \mathbb{C}^k$ is an embedding.*

Then the group G of holomorphic automorphisms generated by elements of the flows of completely integrable algebraic vector fields is infinitely transitive (i.e., it is m-transitive for every natural m).

PROOF. Let us start with the following.

Claim. *Let X^m_* be the subset of points $(x_1, x_2, \ldots, x_m)$ in X^m such that for any $j = 1, \ldots, k$ we have*

$$f_j(x_i) \neq f_j(x_l) \tag{8.2}$$

where $i \neq l$. Then a G-orbit $O \subset X$ through a point $(x_1, \ldots, x_m) \in X^m_*$ contains all points of form $(x, x_2, \ldots, x_m)$ where x runs over some neighborhood U of x_1 in X.

Indeed, let p_j $(j = 1, \ldots, n)$ be a polynomial in one variable such that the function $p_j(f_j)$ vanishes at all points $x_2, \ldots, x_m$ but not at x_1. Then the elements of flows of completely integrable vector fields $p_1(f_1)\sigma_1, \ldots, p_k(f_k)\sigma_k$ keeps points $x_2, \ldots, x_n$ fixed. On the other hand since these fields generate the tangent space $T_{x_1}X$ by (i), compositions of their exponents can send x_1 into any given point in a small neighborhood U of x_1 which yields the claim.

Note that U can chosen so that it depends continuously on $(x_1, x_2, \ldots, x_n)$ in a neighborhood of this point in X^m_*, and we can also repeat this claim with any x_j instead of x_1. This implies immediately that $O \cap X^m_*$ is an open subset of X^m_*. Furthermore, it is closed in X^m_* because if $(x_1', x_2', \ldots, x_m')$ belongs to its closure then the open set $O' \cap X^m_*$ meets O where O' is the G-orbit of $(x_1', x_2', \ldots, x_m')$. Thus O contains X^m_*.

Now let $(x_1, x_2, \ldots, x_m) \in X^m$ be an m-tuple of distinct points that does not satisfy (8.2). Say $f_j(x_1) = f_j(x_2)$ for $j \leq n$ and $f_i(x_1) \neq f_i(x_2)$ for $n + 1 \leq i \leq k$ (note that $n < k$ because of assumption (ii)). It suffices to show that a perturbation of $(x_1, x_2, \ldots, x_m)$ by an element of G close to the identical automorphism destroys at least one of these equalities. Though a priori a perturbation may not destroy the equalities it enables us to assume that σ_k is not tangent at x_1 to the fiber of f_1 (since otherwise $f_1 \in \operatorname{Ker} \sigma_k$ contrary to (ii)). Choose again a polynomial p_k such that $p_k(f_k)$ vanishes at x_2 but not at x_1. Then the flow of $p_k(f_k)\sigma_k$ changes the value of $f_1(x_1)$ while keeping the point x_2 fixed which concludes the proof. □

8.10. Remark. The infinite transitivity as in Proposition 8.9 can be considered not only for the group of all holomorphic automorphisms but for its subgroups. Say, Arzhantsev, Kuyumzhiyan and Zaidenberg [5] studied the infinite transitivity property for a special automorphism group of an affine algebraic variety, i.e., for the group generated by elements of all algebraic $\mathbb{C}_+$-action on this variety. This work will be continued in the coming paper of Arzhantsev, Flenner, Zaidenberg and the authors where in particular the following fact is established: transitivity of a special group on a smooth affine algebraic variety yields its infinite transitivity.

References

1. C. D. Aliprantis and K. C. Border, *Infinite dimensional analysis: A hitchhiker's guide*, 3rd ed., Springer, Berlin, 2006.
2. E. Andersén, *Volume-preserving automorphisms of* $\mathbf{C}^n$, Complex Variables Theory Appl. **14** (1990), no. 1-4, 223–235.
3. ———, *Complete vector fields on* $(\mathbf{C}^*)^n$, Proc. Amer. Math. Soc. **128** (2000), no. 4, 1079–1085.
4. E. Andersén and L. Lempert, *On the group of holomorphic automorphisms of* $\mathbf{C}^n$, Invent. Math. **110** (1992), no. 2, 371–388.
5. I. V. Arzhantsev, K. Kuyumzhiyan, and M. Zaidenberg, *Flag varieties, toric varieties, and suspensions: three instances of infinite transitivity*, available at `arXiv:1003.3164`.
6. T. Asanuma, *Non-linearizable algebraic* k^**-actions on affine spaces*, Invent. Math. **138** (1999), no. 2, 281–306.

7. J. Ax, *Injective endomorphisms of varieties and schemes*, Pacific J. Math. **31** (1969), 1–7.
8. P. Bala and R. W. Carter, *Classes of unipotent elements in simple algebraic groups.* I, Math. Proc. Cambridge Philos. Soc. **79** (1976), no. 3, 401–425; II, **80** (1976), no. 1, 1–17.
9. C. Bisi and F. Polizzi, *Proper polynomial self-maps of the affine space: state of the art and new results*, available at `arXiv:1005.0078`.
10. S. Borell and F. Kutzschebauch, *Non-equivalent embeddings into complex Euclidean spaces*, Internat. J. Math. **17** (2006), no. 9, 1033–1046.
11. N. Bourbaki, *Lie groups and Lie algebras.* Chapters 7 – 9, translated by A. Pressley, Elem. Math. (Berlin), Springer, Berlin, 2005.
12. M. Brunella, *Sur les courbes intégrales propres des champs de vecteurs polynomiaux*, Topology **37** (1998), no. 6, 1229–1246.
13. ———, *Birational geometry of foliations*, Monografías de Matemática, IMPA, Rio de Janeiro, 2000.
14. ———, *Complete polynomial vector fields on the complex plane*, Topology **43** (2004), no. 2, 433–445.
15. ———, *Foliations on complex projective surfaces*, available at `arXiv:math/0212082`.
16. L. Bungart, *Holomorphic functions with values in locally convex spaces and applications to integral formulas*, Trans. Amer. Math. Soc. **111** (1964), 317–344.
17. G. T. Buzzard and F. Forstnerič, *A Carleman type theorem for proper holomorphic embeddings*, Ark. Mat. **35** (1997), no. 1, 157–169.
18. G. T. Buzzard and J. E. Fornæss, *An embedding of* $\mathbf{C}$ *in* $\mathbf{C}^2$ *with hyperbolic complement*, Math. Ann. **306** (1996), no. 3, 539–546.
19. G. T. Buzzard and J. H. Hubbard, *A Fatou – Bieberbach domain avoiding a neighborhood of a variety of codimension* 2, Math. Ann. **316** (2000), no. 4, 699–702.
20. H. Derksen and F. Kutzschebauch, *Nonlinearizable holomorphic group actions*, Math. Ann. **311** (1998), no. 1, 41–53.
21. ———, *Global holomorphic linearization of actions of compact Lie groups on* $\mathbf{C}^n$, Complex Geometric Analysis in Pohang (Pohang, 1997) (K.-T. Kim and S. G. Krantz, eds.), Contemp. Math., vol. 222, Amer. Math. Soc., Providence, RI, 1999, pp. 201–210.
22. H. Derksen, F. Kutzschebauch, and J. Winkelmann, *Subvarieties of* $\mathbf{C}^n$ *with non-extendable automorphisms*, J. Reine Angew. Math. **508** (1999), 213–235.
23. P. G. Dixon and J. Esterle, *Michael's problem and the Poincaré – Fatou – Bieberbach phenomenon*, Bull. Amer. Math. Soc. (N.S.) **15** (1986), no. 2, 127–187.
24. F. Donzelli, A. Dvorsky, and Sh. Kaliman, *Algebraic density property of homogeneous spaces*, Transform. Groups **15** (2011), no. 3, 551–576.
25. J.-M. Drézet, *Luna's slice theorem and applications*, Algebraic Group Actions and Quotients (J. A. Wiśniewski, ed.), Hindawi Publ. Corp., Cairo, 2004, pp. 39–89.
26. A. Dubouloz, L. Moser-Jauslin, and P.-M. Poloni, *Non cancellation for smooth contractible affine threefolds*, available at `arXiv:1004.4723`.
27. A. H. Durfee, *Algebraic varieties which are a disjoint union of subvarieties*, Geometry and Topology (Athens, GA, 1985) (C. McCrory and T. Shifrin, eds.), Lecture Notes in Pure and Appl. Math., vol. 105, Dekker, New York, 1987, pp. 99–102.
28. D. Eisenbud, *Commutative algebra: With a view toward algebraic geometry*, Grad. Texts in Math., vol. 150, Springer, New York, 1995.
29. Ya. Eliashberg and M. Gromov, *Embeddings of Stein manifolds of dimension n into the affine space of dimension* $3n/2 + 1$, Ann. of Math. (2) **136** (1992), no. 1, 123–135.
30. J. E. Fornæss, *An increasing sequence of Stein manifolds whose limit is not Stein*, Math. Ann. **223** (1976), no. 3, 275–277.
31. O. Forster, *Plongements des variétés de Stein*, Comment. Math. Helv. **45** (1970), 170–184.
32. F. Forstnerič, *Equivalence of real submanifolds under volume-preserving holomorphic automorphisms of* $\mathbf{C}^n$, Duke Math. J. **77** (1995), no. 2, 431–445.
33. ———, *The Oka principle for sections of stratified fiber bundles*, Pure Appl. Math. Q. **6** (2010), no. 3, 843–874.
34. F. Forstnerič, J. Globevnik, and J.-P. Rosay, *Nonstraightenable complex lines in* $\mathbf{C}^2$, Ark. Mat. **34** (1996), no. 1, 97–101.
35. F. Forstnerič, B. Ivarsson, F. Kutzschebauch, and J. Prezelj, *An interpolation theorem for proper holomorphic embeddings*, Math. Ann. **338** (2007), no. 3, 545–554.

36. F. Forstnerič and E. Løw, *Global holomorphic equivalence of smooth submanifolds in* $\mathbf{C}^n$, Indiana Univ. Math. J. **46** (1997), no. 1, 133–153.
37. F. Forstnerič, E. Løw, and N. Øvrelid, *Solving the d- and* $\overline{\partial}$*-equations in thin tubes and applications to mappings*, Michigan Math. J. **49** (2001), no. 2, 369–416.
38. F. Forstnerič and J.-P. Rosay, *Approximation of biholomorphic mappings by automorphisms of* $\mathbf{C}^n$, Invent. Math. **112** (1993), no. 2, 323–349.
39. F. Forstnerič and E. F. Wold, *Bordered Riemann surfaces in* $\mathbb{C}^2$, J. Math. Pures Appl. (9) **91** (2009), no. 1, 100–114.
40. T. Fujita, *On the topology of noncomplete algebraic surfaces*, J. Fac. Sci. Univ. Tokyo Sect. IA Math. **29** (1982), no. 3, 503–566.
41. W. Fulton and J. Harris, *Representation theory: A first course*, Grad. Texts in Math., vol. 129, Springer, New York, 1991.
42. P. Griffiths and J. Harris, *Principles of algebraic geometry*, Wiley-Intersci., New York, 1978. Pure Appl. Math. (N. Y.)
43. M. Gromov, *Oka's principle for holomorphic sections of elliptic bundles*, J. Amer. Math. Soc. **2** (1989), no. 4, 851–897.
44. A. Grothendieck, *On the de Rham cohomology of algebraic varieties*, Inst. Hautes Études Sci. Publ. Math. **29** (1966), 95–103.
45. R. Hartshorne, *Algebraic geometry*, Vol. 52, Springer-Verlag, New York, 1977. Grad. Texts in Math.
46. P. Heinzner and F. Kutzschebauch, *An equivariant version of Grauert's Oka principle*, Invent. Math. **119** (1995), no. 2, 317–346.
47. R. B. Holmes, *Geometric functional analysis and its applications*, Vol. 24, Springer, New York, 1975. Grad. Texts in Math.
48. A. Huckleberry and A. Isaev, *Infinite-dimensionality of the automorphism groups of homogeneous Stein manifolds*, Math. Ann. **344** (2009), no. 2, 279–291.
49. Z. Jelonek, *The extension of regular and rational embeddings*, Math. Ann. **277** (1987), no. 1, 113–120.
50. Sh. Kaliman, *Extensions of isomorphisms between affine algebraic subvarieties of* k^n *to automorphisms of* k^n, Proc. Amer. Math. Soc. **113** (1991), no. 2, 325–334.
51. ———, *Some facts about Eisenman intrinsic measures*, Complex Variables Theory Appl. **27** (1995), no. 2, 163–173.
52. ———, *On a theorem of Ax*, Proc. Amer. Math. Soc. **133** (2005), no. 4, 975–977 (electronic).
53. ———, *Actions of* $\mathbb{C}^*$ *and* $\mathbb{C}_+$ *on affine algebraic varieties*, Algebraic Geometry — Seattle 2005. Part 2 (D. Abramovich, A. Bertram, L. Katzarkov, R. Pandharipande, and M. Thaddeus, eds.), Proc. Sympos. Pure Math., vol. 80, Amer. Math. Soc., Providence, RI, 2009, pp. 629–654.
54. S. Kaliman, M. Koras, L. Makar-Limanov, and P. Russell, $\mathbf{C}^*$*-actions on* $\mathbf{C}^3$ *are linearizable*, Electron. Res. Announc. Amer. Math. Soc. **3** (1997), 63–71 (electronic).
55. Sh. Kaliman and F. Kutzschebauch, *Density property for hypersurfaces* $UV = P(\overline{X})$, Math. Z. **258** (2008), no. 1, 115–131.
56. ———, *Criteria for the density property of complex manifolds*, Invent. Math. **172** (2008), no. 1, 71–87.
57. ———, *Algebraic volume density property of affine algebraic manifolds*, Invent. Math. **181** (2010), no. 3, 605–647.
58. Sh. Kaliman and M. Zaidenberg, *Affine modifications and affine hypersurfaces with a very transitive automorphism group*, Transform. Groups **4** (1999), no. 1, 53–95.
59. F. Knop, *Nichtlinearisierbare Operationen halbeinfacher Gruppen auf affinen Räumen*, Invent. Math. **105** (1991), no. 1, 217–220.
60. S. Kobayashi and K. Nomizu, *Foundations of differential geometry*. Vol. II, Wiley Classics Lib., Wiley-Intersci., New York, 1996.
61. H. Kraft and G. W. Schwarz, *Reductive group actions with one-dimensional quotient*, Inst. Hautes Études Sci. Publ. Math. **76** (1992), 1–97.
62. F. Kutzschebauch, *Andersén – Lempert-theory with parameters: a representation theoretic point of view*, J. Algebra Appl. **4** (2005), no. 3, 325–340.
63. ———, *Some results on embedding Stein spaces with interpolation*, Ark. Mat. **43** (2005), no. 2, 419–425.

64. F. Kutzschebauch and S. Lodin, *Families of holomorphic embeddings and group actions on affine space*, preprint.
65. F. Kutzschebauch, E. Løw, and E. F. Wold, *Embedding some Riemann surfaces into* $\mathbb{C}^2$ *with interpolation*, Math. Z. **262** (2009), no. 3, 603–611.
66. J. M. Landsberg, L. Manivel, and B. W. Westbury, *Series of nilpotent orbits*, Experiment. Math. **13** (2004), no. 1, 13–29.
67. I. Majcen, *Embedding certain infinitely connected subsets of bordered Riemann surfaces properly into* $\mathbb{C}^2$, J. Geom. Anal. **19** (2009), no. 3, 695–707.
68. M. McQuillan, *Non-commutative Mori theory*, Technical Report M/01/42, Inst. Hautes Études Sci., Bures-sur-Yvette, 2001.
69. H. Peters and E. F. Wold, *Non-autonomous basins of attraction and their boundaries*, J. Geom. Anal. **15** (2005), no. 1, 123–136.
70. R. Remmert, *Sur les espaces analytiques holomorphiquement séparables et holomorphiquement convexes*, C. R. Acad. Sci. Paris **243** (1956), 118–121.
71. J.-P. Rosay, *Automorphisms of* $\mathbf{C}^n$*, a survey of Andersén – Lempert theory and applications*, Complex Geometric Analysis in Pohang (Pohang, 1997) (K.-T. Kim and S. G. Krantz, eds.), Contemp. Math., vol. 222, Amer. Math. Soc., Providence, RI, 1999, pp. 131–145.
72. J.-P. Rosay and W. Rudin, *Holomorphic maps from* $\mathbf{C}^n$ *to* $\mathbf{C}^n$, Trans. Amer. Math. Soc. **310** (1988), no. 1, 47–86.
73. ———, *Holomorphic embeddings of* $\mathbf{C}$ *in* $\mathbf{C}^n$, Several Complex Variables (Stockholm, 1987/1988) (J. E. Fornæss, ed.), Math. Notes, vol. 38, Princeton Univ. Press, Princeton, NJ, 1993, pp. 563–569.
74. J. Schürmann, *Embeddings of Stein spaces into affine spaces of minimal dimension*, Math. Ann. **307** (1997), no. 3, 381–399.
75. L. Schwartz, *Produits tensoriels topologiques et espaces nucléaires.*
76. ———, *Théorie des distributions à valeurs vectorielles.* I, Ann. Inst. Fourier, Grenoble **7** (1957), 1–141.
77. G. W. Schwarz, *Exotic algebraic group actions*, C. R. Acad. Sci. Paris Sér. I Math. **309** (1989), no. 2, 89–94.
78. V. Srinivas, *On the embedding dimension of an affine variety*, Math. Ann. **289** (1991), no. 1, 125–132.
79. B. Stensønes, *Fatou – Bieberbach domains with* C^∞*-smooth boundary*, Ann. of Math. (2) **145** (1997), no. 2, 365–377.
80. M. Suzuki, *Sur les opérations holomorphes du groupe additif complexe sur l'espace de deux variables complexes*, Ann. Sci. École Norm. Sup. (4) **10** (1977), no. 4, 517–546.
81. ———, *Sur les opérations holomorphes de* $\mathbf{C}$ *et de* $\mathbf{C}^*$ *sur un espace de Stein*, Fonctions de plusieurs variables complexes. III (Sém. François Norguet, 1975) (F. Norguet, ed.), Lecture Notes in Math., vol. 670, Springer, Berlin, 1978, pp. 80–88.
82. Á. Tóth and D. Varolin, *Holomorphic diffeomorphisms of complex semisimple Lie groups*, Invent. Math. **139** (2000), no. 2, 351–369.
83. ———, *Holomorphic diffeomorphisms of semisimple homogeneous spaces*, Compos. Math. **142** (2006), no. 5, 1308–1326.
84. F. Trèves, *Topological vector spaces, distributions and kernels*, Academic Press, New York, 1967.
85. D. Varolin, *The density property for complex manifolds and geometric structures.* II, Internat. J. Math. **11** (2000), no. 6, 837–847.
86. ———, *The density property for complex manifolds and geometric structures*, J. Geom. Anal. **11** (2001), no. 1, 135–160.
87. J. Winkelmann, *On automorphisms of complements of analytic subsets in* $\mathbf{C}^n$, Math. Z. **204** (1990), no. 1, 117–127.
88. ———, *Large discrete sets in Stein manifolds*, Math. Z. **236** (2001), no. 4, 883–901.
89. E. F. Wold, *Embedding Riemann surfaces properly into* $\mathbb{C}^2$, Internat. J. Math. **17** (2006), no. 8, 963–974.
90. ———, *Proper holomorphic embeddings of finitely and some infinitely connected subsets of* $\mathbb{C}$ *into* $\mathbb{C}^2$, Math. Z. **252** (2006), no. 1, 1–9.
91. ———, *Embedding subsets of tori properly into* $\mathbb{C}^2$, Ann. Inst. Fourier (Grenoble) **57** (2007), no. 5, 1537–1555.

92. ———, *A Fatou – Bieberbach domain in $\mathbb{C}^2$ which is not Runge*, Math. Ann. **340** (2008), no. 4, 775–780.
93. ———, *A long $\mathbb{C}^2$ which is not Stein*, Ark. Mat. **48** (2010), no. 1, 207–210.
94. M. G. Zaidenberg, *Isotrivial families of curves on affine surfaces, and the characterization of the affine plane*, Izv. Akad. Nauk SSSR Ser. Mat. **51** (1987), no. 3, 534–567 (Russian); English transl., Math. USSR-Izv. **30** (1988), no. 3, 503–532.

DEPARTMENT OF MATHEMATICS, UNIVERSITY OF MIAMI, CORAL GABLES, FL 33124, USA
E-mail address: kaliman@math.miami.edu

MATHEMATISCHES INSTITUT, UNIVERSITÄT BERN, SIDLERSTRASSE 5, 3012 BERN, SWITZERLAND
E-mail address: frank.kutzschebauch@math.unibe.ch

Centre de Recherches Mathématiques
CRM Proceedings and Lecture Notes
Volume **54**, 2011

Group Actions on Affine Cones

Takashi Kishimoto, Yuri Prokhorov, and Mikhail Zaidenberg

To Peter Russell on the occasion of his 70th birthday

ABSTRACT. We address the following question:

For which smooth projective varieties the corresponding affine cone admits an action of a connected algebraic group different from the standard $\mathbb{G}_m$-action by scalar matrices and its inverse action?

We show in particular that the affine cones over anticanonically embedded smooth del Pezzo surfaces of degree ≥ 4 possess such an action. Besides, we give some examples of rational Fano threefolds which have this property. A question in [29] whether this property holds also for smooth cubic surfaces in $\mathbb{P}^3$, occurs to be out of reach for our methods. We provide nevertheless a general geometric criterion that could be helpful in this case as well.

Introduction

All varieties in this paper are defined over $\mathbb{C}$. By Corollary 1.13 in [29], an isolated Cohen–Macaulay singularity (X, x) of a normal quasiprojective variety X is rational provided that X admits an effective action of the additive group $\mathbb{G}_a = \mathbb{G}_a(\mathbb{C}) = \mathbb{C}_+$, in particular of a connected nonabelian algebraic group. In the opposite direction, let us observe that, for instance, the singularity at the origin of the affine Fermat cubic in $\mathbb{A}^4$

$$x_1^3 + x_2^3 + x_3^3 + x_4^3 = 0$$

is rational. The question was raised [29, Question 2.22] whether it also admits a nondiagonal action of a connected algebraic group, in particular, a $\mathbb{G}_a$-action.

2010 *Mathematics Subject Classification.* Primary 14R20, 14J45; Secondary 14J50, 14R05.

Key words and phrases. Affine cone, del Pezzo surface, Fano threefold, automorphism, additive group, group action.

The first author was supported by a Grant-in-Aid for Scientific Research of JSPS No. 20740004.

The second author was partially supported by an RFBR grant No. 11-01-00336-a and a Leading Scientific Schools grant No. NSh-4713.2010.1.

This work was done during a stay of the second and third authors at the Max Planck Institut für Mathematik at Bonn and a stay of the first and the second authors at the Institut Fourier, Grenoble. The authors thank these institutions for hospitality.

This is the final form of the paper.

So far, we do not know the answer. However, we answer in affirmative a similar question for all del Pezzo surfaces of degree $d \geq 4$.

0.0.1. Theorem. *Let Y_d be a smooth del Pezzo surface of degree d anticanonically embedded into $\mathbb{P}^d$, and let $X_d = \operatorname{Aff\,Cone}(Y_d) \subseteq \mathbb{A}^{d+1}$ be the affine cone over Y_d. If $4 \leq d \leq 9$ then X_d admits a nontrivial $\mathbb{G}_a$-action. Consequently, the automorphism group $\operatorname{Aut}(X_d)$ is infinite-dimensional. Moreover, the Makar-Limanov invariant of X_d is trivial.*

Recall [23, 10.1.1] that for $d \leq 5$ the group $\operatorname{Aut}(Y_d)$ is finite. By definition, the Makar-Limanov invariant of an affine variety X is the subring of $\mathcal{O}(X)$ of common invariants of all $\mathbb{G}_a$-actions on X. It is trivial when it consists of the constants.

One of our main results (Theorem 3.1.9) provides a necessary and sufficient condition for the existence of a nontrivial $\mathbb{G}_a$-action on an affine cone. As a corollary, for affine cones of dimension 3 we obtain the following geometric criterion.

0.0.2. Theorem. *Let Y be a smooth projective rational surface with a polarization $\varphi_{|H|}\colon Y \hookrightarrow \mathbb{P}^n$, and let $X = \operatorname{Aff\,Cone}_H(Y) \subseteq \mathbb{A}^{n+1}$ be the affine cone over $Y \subseteq \mathbb{P}^n$. Then X admits a nontrivial $\mathbb{G}_a$-action if and only if Y contains an H-polar cylinder, i.e., a cylindrical Zariski open set*

$$U = Y \setminus \operatorname{supp} D \simeq Z \times \mathbb{A}^1,$$

where Z is an affine curve and $D \in |\mathrm{d}H|$ is an effective divisor on $\mathbb{P}^n$.

Using this criterion, we show in Proposition 3.1.13 that for every smooth projective rational surface Y there exists a polarization $\varphi_{|H|}\colon Y \hookrightarrow \mathbb{P}^n$ such that Y contains an H-polar cylinder and so the corresponding affine cone possesses an effective action of $\mathbb{G}_a$. It would be interesting to classify in any dimension all pairs (Y, H), where Y is a smooth projective variety and H an ample divisor on Y, such that the affine cone $X = \operatorname{Aff\,Cone}_H(Y)$ admits an effective $\mathbb{G}_a$-action. We recover this classification for $\dim_{\mathbb{C}}(Y) = 1$ and give some concrete examples in higher dimensions, especially in dimensions 2 and 3.

A theorem due to Matsumura, Monsky and Andreotti (see [79], or [42, §I.4], or Section 1 below) claims that any automorphism of a smooth hypersurface Y in $\mathbb{P}^n$ of degree d, where $d, n \geq 3$ and $(d, n) \neq (4, 3)$, is restriction of a unique projective linear transformation, and $\operatorname{Aut}(Y)$ is a finite group. In Corollary 2.1.4 we show that the automorphism group $\operatorname{Aut}(X)$ of the affine cone X over a smooth, nonbirationally ruled projective variety Y is a linear group, and actually a central extension of a finite group by $\mathbb{G}_m = \mathbb{C}^*$. Consequently, among the affine cones over smooth projective surfaces in $\mathbb{P}^3$, only those of degree ≤ 3 can admit a nontrivial action of a connected algebraic group, and their automorphism groups can be infinite-dimensional. Actually, the 3-fold affine quadric cone possesses an effective linear action of the additive group $\mathbb{G}_a^2$, see Example 3.1.3 below or [5, 105].

In Section 1 we give a short overview of the known results on the automorphism groups. In Section 2 we collect generalities on automorphisms of affine cones. Theorems 0.0.1 and 0.0.2 are proven in Section 3. In Section 4 we summarize some geometric facts that could be useful (in view of the criterion of Theorem 0.0.2) in order to answer Question 2.22 in [29] cited above. In the final Section 5 we describe two families of rational Fano threefolds such that the affine cones over their anti-canonical embeddings possess effective $\mathbb{G}_a$-actions.

Acknowledgements. We thank Alvaro Liendo, who participated in discussions on early stages of this work, for his attention and comments. Our thanks also to Dmitry Akhiezer, Michel Brion, and Dimitri Timashev for providing useful information on homogeneous varieties, to Dmitry Akhiezer and Ivan Arzhantsev for reading some chapters and valuable remarks. We are grateful to the referee for valuable comments.

1. Group actions on projective varieties

In this section we recall some well known facts about the automorphism groups of projective or quasiprojective varieties; see, e.g., [42, §I.4; 118, §II.3]. For an algebraic variety Y, we let $\mathrm{Aut}(Y)$ denote the group of all biregular automorphisms of Y and $\mathrm{Bir}(Y)$ the group of all birational transformations of Y into itself. For a projective or an affine embedding $Y \hookrightarrow \mathbb{P}^n$ ($Y \hookrightarrow \mathbb{A}^n$, respectively) we let $\mathrm{Lin}(Y)$ denote the group of all automorphisms of Y which extend linearly to the ambient space.

1.1. Automorphisms of smooth projective hypersurfaces. In the following theorems we gather some results concerning the groups Lin, Aut, and Bir for projective hypersurfaces; see Matsumura and Monsky [79], Iskovskikh and Manin [55], Pukhlikov [95–97], Cheltsov [11], de Fernex, Ein and Mustaţă [19].

1.1.1. Theorem. *Let Y be a smooth hypersurface in $\mathbb{P}^n$ of degree d. Then for all $d, n \geq 3$ except for $(d,n) = (4,3)$,*

$$\mathrm{Aut}(Y) = \mathrm{Lin}(Y)$$

and this group is finite. It is trivial for a general hypersurface of degree $d \geq 3$.

There is a similar result for Schubert hypersurfaces in flag varieties, see Theorem 8.8 in [109]. For the group of birational transformations, the following hold.

1.1.2. Theorem. *For $Y \subseteq \mathbb{P}^n$ as above and for all $d > n \geq 2$ except for $(d,n) = (3,2)$,*

$$\mathrm{Bir}(Y) = \mathrm{Aut}(Y).$$

This group is finite except in the case $(d,n) = (4,3)$ of a smooth quartic surface $Y \subseteq \mathbb{P}^3$, where it is discrete, but can be infinite and different from $\mathrm{Lin}(Y)$ *which is finite. The group* $\mathrm{Bir}(Y)$ *of a very general quartic surface $Y \subseteq \mathbb{P}^3$ is trivial.*

The case $d \leq n$ is much more complicated. However, in this case there are deep partial results, see, e.g., [19, 55, 95].

Let us indicate briefly some ideas used in the proofs. In case $d \neq n+1$ the proof is easy and exploits the fact that the canonical divisor $K_Y = \mathcal{O}_Y(d-n-1)$ is $\mathrm{Aut}(Y)$-stable. In case $d = n+1$ the equalities $\mathrm{Bir}(Y) = \mathrm{Aut}(Y) = \mathrm{Lin}(Y)$ follow since such hypersurfaces represent Mori minimal models. Indeed a birational map between minimal models is an isomorphism in codimension 1, see, e.g., [66], hence it induces an isomorphism of the corresponding Picard groups. If $n \geq 4$ then $\mathrm{Pic}(Y) \simeq \mathbb{Z}$ by the Lefschetz Hyperplane Section Theorem. Therefore any birational transformation φ of Y acts trivially on $\mathrm{Pic}(Y)$ and so preserves the complete linear system of hyperplane sections $|\mathcal{O}_Y(1)|$. Since Y is linearly normal,[1] φ is induced by a projective linear transformation of the ambient projective space $\mathbb{P}^n$. For the

[1] I.e., $H^0(\mathcal{O}_{\mathbb{P}^n}(1)) \to H^0(\mathcal{O}_Y(1))$ is a surjection.

proof of finiteness of the group $\mathrm{Lin}(Y)$ and its triviality for general hypersurfaces, we refer to the classical paper of Matsumura and Monsky [79].

By virtue of the Noether – Lefschetz Theorem, these arguments can be equally applied to very general smooth surfaces in $\mathbb{P}^3$ of degree $d \geq 4$. For an arbitrary smooth surface Y in $\mathbb{P}^3$, the minimality of Y should be combined with the fact that $\mathrm{Pic}(Y)$ is torsion free. Indeed, any smooth surface in $\mathbb{P}^3$ of degree $d \geq 4$ represents a minimal model and so is not birationally ruled, hence its birational automorphisms are biregular; see, e.g., [78, Theorem 1-8-6]. For $d > 4$ the canonical class K_Y defines an equivariant polarization of Y, and $\mathcal{O}_Y(1) \sim (1/(d-4))K_Y$. Since $\mathrm{Pic}(Y)$ is torsion free and $H^0(\mathcal{O}_{\mathbb{P}^3}(1)) \to H^0(\mathcal{O}_Y(1))$ is a surjection, all automorphisms of Y are linear. This is not true, in general, in the case of a smooth quartic surface in $\mathbb{P}^3$. An example of such a surface with infinite automorphism group due to Fano and Severi is discussed in [79, Theorem 4]. A nonlinear biregular involution exists on any smooth quartic in $\mathbb{P}^3$ containing skew lines, for instance, on the Fermat quartic $x^4 + y^4 + z^4 + u^4 = 0$; see Takahashi [108].

For a quadric hypersurface $X \subseteq \mathbb{A}^{n+1}$ of dimension $n \geq 2$, the group $\mathrm{Aut}(X)$ is infinite-dimensional [110, Lemma 1.1], cf. also Example 3.1.3 below. For $n = 2$ this group has an amalgamated product structure [36]; cf. also [73, 74].

For a smooth cubic surface $Y \subseteq \mathbb{P}^3$ the group $\mathrm{Aut}(Y) = \mathrm{Lin}(Y)$ is finite, while the Cremona group $\mathrm{Bir}(Y) \simeq \mathrm{Bir}(\mathbb{P}^2)$ is infinite-dimensional. The automorphism groups of such surfaces were listed by Hosoh [45] who corrected an earlier classification by Segre [103]; see also Manin [75] and Dolgachev [23]. The largest order of such a group is 648. This upper bound is achieved only for the Fermat cubic surface, see [46]. The least common multiple of the orders of all these automorphism groups is $3240 = 2^3 \cdot 3^4 \cdot 5$ (Gorinov [39]; see also Gonzlez-Aguilera and Liendo [37, 38]).

The Fermat quartic $x^4 + y^4 + z^4 + u^4 = 0$ and the smooth quartic $x^4 + y^4 + z^4 + u^4 + 12xyzu = 0$ in $\mathbb{P}^3$ can be also characterized in terms of the orders of their automorphism groups, see Mukai [83], Kondō [69], and Oguiso [87].

1.2. Automorphisms of smooth projective varieties. According to the well-known Matsumura Theorem[2] the group $\mathrm{Bir}(Y)$ of a smooth projective variety Y of general type is finite, hence also the group $\mathrm{Aut}(Y)$ is. In particular, this holds if $c_1(Y) < 0$. See, e.g., Xiao [115, 116] for effective bounds of orders of automorphism groups for the general type surfaces. A vast literature is devoted to automorphism groups of $K3$ and Enriques surfaces. These groups are discrete, and infinite in many cases. Finite groups of automorphism of $K3$-surfaces were classified by Nikulin and Mukai (see, e.g., [22, 57, 68, 76, 83, 84, 116]).

For varieties of nongeneral type we have the following result due to Kalka – Shiffman – Wong [59] and Lin [118, Theorem II.3.1.2].

1.2.1. Theorem. *Let Y be a smooth projective variety of dimension n. Suppose that not all Chern numbers of Y vanish and either $c_1(Y) \leq 0$ or $H^{n,0}(Y) \neq 0$. Then* $\mathrm{Aut}(Y)$ *is a discrete group.*

The first assumption is fulfilled, for instance, if $\mathrm{e}(Y) = c_n(Y) \neq 0$, or $\mathrm{e}(\mathcal{O}_Y) \neq 0$, or $c_1^n(Y) \neq 0$, where e stands for the Euler characteristic. However, this assumption does not hold for an abelian variety $Y = A$. For any projective embedding

[2]Which generalizes earlier results by Andreotti for surfaces of general type; see Kobayashi – Ochiai [63] and Noguchi – Sunada [85] for further generalizations.

$A \hookrightarrow \mathbb{P}^N$, the group $\mathrm{Lin}(A)$ is finite, see [42, §II.6], while the group $\mathrm{Aut}(A) \supseteq A$ is infinite. Conversely, by Théorème Principal I of Blanchard [8] for any finite subgroup $G \subseteq \mathrm{Aut}(A)$ there exists a projective embedding $A \hookrightarrow \mathbb{P}^N$ which linearizes G. A general form of Blanchard's Theorem is as follows (cf. [3, Theorem 3.2.1]).

1.2.2. Theorem. *Let Y be a smooth projective variety and $G \subseteq \mathrm{Aut}(Y)$ a subgroup which acts finitely on $\mathrm{Pic}(Y)$. Then there is a G-equivariant projective embedding $Y \hookrightarrow \mathbb{P}^N$.*

Indeed, such an embedding corresponds to a very ample G-invariant divisor class. However, if G acts finitely on $\mathrm{Pic}(Y)$ then the orbit of any ample class is an ample G-invariant class.

For instance, if $\mathrm{Pic}(Y)$ is discrete and G is connected then G acts trivially on $\mathrm{Pic}(Y)$. Hence there exists a G-equivariant projective embedding $Y \hookrightarrow \mathbb{P}^N$.

As another example, consider a smooth Fano variety Y embedded by a pluri-anticanonical system $\varphi_{|-mK_Y|}\colon Y \hookrightarrow \mathbb{P}^N$ for a suitable $m > 0$. The canonical bundle K_Y being stable under the action of the automorphism group $\mathrm{Aut}(Y)$ on Y, this embedding is equivariant and realizes $\mathrm{Aut}(Y)$ as a closed subgroup of $\mathrm{PGL}_{N+1}(\mathbb{C})$. In particular, this applies to the anticanonical embeddings of del Pezzo surfaces $Y_d \hookrightarrow \mathbb{P}^d$ of degree $d \ge 3$. A rough description of the automorphisms groups of these surfaces is as follows, see Proposition 10.1.1 in [23] (cf., e.g., [6, 7, 18, 20, 24, 25, 44] for more delicate properties).

1.2.3. Theorem. *Let $Y = Y_d$ be a del Pezzo surface of degree $d \ge 3$. Then the automorphism group $\mathrm{Aut}(Y)$ acts on the lattice $Q = (\mathbb{Z}K_Y)^\perp \subseteq \mathrm{Pic}(Y)$ preserving the intersection form. The image of the corresponding homomorphism $\rho\colon \mathrm{Aut}(Y) \to O(Q)$ is contained in the Weyl group $W(Q)$. The kernel of ρ is trivial for $d \le 5$ and is a connected linear algebraic group of dimension $2d - 10$ for $d \ge 6$. More precisely, the following hold.*

(1) *For $d \le 5$ the group $\mathrm{Aut}(Y)$ is finite.*
(2) *For $d \ge 6$ the identity component $\mathrm{Aut}_0(Y) = \ker(\rho)$ contains a 2-torus $\mathbb{T}_2 \simeq \mathbb{G}_m^2$, and $\mathrm{Aut}_0(Y) = \mathbb{T}_2$ for $d = 6$.*
(3) *For $d \ge 7$ besides the 2-torus $\mathbb{T}_2$ the group $\mathrm{Aut}_0(Y)$ contains a subgroup isomorphic to $\mathbb{A}_+^2 = \mathbb{G}_a^2$. In particular, for $d = 7$ there are a decomposition*

$$\mathrm{Aut}(Y) \simeq (\mathbb{A}_+^2 \rtimes \mathbb{T}_2) \rtimes \mathbb{Z}/2\mathbb{Z}$$

and a faithfull presentation $\mathrm{Aut}_0(Y) \hookrightarrow \mathrm{GL}_3(\mathbb{C})$ with image

$$\begin{pmatrix} 1 & 0 & * \\ 0 & * & * \\ 0 & 0 & * \end{pmatrix} = \begin{pmatrix} 1 & 0 & 0 \\ 0 & * & 0 \\ 0 & 0 & * \end{pmatrix} \cdot \begin{pmatrix} 1 & 0 & * \\ 0 & 1 & * \\ 0 & 0 & 1 \end{pmatrix}$$

(4) *For $d = 8$ either $Y \to \mathbb{P}^2$ is a blowup at a point and then $\mathrm{Aut}(Y) \simeq \mathrm{GL}_2 \ltimes \mathbb{A}_+^2$, or $Y \simeq \mathbb{P}^1 \times \mathbb{P}^1$ and then $\mathrm{Aut}(Y) \simeq \left(\mathrm{PGL}_2(\mathbb{C})\right)^2 \rtimes \mathbb{Z}/2\mathbb{Z}$.*
(5) *Finally for $d = 9$, $Y \simeq \mathbb{P}^2$ and $\mathrm{Aut}(Y) \simeq \mathrm{PGL}_3(\mathbb{C})$.*

1.2.4. Remark. An effective $\mathbb{A}_+^2$-action on a del Pezzo surface Y of degree $d = 7$ can be defined via the locally nilpotent derivations

$$\partial_{\alpha,\beta} = \alpha z \frac{\partial}{\partial x} + \beta z \frac{\partial}{\partial y}, \qquad (\alpha, \beta) \in \mathbb{A}_+^2. \tag{1}$$

Indeed, the induced $\mathbb{A}^2_+$-action on $\mathbb{A}^3$:

$$(\alpha, \beta).(x, y, z) = (x + \alpha z, y + \beta z, z)$$

descends to an action on $\mathbb{P}^2$ fixing the line $z = 0$ pointwise. The blowup at two points on this line preserves the action. Likewise one defines an $\mathbb{A}^2_+$-action on Y for $d = 8$ or $d = 9$.

Consult also a recent preprint by Cantat and Dolgachev [10] on rational surfaces with a large automorphism group.

1.3. Homogeneous and almost homogeneous varieties. By the Borel–Remmert Theorem [3, 3.9] any connected, compact, homogeneous Kähler manifold V is biholomorphic to the product $\mathrm{Alb}(V) \times Y$ of the Albanese torus and a (generalized) flag variety $Y = G/P$ (i.e., Y is the quotient of a connected semisimple linear algebraic group by a parabolic subgroup)[3]. It follows that every simply connected homogeneous compact Kähler manifold is a flag variety and the same is true for a rational projective homogeneous variety (for homogeneous compact complex manifolds satisfying both conditions this was established by Goto [40]). Furthermore, Grauert and Remmert [41] carried over a result of Chow [14] from abstract algebraic to Moishezon varieties. Namely, they proved that a homogeneous Moishezon variety is projective algebraic. Thus, if such a variety is simply connected or rational, it is a flag variety.

Every flag variety G/P is a projective rational Fano variety (see [106]). Every ample line bundle L on G/P is very ample (see, e.g., [13; 58; 71, §3.3.2; 109, Theorem 7.52]). The complete linear system $|L|$ defines a G-equivariant embedding $Y \hookrightarrow \mathbb{P}^n$ with a projectively normal image [98, Theorem 1.iii].

For a maximal parabolic subgroup $P_{\max} \subseteq G$, the Picard group $\mathrm{Pic}(G/P_{\max}) \cong \mathbb{Z}$ is generated by the class of a unique Schubert divisorial cycle in $G/P_{\max}$, and this class is very ample. In the case of a Grassmannian this class gives the Plücker embedding. For an arbitrary flag variety G/P, its Picard group $\mathrm{Pic}(G/P)$ is also generated by the classes of the Schubert divisorial cycles; see, e.g., [13, 90]. The set of maximal parabolic subgroups $P_{\max}$ of G which contain P is finite. Every Schubert divisor class in $\mathrm{Pic}(G/P)$ is lifted via a surjection $G/P \to G/P_{\max}$, see, e.g. [70, 106]. A linear combination of these divisors is very ample if and only if its coefficients are all positive (see [9] for the case of a full flag variety; the general case is similar [109, Theorem 7.52]).

For a description of the automorphism groups of flag varieties see, e.g., [3, §3.3].

The following "cone theorem" describes certain *almost homogeneous* complex varieties. It is due to Akhiezer [1, Theorem 3] in algebraic context and to Huckleberry and E. Oeljeklaus [48] in analytic one.[4]

1.3.1. Theorem. *Let X be an irreducible reduced complex space of dimension ≥ 2. Suppose that a connected complex Lie group acts by biholomorphic transformations on X with an open orbit $\Omega \subseteq X$ such that the complement $E = X \setminus \Omega$ is a proper analytic subset with an isolated point, say, $0 \in E$. Then the normalization $\nu\colon \widetilde{X} \to X$ is one-to-one and $\widetilde{X}$ is biholomorphic to a projective or an affine cone over a flag variety G/P of a semisimple linear algebraic group G under a certain*

[3]See also [102] for a more general result in the projective case.

[4]The smooth compact case was done first in [86]. See also [49, Chapter 2, §3, Theorem 1] for real groups.

equivariant projective embedding. The isolated point $0 \in E$ *corresponds to the vertex of the cone. In particular, if* $(X, 0)$ *is smooth then* $X \simeq \mathbb{A}^n$ *or* $X \simeq \mathbb{P}^n$.

Thus the variety X as in the theorem equipped with an appropriate algebraic structure carries a regular almost transitive group action. If the initial group is a complex linear algebraic group, then G is its maximal semisimple subgroup [1]. Given any G-equivariant projective embedding $\varphi_{|H|}\colon Y = G/P \hookrightarrow \mathbb{P}^n$, where $\dim Y \geq 1$, the affine cone $\mathrm{Aff\,Cone}_H(Y)$ over the image admits a regular action transitive off the vertex of a locally direct product $\widetilde{G} \cdot \mathbb{G}_m$ with $\mathbb{G}_m$ acting by homotheties, where $\widetilde{G} \to G$ is a finite group cover.

A similar description exists for the class of quasi-projective G-varieties X, where G is a connected linear algebraic group acting on X with an open orbit Ω, provided that there is an equivariant completion $\overline{X}$ of X with disconnected complement $\overline{X} \setminus \Omega$ [1, Theorem 2]. See also [2] for the case that $\overline{X} \setminus \Omega$ is a G-orbit of codimension 1 in X (in this case it is connected).

An explicit description of almost homogeneous 2-dimensional affine cones over smooth projective curves is due to Popov [89] (see also [30] for an alternative proof). We recall that a Veronese cone V_d is the affine cone over a smooth rational normal curve $\Gamma_d \subseteq \mathbb{P}^d$, i.e., a linearly nondegenerate[5] smooth curve in $\mathbb{P}^d$ of degree d. All such curves in $\mathbb{P}^d$ are projectively equivalent and rational. For normal 2-dimensional cones, Popov's Theorem can be stated as follows.

1.3.2. Theorem (V. Popov). *Let* X *be the affine cone over a smooth projective curve* Y. *If* X *is normal and admits an algebraic group action transitive in* $X \setminus \{0\}$, *then* X *is a Veronese cone* V_d *for some* $d \geq 1$, *and* Y *is a rational normal curve* Γ_d.

Popov [89] actually classified all almost homogeneous cones in dimension 2 with an isolated singularity (not necessarily normal). Every such cone possesses a linear $\mathrm{SL}(2, \mathbb{C})$-action transitive off the vertex. The group $\mathrm{Aut}(X)$ of a Veronese cone is infinite-dimensional and so cannot be linearized under an affine embedding; see Section 2.3 below.

2. Groups acting on affine cones

2.1. Linear automorphisms of affine cones. Let us start with the following result.

2.1.1. Proposition. *Given two affine cones* $X_i = \mathrm{Aff\,Cone}(Y_i) \subseteq \mathbb{A}^{n_i+1}$ *over smooth, linearly nondegenerate, projective varieties* $Y_i \subseteq \mathbb{P}^{n_i}$ $(i = 1, 2)$ *and an isomorphism* $\varphi\colon X_1 \xrightarrow{\simeq} X_2$, *the differential* $\mathrm{d}\varphi(0)$ *provides a linear isomorphism* $\Phi\colon \mathbb{A}^{n_1+1} \xrightarrow{\simeq} \mathbb{A}^{n_2+1}$ *which restricts to an isomorphism* $\Phi|_{X_1}\colon X_1 \xrightarrow{\simeq} X_2$. *In particular* $n_1 = n_2$, *and* Y_1 *and* Y_2 *are projectively equivalent.*

Proof. By the linear nondegeneracy assumption

$$T_0 X_i = \mathbb{A}^{n_i+1}, \quad C_0 X_i = X_i, \quad \text{and} \quad \mathbb{P}(C_0 X_i) = Y_i, \qquad i = 1, 2,$$

where $T_0 X_i$ is the Zariski tangent space to X_i at the vertex $0 \in X_i$, and $C_0 X_i$ is the tangent cone in 0 (see, e.g., [16, §9.7]). Now the assertion follows since $d\varphi(0)$ provides an isomorphism of the Zariski tangent spaces and sends the cone

[5] A projective variety $Y \subseteq \mathbb{P}^n$ is linearly nondegenerate if it is not contained in any hyperplane.

C_0X_1 onto the cone C_0X_2 [17, §7.3]. In fact $\mathrm{d}\varphi(0)$ lifts to an isomorphism of blowups $\mathrm{Bl}_0(X_1) \xrightarrow{\cong} \mathrm{Bl}_0(X_2)$ preserving the exceptional divisors. These divisors are isomorphic to Y_1 and Y_2, respectively, and $\mathrm{d}\varphi(0)$ induces a linear isomorphism $Y_1 \xrightarrow{\cong} Y_2$. □

2.1.2. Remark. The isomorphism φ as in Proposition 2.1.1 does not need to be linear itself. However, this is the case under the additional assumption that Y_1 is not birationally ruled (see Proposition 2.1.3 below). A birationally ruled projective variety is a variety birationally equivalent to a product $Z \times \mathbb{P}^1$. Recall also that a birational map $\bar{f}\colon \overline{X}_1 \dashrightarrow \overline{X}_2$ is said to be *isomorphism in codimension one* if there are subsets $B_i \subseteq \overline{X}_i$ of codimension at least 2 such that

$$\bar{f}|(\overline{X}_1 \setminus B_1) : \overline{X}_1 \setminus B_1 \to \overline{X}_2 \setminus B_2$$

is an isomorphism.

2.1.3. Proposition. *Consider the affine cones $X_i = \operatorname{Aff\,Cone}(Y_i) \subseteq \mathbb{A}^{n_i+1}$ over projective varieties $Y_i \subsetneq \mathbb{P}^{n_i}$, $i = 1, 2$. Suppose that Y_1 and Y_2 are smooth, irreducible, and linearly nondegenerate. If Y_1 is not birationally ruled then every isomorphism $\varphi : X_1 \xrightarrow{\cong} X_2$ extends to a unique linear isomorphism $\mathbb{A}^{n_1+1} \xrightarrow{\cong} \mathbb{A}^{n_2+1}$. In particular $n_1 = n_2$, and Y_1 and Y_2 are projectively equivalent.*

The proposition follows immediately from Lemmas 2.1.7 and 2.1.8 below. Before passing to the lemmas, let us give two corollaries, which are the main results of this subsection.

2.1.4. Corollary. *Let $X = \operatorname{Aff\,Cone}(Y) \subseteq \mathbb{A}^{n+1}$ be the affine cone over a smooth projective variety $Y \subseteq \mathbb{P}^n$. If Y is not birationally ruled then* $\operatorname{Aut}(X) = \operatorname{Lin}(X)$. *Moreover,* $\operatorname{Aut}(X)$ *is a central extension of the group* $\operatorname{Lin}(Y)$ *by* $\mathbb{G}_m$.

Indeed, the exact sequence

$$0 \to \mathbb{G}_m \to \mathrm{GL}(n+1, \mathbb{C}) \to \mathrm{PGL}(n+1, \mathbb{C}) \to 0$$

yields the following one:

$$0 \to \mathbb{G}_m \to \operatorname{Lin}(X) \xrightarrow{\pi} \operatorname{Lin}(Y) \to 0. \tag{2}$$

2.1.5. Corollary. *Let $X = \operatorname{Aff\,Cone}(Y)$ be the affine cone over a smooth projective 3-fold Y. Suppose that Y is rationally connected and nonrational. Then* $\operatorname{Aut}(X) = \operatorname{Lin}(X)$.

Proof. Indeed if Y were birationally ruled, i.e., birational to a product $Z \times \mathbb{P}^1$, then Z would be rationally connected and so a rational surface. Hence Y would be rational too, contrary to our assumption. Thus Corollary 2.1.4 applies and gives the assertion. □

2.1.6. Example. For instance, if $Y \subseteq \mathbb{P}^n$ is a nonrational Fano 3-fold and $X = \operatorname{Aff\,Cone}(Y)$, then $\operatorname{Aut}(X) = \operatorname{Lin}(X)$. As an example, one can consider any smooth cubic or quartic 3-fold $Y \subseteq \mathbb{P}^4$.

For the proof of the next lemma we refer the reader to [15, Proposition 2.7; 110, Theorem 2.2].

2.1.7. Lemma. *Let $X_i = \overline{X}_i \backslash D_i$, $i = 1, 2$, where $\overline{X}_i$ is a projective variety and D_i an irreducible divisor on $\overline{X}_i$. Suppose that X_i is regular near D_i for $i = 1, 2$. If D_1 is not birationally ruled then any isomorphism $f\colon X_1 \to X_2$ extends to a birational map $\bar{f}\colon \overline{X}_1 \dashrightarrow \overline{X}_2$ which is an isomorphism in codimension 1. If in addition the divisors D_1 and D_2 are ample then $\bar{f}\colon \overline{X}_1 \xrightarrow{\simeq} \overline{X}_2$ is an isomorphism.*

Proposition 2.1.3 is now a direct consequence of the following lemma.

2.1.8. Lemma. *Consider two projective varieties $Y_i \subsetneq \mathbb{P}^{n_i}$, where $n_i \geq 2$, $i = 1, 2$. Suppose that Y_1 and Y_2 are smooth, irreducible, and linearly nondegenerate. Consider also the affine cones $X_i = \operatorname{Aff Cone}(Y_i) \subseteq \mathbb{A}^{n_i+1}$ over Y_i and the projective cones $\overline{X}_i \subseteq \mathbb{P}^{n_i+1}$, $i = 1, 2$. Let $\varphi\colon X_1 \xrightarrow{\simeq} X_2$ be an isomorphism such that the induced birational map $\bar{\varphi}\colon \overline{X}_1 \dashrightarrow \overline{X}_2$ is an isomorphism in codimension 1. Then φ extends to a unique linear isomorphism $\Phi\colon \mathbb{A}^{n_1+1} \xrightarrow{\simeq} \mathbb{A}^{n_2+1}$. In particular $n_1 = n_2$, and Y_1 and Y_2 ($\overline{X}_1$ and $\overline{X}_2$, respectively) are projectively equivalent.*

Proof. We let $D_i = \overline{X}_i \setminus X_i$ denote the divisor at infinity; it is a scheme-theoretic hyperplane section. Since D_1 and D_2 are ample then (similarly as in Lemma 2.1.7) φ extends to an isomorphism $\bar{\varphi}\colon \overline{X}_1 \to \overline{X}_2$ which sends $0 \in \mathbb{A}^{n_1+1}$ to $0 \in \mathbb{A}^{n_2+1}$. Indeed, these points are the only singular points of the projective cones $\overline{X}_1$ and $\overline{X}_2$. Moreover, $\bar{\varphi}$ sends the generators of the cone[6] $\overline{X}_1$ into generators of $\overline{X}_2$. Indeed, every generator l_1 of $\overline{X}_1$ meets D_1 transversally in one point. The image $l_2 = \varphi(l_1) \subseteq \overline{X}_2$ possesses similar properties, hence l_2 is again a projective line through the origin, i.e., a generator of the cone $\overline{X}_2$.

It follows that the orbits of the $\mathbb{G}_m$-action on $\overline{X}_1$ are sent to the orbits of the $\mathbb{G}_m$-action on $\overline{X}_2$. Furthermore $\bar{\varphi}$ is $\mathbb{G}_m$-equivariant, hence it induces an isomorphism $\varphi^*\colon \mathcal{O}(Y_2) \xrightarrow{\simeq} \mathcal{O}(Y_1)$ of the homogeneous coordinate rings. These graded rings are the coordinate rings of the affine cones X_1 and X_2, respectively, generated by their first graded pieces.[7] The graded isomorphism φ^* restricts to a linear isomorphism, say, $\Psi\colon \mathbb{A}^{n_2+1} \xrightarrow{\simeq} \mathbb{A}^{n_1+1}$ between these first graded pieces. The dual isomorphism $\Phi = \Psi^\vee\colon \mathbb{A}^{n_1+1} \xrightarrow{\simeq} \mathbb{A}^{n_2+1}$ provides a desired linear extension of φ. The uniqueness of such an extension follows immediately, since Y_1 and Y_2 are assumed to be linearly nondegenerate. □

For a projective variety $Y \subseteq \mathbb{P}^n$ with affine cone $X = \operatorname{Aff Cone}(Y)$ it can happen that $\operatorname{Aut}(Y) \neq \operatorname{Lin}(Y)$, while $\operatorname{Aut}(X) = \operatorname{Lin}(X)$, as in the following examples.

2.1.9. Examples. (1) Let A be an abelian variety. Consider a projective embedding $A \hookrightarrow \mathbb{P}^n$ (for instance, a smooth cubic curve in $\mathbb{P}^2$) with affine cone $X = \operatorname{Aff Cone}(A)$. By Corollary 2.1.4 $\operatorname{Aut}(X) = \operatorname{Lin}(X)$. While $\operatorname{Lin}(A)$ is a finite group (see [42, §II.6] or Section 1 above), the group $\operatorname{Aut}(A)$ contains the subgroup of translations and so is infinite. Thus $\operatorname{Aut}(Y) \neq \operatorname{Lin}(Y)$ (cf. Blanchard's Theorem 1.2.2).

(2) A smooth quartic $Y \subseteq \mathbb{P}^3$ is a $K3$-surface and so is not birationally ruled. Hence again $\operatorname{Aut}(X) = \operatorname{Lin}(X)$, where $X = \operatorname{Aff Cone}(Y) \subseteq \mathbb{A}^4$. Moreover, $\operatorname{Lin}(Y)$ is a finite group, while the group $\operatorname{Aut}(Y)$ can be infinite, see the discussion in §1.1. Clearly, nonlinear automorphisms of Y are not induced by automorphisms of X.

[6]That is the projective lines on $\overline{X}_1$ passing through the origin.

[7]Consisting of the restrictions to X_i of linear functions on $\mathbb{A}^{n_i+1}$, $i = 1, 2$.

2.2. Lifting G-actions to affine cones. In this subsection we address the following questions.

(1) *When a G-action on Y is induced by a G-action on X?*
(2) *When a G-action on Y is induced by a $\widetilde{G}$-action on X?*

A related question is:

Which projective representations can be lifted to linear ones?

Simple examples show that one needs some restrictions on such a projective representation. In the first example below the group G is finite, and is connected algebraic in the second.

2.2.1. Examples. (1) The quaternion group $Q_8 = \{\pm 1, \pm \mathrm{i}, \pm \mathrm{j}, \pm \mathrm{k}\}$ admits a unique irreducible two-dimensional representation. Namely, it acts on the affine plane $\mathbb{A}^2$ via the left multiplications provided that $(z_1, z_2) \in \mathbb{A}^2$ is written as a quaternion $q = z_1 + z_2\mathrm{j}$. This action induces a faithful action of Q_8 on any Veronese cone $V_d \simeq \mathbb{A}^2/\mathbb{Z}_d$ with d odd (cf. Section 2.4 below). The latter descends to an effective linear action on $\mathbb{P}^1$ of the dihedral group (or Klein four-group)

$$D_2 = V_4 = Q_8/Z(Q_8) \simeq (\mathbb{Z}/2\mathbb{Z})^2.$$

However, this D_2-action on $\mathbb{P}^1$ cannot be lifted to a D_2-action on $\mathbb{A}^2$ or on any of the Veronese cones V_d with d odd. Indeed, otherwise the exact sequence

$$0 \to Z(Q_8) \to Q_8 \to D_2 \to 0$$

would split, which is not the case. In other words, the faithful projective representation $D_2 \to \mathrm{PGL}_{n+1}(\mathbb{C})$ induced by the Veronese embedding $\varphi_{|\mathcal{O}_{\mathbb{P}^1}(n)|}\colon \mathbb{P}^1 \to \mathbb{P}^n$ lifts to a linear representation $D_2 \to \mathrm{GL}_{n+1}(\mathbb{C})$ if and only if $n = 2k > 0$ is even and so $\mathcal{O}_{\mathbb{P}^1}(n) = -kK_{\mathbb{P}^1}$.

(2) The standard projective representation of $G = \mathrm{PGL}_2(\mathbb{C})$ on $\mathbb{P}^1$ induces a linear G-action on the rational normal curve $\Gamma_d \subseteq \mathbb{P}^d$. Suppose that the latter action can be lifted to the Veronese cone $V_d = \mathrm{Aff\,Cone}(\Gamma_d) \subseteq \mathbb{A}^{d+1}$. This would give an irreducible representation of $G = \mathrm{PGL}_2(\mathbb{C})$ of dimension $d+1$. However, such a representation does exist only for d even. Indeed, every irreducible representation of $\mathrm{PGL}_2(\mathbb{C})$ yields an irreducible representation of $\mathrm{SL}_2(\mathbb{C})$ trivial on the center, and vice versa.

2.2.2. Remark. Concerning question (2), recall that for any perfect group G there exists a unique universal central extension (or Schur cover) $\widehat{G}$ of G such that every projective representation of G is induced by a linear representation of $\widehat{G}$ (see [107, Section 7]). For a finite perfect group G, the Schur cover $\widehat{G}$ is again finite. For a perfect (e.g., semi-simple) connected linear algebraic group G over $\mathbb{C}$, the Schur cover is just the simply connected universal covering group $\widehat{G}$ of G.

Any linear action $G \to \mathrm{Lin}(X)$ on the affine cone $X = \mathrm{Aff\,Cone}(Y)$ induces (via the exact sequence (2)) a linear action $G \to \mathrm{Lin}(Y)$ on Y. The latter factorizes through the action on Y of the quotient group $G/(G \cap \mathbb{G}_m)$. Answering question (2) above, in the following proposition we provide a simple criterion as to when a G-action on Y is induced by a $\widetilde{G}$-action on $X = \mathrm{Aff\,Cone}(Y)$, where $\widetilde{G}$ is a central extension of G (which is not a Schur cover). The proof is straightforward. Let us remind that any linear action $G \to \mathrm{Lin}(Y)$ of a group G on a projective variety $Y \subseteq \mathbb{P}^n$ stabilizes the very ample divisor class $[\mathcal{O}_Y(1)] \in \mathrm{Pic}(Y)$.

2.2.3. Proposition. (a) *Let Y be a smooth projective variety and $G \to \operatorname{Aut}(Y)$ be a group action on Y. If this action stabilizes a very ample divisor class $|H| \in \operatorname{Pic}(Y)$, then it extends linearly to the ambient projective space $\mathbb{P}^n = \mathbb{P}H^0(Y, \mathcal{O}_Y(H))$.*

(b) *Furthermore, let $X = \operatorname{Aff\,Cone}_H(Y)$ be the affine cone over $\varphi_{|H|}(Y)$. Consider the central extension $\widetilde{G} = \pi^{-1}(G) \subseteq \operatorname{Lin}(X)$ of G by $\mathbb{G}_m$, where $\pi\colon \operatorname{Lin}(X) \twoheadrightarrow \operatorname{Lin}(Y)$ is as in (2). Then the group $\widetilde{G}$ acts linearly on X inducing the given G-action on Y.*

2.2.4. Corollary. *Let G be a connected linear algebraic group. Then any regular G-action on a smooth projective variety $Y \subseteq \mathbb{P}^n$ is induced by a regular $\widetilde{G}$-action on the affine cone* $\operatorname{Aff\,Cone}(Y)$, *where $\widetilde{G} = \pi^{-1}(G) \subseteq \operatorname{Lin}(X)$ is a central extension of G by $\mathbb{G}_m$.*

PROOF. Since G is connected, G acts on $\operatorname{Pic}_0(Y)$. The group G being a rational variety [12], every morphism of G to the abelian variety $\operatorname{Pic}_0(Y)$ is constant. Hence the G-action on $\operatorname{Pic}_0(Y)$ is trivial, and so is the induced action on the Néron–Severi group $\operatorname{NS}(Y) = \operatorname{Pic}(Y)/\operatorname{Pic}_0(Y)$. Thus G acts trivially on $\operatorname{Pic}(Y)$. By Proposition 2.2.3(a) the G-action on Y extends linearly to $\mathbb{P}^n$. Now the result follows. □

2.2.5. Remark. Instead of referring to [12] one can show directly that every morphism $f\colon G \to A$ to an abelian variety A is constant. Clearly, f is constant on any abelian subgroup of G and on its cosets. Hence f is also constant on any solvable subgroup. In particular, it is constant on the solvable radical $\operatorname{Rad}(G)$ and on its cosets. Thus f induces a morphism $G/\operatorname{Rad}(G) \to A$. So we may assume that G is semisimple. Consider a maximal torus $\mathbb{T} \subseteq G$ and the collection of its root vectors $(H_\alpha)_\alpha \subseteq T_eG = \operatorname{lie}(G)$. The subset $T_e\mathbb{T} \cup (H_\alpha)_\alpha$ consists of the tangent vectors of algebraic one-parameter subgroups of G and spans the tangent space T_eG. Hence the differential $df(e)$ vanishes. Now the assertion follows. Indeed, applying left shifts one can produce a similar situation in any point g of G.

A stronger statement holds for pluri-canonical or pluri-anticanonical embeddings.

2.2.6. Proposition. *Let Y be a smooth projective variety. Suppose that for some $m \in \mathbb{Z}$ there is an embedding $\varphi = \varphi_{|mK_Y|}\colon Y \hookrightarrow \mathbb{P}^n$, and let $X = \operatorname{Aff\,Cone}(\varphi(Y))$. Then*

$$\operatorname{Lin}(X) = \mathbb{G}_m \times \operatorname{Lin}(\varphi(Y)) \simeq \mathbb{G}_m \times \operatorname{Aut}(Y),$$

where $\mathbb{G}_m$ acts on the cone X by scalar matrices.

PROOF. Indeed, the group $\operatorname{Aut}(Y)$ acts on the linear system $|mK_Y|$ yielding an isomorphism $\operatorname{Aut}(Y) \simeq \operatorname{Lin}(\varphi(Y))$. Moreover, $\operatorname{Aut}(Y)$ acts on the linear bundle $\mathcal{O}(mK_Y)$. Hence it acts linearly on $H^0(Y, \mathcal{O}(mK_Y))$. The dual action on $H^0(Y, \mathcal{O}(mK_Y))^\vee$ preserves the cone X. This gives an embedding $\operatorname{Aut}(Y) \hookrightarrow \operatorname{Lin}(X)$ and a splitting of the exact sequence

$$0 \to \mathbb{G}_m \to \operatorname{Lin}(X) \to \operatorname{Lin}(\varphi(Y)) \simeq \operatorname{Aut}(Y) \to 0.$$

Since the subgroup $\mathbb{G}_m \subseteq \operatorname{Lin}(X)$ is central, the assertions follow. □

This proposition can be applied to the anticanonical embeddings of del Pezzo surfaces. In the case where there is a $\mathbb{G}_a$-action on X the group $\mathrm{Aut}(X)$ is infinite-dimensional. For instance, this is so for the cones over del Pezzo surfaces of degree $d \geq 4$ (for $d \geq 7$ there exists even a linear $\mathbb{G}_a^2$-action on X). While for $6 \geq d \geq 4$ the group $\mathrm{Aut}(Y)$ is finite or toric, hence any $\mathbb{G}_a$-action on X is nonlinear; cf. Theorem 0.0.1 in the Introduction and also Theorems 1.2.3 and 3.2.2.

2.3. Groups acting on affine cones. Similarly as in Proposition 2.1.1, in the case of a reductive group action a weaker analog of Corollary 2.1.4 holds without the assumption of birational nonruledness.

2.3.1. Lemma. *Suppose that a connected reductive group G acts effectively on the affine cone $X \subseteq \mathbb{A}^{n+1}$ over a smooth linearly nondegenerate projective variety $Y \subsetneq \mathbb{P}^n$. Then there is a faithful representation $\rho\colon G \to \mathrm{GL}(n+1,\mathbb{C})$, which restricts to an effective linear G-action on X inducing a linear action of G on Y.*

PROOF. The vertex $0 \in X$ is an isolated singular point of X, hence a fixed point of G. Since G is reductive, the induced representation ρ of G on the Zariski tangent space T_0X is faithful (see, e.g., [3] or [30, Lemma 2.7(b)]) and descends to Y via the projective representation $\bar{\rho}\colon G \to \mathrm{PGL}(n+1,\mathbb{C}) = \mathrm{GL}(n+1,\mathbb{C})/\mathbb{G}_m$. □

Let us note that for a nonreductive group action, ρ as above can be trivial. For instance, this is the case for the $\mathbb{G}_a$-action $t.(x,y) = (x+ty^2, y)$ on $X = \mathbb{A}^2$.

The following theorem is complementary to Corollary 2.1.4; cf. also [48] for (a).

2.3.2. Theorem. *We let $X \subseteq \mathbb{A}^n$ $(n \geq 2)$ be the affine cone over a smooth projective variety $Y \subseteq \mathbb{P}^{n-1}$. Suppose that*

- *The group $\mathrm{Aut}(Y)$ is finite.*
- *A connected algebraic group G of dimension ≥ 2 acts effectively on X and contains a 1-dimensional torus $\mathbb{T} \simeq \mathbb{G}_m$ acting on $\mathbb{A}^n$ via scalar matrices.*

Then

(a) *G is a solvable group of rank 1.*

Assume further that X is a normal affine threefold. Then

(b) *there exists an $\mathbb{A}^1$-fibration $\theta\colon X \to Z$, where Z is a normal affine surface equipped with a good $\mathbb{G}_m$-action and θ is equivariant with respect to the standard $\mathbb{G}_m$-action on X.*

(c) *Y is uniruled via a family of rational curves parameterized by the curve $(Z \setminus \{\theta(0)\})/\mathbb{G}_m \simeq \mathbb{P}^1$.*

PROOF. Consider a Levi decomposition $G = \mathrm{Rad}_u(G) \rtimes L$, where $L \subseteq G$ is a Levi subgroup (i.e., a maximal connected reductive subgroup) containing $\mathbb{T}$. By Lemma 2.3.1 the induced representation ρ of L on the Zariski tangent space T_0X is faithful. Moreover $\mathbb{T}$ (which acts on T_0X by scalar matrices) is a central subgroup of L. Since the group $\mathrm{Aut}(Y)$ is finite and L is connected, the induced action of the quotient group $L/\mathbb{T}$ on Y is trivial. Thus $L = \mathbb{T}$ is a maximal torus of $G = \mathrm{Rad}_u(G) \rtimes \mathbb{T}$, and so G is solvable of rank 1. This shows (a).

Let us show (b) and (c). By our assumptions $\dim G \geq 2$ and $\dim X = 3$. In particular the unipotent radical $\mathrm{Rad}_u(G)$ is nontrivial and contains a one-parameter subgroup $U \simeq \mathbb{G}_a$. All orbits of U are closed in X, and the one-dimensional orbits are isomorphic to the affine line $\mathbb{A}^1$. Therefore X is affine uniruled. Its coordinate ring $A = \mathcal{O}_X$ is graded by the dual lattice $\mathbb{T}^\vee \simeq \mathbb{Z}$. This grading is actually positive:

$$A = \bigoplus_{k \geq 0} A_k.$$

The infinitesimal generator ∂ of the induced $\mathbb{G}_a$-action on A is a homogeneous locally nilpotent derivation of A (see, e.g., [30, 99]). Since $\dim X = 3$, due to the Zariski Finiteness Theorem [32, Chapter 5, Theorem 6.3.1; 119] the ring of invariants $B = \ker \partial = A^{\mathbb{G}_a}$ is a normal graded finitely generated subalgebra of the normal algebra A with $B_0 = A_0 = \mathbb{C}$. Therefore the normal affine surface $Z = \operatorname{Spec}(B)$ is endowed by a $\mathbb{T}$-action with a unique attractive fixed point $0' = \theta(0)$, where $\theta \colon X \to Z$ is the orbit map of the $\mathbb{G}_a$-action on X. Thus θ is a $\mathbb{T}$-equivariant surjection induced by the inclusion $B \subseteq A$ of graded normal algebras. This yields (b).

Since a general one-dimensional orbit of $U \simeq \mathbb{G}_a$ in X does not pass through the vertex $0 \in X$ and is not contained in an orbit closure of T (i.e., in a generator of the cone), there is a Zariski open subset, say, Ω of Y covered by the images of these orbits. Taking Zariski closures yields a family of rational curves parameterized by the rational normal curve $(Z \setminus \{0'\})/\mathbb{G}_m$. Thus Y is uniruled, as claimed in (c). □

2.4. Group actions on 2-dimensional affine cones. The following corollary is immediate from Proposition 2.1.1.

2.4.1. Corollary. *Consider two smooth linearly nondegenerate curves $Y_i \subseteq \mathbb{P}^{n_i}$ ($i = 1, 2$) of degrees d_i, and let $X_i = \operatorname{Aff\,Cone}(Y_i) \subseteq \mathbb{A}^{n_i+1}$ be the corresponding affine cones. Then $X_1 \simeq X_2$ if and only if these cones are linearly isomorphic, if and only if $n_1 = n_2$, $d_1 = d_2$ and Y_1 and Y_2 are projectively equivalent.*

Similarly, from Corollary 2.1.4 we deduce the following one.

2.4.2. Corollary. *Let $X = \operatorname{Aff\,Cone}(Y) \subseteq \mathbb{A}^{n+1}$ be the affine cone over a smooth, nonrational projective curve $Y \subseteq \mathbb{P}^n$. Then $\operatorname{Aut}(X) = \operatorname{Lin}(X)$, and this group is a central extension of the finite group $\operatorname{Lin}(Y)$ by $\mathbb{G}_m$.*

2.4.3. Remarks. (1) However, $\operatorname{Aut}(Y) \neq \operatorname{Lin}(Y)$ for an elliptic curve $Y \subseteq \mathbb{P}^n$, see Example 2.1.9(1). Consider further a smooth rational curve $Y \subseteq \mathbb{P}^n$ of degree $d > n$. Then Y is neither linearly nor projectively normal. Indeed, Y is a linear projection of the rational normal curve $\Gamma_d \subseteq \mathbb{P}^d$, and $X = \operatorname{Aff\,Cone}(Y)$ is a linear projection of the Veronese cone $V_d = \operatorname{Aff\,Cone}(\Gamma_d)$. The letter projection gives a normalization of X. This is not an isomorphism as it diminishes the dimension of the Zariski tangent space at the vertex.

(2) The normalizations of the affine cones X_1 and X_2 over two smooth rational curves Y_1 and Y_2, respectively, are isomorphic if and only if $\deg(Y_1) = \deg(Y_2)$. While in general the (nonnormal) affine surface $X = \operatorname{Aff\,Cone}(Y)$ admits nontrivial equisingular deformations arising from deformations of the projective embedding $Y \hookrightarrow \mathbb{P}^n$. For instance, smooth rational curves Y of type $(1, a)$ on a quadric $\mathbb{P}^1 \times \mathbb{P}^1 \hookrightarrow \mathbb{P}^3$ vary in a family of projective dimension $2a + 1$. Hence for any $a \geq 3$ the group $\operatorname{PSO}(4, \mathbb{C})$ cannot act transitively on this family.

(3) Since any group action on an affine cone X lifts to the normalization, it is enough to restrict to normal cones. For the normal Veronese cone $X = V_d \subseteq \mathbb{A}^{d+1}$ over a rational normal curve $Y = \Gamma_d \subseteq \mathbb{P}^d$ we have $\operatorname{Aut}(X) \neq \operatorname{Lin}(X)$. Moreover,

$V_d \simeq \mathbb{A}^2/\mathbb{Z}_d$ being a toric surface, for every $d \geq 1$ the group $\mathrm{Aut}(V_d)$ is infinite-dimensional. In particular this is not an algebraic group. Indeed, the graded coordinate ring $\mathcal{O}(V_d)$ admits a nonzero locally nilpotent derivation ∂ corresponding to an effective $\mathbb{G}_a$-action on V_d [30]. The kernel $\ker(\partial) \subseteq \mathcal{O}(V_d)$ is isomorphic to the polynomial ring $\mathbb{C}[t]$. For any $p \in \mathbb{C}[t]$, the derivation $p \cdot \partial$ is again locally nilpotent. Thus $\mathbb{C}[t]\cdot\partial$ is the Lie algebra of an infinite-dimensional abelian subgroup $G \subseteq \mathrm{Aut}(V_d)$.

(4) There are actually two independent $\mathbb{G}_a$-actions on V_d with different orbits, and even a continuous family of such actions; see, e.g., [28, 30]. Gizatullin and Danilov [36] studied the structure of an amalgamated product on the group $\mathrm{Aut}(V_d)$, while Makar-Limanov [73, 74] provided an explicit description of this group.

(5) Similarly, independent $\mathbb{G}_a$-actions, and an amalgamated product structure, exist on any normal affine toric surface different from $\mathbb{A}^1_* \times \mathbb{A}^1$ or $\mathbb{A}^1_* \times \mathbb{A}^1_*$. Every such surface is of the form $V_{d,e} = \mathbb{A}^2/\mathbb{Z}_d$ for an appropriate diagonal action

$$\zeta.(x,y) = (\zeta x, \zeta^e y), \quad \text{where } \zeta^d = 1 \text{ and } \gcd(e,d) = 1,$$

of the cyclic group $\mathbb{Z}_d = \langle \zeta \rangle$ on the affine plane $\mathbb{A}^2$. Choosing a system of homogeneous generators in the graded coordinate ring $\mathcal{O}(V_{d,e})$ yields an embedding $V_{d,e} \hookrightarrow \mathbb{A}^{N+1}$, which is equivariant with respect to a suitable diagonal $\mathbb{G}_m$-action on $\mathbb{A}^{n+1}$ with positive weights $w = (w_0, \ldots, w_N)$. In this way $V_{d,e}$ can be realized as the affine cone over a smooth rational curve in the corresponding weighted projective space $\mathbb{P}^N_w = (\mathbb{A}^{N+1} \setminus \{0\})/\mathbb{G}_m$.

(6) By Popov's Theorem [89][8] the Veronese cones $V_d = V_{d,1}$ can be characterized as normal affine surfaces on which an algebraic group acts with just two orbits.

Let us construct an explicit example of a nonlinear biregular automorphism of a Veronese cone V_d for every $d \geq 1$.

2.4.4. Example. Consider as before the Veronese cone $V_d \subseteq \mathbb{A}^{d+1}$ over a rational normal curve $\Gamma_d \subseteq \mathbb{P}^d$. Let $\overline{V}_d \subset \mathbb{P}^{d+1}$ be the Zariski closure, and $\widetilde{V}_d$ be the blowup of $\overline{V}_d$ at the vertex $0 \in V_d$. It is well known [36] that $\widetilde{V}_d \simeq \Sigma_d$, where Σ_d denotes a Hirzebruch surface with the exceptional section S_0 and a disjoint section at infinity, say, S_∞ with $S_0^2 = -d$ and $S_\infty^2 = d$. Therefore

$$V_d \simeq (\Sigma_d \setminus S_\infty)/S_0.$$

To exhibit a nonlinear automorphism of V_d is the same as to exhibit an automorphism of $\Sigma_d \setminus S_\infty$, which extends to a birational transformation of Σ_d preserving the exceptional section S_0 but not the ruling $\pi\colon \Sigma_d \to \mathbb{P}^1$ (or, equivalently, which blows down the curve S_∞). On the level of dual graphs, such a birational transformation consists e.g., in the following sequence of blowups and blowdowns [27]:

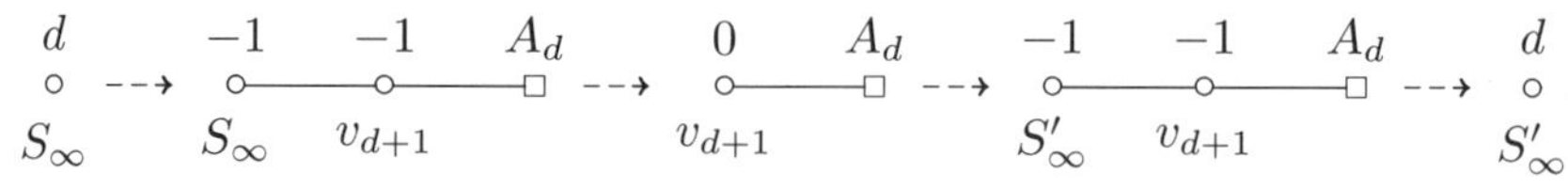

Here a box marked A_d represents the linear chain $[\![-2, \ldots, -2]\!]$ of length d. The centers of blowups on the curves S_∞ and v_{d+1} can vary. Anyhow, the section S_∞ being contracted, the resulting biregular transformation of the Veronese cone V_d is nonlinear.

[8]This is actually an earlier version of the Cone Theorem in dimension 2; see Section 1.

3. Group actions on 3-dimensional affine cones

The main result of this section is the existence of a $\mathbb{G}_a$-actions on the affine cones over any smooth del Pezzo surface of degree ≥ 4. The proof exploits a general geometric criterion for the existence of such an action.

3.1. Existence of $\mathbb{G}_a$-actions on affine cones: a geometric criterion.

3.1.1. Let Y be a smooth projective variety, and let $H \in \operatorname{Div}(Y)$ be an ample polarization of Y. Consider the total space $\widehat{X}$ of the line bundle $\mathcal{O}_Y(H)$ with the zero section $S_0 \subseteq \widehat{X}$. Under the natural identification $S_0 \simeq Y$, we have $\mathcal{O}_{S_0}(S_0) = \mathcal{O}_Y(-H)$. Hence S_0 is contractible, i.e., there is a birational contraction $\upsilon\colon \widehat{X} \to X$, where X is a normal affine variety and $\upsilon(S_0)$ is a point. In this situation, we call X a *generalized cone* over (Y, H). If H is very ample, then X coincides with the normalization of the usual affine cone $\operatorname{Aff\,Cone}(Y)$ over $Y \hookrightarrow \mathbb{P}^n$, where the embedding is given by the linear system $|H|$. So we write $X = \operatorname{Aff\,Cone}_H(Y)_{\mathrm{norm}}$. In this section we provide a criterion of existence of a $\mathbb{G}_a$-action on a generalized cone.

Let us note that X can be compactified to the projective cone $\overline{X}$ over Y by adding a divisor at infinity $S_\infty \simeq Y$. The divisor S_∞ on $\overline{X}$ being ample, the variety $X = \overline{X} \setminus S_\infty$ is affine.

3.1.2. For instance, the affine cone over $\mathbb{P}^2$ coincides with $\mathbb{A}^3$ and so admits a transitive action of the additive group $\mathbb{G}_a^3$. In the following example we exhibit an effective $\mathbb{G}_a^2$-action on the affine cone $X \subseteq \mathbb{A}^4$ over a smooth quadric $Y \subseteq \mathbb{P}^3$ (cf. another constructions in [105]). The automorphism groups of affine quadrics were studied, e.g., in [21, 36, 110]. Over a general base field, this group is infinite-dimensional as soon as the corresponding quadratic form is isotropic [110, Lemma 1.1]. The proof of Lemma 1.1 in [110] provides a nontrivial linear $\mathbb{G}_a$-action on any quadric over $\mathbb{C}$. In the following example we exhibit an explicit effective $\mathbb{G}_a^2$-action on the affine cone over a smooth quadric in $\mathbb{P}^3$.

3.1.3. Example. All smooth quadrics in $\mathbb{P}^3$ are projectively equivalent. Choosing for instance the quadric

$$Y = \{xy = zu\}$$

we can define a linear $\mathbb{G}_a^2$-action on $X = \operatorname{Aff\,Cone}(Y) \subseteq \mathbb{A}^4$ by the following pair of commuting locally nilpotent derivations on the ring $A = \mathcal{O}(X)$:

$$\partial_1 = u\partial/\partial x + y\partial/\partial z \quad \text{and} \quad \partial_2 = u\partial/\partial y + x\partial/\partial z.$$

See [5, 105] for a more thorough treatment on the subject.

Let us introduce the following notion.

3.1.4. Definition. Let X be an affine variety. For a function $f \in \mathcal{O}(X)$ we let

$$\mathbb{D}_+(f) = X \setminus \mathbb{V}_+(f), \quad \text{where } \mathbb{V}_+(f) = f^{-1}(0).$$

We say that X is *cylindrical* if X contains a dense principal Zariski open subset $U = \mathbb{D}_+(f)$ isomorphic to the cylinder $Z \times \mathbb{A}^1$ over an affine variety Z.

The following proposition generalizes Lemma 1.6 in [30].

3.1.5. Proposition. *For an irreducible affine variety X, the following conditions are equivalent*:

(i) *X possesses an effective $\mathbb{G}_a$-action.*

(ii) *X is cylindrical.*

PROOF. First we suppose that X possesses an effective $\mathbb{G}_a$-action ψ with the associate locally nilpotent derivation $\partial \neq 0$. The filtration

$$0 \in \ker \partial \subsetneq \ker \partial^{(2)} \subsetneq \ker \partial^{(3)} \cdots \tag{3}$$

being strictly increasing, we can find $g \in \mathcal{O}(X)$ such that $\partial^{(2)} g = 0$ but $h := \partial g \neq 0$. Thus $\partial h = 0$ and so $h \in \mathcal{O}(X)$ is ψ-invariant. Letting $s = g/h$ and $U = \mathbb{D}_+(h)$ the function $s \in \mathcal{O}(U)$ gives a *slice* of ∂ that is, $\partial(s) = 1$. Consequently, the restriction of s to any 1-dimensional orbit O of ψ in U is an affine coordinate on $O \simeq \mathbb{A}^1$. By the Slice Theorem ([32, Corollary 1.22]), $\mathcal{O}(U) \simeq \ker(\partial)[s]$ and $\partial = \partial/\partial s$. Therefore $U \simeq Z \times \mathbb{A}^1$, where $Z = \operatorname{Spec}(\ker \partial) \simeq s^{-1}(0)$. This yields (ii).

To show the converse, assume that X is cylindrical. Let $U = \mathbb{D}_+(f) \simeq Z \times \mathbb{A}^1$ be a principal cylinder in X as in Definition 3.1.4. We consider the natural $\mathbb{G}_a$-action ϕ on U by translations along the second factor. Since $f|U$ does not vanish it is constant along any orbit of ϕ and so ϕ-invariant. Letting ∂ denote the locally nilpotent derivation on $\mathcal{O}(U)$ associated to ϕ, the derivation $\partial_n = f^n \partial \in \operatorname{Der}(\mathcal{O}(U))$ is again locally nilpotent for any $n \in \mathbb{N}$. Let $a_1, \ldots, a_k$ be a system of generators of $\mathcal{O}(X)$, and let $N \in \mathbb{N}$ be sufficiently large so that $f^N \partial a_i \in \mathcal{O}(X)$ for any $i = 1, \ldots, k$. Then $\partial_N(a_i) \in \mathcal{O}(X)$ for any $i = 1, \ldots, k$, hence $\partial_N(\mathcal{O}(X)) \subseteq \mathcal{O}(X)$. Thus the derivation $\partial_N|\mathcal{O}(X) \in \operatorname{Der}(\mathcal{O}(X))$ is locally nilpotent and so generates an effective $\mathbb{G}_a$-action ψ on X. Therefore (i) holds. □

3.1.6. Remark. Clearly the $\mathbb{G}_a$-actions ϕ and $\psi|U$ as in the proof have the same orbits, and $\mathbb{V}_+(f) = \{f = 0\}$ consists of fixed points of ψ.

In the case of affine cones, Theorem 3.1.9 below gives a more practical criterion. We need the following definition.

3.1.7. Definition. For a projective variety Y with a very ample polarization $\varphi_{|H|}\colon Y \hookrightarrow \mathbb{P}^n$, we call an *$H$-polar subset* any Zariski open subset of the form $U = Y \setminus \operatorname{supp} D$, where $D \in |dH|$ is an effective divisor on $\mathbb{P}^n$.

3.1.8. Recall that an affine ruling on a variety U is a morphism $\pi\colon U \to Z$ such that every scheme theoretic fiber of π is isomorphic to the affine line $\mathbb{A}^1$. By a theorem due to Kambayashi, Miyanishi and Wright [60, 61] (see also [26, 100]), every affine ruling $\pi\colon U \to Z$ on a normal variety U over a normal base Z is a locally trivial $\mathbb{A}^1$-bundle.

3.1.9. Theorem. *Let Y be a smooth projective variety with a very ample polarization $\varphi_{|H|}\colon Y \hookrightarrow \mathbb{P}^n$. Then the following hold.*

(a) *If the affine cone $X = \operatorname{Aff\,Cone}_H(Y)$ admits an effective $\mathbb{G}_a$-action, then Y possesses an H-polar open subset U which is the total space of a line bundle $U \to Z$.*

(b) *Conversely, if Y possesses an H-polar open subset U equipped with an affine ruling $U \xrightarrow{\mathbb{A}^1} Z$ and a section $Z \to U$, where Z is smooth and $\operatorname{Pic}(Z) = 0$, then the affine cone $X = \operatorname{Aff\,Cone}_H(Y)$ admits an effective $\mathbb{G}_a$-action.*

PROOF. (a) Let ψ' be an effective $\mathbb{G}_a$-action on X with associate locally nilpotent derivation $\partial' \neq 0$. Using the natural grading of the coordinate ring

$$A = \mathcal{O}(X) = \bigoplus_{i \geq 0} A_i,$$

∂' can be decomposed into a finite sum of homogeneous derivations $\partial' = \sum_{i=1}^{n} \partial'_i$, where the principal component $\partial = \partial'_n \neq 0$ is again locally nilpotent. The $\mathbb{G}_a$-action ψ on X generated by ∂ extends to an effective action of a semi-direct product $G = \mathbb{G}_a \rtimes \mathbb{G}_m$ on X.

The filtration (3) from the proof of Proposition 3.1.5 consists now of graded subrings. Hence we can find homogeneous elements $\hat{g}$, $\hat{h} \in A$ such that $\partial\hat{g} = \hat{h}$ and $\partial\hat{h} = 0$. In the notation of 3.1.4 we let

$$\widehat{U} = \mathbb{D}_+(\hat{h}) \subseteq X \quad \text{and} \quad \widehat{Z} = \mathbb{V}_+(\hat{g}) \setminus \mathbb{V}_+(\hat{h}) \subseteq \widehat{U}.$$

Likewise in the proof of Proposition 3.1.5, we obtain a decomposition $\widehat{U} \simeq \widehat{Z} \times \mathbb{A}^1$.

Furthermore, G acts on $\widehat{U} \simeq \widehat{Z} \times \mathbb{A}^1$ respecting the product structure. More precisely, $\mathbb{G}_a$ acts by shifts on the second factor, i.e., along the fibers of the morphism $\hat{\pi} : \widehat{U} \to \widehat{Z}$. Since $\hat{g}$, $\hat{h}$, and ∂ are homogeneous, $\mathbb{G}_m$ acts on $\widehat{U}$ stabilizing $\widehat{Z}$ and sending the fibers of $\hat{\pi}$ into fibers. The factorization by the $\mathbb{G}_m$-action on $\widehat{U}$ yields a Zariski open subset $U = \hat{U}/\mathbb{G}_m \subseteq Y$ and a divisor $Z = \widehat{Z}/\mathbb{G}_m$ on U so that

$$\widehat{U} = \operatorname{Aff\,Cone}(U) \setminus \{0\} \quad \text{and} \quad \widehat{Z} = \operatorname{Aff\,Cone}(Z) \setminus \{0\}. \tag{4}$$

The map $\hat{\pi}$ defines an affine ruling $\pi\colon U \xrightarrow{\mathbb{A}^1} Z$ with a section $Z \hookrightarrow U$. Each fiber of π is the quotient of a G-orbit in U by the $\mathbb{G}_m$-action. Since Y and Z are smooth, by the Kambayashi – Miyanishi – Wright theorem cited in 3.1.8, π is an $\mathbb{A}^1$-bundle and, moreover, a vector bundle since it possesses a section.

Finally, since $D = \hat{h}^*(0) \in |dH|$, where $d = \deg(\hat{h})$, the open set $U \subseteq Y$ is H-polar (see Definition 3.1.7). This shows (a).

To show (b), suppose that Y possesses an H-polar open subset U with an affine ruling $\pi\colon U \xrightarrow{\mathbb{A}^1} Z$ and a section $Z \to U$, where Z is smooth and $\operatorname{Pic}(Z) = 0$. Since both U and Z are smooth, π is locally trivial by the Kambayashi – Miyanishi – Wright theorem. Since π has a section Z, $\pi\colon U \to Z$ is a line bundle and Z is the zero section. This bundle is trivial since $\operatorname{Pic}(Z) = 0$. Thus $U \simeq Z \times \mathbb{A}^1$. In particular $\operatorname{Pic}(U) = 0$.

Let further $\sigma\colon \widetilde{X} \to X$ be the blowup of the vertex $0 \in X$. The induced morphism $\rho\colon \widetilde{X} \to Y$ has a natural structure of a line bundle with the exceptional divisor $E = \sigma^{-1}(0)$ as the zero section. Since $\operatorname{Pic}(U) = 0$, the restriction $\rho|\widetilde{U}\colon \widetilde{U} \to U$ to $\widetilde{U} = \rho^{-1}(U) \subseteq \widetilde{X}$ yields a trivial line bundle. Hence

$$\widetilde{U} \simeq U \times \mathbb{A}^1 \simeq Z \times \mathbb{A}^1 \times \mathbb{A}^1 \quad \text{and, similarly,} \quad \widetilde{Z} = \rho^{-1}(Z) \simeq Z \times \mathbb{A}^1.$$

Under this isomorphism $E \cap \widetilde{U}$ is sent to $U \times \{0\} \simeq Z \times \mathbb{A}^1 \times \{0\}$ and $E \cap \widetilde{Z}$ to $Z \times \{0\}$. For $\widehat{U} = \widetilde{U} \setminus E$ and $\widehat{Z} = \widetilde{Z} \setminus E$ as in (4) we obtain

$$\widehat{U} \simeq Z \times \mathbb{A}^1 \times \mathbb{A}^1_* \quad \text{and} \quad \widehat{Z} \simeq Z \times \{0\} \times \mathbb{A}^1_*. \tag{5}$$

Thus $\widehat{U} \simeq \widehat{Z} \times \mathbb{A}^1$ is a cylinder in X. Since $U \subseteq Y$ is H-polar, $\hat{U} \subseteq X$ is a principal Zariski open subset and so X is cylindrical. Now (b) follows by Proposition 3.1.5. □

3.1.10. Remarks. (1) This theorem, with the same proof, holds also for generalized cones (see 3.1.1). In particular, we may assume that H is just an ample divisor.

(2) It is easily seen that if a cone $X = \operatorname{Aff Cone}_H(Y)$ admits an effective $\mathbb{G}_a$-action, then also the cone $X_k = \operatorname{Aff Cone}_{kH}(Y)$ admits such an action for any $k \geq 1$. Moreover, the cone X_k is normal for $k \gg 1$, see [43, Chapter II, Ex. 5.14].

3.1.11. Remark. The construction of a $\mathbb{G}_a$-action on X as in the proof of (b) can be made more explicite. The product $G = \mathbb{G}_a \times \mathbb{G}_m$ acts on $\widehat{U}$ preserving the product structure in (5):

$$G \ni (a, \lambda) \colon \widehat{U} \to \widehat{U}, \quad (z, x, y) \mapsto (z, x + a, \lambda y).$$

The infinitesimal generators $\partial/\partial x$ and $y\partial/\partial y$ of the $\mathbb{G}_a$- and $\mathbb{G}_m$-actions commute. Letting $D = X \setminus \widehat{U}$, there is a regular function $f \in \mathcal{O}(X)$ such that $\operatorname{div}(f) = nD$. Moreover, we can choose f of the form $f = y^k g(z)$, where $g \neq 0$. For $N \gg 1$ the $\mathbb{G}_a$-action generated by $\partial = f^N \partial/\partial x$ extends to the cone X, see the proof of Proposition 3.1.5. With this new $\mathbb{G}_a$-action, a semidirect product $\mathbb{G}_a \rtimes \mathbb{G}_m$ acts effectively on X. However, the factors do not commute any more.

Theorem 3.1.9 yields the following criterion of existence of a $\mathbb{G}_a$-action on certain 3-dimensional affine cones.

3.1.12. Corollary. *Let Y be a rational smooth projective surface with a polarization $\varphi_{|H|} \colon Y \hookrightarrow \mathbb{P}^n$, and let $X = \operatorname{Aff Cone}_H(Y) \subseteq \mathbb{A}^{n+1}$ be the affine cone over Y. Then X admits a nontrivial $\mathbb{G}_a$-action if and only if Y possesses an H-polar cylinder $U \simeq Z \times \mathbb{A}^1$, where Z is a smooth affine curve.*

PROOF. Since Y is smooth and rational, Z as in Theorem 3.1.9 is a noncomplete smooth rational curve. Thus $\operatorname{Pic}(Z) = 0$. Hence the affine rulings from Theorem 3.1.9 and its proof are actually direct products. Our assertion can be easily deduced now from Theorem 3.1.9. □

Using this criterion we can show that, given an arbitrary smooth rational surface Y, some affine cone over Y admits a nontrivial $\mathbb{G}_a$-action.

3.1.13. Proposition. *Let Y be a rational smooth projective surface. Then there is an embedding $\varphi \colon Y \hookrightarrow \mathbb{P}^n$ such that the affine cone $X = \operatorname{Aff Cone}(\varphi(Y))$ is normal and admits an effective $\mathbb{G}_a$-action.*

PROOF. By Corollary 3.1.12 it is enough to find an effective ample divisor D on Y with cylindrical complement $Y \setminus \operatorname{supp} D$. If Y is isomorphic either to the projective plane $\mathbb{P}^2$ or to a Hirzebruch surface, the existence of such a divisor D is easy. Hence we may assume that Y possesses a $\mathbb{P}^1$-fibration, say, $f : Y \to \mathbb{P}^1$ with singular fibers $F_1, \ldots, F_r$.[9] Let S be a section of f and $F_{i,1}, \ldots, F_{i,r_i}$ be the irreducible components supporting the fiber F_i. We claim that for suitable positive integers m and $m_{i,1}, \ldots, m_{i,r_i}$ the divisor

$$D = mS + \sum_{\substack{1 \leq i \leq r \\ 1 \leq j \leq r_i}} m_{i,j} F_{i,j}$$

is ample.

[9] Since Y is rational, the base curve is rational too.

Indeed, the divisors $F_{i,j}$ and S generate the group $\mathrm{Pic}(Y)$. Hence there exists an ample divisor A whose support is contained in $S+\sum F_{i,j}$. Let $\widehat{F}_1,\dots,\widehat{F}_r$ be scheme theoretic fibers so that $\widehat{F}_i=\sum_{j=1}^{r_i}a_{i,j}F_{i,j}$, where $a_{i,j}>0\ \forall i,j$. Clearly, the divisor $\widehat{F}=\sum\widehat{F}_i$ is nef and $B=S+t\widehat{F}$ is also nef for $t\gg 0$. Furthermore, B is an effective nef divisor with support $S+\sum F_{i,j}$. Therefore $D=A+nB$ is an effective ample divisor with $\operatorname{supp} D=S+\sum F_{i,j}$ for $n\gg 1$.

By construction, the complement $Y\setminus\operatorname{supp} D$ is an $\mathbb{A}^1$-cylinder over the affine rational curve obtained from $\mathbb{P}^1$ by deleting r points which correspond to singular fibers.[10] So for $m\gg 1$ the linear system $|mD|$ defines an embedding $Y\hookrightarrow\mathbb{P}^n$ with a projectively normal image, see Exercise 5.14 in [43, Chapter II]. Since Y admits an $|mD|$-polar cylinder, X is normal and cylindrical. By Corollary 3.1.12, X admits an effective $\mathbb{G}_a$-action, as required. Now the proof is completed. □

The following question arises.

3.1.14. Question. Does there exist a polarized smooth rational surface (Y,H) without any H-polar cylinder?

3.1.15. Remark. If U is an H-polar cylinder on Y then it is also kH-polar for any $k\in\mathbb{N}$, and vice versa. Thus the existence of an H-polar cylinder depends only on the ray of H in the ample cone of Y. Moreover, since the irreducible components of the divisor $D=Y\setminus U$ span the Picard group $\mathrm{Pic}(Y)$ and the ample cone is open, the property of a cylinder U to be H-polar is stable under small perturbations of H.

For any smooth rational projective surface, the H-polar cylinder from Proposition 3.1.13 can be chosen to be isomorphic to the affine plane. Let us provide similar examples in higher dimensions.

3.1.16. Example. Consider a flag variety G/P with an ample polarization H (see Section 1.3). By Corollary 2.2.4 the G-action on G/P lifts to a $\widetilde{G}$-action on the cone $\operatorname{Aff Cone}_H(G/P)$, where $\widetilde{G}$ is the universal cover of G. The actions of one-parameter unipotent subgroups of $\widetilde{G}$ yield effective $\mathbb{G}_a$-actions on the cone. Actually G/P contains an H-polar open cylinder U isomorphic to an affine space $\mathbb{A}^n$ (cf. Theorem 3.1.9(a)).

Indeed, let $B_+\subseteq P$ be a Borel subgroup of G, and let B_- be the opposite Borel subgroup so that $B_+\cap B_-$ is a Cartan subgroup. Then $B_-\cdot P$ is open in G and so the B_--orbit U of $e\cdot P$ is open in G/P. Thus U is a big Schubert cell. Since U is also an orbit of the maximal unipotent subgroup $B_u\subseteq B_-$, it is isomorphic to $\mathbb{A}^n$. In particular, U is a cylinder in Y. Letting $D=Y\setminus U=\bigcup_i D_i$, the Schubert divisors D_i form a basis in $\mathrm{Pic}(G/P)$. In this basis $H=\sum_i\alpha_iD_i$, where $\alpha_i>0$ for all i since the divisor H is ample, see [106; 109, Theorem 7.53]. Hence U is an H-polar cylinder in G/P.

We note that the action of $\widetilde{G}$ on the affine cone $\widetilde{X}=\operatorname{Aff Cone}_H(G/P)$ is transitive off the vertex $0\in\widetilde{X}$. Indeed, we may suppose that $X=\widetilde{G}/\widetilde{P}$, where $\widetilde{G}$ is semisimple, simply connected, and $\widetilde{P}\subseteq\widetilde{G}$ is parabolic. Since $\widetilde{X}$ is affine, the stabiliser $\mathrm{Stab}_{\widetilde{G}}(x)$ of a point $x\in\widetilde{X}\setminus\{0\}$ cannot contain a parabolic subgroup.

[10] Alternatively, we can use the fact that any point of Y admits a neighborhood U isomorphic to the affine plane $\mathbb{A}^2$ so that $Y\setminus U$ supports the Picard group $\mathrm{Pic}(Y)$.

Hence the stabilizer $\mathrm{Stab}_{\widetilde{G}}([x])$ (conjugate to $\widetilde{P}$) acts nontrivially on the generator of the cone through x. By Corollary 1.5 in [92] the Makar-Limanov invariant $\mathrm{ML}(\widetilde{X})$ is trivial (cf. Theorem 3.3.2; see Section 3.3 below for the definition of the Makar-Limanov invariant). Moreover, the cone $\widetilde{X}$ is flexible and the group $\mathrm{Aut}(\widetilde{X})$ acts m-transitively on $\widetilde{X}$ for any $m \geq 1$ [4].

3.1.17. Remark. The existence in a projective variety of an open subset isomorphic to an affine space is rather exceptional. For instance, none of smooth rational cubic 4-folds in $\mathbb{P}^5$, and none of smooth 3-fold intersections of a pair of quadrics in $\mathbb{P}^5$ contains a Zariski open set isomorphic to an affine space, see [88, 94]. At the same time, every smooth intersection Y of a pair of quadrics in $\mathbb{P}^5$ contains a $(-K_Y)$-polar cylinder, see Proposition 5.0.1 below and its proof.

3.2. $\mathbb{G}_a$-actions on affine cones over del Pezzo surfaces. Let us explain first the reason of our special interest in the affine cones over del Pezzo surfaces.

3.2.1. A normal variety X is called $\mathbb{Q}$-*Gorenstein* if some multiple nK_X is Cartier, where K_X is the canonical (Weil) divisor. This notion is important in the Mori minimal model program (MMP). It is easily seen that the generalized cone $X = \mathrm{Aff\,Cone}_H(Y)$ over a smooth polarized variety (Y, H) is $\mathbb{Q}$-Gorenstein if and only if $aH \sim -bK_Y$ for some $a \in \mathbb{N}$, $b \in \mathbb{Z}$ ([65, Example 3.8]). If, moreover, $H \sim -K_Y$, then X is Gorenstein and has at most canonical singularity at the origin, see *loc. cit.*

On the other hand, if $\mathrm{Aut}(X) \neq \mathrm{Lin}(X)$ then by Corollary 2.1.4 Y is birationally ruled. Hence the Kodaira dimension of Y is negative, see [64]. Thus $b > 0$, i.e, $-K_Y$ is ample. Consequently, Y is a Fano variety.

Therefore, if the affine cone X over (Y, H) is $\mathbb{Q}$-Gorenstein and admits an effective nonlinear $\mathbb{G}_a$-action, then Y is a Fano variety and $H \in \mathbb{Q}_{>0}[-K_Y]$. In particular, if $\dim(Y) = 2$ then (Y, H) is a pluri-anticanonically polarized del Pezzo surface.[11]

From now on we assume that Y is a del Pezzo surface of degree $d \geq 3$ and $H = -K_Y$ is the anti-canonical polarization. Thus the linear system $|-K_Y|$ is very ample and provides an embedding $Y \hookrightarrow \mathbb{P}^d$ onto a projectively normal smooth surface of degree d, see, e.g., [23]. The affine cone $X = \mathrm{Aff\,Cone}_{-K_Y}(Y)$ has a normal, canonical, Gorenstein (hence also Cohen – Macaulay) singularity at the vertex.

The following theorem is the main result of this subsection (see Theorem 0.0.1 in the Introduction).

3.2.2. Theorem. *Let Y_d be a smooth del Pezzo surface of degree d anticanonically embedded into $\mathbb{P}^d$, where $4 \leq d \leq 9$, and let $X_d \subseteq \mathbb{A}^{d+1}$ be the affine cone over Y_d. Then X_d admits a nontrivial $\mathbb{G}_a$-action.*

PROOF. Consider a pencil $\mathscr{L}_{\mathbb{P}^2} = \langle C_1, C_2 \rangle$ on $\mathbb{P}^2$ generated by a smooth conic C_1 and a double line $C_2 = 2l$, where l is tangent to C_1 at a point $P_0 \in C_1$. Then $L \setminus \{P_0\} \simeq \mathbb{A}^1$, where L is a general member of $\mathscr{L}_{\mathbb{P}^2}$. Moreover, $U = \mathbb{P}^2 \setminus (C_1 \cup C_2)$ is a cylinder over $\mathbb{A}^1_*$. Blowing up at $9 - d$ distinct points Q_i on $C_1 \setminus \{P_0\}$, where $9 \geq d \geq 4$, we obtain a del Pezzo surface Y of degree d with a contraction $\sigma \colon Y \to \mathbb{P}^2$, and any such surface can be obtained in this way, except for $\mathbb{P}^1 \times \mathbb{P}^1$.

---

[11] Cf. Remark 2.1.9(3).

The cylinder $U' = \sigma^{-1}(U) \simeq U$ is $(-K_Y)$-polar (see Definition 3.1.7). Indeed, let $E_i = \sigma^{-1}(Q_i)$, $i = 1, \ldots, 9-d$. For any $1 \gg \varepsilon > 0$ we have

$$-K_{\mathbb{P}^2} \equiv (1+\varepsilon)C_1 + (1-2\varepsilon)l.$$

Hence,

$$-K_Y = \sigma^*(-K_{\mathbb{P}^2}) - \sum_{i=1}^{9-d} E_i \equiv (1+\varepsilon)\Delta_1 + (1-2\varepsilon)\Delta_2 + \varepsilon \sum_{i=1}^{9-d} E_i,$$

where Δ_1 and Δ_2 are the proper transforms in Y of C_1 and l, respectively. Thus U' is a $(-K_Y)$-polar cylinder on Y.

In the remaining case where $Y = \mathbb{P}^1 \times \mathbb{P}^1$, the natural embedding of $\mathbb{A}^2 = \mathbb{A}^1 \times \mathbb{A}^1$ into Y yields a $(-K_Y)$-polar cylinder on Y. Applying now Corollary 3.1.12 ends the proof. □

The proof exploits a $(-K_{\mathbb{P}^2})$-polar cylinder on $\mathbb{P}^2$ made of a pencil of conics with a common tangent line. Based on the same idea, we give below some alternative constructions of polar cylinders on anticanonically polarized del Pezzo surfaces of degree ≥ 4. Due to Corollary 3.1.12, this leads to new $\mathbb{G}_a$-actions on the cones over del Pezzo surfaces of degree ≥ 4 under their anticanonical embeddings. These examples will be useful in the sequel.

3.2.3. Example. Consider a pencil of rational curves on $\mathbb{P}^2$ with a unique base point P. (Similarly, one can find such a pencil on the quadric $\mathbb{P}^1 \times \mathbb{P}^1$.) Then the complement of the union of its degenerate members (or of a general one, if all members are nondegenerate) is a $(-K_{\mathbb{P}^2})$-polar cylinder on $\mathbb{P}^2$. In [81] an example was proposed of such a pencil of quintic curves. Moreover, there is a smooth conic C_1 and a rational unicuspidal quintic C_2 from the pencil as in [81] that meet in one point, the cuspidal point of the quintic.

These two curves generate a pencil $\mathscr{L}_{\mathbb{P}^2} = \langle 5C_1, 2C_2 \rangle$ of rational curves of degree 10 with a unique base point such that $\mathbb{P}^2 \setminus (C_1 \cup C_2)$ is a cylinder. Similarly as in the proof above, every del Pezzo surface Y of degree $d \geq 4$ can be obtained, along with a $(-K_Y)$-polar cylinder, by blowing up a certain set of $9-d$ points on C_1. Indeed, we can write

$$-K_{\mathbb{P}^2} \equiv (\tfrac{3}{2} - \varepsilon)C_1 + \tfrac{2}{5}\varepsilon C_2$$

with an appropriate $\varepsilon > 0$, and then proceed in the same fashion as in the proof.

3.2.4. Example. Picking up four points $P_1, \ldots, P_4$ in $\mathbb{P}^2$ in general position, we consider the pencil of lines $\mathscr{L}_{\mathbb{P}^2}$ on $\mathbb{P}^2$ generated by $l_1 = (P_1P_2)$ and $l_2 = (P_3P_4)$. The blowup $\sigma \colon Y \to \mathbb{P}^2$ of these points yields a del Pezzo surface Y of degree 5. We have

$$-K_{\mathbb{P}^2} \equiv \frac{3}{2}l_1 + \frac{3}{2}l_2 \quad \text{and so } -K_Y \equiv \frac{3}{2}l_1' + \frac{3}{2}l_2' + \frac{1}{2}\sum_{i=1}^{4} E_i,$$

where l_i' is the proper transform of l_i, $i = 1, 2$, and E_i is the exceptional (-1)-curve over P_i, $i = 1, \ldots, 4$. Then $L^{(1)} = l_1' + E_1 + E_2$ and $L^{(2)} = l_2' + E_3 + E_4$ are the only degenerate fibers of the pencil $\mathscr{L} = \sigma_*^{-1}(\mathscr{L}_{\mathbb{P}^2})$ on Y. Since

$$D = \tfrac{1}{2}(L^{(1)} + L^{(2)}) + (l_1' + l_2') \equiv -K_Y,$$

the open set

$$Y \setminus \operatorname{supp} D = \mathbb{P}^2 \setminus (l_1 \cup l_2) \simeq \mathbb{A}^1_* \times \mathbb{A}^1$$

is a $(-K_Y)$-polar cylinder on Y. A similar construction can be applied to any del Pezzo surface of degree $d \geq 5$.

3.2.5. Example. Consider the pencil $\mathscr{L}_{\mathbb{P}^2}$ of unicuspidal rational curves $\alpha y z^{n-1} + \beta x^n = 0$ in $\mathbb{P}^2$, where $n \geq 1$. Blowing up $k \leq 4$ points in $\mathbb{P}^2$, at most two on each of the lines $x = 0$ and $y = 0$ off their common point $(0:0:1)$, we obtain examples of $(-K_Y)$-polar cylinders on an arbitrary del Pezzo surface of degree $d \geq 5$. For $n = 1$ and $k = 4$ this gives again the cylinder from Example 3.2.4.

3.2.6. Remark. The idea to start with a $(-K_{\mathbb{P}^2})$-polar cylinder on $\mathbb{P}^2$ cannot be carried out any more in case of a smooth cubic surface $Y \subseteq \mathbb{P}^3$. Indeed, suppose we are given a cylinder $\mathbb{P}^2 \setminus C \simeq Z \times \mathbb{A}^1$, where C is a reduced plane curve of degree d, not necessarily smooth or irreducible, and let $\mathscr{L}_{\mathbb{P}^2}$ be the corresponding pencil. Then $\mathrm{Bs}(\mathscr{L}_{\mathbb{P}^2})$ consists of one point $P_0 \in C$, and $C \setminus \{P_0\}$ is a disjoint union of components isomorphic to $\mathbb{A}^1$. Performing a blowup $\sigma\colon Y \to \mathbb{P}^2$ of m points $P_i \in C \setminus \{P_0\}$ with exceptional curves $E_j = \sigma^{-1}(p_j)$, $j = 1, \dots, m$, from the equalities

$$3\sigma^*(C) \sim d\sigma^*(-K_{P^2}) = -dK_Y + d\sum_{j=1}^{m} E_j \quad \text{and} \quad 3\sigma^*(C) = 3C' + 3\sum_{j=1}^{m} E_j,$$

where C' is the proper transform of C on Y, we obtain

$$-dK_Y \sim 3C' + (3-d)\sum_{j=1}^{m} E_j =: D.$$

The divisor D is effective and $\operatorname{supp} D = C' + \sum_{j=1}^{m} E_j$ if and only if $d \leq 2$ i.e., C is a line or a conic. Since the centers of blowup P_i, $i = 1, \dots, m$, are situated on C and Y must be del Pezzo, we have $m \leq 5$ and so $\deg(Y) \geq 4$.

In the next example, starting with a pencil on $\mathbb{P}^2$ with five base points, we construct a $(-K_Y)$-polar cylinder of a different type on arbitrary del Pezzo surface Y of degree $d = 5$, by resolving all the base points but one.

3.2.7. Example. Consider the following pencil $\mathscr{L}_{\mathbb{P}^2}$ of rational plane sextics:

$$\alpha(y^2 z - x^3)^2 + \beta(y^2 - xz)(y^4 - x^4) = 0.$$

The base locus of $\mathscr{L}_{\mathbb{P}^2}$ consists of the points $P_0, \dots, P_4$, where $P_0 = (0:0:1)$ and $\{P_1, \dots, P_4\} = (x^2 = z^2, \quad xz = y^2)$. Furthermore, $\mathscr{L}_{\mathbb{P}^2}$ has no fixed component and so its general member L is irreducible. Since $\operatorname{mult}_{P_0}(L) = 4$ and $\operatorname{mult}_{P_i}(L) = 2$ for $i = 1, \dots, 4$, the curve L is rational. Any singular point P_i of L is resolved by one blowup, and the singularity of L at P_0 is cuspidal. No three of the points $P_1, \dots, P_4$ are collinear. Therefore the blowup $\sigma : Y \to \mathbb{P}^2$ of the latter points yields a del Pezzo surface Y of degree 5, and any such surface arises in this way. Let $\mathscr{L}$ be the proper transform of $\mathscr{L}_{\mathbb{P}^2}$ on Y, and let $P = \sigma^{-1}(P_0)$. Then $\mathscr{L}$ is a pencil of rational curves with a cuspidal singularity at the unique base point P, smooth and disjoint outside P.

There are exactly two degenerate members of $\mathscr{L}_{\mathbb{P}^2}$, namely the double cuspidal cubic $C' = 2(y^2 z = x^3)$ and the union C'' of the conic $(y^2 = xz)$ and the four lines $(y^4 = x^4)$. We have $-K_{\mathbb{P}^2} \equiv D_{\mathbb{P}^2} = \frac{1}{4}C' + \frac{1}{4}C''$. Let D be the proper transform of $D_{\mathbb{P}^2}$ on Y. Since $\operatorname{mult}_{P_i}(D_{\mathbb{P}^2}) = 1$ for $i = 1, \dots, 4$, we have $K_Y + D = \sigma^*(K_{\mathbb{P}^2} + D_{\mathbb{P}^2}) \equiv 0$. Hence $U = Y \setminus (C' \cup C'')$ is a $(-K_Y)$-polar cylinder on Y (see also Example 4.1.16 below).

3.3. The Makar-Limanov invariant on affine cones over del Pezzo surfaces.

3.3.1. For an algebra A over a field k, its Makar-Limanov invariant $\mathrm{ML}(A)$ is defined as the intersection of the kernels of all locally nilpotent derivations on A. It is trivial if $\mathrm{ML}(A) = k$. Following [77] we say that A is of class ML_i if the quotient field $\mathrm{Frac}(\mathrm{ML}(A))$ has transcendence degree i. If $\mathrm{ML}(A)$ is finitely generated then $i = \dim(Z)$, where $Z = \mathrm{Spec}(\mathrm{ML}(A))$. Thus $A \in \mathrm{ML}_0$ whenever A has trivial Makar-Limanov invariant. For instance, $\mathbb{A}^3 \in \mathrm{ML}_0$ (regarded as the affine cone over $\mathbb{P}^2$).

For A graded there are graded versions $\mathrm{ML}^{(h)}(A)$ and $\mathrm{ML}_i^{(h)}$ of $\mathrm{ML}(A)$ and ML_i, respectively [30], where one restricts to homogeneous locally nilpotent derivations. Clearly, $\mathrm{ML}(A) \subseteq \mathrm{ML}^{(h)}(A)$. Hence the usual ML invariant is trivial if the homogeneous is.

3.3.2. Theorem. *Let X be the affine cone over a smooth, anticanonically embedded del Pezzo surface $Y \subseteq \mathbb{P}^d$ of degree $d \geq 4$. Then the homogeneous Makar-Limanov invariant $\mathrm{ML}^{(h)}(X)$ is trivial, i.e., $X \in \mathrm{ML}_0^{(h)}$.*

PROOF. By Theorem 1.2.3 and Proposition 2.2.6, for $d \geq 6$ the surface Y and the cone $X = \mathrm{Aff\,Cone}_{-K_Y}(Y)$ are toric. Since X is not isomorphic to a product $X' \times \mathbb{A}^1_*$, by Lemma 4.5 in [72] the homogeneous Makar-Limanov invariant of X is trivial.

It remains to show that $X \in \mathrm{ML}_0^{(h)}$ for $d = 4, 5$. Note that for an arbitrary graded algebra $A = \bigoplus_i A_i$, the graded subalgebra $\mathrm{ML}^{(h)}(A)$ is nontrivial if and only if there exists a nonconstant homogeneous element $h \in A_n \cap \mathrm{ML}^{(h)}(A)$ (so h is annihilated by all homogeneous locally nilpotent derivations on A). In the case of an affine cone X, the degree $n = \deg(h)$ is positive. Hence $\Gamma = \mathbb{V}(h) \in |n(-K_Y)|$ is an effective ample divisor on Y.

Let as before $Y \subseteq \mathbb{P}^d$ be a del Pezzo surface of degree $d \geq 4$. Suppose to the contrary that $\mathrm{ML}^{(h)}(X) \neq \mathbb{C}$, and let $h \in A_n \cap \mathrm{ML}^{(h)}(A)$ be nonconstant. Then the affine cone over the curve $\mathrm{supp}(\Gamma)$ is a divisor on X stable under any $\mathbb{G}_a$-action defined by a homogeneous locally nilpotent derivation on $\mathcal{O}(X)$. Hence for every $(-K_Y)$-polar cylinder U on Y, the curve $\mathrm{supp}(\Gamma)$ consists of components of the members of the linear pencil $\mathscr{L}$ on Y associated with U. In particular, for all $(-K_Y)$-polar cylinders on Y there must be a common component of the associated linear pencils $\mathscr{L}$.

For $d = 5$, we let $\sigma\colon Y \to \mathbb{P}^2$ be the blowup of four points $P_1, \ldots, P_4$ in $\mathbb{P}^2$ with exceptional curves $E_i = \sigma^{-1}(P_i)$. There are exactly ten lines on Y. Besides $E_1, \ldots, E_4$ these are the proper transforms l_{ij} of the lines (P_iP_j) on $\mathbb{P}^2$, where $1 \leq i < j \leq 4$. For every pair of lines $(l_{ij}, l_{i'j'})$ with distinct indices i, j, i', j', the curves

$$L_1 = l_{ij} + E_i + E_j \quad \text{and} \quad L_2 = l_{i'j'} + E_{i'} + E_{j'}$$

are the only degenerate members of a cylindrical linear pencil on Y (cf. Example 3.2.4). Three such pencils have no common component except for the lines $E_1, \ldots, E_4$.

Let us replace the lines $E_1, \ldots, E_4$ on Y by some other four disjoint lines, e.g., by $l_{12}, l_{13}, l_{23}, E_4$. We consider also three associated cylindrical pencils on Y e.g., those with degenerate members

$$L'_1 = E_2 + l_{12} + l_{23} \quad \text{and} \quad L'_2 = E_4 + l_{13} + l_{24}.$$

These pencils have no common component except for the lines $l_{12}, l_{13}, l_{23}, E_4$. The line E_4 is the only common component of all six pencils. With yet another choice of a pencil, we can eliminate also this latter line. Thus the homogeneous Makar-Limanov invariant of Y is trivial, as stated.

Let further $d = 4$, and let $\sigma_0 \colon Y \to \mathbb{P}^2$ be the blowup of five points $P_1, \ldots, P_5$ in general position in $\mathbb{P}^2$, with exceptional curves $E_i = \sigma^{-1}(P_i)$, $i = 1, \ldots, 5$. We let C denote the unique smooth conic through the points P_i. Given a point $Q \in C$ different from all the points P_i, similarly as in the proof of Theorem 3.2.2 we consider the pencil of conics on $\mathbb{P}^2$ generated by C and $2l_Q$, where l_Q is the tangent line to C at Q. Two different such pencils on $\mathbb{P}^2$ have no common member except for the conic C itself. Thus C' and the lines E_i, $i = 1, \ldots, 5$, are the only common components of the induced cylindrical pencils on the del Pezzo surface Y of degree 4.

Consider next the contraction $\sigma_1 \colon Y \to \mathbb{P}^2$ of the five disjoint lines $C', l_{12}, l_{13}, l_{14}, l_{15}$ on Y. Then $\sigma_1(E_1)$ is a conic in $\mathbb{P}^2$, which plays now the role of C. Once again, the only common components of the induced cylindrical pencils on Y are E_1 and the five disjoint lines as above meeting E_1.

Likewise, for the six different contractions $\sigma_i \colon Y \to \mathbb{P}^2$, $i = 0, \ldots, 5$, the only common component of the induced cylindrical pencils on Y is C'. However, the ample divisor Γ from the beginning of the proof cannot be supported by C'. This contradiction finishes the proof. □

3.3.3. Problem. Classify all affine cones over projective varieties whose homogeneous Makar-Limanov invariant is trivial.

4. On existence of $\mathbb{G}_a$-actions on cones over cubic surfaces

In this section we analyze in detail the case of a smooth cubic surface $Y \subseteq \mathbb{P}^3$. We do not know so far whether or not the affine cone $X = \operatorname{Aff\,Cone}(Y)$ carries a $\mathbb{G}_a$-action. However, we obtain in Proposition 4.1.20 and Theorem 4.2.2 below a detailed information on an eventual anticanonical polar cylinder in Y. This makes the criterion 3.1.12 of the existence of a $\mathbb{G}_a$-action more concrete in this particular case. We adopt the following convention.

4.0.1. Convention. Let Y be a smooth cubic surface in $\mathbb{P}^3$. Suppose that the affine cone $X = \operatorname{Aff\,Cone}(Y) \subseteq \mathbb{A}^4$ admits an effective $\mathbb{G}_a$-action. By Corollary 3.1.12, Y possesses a $(-K_Y)$-polar cylinder $U \simeq \mathbb{A}^1 \times Z$, where Z is an affine smooth rational curve. In other words $Y \setminus U = \operatorname{supp} D$, where

$$D = \sum_{i=1}^{n} \delta_i \Delta_i \equiv -K_Y \tag{6}$$

is an effective $\mathbb{Q}$-divisor on Y with $\delta_i \in \mathbb{Q}_{>0}$ $\forall i = 1, \ldots, n$.

4.1. The linear pencil on a cubic surface compatible with a cylinder. Let $\mathscr{L}$ be the pencil on Y with general member $L_z = \overline{\operatorname{pr}_2^{-1}(z)}$ for $z \in Z$. It is easily seen that $\mathscr{L}$ has at least one degenerate member. In what follows we suppose that $\operatorname{supp} D$ does not contain a nondegenerate member of $\mathscr{L}$ (otherwise, up to numerical equivalence, we replace such a member by a degenerate one). Under these assumptions, the following hold.

4.1.1. Lemma. *The support of D is connected and simply connected, and contains at least 7 irreducible components.*

PROOF. The projection $\mathrm{pr}_2 \colon U \to Z$ extends to a rational map $Y \dashrightarrow \mathbb{P}^1$ defined by the pencil $\mathscr{L}$ as above. A general member L of $\mathscr{L}$ is a rational curve smooth off a unique point P, where $\{P\} = L \cap \operatorname{supp} D$, and $L \setminus \{P\} = L \cap U \simeq \mathbb{A}^1$. Thus L is unibranch at P. The only base point of $\mathscr{L}$ (if exists) is contained in $\operatorname{supp} D$.

Since D is ample the support $\operatorname{supp} D$ is connected, see, e.g., Corollary 7.9 in [43, Chapter III, §7]. Resolving, if necessary, the base point of $\mathscr{L}$ by a modification $p : W \to Y$ yields a rational surface W with a pencil of rational curves $\mathscr{L}_W = {p_*}^{-1}\mathscr{L}$ that are fibers of $q = q_{|\mathscr{L}_W|} \colon W \to \mathbb{P}^1$. By Zariski's Main Theorem, the total transform $p^{-1}(\operatorname{supp} D)$ is still connected, and is a union of a (-1)-section, say, S of $\mathscr{L}_W$ and of some number of rational trees contained in fibers of $\mathscr{L}_W$. Hence $p^{-1}(\operatorname{supp} D)$ is also a tree of rational curves, i.e., is connected and simply connected. The exceptional divisor E of p being a subtree of $p^{-1}(\operatorname{supp} D)$, the contraction of E does not affect the simply-connectedness. This proves the first assertion.

Since Z is a rational smooth affine curve, we have $\operatorname{Pic}(U) = \operatorname{Pic}(\mathbb{A}^1) \times \operatorname{Pic}(Z) = 0$. By virtue of the exact sequence

$$G := \sum_{i=1}^{n} \mathbb{Z}\Delta_i \to \operatorname{Pic}(Y) \to \operatorname{Pic}(U) = 0 \tag{7}$$

the free abelian group G generated by the components Δ_i of D surjects onto $\operatorname{Pic}(Y) \simeq \mathbb{Z}^7$. Therefore $\operatorname{rk}(G) \geq 7$, which proves the second assertion. □

4.1.2. Lemma. *The pencil $\mathscr{L}$ has a unique base point, say, P, and* $\deg \mathscr{L} \geq 3$.

PROOF. If to the contrary $\operatorname{Bs} \mathscr{L} = \varnothing$, then the pencil of conics[12] $\mathscr{L}$ on Y with a section, say, $S = \Delta_0$ defines a morphism $\varphi_{|C|} \colon Y \to \mathbb{P}^1$ (extending the projection pr_2 of the cylinder) with exactly five degenerate fibers $L_1, \ldots, L_5$. Each degenerate fiber consists of two lines on Y intersecting transversally at one point. At most one of these two lines, say, l_i meets the cylinder U, while the other one, say, Δ_i is a components of D. Since D is connected we have $\Delta_i \cdot S = 1$ and $l_i \cdot S = 0$, $i = 1, \ldots, 5$. By the Adjunction Formula we get

$$1 = (-K_Y) \cdot l_i = D \cdot l_i = \delta_i - x,$$

where $x = 0$ if $l_i \neq \Delta_j$ $\forall j$ and $x = \delta_j > 0$ otherwise. Hence $\delta_i = 1 + x \geq 1$. Similarly, for a general fiber L of $\mathscr{L}$,

$$2 = (-K_Y) \cdot L = D \cdot L = \delta_0 S \cdot L = \delta_0$$

and so $\delta_0 = 2$.

On the other hand, by the Adjunction Formula

$$(-K_Y) \cdot S = 2 + S^2 = D \cdot S = 2S^2 + \sum_{i=1}^{5} \delta_i.$$

[12]For a curve C on a cubic surface Y, $\deg C$ stands for the degree of C with respect to the given embedding $Y \hookrightarrow \mathbb{P}^3$.

Therefore

$$2 = S^2 + \sum_{i=1}^{5} \delta_i \geq 4,$$

a contradiction. The inequality $(-K_Y) \cdot \mathscr{L} \geq 3$ is now immediate. □

4.1.3. Remarks. (1) Actually the degree of $\mathscr{L}$ must be essentially higher, since by Lemma 4.1.8 below D has 8 irreducible components.

(2) The assertion of the lemma holds also for any del Pezzo surface of degree 4 or 5, with a similar proof. However, it fails for degree 6. Indeed, pick three points P_0, P_1, P_2 in general position in $\mathbb{P}^2$, and consider the pencil generated by the lines $l_i = (P_0P_i)$, $i = 1, 2$. Blowing up these points we get a del Pezzo surface Y of degree 6 with a base point free pencil. Then the complement in Y of the total transform of $l_1 \cup l_2$ is a $(-K_Y)$-polar cylinder.

4.1.4. In the sequel we frequently use the following commutative diagram:

(8)

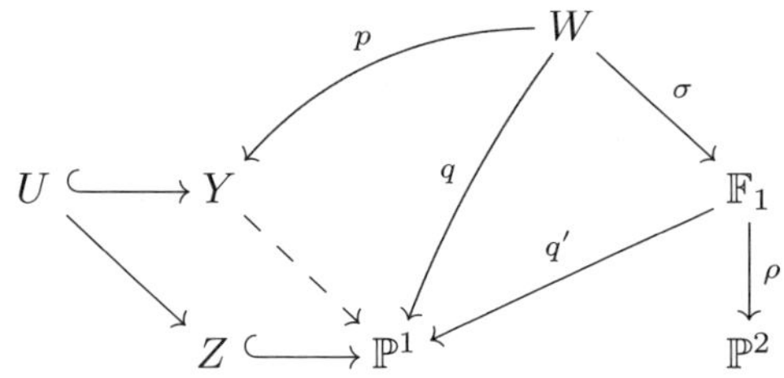

where $p\colon W \to Y$ is the minimal resolution of the base point P of $\mathscr{L}$, $q\colon W \to \mathbb{P}^1$ is the induced pencil, $\sigma\colon W \to \mathbb{F}_1$ is composed of the contraction of all components of degenerate fibers of q except those which meet the exceptional (-1)-section S_W of q, and $\rho\colon \mathbb{F}_1 \to \mathbb{P}^2$ is the contraction of this exceptional section.

4.1.5. Lemma. *$\delta_i < 1$ in (6) for all $i = 1, \ldots, n$.*

Proof. It is enough to show that $\delta_1 < 1$. By symmetry then $\delta_i < 1$ $\forall i$. Since the anticanonical divisor of Y is ample, we have $(-K_Y) \cdot \Delta_i > 0$ $\forall i = 1, \ldots, n$. We distinguish between the following cases :

(i) $(-K_Y) \cdot \Delta_1 \geq 3$,
(ii) $(-K_Y) \cdot \Delta_1 = 2$, and
(iii) $(-K_Y) \cdot \Delta_1 = 1$.

In case (i) suppose to the contrary that $\delta_1 \geq 1$. Since $n \geq 7$ by Lemma 4.1.1 and the divisor $-K_Y \equiv D$ is ample, we obtain

$$3 = (-K_Y) \cdot D = \sum_{i=1}^{n} \delta_i (-K_Y) \cdot \Delta_i > \delta_1 (-K_Y) \cdot \Delta_1 \geq 3,$$

which gives a contradiction.

In case (ii) $\Delta_1 \subseteq Y$ is a conic, $\Delta_1^2 = 0$, and $-K_Y \equiv \Delta_1 + E$, where E is the residual line cut out on Y by a plane in $\mathbb{P}^3$ through Δ_1.

Let $E = \Delta_i$ for some $i > 1$; we may assume that $i = 2$. Then $-K_Y \equiv \Delta_1 + \Delta_2$, $\Delta_1^2 = 0$ and $\Delta_1 \cdot \Delta_2 = 2$. Thus

$$2 = (-K_Y) \cdot \Delta_1 = D \cdot \Delta_1 \geq \delta_2 \Delta_1 \cdot \Delta_2 = 2\delta_2, \tag{9}$$

so $\delta_2 \le 1$. Furthermore,

$$1 = \Delta_2 \cdot D = \sum_{i=1}^{n} \delta_i \Delta_2 \cdot \Delta_i \ge 2\delta_1 - \delta_2.$$

Hence $\delta_1 \le \frac{1}{2}(\delta_2 + 1) \le 1$.

If $\delta_1 = 1$ then also $\delta_2 = 1$. Since $\Delta_1 + \Delta_2 \equiv -K_Y \equiv D$ we get $D = \Delta_1 + \Delta_2$. Thus $n = 2$, which contradicts Lemma 4.1.1. Therefore in this case $\delta_1 < 1$, as stated.

If further $E \ne \Delta_i\ \forall i$ then

$$(10) \qquad 1 = (-K_Y) \cdot E = D \cdot E \ge \delta_1 \Delta_1 \cdot E = 2\delta_1,$$

hence $\delta_1 \le \frac{1}{2}$. Thus anyway, $\delta_1 < 1$ in case (ii).

In case (iii) Δ_1 is a line on Y. Let C be a residual conic of Δ_1, so that $\Delta_1 + C \equiv -K_Y$ is a hyperplane section. We have as before

$$2 = (-K_Y) \cdot C = D \cdot C \ge \delta_1 \Delta_1 \cdot C = 2\delta_1,$$

hence $\delta_1 \le 1$.

If $\delta_1 = 1$ then $D \cdot C = \Delta_1 \cdot C = 2$ and so $(D - \Delta_1) \cdot C = 0$. Therefore the divisor $\operatorname{supp}(D - \Delta_1)$ is supported on the members of the pencil of conics $|C|$ on Y. The curve Δ_1 meets each fiber twice, and so the morphism $\varphi_{|C|}$ restricted to Δ_1 has two branch points.

By Lemma 4.1.1 the curve $\operatorname{supp} D$ is simply connected, hence it cannot contain the whole fiber of $\varphi_{|C|}$ which meets the component Δ_1 of D at two distinct points. We claim however that if a degenerate fiber $l_1 + l_2$ of the pencil $|C|$ contains a component, say, $\Delta_i = l_1$ of D, then its second component $l_2 = \Delta_j$ is also contained in $\operatorname{supp} D$ and, moreover, $\delta_i = \delta_j$. Indeed, since $\delta_1 = 1$ and Δ_1 is a line on Y we have

$$(11) \qquad 1 = \Delta_i \cdot D = \Delta_i \cdot \Delta_1 + \Delta_i \cdot (D - \Delta_1) = 1 - \delta_i + \sum_{k \ne 1,i} \delta_k \Delta_i \cdot \Delta_k.$$

The only component of the latter sum that meets Δ_i can be the line l_2. Hence $l_2 = \Delta_j$ for some $j \ne 1, i$. Now (11) shows that $\delta_i = \delta_j$, as claimed.

Furthermore, since $\Delta_i \cup \Delta_j$ meets Δ_1 twice and $\operatorname{supp} D$ is a tree, the line Δ_1 passes through the intersection point $\Delta_i \cap \Delta_j$. On the other hand, Δ_1 is tangent to exactly two members of the pencil $|C|$, which are either smooth or consist of two lines Δ_i and Δ_j meeting Δ_1 at their common point (an Eckardt point of Y). By the simply connectedness of $\operatorname{supp} D$, none of the other components of members of $|C|$ can be contained in $\operatorname{supp} D$. Hence $\operatorname{supp} D$ can contain at most 5 components, namely, Δ_1 and the components of two degenerate members tangent to Δ_1. However, this contradicts Lemma 4.1.1, since by this lemma $\operatorname{supp} D$ consists of at least 7 components. Now the proof is completed. □

4.1.6. Lemma. *Every component of the degenerate members of the pencil $\mathscr{L}$ on Y passes through the base point P of $\mathscr{L}$.*

PROOF. Assume to the contrary that there is a component C_0 of a degenerate member $L^{(0)}$ of $\mathscr{L}$ such that $P \notin C$. By Zariski's Lemma $C_0^2 < 0$. Hence C_0 is a (-1)-curve on Y and so $D \cdot C_0 = (-K_Y) \cdot C_0 = 1$. By an easy argument (cf. the proof of Lemma 4.1.1) the curve $\operatorname{supp} L^{(0)}$ is connected and simply connected (this is a tree of rational curves outside P). If there is another component C_1 of $L^{(0)}$ which meets C_0 and does not pass through P, then $L^{(0)}$ contains the configuration of two crossing lines $C_0 + C_1$ which do not pass through P. Then $\mathscr{L}$ must be the linear system of conics $|C_0 + C_1|$. By Lemma 4.1.2 this leads to a contradiction. Hence C_0 cannot separate $\operatorname{supp} L^{(0)}$ and so C_0 meets the complement $\operatorname{supp}(L^{(0)} - C_0)$ at one point transversally.

It follows that $\Delta_i \cdot C_0 = 1$ for a unique index i. If C_0 is not a component of D, then $1 = D \cdot C_0 = \delta_i$, which contradicts Lemma 4.1.5. Similarly, if $C_0 = \Delta_j$, then $1 = D \cdot C_0 = \delta_i - \delta_j$. Hence $\delta_i = 1 + \delta_j > 1$. Once again, this contradicts Lemma 4.1.5. □

4.1.7. Lemma. *The pencil $\mathscr{L}$ has at most two degenerate members.*

PROOF. Recall (see 4.1.4) that $p\colon W \to Y$ stands for the minimal resolution of the base locus of $\mathscr{L}$ and $q\colon W \to \mathbb{P}^1$ for the fibration given by $p_*^{-1}\mathscr{L}$. Write p as a composition of blowups of points over P:

$$p\colon W \xrightarrow{p_1} W_1 \xrightarrow{p_2} \cdots \xrightarrow{p_N} W_N = Y, \tag{12}$$

where the exceptional divisor S_W of p_1 is a q-horizontal (-1)-curve on W. A general fiber L of q is a smooth rational curve meeting S_W at one point. Indeed, $L \setminus S_W \simeq p(L) \setminus P \simeq \mathbb{A}^1$. Therefore S_W is a section of q.

Let $C_1, \ldots, C_m$ be the components of degenerate fibers $F_1, \ldots, F_m$ of q meeting S_W. We claim that all the curves C_i are p-exceptional. Indeed, otherwise for some i, the image $p(\overline{(F_i \setminus C_i)})$ would be a component of a degenerate member of $\mathscr{L}$ which does not pass through P. The latter contradicts Lemma 4.1.6.

Note that, on each step, the exceptional divisor of $p_k \circ \cdots \circ p_N$ is an SNC tree of rational curves. On the other hand, all the curves $p_1(C_i)$ on W_1 pass through the point $p_1(S_W)$. Therefore $m \leq 2$. □

4.1.8. Lemma. *The pencil $\mathscr{L}$ has exactly two degenerate members, say, $L^{(1)}$ and $L^{(2)}$. Furthermore, $\operatorname{supp} D = \operatorname{supp}(L^{(1)} + L^{(2)})$ consists of 8 irreducible components, i.e., $n = 8$ in (6).*

PROOF. Assume that the only degenerate member of $\mathscr{L}$ is $L^{(1)}$. In this case, $\operatorname{supp} D \subseteq \operatorname{supp} L^{(1)}$, $Z \supseteq \mathbb{A}^1$, and $U \supseteq \mathbb{A}^2$.

If $Z \simeq \mathbb{A}^1$ then $\operatorname{Pic}(U) = 0$ and $H^0(U, \mathcal{O}_U) = \mathbb{C}$, hence $\operatorname{Pic}(Y) \simeq \sum_{i=1}^n \mathbb{Z} \cdot \Delta_i$. Thus in this case $n = 7$ and $-K_Y = \sum_{i=1}^7 m_i \Delta_i$ for some $m_i \in \mathbb{Z}$. On the other hand, $-K_Y \equiv D = \sum_{i=1}^7 \delta_i \Delta_i$, where $0 < \delta_i < 1$ $\forall i$ according to Lemma 4.1.5. Since the decomposition of $-K_Y$ in $\operatorname{Pic}(Y) \otimes \mathbb{Q}$ is unique, this yields a contradiction.

If further $Z \simeq \mathbb{P}^1$ then $\operatorname{Pic}(U) \simeq \mathbb{Z}$ and $H^0(U, \mathcal{O}_U) = \mathbb{C}$. From the exact sequence (7), where $\operatorname{Pic}(U) = 0$ is replaced by $\operatorname{Pic}(U) \simeq \mathbb{Z}$, we obtain $n = 6$. By Lemma 4.1.1, this leads again to a contradiction. Therefore $\mathscr{L}$ has indeed two degenerate members. This fixes the first assertion.

As for the second one, assuming to the contrary that $\operatorname{supp} D \neq \operatorname{supp}(L^{(1)} + L^{(2)})$ we would have $Z \supseteq \mathbb{A}^1$. Now the same argument as before yields a contradiction. Since the Picard group $\operatorname{Pic}(Y) \cong \mathbb{Z}^7$ is generated by the irreducible components of $L^{(1)} + L^{(2)}$ and $L^{(1)} \equiv L^{(2)}$ is the only relation between these components, we obtain that $n = 8$. □

4.1.9. Corollary. *The pencil $\mathscr{L}$ is ample. Furthermore, for every irreducible component C of a member of the pencil $\mathscr{L}$ on Y we have $C \setminus \{P\} \simeq \mathbb{A}^1$, and two such components have just the point P in common.*

PROOF. The first assertion follows immediately from Lemma 4.1.6 by the Nakai–Moishezon criterion. As for the second one, this follows from the well known fact that on an affine surface $V = Y \setminus L$, where L is a general fiber of $\mathscr{L}$, every degenerate fiber of the $\mathbb{A}^1$-fibration $\varphi_{|\mathscr{L}|}\colon V \to \mathbb{A}^1$ is a disjoint union of affine lines, see, e.g., [80, 117]. □

4.1.10. Lemma. *The pair (Y, D) is not log canonical at P.*

PROOF. Let D_W denote the crepant pull-back of D on W as in (8), i.e., a $\mathbb{Q}$-divisor on W such that

$$K_W + D_W = p^*(K_Y + D) \quad \text{and} \quad p_* D_W = D.$$

The exceptional (-1)-section S on W is the only q-horizontal component of D_W. For a general fiber L of q we have $2 = (-K_W) \cdot L = D_W \cdot L$. Therefore the discrepancy $a(S, D)$ (i.e., the coefficient of S in D_W with the opposite sign) equals -2. This proves our assertion. □

4.1.11. Corollary. $\operatorname{mult}_P(D) > 1$.

PROOF. If $\operatorname{mult}_P(D) \leq 1$, then the pair (Y, D) is canonical at P, because $P \in Y$ is a smooth point. In particular, it is log canonical. This contradicts Lemma 4.1.10. □

4.1.12. Lemma. *Any line l on Y through P is contained in* $\operatorname{supp} D$.

PROOF. Assuming the contrary we obtain $1 = (-K_Y) \cdot l = D \cdot l \geq \operatorname{mult}_P(D) > 1$, a contradiction. □

4.1.13. Lemma. *P cannot be an Eckardt point on Y.*

PROOF. Suppose the contrary. Then by Lemma 4.1.12, up to a permutation we may assume that $\Delta_1, \Delta_2, \Delta_3$ are lines through P, where $\delta_1 \leq \delta_2 \leq \delta_3$. Since $D \cdot \Delta_1 = (-K_Y) \cdot \Delta_1 = 1$, for an effective $\mathbb{Q}$-divisor on Y

$$D' = \frac{1}{1 - \delta_1}\left(D - \delta_1(\Delta_1 + \Delta_2 + \Delta_3)\right) \tag{13}$$

we obtain $D' \cdot \Delta_1 = 1$ and by (13)

$$D \equiv -K_Y \equiv \Delta_1 + \Delta_2 + \Delta_3 \equiv D'.$$

Now the proof of Lemma 4.1.10 works equally for D'. Hence the pair (Y, D') is not canonical at P and so $\operatorname{mult}_P(D') > 1$. This contradicts the inequality $\operatorname{mult}_P(D') \leq D' \cdot \Delta_1 = 1$. □

4.1.14. Lemma. *The fibration $q\colon W \to \mathbb{P}^1$ in (8) has exactly two degenerate fibers. The general member L of $\mathscr{L}$ is singular at P.*

PROOF. The first assertion follows from Lemma 4.1.8. Let us show the second one. Assuming the contrary, for a smooth rational curve L on Y by the adjunction formula we obtain $(K_Y + L) \cdot L = -2$. The Mori cone of Y being spanned by the (-1)-curves $E_1, \ldots, E_{27}$, there is a decomposition

$$L \equiv \sum_{i=1}^{27} \alpha_i E_i,$$

where $\alpha_i \in \mathbb{Q}_{\geq 0}$. Hence $(K_Y + L) \cdot E_i < 0$ for some i. Thus $L \cdot E_i < (-K_Y) \cdot E_i = 1$, and so L cannot be ample. This contradicts Corollary 4.1.9. □

4.1.15. Remark. The minimal resolution $p : W \to Y$ of the base point P of $\mathscr{L}$ dominates the embedded minimal resolution of the cusp at P of a general member L of $\mathscr{L}$. The exceptional divisor $E = p^{-1}(P) \subseteq W$ is a rational comb with the number of teeth equal the length of the Puiseaux sequence of the cusp. The only (-1)-curve S_W in E is sitting on the handle of the comb. Hence $E = E^{(1)} + S_W + E^{(2)}$, where $E^{(k)} \subseteq p^{-1}(L^{(k)})$, $k = 1, 2$, and exactly one of the $E^{(k)}$ is a negatively definite linear chain of rational curves.

The degenerate members $L_W^{(1)}, L_W^{(2)}$ of the induced linear system $\mathscr{L}_W$ on W have the following structure: $L_W^{(k)}$ consists of $E^{(k)}$ and the proper transforms Δ_i' of the components $\Delta_i \subseteq L^{(k)}$, called *feathers*. The feathers are disjoint in W, and each of them meets $E^{(k)}$ at one point transversally. Let us illustrate this for the surface Y from Example 3.2.7, with the pullback $\mathscr{L}_W$ of the pencil $\mathscr{L}$ on Y.

4.1.16. Example. Up to interchanging the degenerate fibers $L_W^{(1)}$ and $L_W^{(2)}$ of $\mathscr{L}_W$, their structure is as follows. The fiber $L_W^{(1)}$ contains a component E_1 with self-intersection -5 joint with the (-1)-section S_W, and also the (-1)-feathers $\Delta_1', \dots, \Delta_5'$ meeting E_1. The second fiber $L_W^{(2)}$ contains the only feather Δ_0' of multiplicity 2. The weighted dual graph of the configuration $L_W^{(1)} + S_W + L_W^{(2)}$ is as follows:

(14)

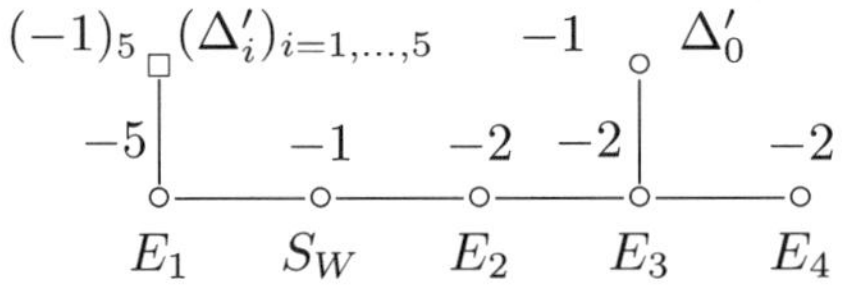

where the box denotes a disjoint union of five (-1)-feathers $\Delta_1', \dots, \Delta_5'$ joint with E_1. The exceptional divisor of $p \colon W \to Y$ is $E = E_1 + \dots + E_4$, and $\Delta_i = p(\Delta_i')$ $(i = 0, \dots, 5)$ are the components of the degenerate fibers $L^{(j)} = p(L_W^{(j)})$ $(j = 1, 2)$ of the pencil $\mathscr{L}$ on Y. More precisely, Δ_0 is the proper transform in Y of the cuspidal cubic C', Δ_1 is that of the conic C'', and $\Delta_2, \dots, \Delta_5 \subseteq Y$ are the proper transforms of the four lines $(y^4 = x^4)$ in $\mathbb{P}^2$.

4.1.17. Let $Q = (\rho \circ \sigma)(S_W) \in \mathbb{P}^2$, where as in 4.1.4 S_W stands for a (-1)-section of q contained in $\mathrm{Exc}(p)$. Then $\mathscr{L}_{\mathbb{P}^2} = \rho_*(\sigma_* \mathscr{L}_W)$ is the linear pencil of lines through Q on $\mathbb{P}^2$. We let $\varphi = \rho \circ \sigma \circ p^{-1} \colon Y \dashrightarrow \mathbb{P}^2$, and we let $\mathscr{H}$ denote the proper transform of $\mathscr{H}_{\mathbb{P}^2} = |\mathcal{O}_{\mathbb{P}^2}(1)|$ on Y via φ. With this notation, the following holds.

4.1.18. Lemma. $\mathscr{L} \subseteq \mathscr{H}$.

Proof. We have

$$\mathscr{H}_W = (\rho \circ \sigma)_*^{-1} \mathscr{H}_{\mathbb{P}^2} = (\rho \circ \sigma)^* \mathscr{H}_{\mathbb{P}^2} = \sigma^*(\rho^* \mathscr{H}_{\mathbb{P}^2}) \supseteq \sigma^*(\rho^* \mathscr{L}_{\mathbb{P}^2}).$$

Indeed, $\mathscr{H}_{\mathbb{P}2}$ is base point free. It is clear that $\rho^* \mathscr{L}_{\mathbb{P}^2} = \mathscr{L}_{\mathbb{F}_1} + S$, where S is the exceptional curve of ρ. Note that the centers of subsequent blowups in σ (including infinitesimally near centers) lie neither on the proper transform of S nor on that of general members of $\mathscr{L}_{\mathbb{P}^2}$. Hence,

$$\sigma^*(\rho^* \mathscr{L}_{\mathbb{P}^2}) = \sigma^*(\mathscr{L}_{\mathbb{F}_1} + S) = \mathscr{L}_W + S_W.$$

Since S_W is p-exceptional, applying p_* yields the assertion. □

4.1.19. Corollary. *$\mathscr{L}$ cannot be contained in $|-mK_Y|$ for any $m \in \mathbb{N}$.*

PROOF. Assume to the contrary that $\mathscr{L} \subseteq |-mK_Y|$ for some $m \in \mathbb{N}$. Then the mobile linear system $\mathscr{H}$, which defines the birational map $\varphi\colon Y \dashrightarrow \mathbb{P}^2$ (see 4.1.17), is also contained in $|-mK_Y|$. This contradicts the Segre–Manin theorem as stated in [67, Theorem 2.13]. □

The results of this subsection can be summarized as follows.

4.1.20. Proposition. *Let $Y \subseteq \mathbb{P}^3$ be a smooth cubic surface, and $X \subseteq \mathbb{A}^4$ be the affine cone over Y. Then X admits an effective $\mathbb{G}_a$-action if and only if Y admits a linear pencil $\mathscr{L}$ with the following properties:*

(1) *The base locus $\operatorname{Bs}\mathscr{L}$ consists of a single point, say, P, which is not an Eckardt point on Y.*
(2) *A general member L of $\mathscr{L}$ is singular at P, and $L \setminus \{P\} \simeq \mathbb{A}^1$.*
(3) *$\mathscr{L}$ has exactly two degenerate members, say $L^{(1)}$ and $L^{(2)}$, where the curve $L^{(1)} \cup L^{(2)}$ consist of 8 irreducible components $\Delta_1, \ldots, \Delta_8$.*
(4) *All curves Δ_i, $i = 1, \ldots, 8$, pass through P and are pairwise disjoint off P. Furthermore, $\Delta_i \setminus \{P\} \simeq \mathbb{A}^1$ $\forall i$.*
(5) *Every line on Y passing through P is one of the Δ_i.*
(6) *$-K_Y \equiv D = \sum_{i=1}^{8} \delta_i \Delta_i$, where $\delta_i \in \mathbb{Q}$ and $0 < \delta_i < 1$ $\forall i$.*
(7) *The pair (Y, D) is not log canonical at P.*
(8) *For every $m > 0$, $\mathscr{L}$ is not contained in $|-mK_Y|$.*

We do not know so far any example of a cubic surface with such a pencil $\mathscr{L}$. For the pencils on del Pezzo surfaces from Examples 3.2.3–3.2.7, not all of the properties (1)–(8) are fulfilled. For instance, the pencil of Example 3.2.7 satisfies (1)–(7), however (8) fails, since $\mathscr{L} \sim -2K_Y$.

4.2. The inverse nef value.

4.2.1. The nef value plays an important role in the adjunction-theoretic classification of polarized varieties. Let Y be a projective variety such that the canonical divisor K_Y is not nef, and let H be an ample divisor on Y. We define the *inverse nef value* $t_0 = t_0(Y, H)$ to be the supremum of t such that the divisor $H + tK_Y$ is nef, i.e.,

$$H \cdot C \geq t(-K_Y) \cdot C = t \deg(C) \tag{15}$$

for every curve C on Y. By the Kawamata rationality theorem [78, Theorem 7.1.1], t_0 is achieved and is rational. By the Kawamata–Shokurov base-point-freeness theorem [78, Theorem 6.2.1], the divisor $H + t_0K_Y$ is semiample, i.e., the complete linear system $|m(H+t_0K_Y)|$ has no base point for all $m \gg 0$ and defines a surjective morphism $\varphi\colon Y \to Y'$ with connected fibers onto a normal projective variety Y'. In particular, $\kappa(H+t_0K_Y) \geq 0$, where κ stands for the Iitaka–Kodaira dimension.

For a smooth cubic surface Y in $\mathbb{P}^3$ satisfying Convention 4.0.1, we let $\mathscr{H} = \varphi_*^{-1}(\mathcal{O}_{\mathbb{P}^2}(1))$ be the mobile linear system on Y constructed in 4.1.17. In this setting the inverse nef value $t_0 = t_0(Y, \mathscr{H})$ is a positive integer (indeed, for $t = t_0$ the equality in (15) is achieved on a (-1)-curve). Moreover, $\kappa(\mathscr{H} + t_0K_Y) = 0$ if and only if $\mathscr{H} + t_0K_Y \equiv 0$. However, in the latter case by Corollary 4.19 a $(-K_Y)$-polar cylinder on Y cannot exist.

By virtue of Theorem 4.2.2 below, the same conclusion holds in the case where $\kappa(\mathscr{H} + t_0K_Y) = 1$. In the latter case the linear system $|m(\mathscr{H} + t_0K_Y)|$ defines for $m \gg 1$ a conic bundle $Y \to \mathbb{P}^1$. Indeed, the image curve is rational since Y is, and an irreducible general fiber F with $F^2 = 0$ and $-K_F = -K_Y|_F$ ample is a smooth conic. Actually $|\mathscr{H} + t_0 K_Y|$ defines already a conic bundle. For assuming that $\mathscr{H} + t_0 K_Y \equiv \beta F$, where $\beta \in \mathbb{Q}$, and taking intersection with a line l on Y such that $F \cdot l = 1$, we obtain $\beta \in \mathbb{N}$.

4.2.2. Theorem. *Let $\chi\colon Y \dashrightarrow \mathbb{P}^2$ be a birational map and $\mathscr{H} = \chi_*^{-1}(|\mathcal{O}_{\mathbb{P}^2}(1)|)$ be the proper transform on Y of the complete linear system of lines on $\mathbb{P}^2$. Then there is no conic F on Y such that*

$$\mathscr{H} \sim -aK_Y + bF \quad \text{for some } a \in \mathbb{N} \text{ and } b \in \mathbb{Z}. \tag{16}$$

PROOF. We use the methods developed in [52, 53]. Consider a resolution of indeterminacies of χ:

$$\begin{array}{ccc} & \widetilde{Y} & \\ {\scriptstyle p}\swarrow & & \searrow{\scriptstyle q} \\ Y & \overset{\chi}{\dashrightarrow} & \mathbb{P}^2. \end{array}$$

Decomposing p into a sequence of blowups with exceptional curves $E_1, \ldots, E_n$, the linear system $\widetilde{\mathscr{H}} = q^*(|\mathcal{O}_{\mathbb{P}^2}(1)|)$ on $\widetilde{Y}$ and the line bundle $K_{\widetilde{Y}}$ can be written in $\operatorname{Pic}(\widetilde{Y})$ as

$$\widetilde{\mathscr{H}} = p^*(\mathscr{H}) - \sum_{i=1}^{n} m_i E_i^* \quad \text{and} \quad K_{\widetilde{Y}} = p^*(K_Y) + \sum_{i=1}^{n} E_i^*. \tag{17}$$

Computing the intersection numbers $\widetilde{\mathscr{H}}^2$ and $\widetilde{\mathscr{H}} \cdot K_{\widetilde{Y}}$, by (17) we obtain

$$1 = \mathscr{H}^2 - \sum_{i=1}^{n} m_i^2 \quad \text{and} \quad -3 = K_Y \cdot \mathscr{H} + \sum_{i=1}^{n} m_i. \tag{18}$$

Suppose to the contrary that (16) holds for some conic F on Y. We choose the minimal possible value of $a > 0$. Since $F^2 = 0$ on Y, from (16) and (18) we deduce

$$\sum_{i=1}^{n} m_i^2 = 3a^2 + 4ab - 1 \quad \text{and} \quad \sum_{i=1}^{n} m_i = 3a + 2b - 3. \tag{19}$$

In the rest of the proof we use the following Claims 1–4.

Claim 1. *There is a birational transformation $Y \dashrightarrow Y'$, where Y' is again a smooth cubic surface in $\mathbb{P}^3$, such that for the proper transforms $\mathscr{H}'$ of $\mathscr{H}$ and F' of F on Y' we have*

$$\mathscr{H}' \sim -aK_Y + b'F'$$

with the same a as in (16), and additionally with

$$m' := \max_i \{m_i'\} \leq a \quad \forall i,$$

where the integers m_i' have the same meaning on Y' as the m_i have on Y.

PROOF OF CLAIM 1. Suppose that $m_i > a$ for some value of i. Consider the conic bundle $\varphi = \varphi_{|F|}\colon Y \to \mathbb{P}^1$. Let us perform an elementary transformation at the point $P = p(E_i) \in Y$. First we apply the blowup $\sigma\colon \widehat{Y} \to Y$ of P with exceptional divisor E. Assuming that $m_i = \mathrm{mult}_P(\mathscr{H})$, on the new surface $\widehat{Y}$ we have

(20) $$\widehat{\mathscr{H}} := \sigma_*^{-1}(\mathscr{H}) = \sigma^*(\mathscr{H}) - m_i E \quad \text{and} \quad K_{\widehat{Y}} = \sigma^*(K_Y) + E.$$

Modulo the linear equivalence we may choose the conic F passing through the point P. Then F is irreducible. Indeed, otherwise $F = F_1 + F_2$, where F_1, F_2 are two lines on Y and F_1 passes through P. So

$$a < m_i \le (F_1 \cdot \mathscr{H})_P \le F_1 \cdot \mathscr{H} = F_1 \cdot \big(-aK_Y + b(F_1 + F_2)\big) = a,$$

which is impossible. Thus the proper transform $\widehat{F} \sim \sigma^*(F) - E$ of F on $\widehat{Y}$ is a (-1)-curve.

The contraction $\sigma' \colon \widehat{Y} \to Y'$ of $\widehat{F}$ to a point $P' \in Y'$ yields a smooth conic $F' = \sigma'_*(E)$ passing through P' on the resulting cubic surface Y', such that

$$\mathscr{H}' := \sigma'_*(\widehat{\mathscr{H}}) \sim -aK_Y + b'F' \qquad \text{for some } b' \in \mathbb{Z}.$$

Using (16) and (20), on Y' we obtain

$$\begin{aligned}\mathrm{mult}_{P'}(\mathscr{H}') = \widehat{\mathscr{H}} \cdot \widehat{F} &= \sigma^*(\mathscr{H}) \cdot \widehat{F} - m_i E \cdot \widehat{F} \\ &= \mathscr{H} \cdot F - m_i = -aK_Y \cdot F - m_i = 2a - m_i < a.\end{aligned}$$

Iterating this procedure we achieve finally that $m'_i \le a$ for all values of i, as required. □

So we assume in the sequel that

(21) $$m = \max_i \{m_i\} \le a \quad \forall i.$$

Claim 2. *Under the assumption* (21) *we have* $b < 0$.

PROOF OF CLAIM 2. From (19) and (21) we obtain

(22) $$3a^2 + 4ab = 1 + \sum_{i=1}^n m_i^2 \le 1 + m \sum_{i=1}^n m_i \le 1 + m(3a + 2b - 3) \le 1 + a(3a + 2b - 3).$$

It follows by (22) that $2ab \le 1 - 3a$. Since $a \ge 1$ then $b \le 1/(2a) - \frac{3}{2} \le -1$. □

Claim 3 (the Noether – Fano inequality). *For m as in* (21) *we have*

$$a \ge m > a + b.$$

PROOF OF CLAIM 3. The first inequality follows by (21). To show the second, suppose to the contrary that

(23) $$m \le a + b.$$

From (22) and (23) we obtain

(24) $$3a^2 + 4ab \le 1 + m(3a + 2b - 3) \le 1 + (a + b)(3a + 2b - 3).$$

Thus by (23) and (24)

(25) $$3 \le 3m \le 3(a + b) \le 1 + b(a + 2b).$$

We claim that $a + 2b \ge 0$. Indeed, let C be the residual line of the conic F on Y so that $F + C \sim -K_Y$. Then by (16)

$$0 \le \mathscr{H} \cdot C = (-aK_Y + bF) \cdot C = a + 2b.$$

Now (25) leads to a contradiction, since $b < 0$ by Claim 2. □

Claim 4. *Consider the morphism $\varphi\colon Y \to \mathbb{P}^1$ defined by the pencil of conics $|F|$ on Y. Let $m = m_i$, and let $Q = p(E_i) \in Y$. If a line l on Y passes though Q, then l is a component of the fiber of φ through Q.*

PROOF OF CLAIM 4. We have

$$(26)\qquad (\mathscr{H} \cdot l)_Q \geq \operatorname{mult}_Q(\mathscr{H}) = m > a + b.$$

On the other hand,

$$(27)\qquad (\mathscr{H} \cdot l)_Q \leq \mathscr{H} \cdot l = (-aK_Y + bF) \cdot l = a + bF \cdot l.$$

By (26) and (27), $b < bF \cdot l$, where $b < 0$ by Claim 2. Therefore $F \cdot l = 0$, and the claim follows. □

Let again C be the residual (-1)-curve of the conic F on Y so that $-K_Y \sim F + C$. We let V denote a del Pezzo surface of degree 4 obtained by the contraction $\pi\colon Y \to V$ of C, and $\mathscr{H}_V, F_V$, etc. denote the images on V of $\mathscr{H}, F$, etc. Due to (16)

$$(28)\qquad -K_V \sim F_V \quad \text{and} \quad \mathscr{H}_V \sim -aK_V + bF_V \sim -(a+b)K_V.$$

By Claim 4 there is no line on V through the point $Q_V = \pi(Q)$. The blowup $\sigma\colon Y' \to V$ at Q_V yields yet another cubic surface Y' with the exceptional (-1)-curve $E = \sigma^{-1}(Q_V)$. For the proper transform $\mathscr{H}' = \sigma_*^{-1}(\mathscr{H}_V)$ on Y' we obtain by (28):

$$(29)\quad \mathscr{H}' \sim \sigma^*(\mathscr{H}_V) - mE \sim (a+b)\sigma^*(-K_V) - mE \sim (a+b)(-K_{Y'}) + (a+b-m)E.$$

The linear system

$$|F'| = |-K_{Y'} - E|$$

defines a conic bundle on Y'. Plugging $E \sim -K_{Y'} - F'$ into (29) we deduce:

$$(30)\qquad \mathscr{H}' \sim (2a + 2b - m)(-K_{Y'}) - (a + b - m)F'.$$

Using Claims 2 and 3,

$$(31)\qquad 2a + 2b - m = (a + b) + (a + b - m) < a.$$

By virtue of (30) the latter inequality contradicts the minimality of a. Now the proof of Theorem 4.2.2 is completed. □

4.2.3. Corollary. *Under Convention* 4.0.1 *the divisor $\mathscr{H} + t_0 K_Y$ in* 4.2.1 *is big, i.e.,*

$$\kappa(\mathscr{H} + t_0 K_Y) = 2.$$

5. Cones over some rational Fano threefolds

In this section we provide examples of two families of rational Fano 3-folds such that the affine cones over their anti-canonical embeddings admit nontrivial $\mathbb{G}_a$-actions.

See also [62] for more such examples.

5.0.1. Proposition. *Consider a smooth intersection $Y = Y_{2,2}$ of two quadric hypersurfaces in $\mathbb{P}^5$. Then the affine cone X over Y admits an effective $\mathbb{G}_a$-action.*

PROOF. According to the criterion of Theorem 3.1.9, it is enough to construct a $(-K_Y)$-polar open cylinder on Y. Fix a line $l \subseteq Y$. Consider the diagram

$$\begin{array}{ccccc} & & \widetilde{Y} & & \\ & \overset{\sigma}{\swarrow} & & \overset{\varphi}{\searrow} & \\ Y & & \overset{\psi}{\dashrightarrow} & & \mathbb{P}^3 \end{array}$$

where ψ is the projection with center l, σ is the blowup of l, and φ is the blowdown of the divisor D which is swept out by lines meeting l (i.e., $\sigma(D)$ is the union of lines meeting l; see [42, Chapter 6]). It is easily seen that $\Gamma = \varphi(D)$ is a smooth quintic curve in $\mathbb{P}^3$ of genus 2. The image $Q = \varphi(E)$ of the exceptional divisor E of σ is a quadric in $\mathbb{P}^3$. For a line $l \subseteq Y$ the following alternative holds: either

(1) $N_{l/X} = \mathcal{O}_l \oplus \mathcal{O}_l$ and Q is smooth, or
(2) $N_{l/X} = \mathcal{O}_l(1) \oplus \mathcal{O}_l(-1)$ and Q is singular.

Anyhow

$$Y \setminus \sigma(D) \simeq \mathbb{P}^3 \setminus Q.$$

Suppose that Q is singular; then Q is a quadratic cone. Let Π be a plane in $\mathbb{P}^3$ passing through the vertex P of Q. We claim that $\mathbb{P}^3 \setminus (Q \cup \Pi)$ is a principal cylinder. Indeed, consider the projection π_P with center P and its resolution:

$$\begin{array}{ccccc} & & \widetilde{\mathbb{P}}^3 & & \\ & \overset{\sigma'}{\swarrow} & & \overset{\varphi'}{\searrow} & \\ \mathbb{P}^3 & & \overset{\pi_P}{\dashrightarrow} & & \mathbb{P}^2. \end{array}$$

Let $E' \subseteq \widetilde{\mathbb{P}}^3$ be the exceptional divisor of σ', and let $Q' \subseteq \widetilde{\mathbb{P}}^3$ be the proper transform of Q. Then $C = \varphi'(Q') \subseteq \mathbb{P}^2$ is a conic, and E' is a section of the $\mathbb{P}^1$-bundle $\widetilde{\mathbb{P}}^3 \to \mathbb{P}^2$. Furthermore,

$$\widetilde{\mathbb{P}}^3 \simeq \mathbb{P}\big(\mathcal{O}_{\mathbb{P}^2} \oplus \mathcal{O}_{\mathbb{P}^2}(1)\big).$$

Letting $\Pi' \subseteq \widetilde{\mathbb{P}}^3$ be the proper transform of Π, the image $\varphi'(\Pi') = H \subseteq \mathbb{P}^2$ is a line. We have

$$\mathbb{P}^3 \setminus (Q \cup \Pi) \simeq \widetilde{\mathbb{P}}^3 \setminus (Q' \cup \Pi' \cup E')$$

is an $\mathbb{A}^1$-bundle over $\mathbb{P}^2 \setminus (C \cup H)$. Since $\operatorname{Pic}\big(\mathbb{P}^2 \setminus (C \cup H)\big) = 0$ we obtain

$$\mathbb{P}^3 \setminus (Q \cup \Pi) \simeq \mathbb{A}^1 \times \big(\mathbb{P}^2 \setminus (C \cup H)\big),$$

as required. □

Let us exhibit yet another family of Fano threefolds with Picard rank one. Their moduli space is six-dimensional. Every member Y of this family admits a $(-K_Y)$-polar cylinder, whereas the subfamily of compactifications of $\mathbb{A}^3$ is only four-dimensional [34, 93].

5.0.2. Proposition. *Let $Y = Y_{22} \subseteq \mathbb{P}^{13}$ be a Fano variety of genus 12 with $\operatorname{Pic}(Y) = \mathbb{Z} \cdot [-K_Y]$, anticanonically embedded into $\mathbb{P}^{13}$. Then the affine cone X over Y admits an effective $\mathbb{G}_a$-action.*

PROOF. Again, it is enough to construct a $(-K_Y)$-polar open cylinder on Y. Then the result follows by applying Theorem 3.1.9. Picking a line $l_1 \subseteq Y$ we consider the commutative diagram

$$\begin{array}{ccc} \widetilde{Y} & \overset{\chi}{\dashrightarrow} & \widetilde{Y}^+ \\ {\scriptstyle\sigma_1}\big\downarrow & & \big\downarrow{\scriptstyle\varphi_1} \\ \mathbb{P}^{13} \supseteq Y = Y_{22} & \overset{\psi_1}{\dashrightarrow} & Y_5 \subseteq \mathbb{P}^6 \end{array}$$

where σ_1 is the blowup of l_1, ψ_1 is the double projection with center[13] l_1 onto a Fano threefold Y_5 of degree 5 and of Fano index 2 half-anticanonically embedded into $\mathbb{P}^6$, φ_1 is the blowup of a smooth rational curve $\Gamma \subseteq Y_5$ of degree 5, and χ is a flop; see [56, §4.3]. We have

$$Y \setminus H_1 \simeq Y_5 \setminus H_2,$$

where $H_1 \subseteq Y$ is a hyperplane section with $\operatorname{mult}_{l_1}(H_1) = 3$, and $H_2 \subseteq Y_5$ is a hyperplane section passing through Γ. Thus it suffices to show that $Y_5 \setminus H_2$ contains an H_2-polar cylinder.

Recall that the family of all lines on Y_5 is parameterized by $\mathbb{P}^2$. For such a line l either $N_{l/Y_5} \simeq \mathcal{O}_l \oplus \mathcal{O}_l$ or $N_{l/Y_5} \simeq \mathcal{O}_l(1) \oplus \mathcal{O}_l(-1)$. The lines of the second type are parameterized by a smooth conic on $\mathbb{P}^2$; see [35]. There exists a line l_2 on Y_5 contained in H_2. Consider the projection ψ_2 with center l_2 and its resolution:

$$\begin{array}{ccccc} & & \widetilde{Y}_5 & & \\ & {\scriptstyle\sigma_2}\swarrow & & \searrow{\scriptstyle\varphi_2} & \\ \mathbb{P}^6 \supseteq Y_5 & & \overset{\psi_2}{\dashrightarrow} & & Q \subseteq \mathbb{P}^4 \end{array}$$

where σ_2 is the blowup of l_2, and $Q \subseteq \mathbb{P}^4$ is a smooth quadric. We have

$$Y_5 \setminus H_2' \simeq Q \setminus H_3',$$

where $H_2' \subseteq Y_5$ and $H_3' \subseteq Q$ are hyperplane sections such that $\operatorname{mult}_{l_2}(H_2') = 2$, H_2' is swept out by lines meeting l_2, and H_3' is the image of the σ_2-exceptional divisor. Since ψ_2 is a projection,

$$Y_5 \setminus (H_2 \cup H_2') \simeq Q \setminus (H_3 \cup H_3'),$$

where the image $H_3 \subseteq Q$ of H_2 is again a hyperplane section (possibly $H_3 = H_3'$). It remains to show that the complement $Q \setminus (H_3 \cup H_3')$ contains an H_3'-polar cylinder.

We may assume that $H_3' \neq H_3$. The projection $\pi_P \colon \mathbb{P}^4 \dashrightarrow \mathbb{P}^3$ with center at a general point $P \in H_3 \cap H_3'$ yields an isomorphism

$$Q \setminus (H_3 \cup H_3') \simeq \mathbb{P}^3 \setminus (\Pi_1 \cup \Pi_2 \cup \Pi_3),$$

where Π_1, Π_2, Π_3 are three planes in $\mathbb{P}^3$. So the existence of an H_3'-polar cylinder on $Q \setminus (H_3 \cup H_3')$ is equivalent to the existence of a Π_1-polar cylinder on $\mathbb{P}^3 \setminus (\Pi_1 \cup \Pi_2 \cup \Pi_3)$. Now the assertion easily follows. □

---

[13]That is, $\psi_1 \colon Y \dashrightarrow \mathbb{P}^6$ is defined by the linear system $|\mathcal{O}_Y(1) - 2l_1|$ of hyperplane sections on Y passing through l_1 and singular along l_1.

References

1. D. N. Akhiezer, *Dense orbits with two ends*, Izv. Akad. Nauk SSSR Ser. Mat. **41** (1977), no. 2, 308–324 (Russian); English transl., Math. USSR-Izv. **11** (1977), no. 2, 293–307.
2. ———, *Algebraic groups that are transitive in the complement to a homogeneous hypersurface*, Trudy Moskov. Mat. Obshch. **48** (1985), 77–96 (Russian); English transl., Trans. Moscow Math. Soc. **1986** (1986), 83–103.
3. ———, *Lie group actions in complex analysis*, Aspects Math., vol. E27, Friedr. Vieweg, Braunschweig, 1995.
4. I. V. Arzhantsev, K. Kuyumzhiyan, and M. Zaidenberg, *Flag varieties, toric varieties, and suspensions: three instances of infinite transitivity*, available at `arXiv:1003.3164`.
5. I. V. Arzhantsev and E. V. Sharoiko, *The Hassett – Tschinkel correspondence: modality and projective hypersurfaces*, available at `arXiv:0912.1474`.
6. A. Beauville and J. Blanc, *On Cremona transformations of prime order*, C. R. Math. Acad. Sci. Paris **339** (2004), no. 4, 257–259.
7. J. Blanc, *Sous-groupes algébriques du groupe de Cremona*, Transform. Groups **14** (2009), no. 2, 249–285.
8. A. Blanchard, *Sur les variétés analytiques complexes*, Ann. Sci. École Norm. Sup. (3) **73** (1956), 157–202.
9. M. Brion, *Lectures on the geometry of flag varieties*, Topics in Cohomological Studies of Algebraic Varieties (P. Pragacz, ed.), Trends Math., Birkhäuser, Basel, 2005, pp. 33–85.
10. I. Cantat and K. Dolgachev, *Rational surfaces with a large group of automorphisms*, available at `arXiv:1106.0930`.
11. I. A. Cheltsov, *Birationally rigid Fano varieties*, Uspekhi Mat. Nauk **60** (2005), no. 5(365), 71–160 (Russian); English transl., Russian Math. Surveys **60** (2005), no. 5, 875–965.
12. C. Chevalley, *On algebraic group varieties*, J. Math. Soc. Japan **6** (1954), 303–324.
13. ———, *Classification des groupes algébriques semi-simples*, Collected Works, vol. 3, Springer, Berlin, 2005.
14. W.-L. Chow, *On the projective embedding of homogeneous varieties*, Algebraic Geometry and Topology. A Symposium in Honor of S. Lefschetz (R. H. Fox, D. C. Spencer, and A. W. Tucker, eds.), Princeton Univ. Press, Princeton, NJ, 1957, pp. 122–128.
15. A. Corti, *Factoring birational maps of threefolds after Sarkisov*, J. Algebraic Geom. **4** (1995), no. 2, 223–254.
16. D. Cox, J. Little, and D. O'Shea, *Ideals, varieties, and algorithms*, 3rd ed., Undergrad. Texts Math., Springer, New York, 2007.
17. V. I. Danilov, *Algebraic varieties and schemes*, Algebraic Geometry. I, Encyclopaedia Math. Sci., vol. 23, Springer, Berlin, 1994, pp. 167–297.
18. T. de Fernex, *On planar Cremona maps of prime order*, Nagoya Math. J. **174** (2004), 1–28.
19. T. de Fernex, L. Ein, and M. Mustaţă, *Bounds for log canonical thresholds with applications to birational rigidity*, Math. Res. Lett. **10** (2003), no. 2-3, 219–236.
20. M. Demazure, *Surfaces de Del Pezzo.* II–V, Séminaire sur les singularités des surfaces (Palaiseau, 1976) (M. Demazure, H. C. Pinkham, and B. Teissier, eds.), Lecture Notes in Math., vol. 777, Springer, Berlin, 1980, pp. 21–69.
21. M. Doebeli, *Linear models for reductive group actions on affine quadrics*, Bull. Soc. Math. France **122** (1994), no. 4, 505–531.
22. I. V. Dolgachev, *Integral quadratic forms: applications to algebraic geometry (after V. Nikulin)*, Séminaire Bourbaki. Vol. 1982/83, Astérisque, vol. 105, Soc. Math. France, Paris, 1983, pp. 251–278.
23. ———, *Topics in classical algebraic geometry*, available at `http://www.math.lsa.umich.edu/~idolga/topics.pdf`.
24. I. V. Dolgachev and V. A. Iskovskikh, *Finite subgroups of the plane Cremona group*, Algebra, Arithmetic, and Geometry: in Honor of Yu. I. Manin. Vol. I (Yu. Tschinkel and Yu. Zarhin, eds.), Progr. Math., vol. 269, Birkhäuser, Boston, MA, 2009, pp. 443–548.
25. ———, *On elements of prime order in the plane Cremona group over a perfect field*, Int. Math. Res. Not. IMRN **18** (2009), 3467–3485.
26. A. K. Dutta, *On* $\mathbf{A}^1$*-bundles of affine morphisms*, J. Math. Kyoto Univ. **35** (1995), no. 3, 377–385.

27. H. Flenner, Sh. Kaliman, and M. Zaidenberg, *Birational transformations of weighted graphs*, Affine Algebraic Geometry (T. Hibi, ed.), Osaka Univ. Press, Osaka, 2007, pp. 107–147; Corrigendum, in this volume.
28. ———, *Smooth affine surfaces with non-unique* $\mathbb{C}^*$*-actions*, J. Algebraic Geom. **20** (2011), no. 2, 329–398.
29. H. Flenner and M. Zaidenberg, *Rational curves and rational singularities*, Math. Z. **244** (2003), no. 3, 549–575.
30. ———, *Locally nilpotent derivations on affine surfaces with a* $\mathbb{C}^*$*-action*, Osaka J. Math. **42** (2005), no. 4, 931–974.
31. ———, *On the uniqueness of* $\mathbb{C}^*$*-actions on affine surfaces*, Affine Algebraic Geometry (Seville, 2003) (J. Gutierrez, V. Shpilrain, and J.-T. Yu, eds.), Contemp. Math., vol. 369, Amer. Math. Soc., Providence, RI, 2005, pp. 97–111.
32. G. Freudenburg, *Algebraic theory of locally nilpotent derivations*, Encyclopaedia Math. Sci., vol. 136, Springer, Berlin, 2006.
33. T. Fujita, *On the topology of noncomplete algebraic surfaces*, J. Fac. Sci. Univ. Tokyo Sect. IA Math. **29** (1982), no. 3, 503–566.
34. M. Furushima, *The complete classification of compactifications of* $\mathbf{C}^3$ *which are projective manifolds with the second Betti number one*, Math. Ann. **297** (1993), no. 4, 627–662.
35. M. Furushima and N. Nakayama, *The family of lines on the Fano threefold* V_5, Nagoya Math. J. **116** (1989), 111–122.
36. M. H. Gizatullin and V. I. Danilov, *Automorphisms of affine surfaces.* I, Izv. Akad. Nauk SSSR Ser. Mat. **39** (1975), no. 3, 523–565; II, **41** (1977), no. 1, 54–103; English transl. in I, Math USSR-Izv. **9** (1975), no. 3, 493–534; II, **11** (1977), no. 1, 51–98.
37. V. Gonzlez-Aguilera and A. Liendo, *Automorphisms of prime order of smooth cubic n-folds*, available at `arXiv:1002.4136`.
38. ———, *On the order of an automorphism of a smooth hypersurface*, available at `arXiv:1102.3443`.
39. A. G. Gorinov, *Division theorems for the rational cohomology of some discriminant complements*, available at `arXiv:math/0511593`.
40. M. Goto, *On algebraic homogeneous spaces*, Amer. J. Math. **76** (1954), 811–818.
41. H. Grauert and R. Remmert, *Über kompakte homogene komplexe Mannigfaltigkeiten*, Arch. Math. (Basel) **13** (1962), 498–507.
42. P. Griffiths and J. Harris, *Principles of algebraic geometry*, Wiley Classics Lib., Wiley, New York, 1994.
43. R. Hartshorne, *Algebraic geometry*, Grad. Texts in Math., vol. 52, Springer, New York, 1977.
44. T. Hosoh, *Automorphism groups of quartic del Pezzo surfaces*, J. Algebra **185** (1996), no. 2, 374–389.
45. ———, *Automorphism groups of cubic surfaces*, J. Algebra **192** (1997), no. 2, 651–677.
46. ———, *Equations of specific cubic surfaces and automorphisms*, SUT J. Math. **38** (2002), no. 2, 127–134.
47. A. T. Huckleberry, *Actions of groups of holomorphic transformations*, Complex Manifolds, Springer, Berlin, 1998, pp. 143–196.
48. A. T. Huckleberry and E. Oeljeklaus, *A characterization of complex homogeneous cones*, Math. Z. **170** (1980), no. 2, 181–194.
49. A. Huckleberry and E. Oeljeklaus, *Classification theorems for almost homogeneous spaces*, Inst. élie Cartan, vol. 9, Univ. Nancy I, Inst. élie Cartan, Nancy, 1984.
50. J.-M. Hwang, S. Kebekus, and T. Peternell, *Holomorphic maps onto varieties of non-negative Kodaira dimension*, J. Algebraic Geom. **15** (2006), no. 3, 551–561.
51. S. Iitaka, *Algebraic geometry*, Grad. Texts in Math., vol. 76, Springer, New York, 1982.
52. V. A. Iskovskih, *Rational surfaces with a pencil of rational curves and with positive square of the canonical class*, Math. USSR-Sb. **12** (1970), no. 1, 91–117.
53. V. A. Iskovskikh, *Factorization of birational mappings of rational surfaces from the point of view of Mori theory*, Uspekhi Mat. Nauk **51** (1996), no. 4(310), 3–72 (Russian); English transl., Russian Math. Surveys **51** (1996), no. 4, 585–652.
54. ———, *Birational rigidity of Fano hypersurfaces in the framework of Mori theory*, Uspekhi Mat. Nauk **56** (2001), no. 2(338), 3–86 (Russian); English transl., Russian Math. Surveys **56** (2001), no. 2, 207–291.

55. V. A. Iskovskih and Yu. I. Manin, *Three-dimensional quartics and counterexamples to the Lüroth problem*, Mat. Sb. (N.S.) **86(128)** (1971), 140–166 (Russian); English transl., Math. USSR-Sb. **15** (1971), 141–166.
56. V. A. Iskovskikh and Yu. G. Prokhorov, *Fano varieties*, Algebraic Geometry. V (A. N. Parshin and I. R. Shafarevich, eds.), Encyclopaedia Math. Sci., vol. 47, Springer, Berlin, 1999, pp. 1–247.
57. V. A. Iskovskikh and I. R. Shafarevich, *Algebraic surfaces*, Algebraic Geometry. II (I. R. Shafarevich, ed.), Encyclopaedia Math. Sci., vol. 35, Springer, Berlin, 1996, pp. 127–262.
58. J. C. Jantzen, *Representations of algebraic groups*, 2nd ed., Math. Surveys Monogr., vol. 107, Amer. Math. Soc., Providence, RI, 2003.
59. M. Kalka, B. Shiffman, and B. Wong, *Finiteness and rigidity theorems for holomorphic mappings*, Michigan Math. J. **28** (1981), no. 3, 289–295.
60. T. Kambayashi and M. Miyanishi, *On flat fibrations by the affine line*, Illinois J. Math. **22** (1978), no. 4, 662–671.
61. T. Kambayashi and D. Wright, *Flat families of affine lines are affine-line bundles*, Illinois J. Math. **29** (1985), no. 4, 672–681.
62. T. Kishimoto, Yu. Prokhorov, and M. Zaidenberg, *Affine cones over Fano threefolds and additive group actions*, available at `arXiv:1106.1312`.
63. S. Kobayashi and T. Ochiai, *Meromorphic mappings onto compact complex spaces of general type*, Invent. Math. **31** (1975), no. 1, 7–16.
64. J. Kollár, *Rational curves on algebraic varieties*, Ergeb. Math. Grenzgeb. (3), vol. 32, Springer, Berlin, 1996.
65. ———, *Singularities of pairs*, Algebraic Geometry — Santa Cruz 1995 (J. Kollár, R. Lazarsfeld, and D. R. Morrison, eds.), Proc. Sympos. Pure Math., vol. 62, Amer. Math. Soc., Providence, RI, 1997, pp. 221–287.
66. J. Kollár and S. Mori, *Birational geometry of algebraic varieties*, Cambridge Tracts in Math., vol. 134, Cambridge Univ. Press, Cambridge, 1998.
67. J. Kollár, K. E. Smith, and A. Corti, *Rational and nearly rational varieties*, Cambridge Stud. Adv. Math., vol. 92, Cambridge Univ. Press, Cambridge, 2004.
68. S. Kondō, *Niemeier lattices, Mathieu groups, and finite groups of symplectic automorphisms of $K3$ surfaces*, with an appendix by S. Mukai, Duke Math. J. **92** (1998), no. 3, 593–603.
69. ———, *The maximum order of finite groups of automorphisms of $K3$ surfaces*, Amer. J. Math. **121** (1999), no. 6, 1245–1252.
70. V. Lakshmibai and P. Littelmann, *Richardson varieties and equivariant K-theory*, J. Algebra **260** (2003), no. 1, 230–260.
71. S. Lamy, *Automorphismes polynomiaux préservant une action de groupe*, Bol. Soc. Mat. Mexicana (3) **9** (2003), no. 1, 1–19.
72. A. Liendo, *Affine $\mathbb{T}$-varieties of complexity one and locally nilpotent derivations*, Transform. Groups **15** (2010), no. 2, 389–425.
73. L. Makar-Limanov, *On groups of automorphisms of a class of surfaces*, Israel J. Math. **69** (1990), no. 2, 250–256.
74. ———, *On the group of automorphisms of a surface $x^n y = P(z)$*, Israel J. Math. **121** (2001), 113–123.
75. Yu. I. Manin, *Cubic forms: Algebra, geometry, arithmetic*, 2nd ed., North-Holland Math. Library, vol. 4, North-Holland, Amsterdam, 1986.
76. G. Mason, *Symplectic automorphisms of $K3$-surfaces (after S. Mukai and V. V. Nikulin)*, CWI Newslett. **13** (1986), 3–19.
77. K. Masuda and M. Miyanishi, *The additive group actions on $\mathbb{Q}$-homology planes*, Ann. Inst. Fourier (Grenoble) **53** (2003), no. 2, 429–464.
78. K. Matsuki, *Introduction to the Mori program*, Universitext, Springer, New York, 2002.
79. H. Matsumura and P. Monsky, *On the automorphisms of hypersurfaces*, J. Math. Kyoto Univ. **3** (1963/1964), 347–361.
80. M. Miyanishi, *Open algebraic surfaces*, CRM Monogr. Ser., vol. 12, Amer. Math. Soc., Providence, RI, 2001.
81. M. Miyanishi and T. Sugie, *On a projective plane curve whose complement has logarithmic Kodaira dimension $-\infty$*, Osaka J. Math. **18** (1981), no. 1, 1–11.
82. S. Mori, *Threefolds whose canonical bundles are not numerically effective*, Ann. of Math. (2) **116** (1982), no. 1, 133–176.

83. S. Mukai, *Finite groups of automorphisms of $K3$ surfaces and the Mathieu group*, Invent. Math. **94** (1988), no. 1, 183–221.
84. V. V. Nikulin, *Finite groups of automorphisms of Kähler $K3$ surfaces*, Trudy Moskov. Mat. Obshch. **38** (1979), 75–137 (Russian); English transl., Trans. Moscow Math. Soc. **38** (1980), 71–135.
85. J. Noguchi and T. Sunada, *Finiteness of the family of rational and meromorphic mappings into algebraic varieties*, Amer. J. Math. **104** (1982), no. 4, 887–900.
86. E. Oeljeklaus, *Ein Hebbarkeitssatz für Automorphismengruppen kompakter komplexer Mannigfaltigkeiten*, Math. Ann. **190** (1970/1971), 154–166.
87. K. Oguiso, *A characterization of the Fermat quartic $K3$ surface by means of finite symmetries*, Compos. Math. **141** (2005), no. 2, 404–424.
88. T. Peternell and M. Schneider, *Compactifications of $\mathbf{C}^3$*. I, Math. Ann. **280** (1988), no. 1, 129–146.
89. V. L. Popov, *Classification of affine algebraic surfaces that are quasihomogeneous with respect to an algebraic group*, Izv. Akad. Nauk SSSR Ser. Mat. **37** (1973), 1038–1055 (Russian); English transl., Math. USSR-Sb. **7** (1973), no. 5, 1039–1055.
90. ———, *Picard groups of homogeneous spaces of linear algebraic groups and one-dimensional homogeneous vector fiberings*, Izv. Akad. Nauk SSSR Ser. Mat. **38** (1974), 294–322 (Russian); English transl., Math. USSR-Sb. **8** (1974), no. 2, 301–327.
91. ———, *Classification of three-dimensional affine algebraic varieties that are quasihomogeneous with respect to an algebraic group*, Izv. Akad. Nauk SSSR Ser. Mat. **39** (1975), no. 3, 566–609 (Russian); English transl., Math. USSR-Sb. **9** (1975), no. 3, 535–576.
92. ———, *On the Makar-Limanov, Derksen invariants, and finite automorphism groups of algebraic varieties*, in this volume, available at `arXiv:1001.1311`.
93. Yu. G. Prokhorov, *Fano threefolds of genus* 12 *and compactifications of* $\mathbf{C}^3$, Algebra i Analiz **3** (1991), no. 4, 162–170 (Russian); English transl., St. Petersburg Math. J. **3** (1992), no. 4, 855–864.
94. ———, *Compactifications of* $\mathbf{C}^4$ *of index* 3, Algebraic Geometry and Its Applications (Yaroslavl′, 1992) (A. Tikhomirov and A. Tyurin, eds.), Aspects Math., vol. E25, Vieweg, Braunschweig, 1994, pp. 159–169.
95. A. V. Pukhlikov, *Birational isomorphisms of four-dimensional quintics*, Invent. Math. **87** (1987), no. 2, 303–329.
96. ———, *Birationally rigid Fano hypersurfaces*, Izv. Ross. Akad. Nauk Ser. Mat. **66** (2002), no. 6, 159–186 (Russian); English transl., Izv. Math. **66** (2002), no. 6, 1243–1269.
97. ———, *Birationally rigid Fano varieties*, The Fano Conference (Torino, 2002) (A. Collino, A. Conte, and M. Marchisio, eds.), Univ. Torino, Torino, 2004, pp. 659–681.
98. S. Ramanan and A. Ramanathan, *Projective normality of flag varieties and Schubert varieties*, Invent. Math. **79** (1985), no. 2, 217–224.
99. R. Rentschler, *Opérations du groupe additif sur le plan affine*, C. R. Acad. Sci. Paris Sér. A-B **267** (1968), A384–A387.
100. P. Russell and A. Sathaye, *On finding and cancelling variables in* $k[X,Y,Z]$, J. Algebra **57** (1979), no. 1, 151–166.
101. F. Sakai, *Kodaira dimensions of complements of divisors*, Complex Analysis and Algebraic Geometry (W. L. Baily Jr. and T. Shioda, eds.), Iwanami Shoten, Tokyo, 1977, pp. 239–257.
102. C. Sancho de Salas, *Complete homogeneous varieties: structure and classification*, Trans. Amer. Math. Soc. **355** (2003), no. 9, 3651–3667.
103. B. Segre, *The non-singular cubic surfaces*, Oxford Univ. Press, Oxford, 1942.
104. A. Seidenberg, *Derivations and integral closure*, Pacific J. Math. **16** (1966), 167–173.
105. E. V. Sharoiko, *The Hassett – Tschinkel correspondence and automorphisms of a quadric*, Mat. Sb. **200** (2009), no. 11, 145–160 (Russian); English transl., Sb. Math **200** (2009), no. 11, 1715–1729.
106. D. M. Snow, *The nef value and defect of homogeneous line bundles*, Trans. Amer. Math. Soc. **340** (1993), no. 1, 227–241.
107. R. Steinberg, *Lectures on Chevalley groups*, Yale University, New Haven, CT, 1968.
108. N. Takahashi, *An application of Noether – Fano inequalities*, Math. Z. **228** (1998), no. 1, 1–9.
109. E. A. Tevelev, *Projective duality and homogeneous spaces*, Encyclopaedia Math. Sci., vol. 133, Springer, Berlin, 2005.

110. B. Totaro, *The automorphism group of an affine quadric*, Math. Proc. Cambridge Philos. Soc. **143** (2007), no. 1, 1–8.
111. R. Tsushima, *Rational maps to varieties of hyperbolic type*, Proc. Japan Acad. Ser. A Math. Sci. **55** (1979), no. 3, 95–100.
112. T. Vust, *Opération de groupes réductifs dans un type de cônes presque homogènes*, Bull. Soc. Math. France **102** (1974), 317–333.
113. J. M. Wahl, *Derivations, automorphisms and deformations of quasihomogeneous singularities*, Singularities. Part 2 (Arcata, CA, 1981) (P. Orlik, ed.), Proc. Sympos. Pure Math., vol. 40, Amer. Math. Soc., Providence, RI, 1983, pp. 613–624.
114. W. V. Vasconcelos, *Derivations of commutative noetherian rings*, Math. Z. **112** (1969), 229–233.
115. G. Xiao, *Bound of automorphisms of surfaces of general type*. I, Ann. of Math. (2) **139** (1994), no. 1, 51–77; II, J. Algebraic Geom. **4** (1995), no. 4, 701–793.
116. ———, *Galois covers between $K3$ surfaces*, Ann. Inst. Fourier (Grenoble) **46** (1996), no. 1, 73–88.
117. M. G. Zaidenberg, *Isotrivial families of curves on affine surfaces, and the characterization of the affine plane.*, Izv. Akad. Nauk SSSR Ser. Mat. **51** (1987), no. 3, 534–567; Addendum, **55** (1991), no. 2, 444–446 (Russian); English transl. in Math. USSR-Izv. **30** (1988), no. 3, 503–532; Addendum, **38** (1992), no. 2, 435–437.
118. M. G. Zaidenberg and V. Ya. Lin, *Finiteness theorems for holomorphic mappings*, Several Complex Variables. III (G. M. Khenkin, ed.), Encyclopaedia Math. Sci., vol. 9, Springer, Berlin, 1989, pp. 113–172.
119. O. Zariski, *Interprétations algébrico-géométriques du quatorzième problème de Hilbert*, Bull. Sci. Math. (2) **78** (1954), 155–168.

Department of Mathematics, Faculty of Science, Saitama University, Saitama 338-8570, Japan

E-mail address: tkishimo@rimath.saitama-u.ac.jp

Department of Algebra, Faculty of Mathematics, Moscow State University, Moscow 117234, Russia and Laboratory of Algebraic Geometry, SU-HSE, 7 Vavilova Str., Moscow 117312, Russia

E-mail address: prokhoro@gmail.com

Institut Fourier, Université Joseph Fourier, UMR 5582 CNRS-UJF, BP 74, 38402 Saint-Martin-d'Hères cedex, France

E-mail address: zaidenbe@ujf-grenoble.fr

Centre de Recherches Mathématiques
CRM Proceedings and Lecture Notes
Volume **54**, 2011

$\mathbb{C}^*$ in $\mathbb{C}^2$ is Birationally Equivalent to a Line

Mariusz Koras

To Peter

0. Introduction

We prove the following.

0.1. Theorem. *Let U be a closed curve in $\mathbb{C}^2$ isomorphic to $\mathbb{C}^*$. Then there exists a birational map $\sigma\colon \mathbb{P}^2 \dashrightarrow \mathbb{P}^2$ such that the proper transform of $\overline{U}$ is a line in $\mathbb{P}^2$.*

Peter Russell posed the problem some 15 years ago and has been working on it. In fact the main ideas used in the paper come from our joint work with Peter.

If both branches of $\overline{U}$ at infinity are disjoint then $\overline{U}$ is a cuspidal curve. In this case the theorem gives a partial answer to the famous Coolidge±dash Nagata conjecture

Every cuspidal rational curve in $\mathbb{P}^2$ can be transformed into a line by a Cremona transformation.

Embeddings of $\mathbb{C}^*$ into $\mathbb{C}^2$ can be divided into two classes; the first class consists of embeddings which admit *a good asymptote*, see Definition 0.3, the second class consists of those with no good asymptote. The embeddings from the first class are completely classified, [1]. By examining the way in which they are constructed one has the following.

0.2. Theorem. *If U admits a good asymptote then $\overline{U}$ can be transformed into a line by a Cremona transformation.*

Hence throughout the paper we will assume that U has no a good asympote.

0.3. Definition. We say that a rational curve $L \in \mathbb{P}^2$ is a *good asymptote* of U if $L \cap \mathbb{C}^2 \backsimeq \mathbb{C}^1$, and L meets $\overline{U}$ at most once at finite distance, i.e., $L \cdot U \leq 1$.

Notice that this definition differs slightly from the definition in [1]. Both definitions are equivalent up to isomorphism of $\mathbb{C}^2$.

2010 *Mathematics Subject Classification.* Primary 14R10; Secondary 14H50.

Key words and phrases. Embedding of $\mathbb{C}^*$, Cremona transformation, Kodaira dimension.

The author was supported by Polish grant NCN 201 608 640.

This is the final form of the paper.

The key point in the proof of the Theorem 0.1 is the Proposition 2.3. It provides a bound for E^2 and for $(K+D+E)^2$ where E is the proper transform of $\overline{U}$ under a smooth resolution of $\overline{U}$ and D is the resolution divisor, see Setup 0.4. The Proposition 2.3 is true for all embeddings with no good asymptote and will be a basic fact used in the classification of such embeddings. The classification is not completed yet, the author and Peter Russell expect to finish it in near future. The Proposition 2.3 follows from the Proposition 2.2.

0.4. Setup. Let U be a closed algebraic curve in $\mathbb{C}^2$ isomorphic to $\mathbb{C}^*$. Let $\overline{U}$ be the closure of U in $\mathbb{P}^2$. By L_∞ we denote the line at infinity in $\mathbb{P}^2$. Let $\Phi\colon \overline{S}' \to \mathbb{P}^2$ be the resolution of $\overline{U}$, i.e., Φ^{-1} is the minimal sequence of blow ups such that the reduced inverse image of the divisor $\overline{U}+L_\infty$ is an SNC-divisor. Let E' be the proper transform of $\overline{U}$ and let $D' = \Phi^{-1}_{red}(L_\infty)$. Let L'_∞ denotes the proper transform of the line L_∞ in $\overline{S}'$. Let $\Psi\colon \overline{S}' \to \overline{S}$ be the NC-minimalization of the divisor D' with respect to E' i.e. Ψ is the successive contraction of L'_∞ and possibly more D'−components such that $\Psi(D'+E')$ is an SNC-divisor and each (-1)-component of D' is a branching component of $\Psi(D'+E')$. We put $D=\Psi(D')$, $E=\Psi(E')$. Let $S=\overline{S}\setminus D$. Of course $S\simeq\mathbb{C}^2$.

We use the classical Coolidge theorem.

0.5. Theorem (Coolidge, see e.g., [11, Theorem 2.6]**).** *$\overline{U}$ can be transformed into a line by a Cremona transformation if and only if $\bar\kappa(\overline{S}\setminus E) = -\infty$.*

From now on we assume that $\overline{U}$ is not transformable into a line by a Cremona transformation and finally come to a contradiction.

The author thanks the referee for very careful reading and pointing out several mistakes.

1. Preliminaries

In the article we use several notions and results from the theory of open algebraic surfaces. In order to keep the length of the paper to a minimum we refer the reader to [9] for any undefined terms here. We will also use some results from T. Fujita's paper [2], particularly Section 3.

1.1. Let $\overline{M}$ be a complete, non-singular surface and $T=\sum_{i=1}^n m_iT_i$ a divisor on $\overline{M}$ with $T_1,\dots,T_n$ distinct, irreducible curves.

- (i) We call the T_j with $m_j\neq 0$ the components of T and denote by $\operatorname{Supp} T$ their union.
- (ii) We write $\sim$ for linear equivalence of integral divisors. We write $\equiv$ for numerical equivalence of divisors, both over $\mathbb{Z}$ and over $\mathbb{Q}$.
- (iii) We call T reduced if $m_j=1$ for each component T_j of T and then usually write T instead of $\operatorname{Supp} T$.
- (iv) We call T a simple normal crossing divisor (an SNC-divisor) if T is reduced, all its components are smooth and at most two of them meet at any point, and if so, transversally.
- (v) A (b)-curve on $\overline{M}$ is a curve $L\simeq\mathbb{P}^1$ with $L^2=b$.
- (vi) An SNC-divisor T is NC-minimal if every (-1)-component of T is a branching component.

(vii) We call $Q(T) = (T_i \cdot T_j)_{1 \le i,j \le n}$ the intersection matrix of a reduced T and put $d(T) = \det\bigl(-Q(T)\bigr)$. We put $d(T) = 1$ if $\operatorname{Supp}(T) = \varnothing$.

1.2. Let T be a tree, i.e., T has no a cycle. We call T *a rational tree* if the components of T are smooth rational curves. The following two elementary formulas will be used in the calculation of $d(T)$. They correspond to splitting the dual graph of T by deleting respectively a vertex or an edge.

(i) Let C be a component of T and let $T_1', \dots, T_\beta'$ be the connected components of $T - C$. Let C_i be the irreducible component of T_i' meeting C. Then

$$d(T) = -C^2 \prod_i d(T_i') - \sum_i d(T_i' - C_i) \prod_{j \ne i} d(T_j').$$

(ii) Let $T = T_1' + T_2'$ where T_1', T_2' are connected and have no component in common. There are unique components C_1, C_2 of T_1', T_2' respectively such that $C_1 \cdot C_2 = 1$ and we have

$$d(T) = d(T_1')d(T_2') - d(T_1' - C_1)d(T_2' - C_2).$$

1.3. For the definition of *twig*, *tip*, *bark* of a divisor and their properties we refer to [2, Section 3; 9, Chapter 2, Section 3]. We recall only the definition of a *capacity* of a rational chain. Let

$$R = R_1 + \cdots + R_s$$

be a chain of smooth rational curves with dual graph

$$\underset{b_1}{\otimes} \!\!-\!\!-\!\!-\!\!- \cdots -\!\!-\!\!-\!\!-\!\! \underset{b_s}{\otimes}$$

Suppose that R is admissible, i.e., that $b_i = R_i^2 \le -2$, $i = 1, \dots, s$. We put $e(R) = d(R_2 + \cdots + R_s)/d(R)$.

1.3.1. *If $R_1^2 = -k$ then $e(R) \ge 1/k$.*

We recall

1.4. *Let T be a connected NC-minimal divisor consisting of rational curves and let $T_1, \dots, T_s$ be all the maximal twigs of T. Then* $\operatorname{Bk}(T)^2 = -\sum e(T_i)$.

We need the following result on the Kodaira dimension on the complement of a single smooth rational curve from [11]. Here $\kappa(K_{\overline{M}} + B)$ denotes the Iitaka dimension of the divisor $K_{\overline{M}} + B$. It is equal to the logarithmic Kodaira dimension of the complement of B.

1.5. Proposition. *Let B be a smooth rational curve on a smooth complete rational surface $\overline{M}$. Suppose that each (-1)-curve on $\overline{M}$ meets B at least twice. Then*

(a) *if $\kappa(K_{\overline{M}} + B) \ge 0$, then $B^2 \le -4$* ([11, Corollary 2.5]),
(b) *$\kappa(K_{\overline{M}} + B) = 0$ or 1 if and only if $h^0(2K_{\overline{M}} + B) = 1$. Then $B^2 = -4, K_{\overline{M}}^2 = -1$* ([11, Corollary 3.2]).
(c) *If $\kappa(K_{\overline{M}} + B) \ge 0$ then $|B + 2K_{\overline{M}}| \ne \varnothing$.*

1.6. Lemma. *Let $\alpha\colon \overline{S} \to \overline{M}$ be a birational map such that the proper transform of E is a smooth curve $G \subset \overline{M}$. Then*

(a) $\kappa(K_{\overline{M}} + G) \ge 0$.
(b) $G^2 \le -4$.

(c) $\overline{M}$ *is not isomorphic to* $\mathbb{P}^2$ *or to a Hirzebruch surface* F_n.
(d) *There is no a* $\mathbb{P}^1$*-ruling* $F\colon \overline{M} \to \mathbb{P}^1$ *such that* $G \cdot F_t \leq 3$ *where* F_t *is a general fiber of* F.

PROOF. (a) We have $\kappa(K_{\overline{S}}+E) \geq 0$ by our assumption that $\overline{U}$ is not transformable into a line by a Cremona transformation and by Coolidge Theorem 0.5. Since $\kappa(K_{\overline{S}} + E) = \kappa(K_{\overline{M}} + G)$ the statement follows.

(b) It follows from (a) and Proposition 1.5(a).

(c) It follows from the fact that the complement of the negative curve in F_n has negative Kodaira dimension.

(d) Suppose that such a ruling exists. By contracting (-1)-curves in fibers which meet G at most once we may assume that every exceptional curve in a fiber meets G at least twice. Consider a singular fiber F_0 and let L be a (-1)-curve in F_0. Then $L \cdot E \geq 2$. It follows that L is an unique (-1)-curve in F_0. Thus L is a multiple component of F_0 but then $E \cdot F_0 \geq 4$; contradiction. Hence each fiber of F is a smooth rational curve. Therefore $\overline{M}$ is isomorphic to a Hirzebruch surface; contradiction by (b). □

1.7. Corollary. (1) *If* D_j *is not a branching component of* $D + E$ *then* $D_j^2 \leq -2$.
(2) *For every component* $D_i \subset D$, $D_i^2 \leq -1$.

1.8. Lemma. (i) $\bar{\kappa}\big(\overline{S} \setminus (D + E)\big) \geq 0$.
(ii) $(K_{\overline{S}} + D + E)^2 < 3$.

PROOF. (i) If $\bar{\kappa}\big(\overline{S} \setminus (D+E)\big) = -\infty$ then $|K_{\overline{S}}+D+E| = \varnothing$ which implies $E \cdot D \leq 1$ but $E \cdot D = 2$.

(ii) By [10] $(K_{\overline{S}}+D+E)^2 \leq 3(\chi(S \setminus E) + N^2/12)$ where $N = (K_{\overline{S}}+D+E)^-$ is the negative part in the Zariski decomposition of the divisor $K_{\overline{S}} + D + E$. If $N \neq 0$ we are done since $\chi(S \setminus E) = 1$. Suppose that $N = 0$. It follows that the divisor $D+E$ has no twigs (see [2, Section 3; 9, Chapter 2, Section 3]). Hence D is a chain and E meets the tips of D. Suppose that D has more than one component. Then both tips of D are not (-1)-curves since $D + E$ is NC-minimal. D is not an admissible chain since $d(D) = -1 < 0$. Thus D contains a nonnegative component; contradiction by Corollary 1.7(2). If D consists of only one component then $\overline{S} = \mathbb{P}^2$ and E is a conic so E is transformable into a line; contradiction. □

1.9. Write

$$(K_{\overline{S}} + D + E)^2 = 2 - \varepsilon$$

where $\varepsilon \geq 0$. We have $K_{\overline{S}} \cdot (K_{\overline{S}} + D + E) = 2 - \varepsilon$. We put $\gamma = -E^2$.

1.10. Lemma. *There is no a curve* $C \subset \overline{S}$ *such that* $C \cap S \simeq \mathbb{C}^1$.

PROOF. Let L be the proper transform of C in $\mathbb{P}^2$. Then L is a good asymptote of $\overline{U}$; contradiction. □

1.11. Corollary. *The pair* $(\overline{S}, D+E)$ *is relatively minimal* (*see* [9, Chapter 2, Section 3]), *i.e.*, $K_{\overline{S}}+D+E \equiv P + \mathrm{Bk}(D+E)$ *is the Zariski decomposition, where* $P = (K_{\overline{S}} + D + E)^+$.

1.12. In view of Corollary 1.11 and the Kobayashi inequality [,6; 9, Theorem 6.6.2] $P^2 \leq 3\chi(S \setminus E) = 3$. Let R_i, $i = 1, \dots, s$ be all maximal twigs of $D + E$. Let $e_i = e(R_i)$. We obtain that $2 - \varepsilon = P^2 - \sum e_i \leq 3 - \sum e_i$.

1.13. Corollary. $-\big(\mathrm{Bk}(D+E)\big)^2 = \sum e_i \leq 1 + \varepsilon$.

1.14. Lemma. *Let $\overline{M}$ be a smooth projective surface. Let $r = \mathrm{rank}\big(\mathrm{NS}(\overline{M})\big)$ (the Néron – Severi group). Then for any set $C_1, \dots, C_r$ of distinct irreducible curves in $\overline{M}$ the matrix $[C_i \cdot C_j]$ is not negative definite.*

PROOF. Suppose it is. Then in particular $C_1, \dots, C_r$ are independent in $\mathrm{NS}(\overline{M}) \otimes \mathbb{Q}$ hence they form a basis of $\mathrm{NS}(\overline{M}) \otimes \mathbb{Q}$. We reach a contradiction with the Hodge Index Theorem. □

1.15. Lemma. *Let a rational tree Y be a boundary divisor of $\mathbb{C}^2$ on a smooth rational surface $\overline{M}$, i.e., $\overline{M} \setminus Y \simeq \mathbb{C}^2$. Let $Y' \subset Y$ be a connected divisor such that the intersection matrix $[Y_i \cdot Y_j]_{Y_i, Y_j \subset Y'}$ is not negative definite. Suppose that all components of $Y - Y'$ are (≤ -2)-curves. Then any component of $Y - Y'$ is not a branching component of Y.*

PROOF. Suppose that Y_0, a component of $Y - Y'$, is a branching component of Y. Write $Y - Y_0 = Y^{(1)} \cup \dots \cup Y^{(s)}$ where $Y^{(i)}$, $i = 1, \dots, s$, are pairwise disjoint and connected. Let $Y' \subset Y^{(1)}$. There exists a modification $\Phi\colon \overline{M}' \to \overline{M}$ over Y and $\Psi\colon \overline{M}' \to \mathbb{P}^2$ such that $\Psi\big(\Phi^{-1}(Y)\big)$ is a line L_∞ in $\mathbb{P}^2$. We think of Ψ as a sequence of contractions of (-1)-curves. The intersection matrix of $\Phi^{-1}(Y^{(1)})$ is not negative definite hence $\Phi^{-1}(Y^{(1)})$ cannot be mapped to a point by Ψ. Therefore Y_0 must be mapped to a point. So Y_0 at some stage becomes a non branching (-1)-curve. At this stage at least one connected component of $\Phi^{-1}(Y - Y_0)$, say $\Phi^{-1}(Y^{(2)})$, is contracted to a point by a sequence of contractions which take place inside $\Phi^{-1}(Y^{(2)})$. We may assume that Φ, Ψ are minimal in the sense that $\Psi \circ \Phi^{-1}$ has a minimal number of blowing ups and down. The only (-1)-curves in $\Phi^{-1}(Y^{(2)})$ are those produced by Φ^{-1}. Hence, by minimality, none blowing up in Φ^{-1} is performed over $Y^{(2)}$. But all components of $Y^{(2)}$ are (≤ -2)-curves hence there is no possibility of contracting $Y^{(2)}$ to a point. □

1.16. Lemma. *In notation of Lemma 1.15 assume that there exists a component Y_1 of Y such that $Y_1^2 \geq 0$ and for any component $Y_i \neq Y_1$, $Y_i^2 \leq -2$. Then $Y_1^2 = 0$ and $Y = Y_1 + Y_2$ or $Y = Y_1, Y^2 = 1$.*

PROOF. By Lemma 1.15 Y_1 is the only possible branching component of Y. Let $Y_1^2 = k$. Suppose that $Y \neq Y_1$. Let $Y - Y_1 = Y^{(1)} \cup \dots \cup Y^{(s)}$, be the decomposition into connected components; $Y^{(i)}$, $i = 1, \dots, s$, are chains and Y_1 meets them at tips. Let $d_i = d(Y^{(i)})$, $\bar{d}_i$ denotes the determinant of the chain obtained from $Y^{(i)}$ by removing the tip of $Y^{(i)}$ meeting Y_1. Then $d(Y) = (\prod_{i=1}^s d_i)(-k - \sum_{i=1}^s \bar{d}_i/d_i) = -1$ (see 1.2(i)). So

$$k + \sum_i \frac{\bar{d}_i}{d_i} = \frac{1}{\prod_i d_i}.$$

We get $\sum_i (\bar{d}_i \prod_{j \neq i} d_j) \leq 1$. It follows that $s = 1$. The statement easily follows. □

1.17. Lemma. *In notation of Lemma 1.15 assume that Y contains two irreducible components Y_1, Y_2 such that $Y_1^2 = -1$, $Y_2^2 \geq -1$, $Y_1 \cdot Y_2 = 1$ and $Y_i^2 \leq -2$*

for any $Y_i \neq Y_1, Y_2$. *Then* Y *is a chain and* $Y_1^2 = 0$, $Y = Y_1 + Y_2$ *or* $Y_1^2 = -1 = Y_2^2$ *and* $Y = Y_1 + Y_2 + Y_3$ *where* $Y_3 \cdot Y_2 = 1$.

PROOF. Suppose Y is not a chain. By Lemma 1.15 the only possible branching components of Y are Y_1 or Y_2. Suppose that Y_{i_1}, $i_1 \in \{1,2\}$, is a branching component of Y. Let $i_2 \in \{1,2\}$ be different from i_1. Let $Y - Y_{i_1} = Y^{(2)} \cup \cdots \cup Y^{(s)}$, $s \geq 4$ be the decomposition into connected components. We may assume that $Y_{i_2} \subset Y^{(2)}$. Suppose that Y_{i_2} is also a branching component of Y or that $Y_{i_2}^2 \geq 0$. Let $\Phi \colon \overline{M}' \to \overline{M}$ and $\Psi \colon \overline{M}' \to \mathbb{P}^2$ be modifications such that $\Psi(\Phi^{-1}(Y))$ is a line L_∞ in $\mathbb{P}^2$. We take Φ with smallest possible number of blowing ups. It is now clear that none of proper transforms of components of Y is contracted as a first curve by Ψ. It means that Ψ contracts as a first curve one of curves produced by Φ. We may replace Φ by another modification with smaller number of blowings up. It contradicts minimality of Φ. Thus $Y^{(2)}$ is a chain, Y_{i_2} is a tip of $Y^{(2)}$ and $Y_{i_2}^2 = -1$. Let us contract successively Y_{i_2} inside $Y^{(2)}$. Let $\widetilde{Y}$ be the image of Y. $\widetilde{Y}$ satisfies assumptions of Lemma 1.16. It follows that $\widetilde{Y}$ is a chain and that the image of Y_{i_1} is a tip of $\widetilde{Y}$. It follows that Y is a chain; contradiction.

Hence Y is a chain. We contract successively Y_1. Let $\widetilde{Y}$ denotes the image of Y and $\widetilde{Y}_2$ denotes the image of Y_1. By Lemma 1.16 $\widetilde{Y}_2^2 = 0$ or 1 and $\#\widetilde{Y}$=2 or 1 respectively. The statement follows. □

2. Basic inequality

2.1. Let $\psi \colon \bar{\mathbb{S}} \to \bar{\mathbb{N}}$ be the 2-reduction of the divisor D with respect to E, i.e., ψ is a sequence of successive contractions of (-1)-curves in D meeting E once such that the divisors $T = \psi(D)$ and $E_0 = \psi(E)$ satisfy the following:

(i) T is an NC-divisor.

(ii) for any (-1)-component T_i of T, $T_i \cdot E_0 \geq 2$ or T_i is a branching component of T.

2.2. Proposition. *Suppose that* $(E_0 + 2K_{\overline{N}})(K_{\overline{N}} + T) \leq 0$. *Then there exists a* (-1)*-curve* A *in* $\bar{\mathbb{N}}$ *such that* $A \cdot E_0 \leq 1$.

PROOF. Suppose that such a curve does not exist. Let C_1, C_2 be the components of D which meet E. It may happen that $C_1 = C_2$.

2.2.1. Lemma. *There is no a curve* B *in* $\bar{\mathbb{N}}$ *such that* $(E_0 + 2K_{\overline{N}}) \cdot B < 0$.

PROOF. Suppose B exists. Suppose first that $|B + K_{\overline{N}} + T| \neq \varnothing$. Let m be the greatest integer such that $|B + m(K_{\overline{N}} + T)| \neq \varnothing$, see [3] for the existence of m. Write $B + m(K_{\overline{N}} + T) = \sum B_i$. Then $|B_i + K_{\overline{N}} + T| = \varnothing$ for every i. Since $0 > B \cdot (E_0 + 2K_{\overline{N}}) \geq \sum B_i \cdot (E_0 + 2K_{\overline{N}})$, there exists B_i such that $B_i \cdot (E_0 + 2K_{\overline{N}}) < 0$. Thus we may assume that $|B + K_{\overline{N}} + T| = \varnothing$. It follows that B is a smooth rational curve and that $B \cdot T \leq 1$, see [14, 2.1, 2.2] for example. In particular $B \neq E_0$ since $E_0 \cdot T \geq 2$. Hence $K_{\overline{N}} \cdot B < 0$, i.e., $B^2 \geq -1$. Suppose that $B^2 \geq 0$. Since $\overline{N}$ is not a minimal rational surface by Lemma 1.6(c) there exists a singular member $\sum B_j$ of $|B|$ such that $B_j^2 < 0$ for every j. There exists B_j such that $B_j \cdot (E_0 + 2K_{\overline{N}}) < 0$. It follows that $B_j \cdot K_{\overline{N}} < 0$ hence $B_j^2 = -1$. We may assume therefore that $B^2 = -1$. But then $B \cdot E_0 < 2$ and we have contradiction. The lemma is proved. □

2.2.2. Lemma. (a) *$K_{\overline{S}} \cdot D_i \geq 0$ for every $D_i \subset D$, $D_i \neq C_1, C_2$.*
(b) *$K_{\overline{S}} \cdot C_i \geq -1$, i.e., $C_i^2 \leq -1$ for $i = 1, 2$.*

PROOF. We will use the following, easily verifiable, fact:

2.2.3. *Let C be a (-1)-curve on a smooth complete surface $\overline{M}$. Let A, E be irreducible curves on $\overline{M}$ such that C is not a component of $A + E$. Let A', E' be their images under the contraction $\overline{M} \to \overline{M}'$ of C and let r be a non-negative integer. Then*

$$A' \cdot (E' + rK_{\overline{M}'}) = A \cdot (E + rK_{\overline{M}}) + A \cdot C(E \cdot C - r).$$

Suppose that $K_{\overline{S}} \cdot D_i < 0$, $D_i \neq C_1, C_2$. Then $D_i^2 \geq -1$. Clearly D_i is not contracted under ψ (if $D_i^2 = -1$ then it is a branching component of D and at the moment it is touched during the contraction process ψ it becomes a 0-curve). Let $T_i = \psi(D_i)$. Then, by 2.2.3 since ψ contracts curves C such that $E \cdot C \leq 1$, $T_i \cdot (E_0 + 2K_{\overline{N}}) \leq D_i \cdot (E + 2K_{\overline{S}}) < 0$ and we reach contradiction with Lemma 2.2.1. This proves (a). If $C_i^2 \geq 0$ then C_i is not contracted by ψ and $C_i \cdot (E + 2K_{\overline{S}}) < 0$. We again reach contradiction by 2.2.3 and Lemma 2.2.1. This proves (b). □

2.2.4. Lemma. (i) *For every $D_i \subset D$ which is not contracted by ψ $D_i \cdot (E + K_{\overline{S}}) = T_i \cdot (E_0 + K_{\overline{N}})$ where $T_i = \psi(D_i)$.*
(ii) *C_i is contracted iff $C_i^2 = -1$.*
(iii) *$T \cdot (E_0 + K_{\overline{N}}) = D \cdot (E + K_{\overline{S}})$.*

PROOF. (1) Let D_i be a component of D. If ψ contracts a component which meets D_i then $K \cdot D_i$ decreases by 1 and $D_i \cdot E$ increases by 1 so the quantity $D_i \cdot (E + K_{\overline{S}})$ doesn't change.

(2) Suppose that C_i is contracted and $C_i^2 \neq -1$. At some stage we have to contract an adjacent component of C_i. After this contraction C_i meets E_0 at least twice so it cannot be contracted. Suppose that $C_i^2 = -1$. Since $C_i \cdot (E + 2K_{\overline{S}}) < 0$ C_i must be contracted otherwise, using 2.2.3, we have contradiction with Lemma 2.2.1.

(3) Suppose that $D_j \neq C_i$ is contracted. At some stage ψ contracts an adjacent component of D_j and then $D_j' \cdot E' = 1$ where D_j', E' denote the images of D_j, E. If $D_j'^2 \neq -1$ then an adjacent component of D_j' must be contracted but after this contraction $D_j'' \cdot E'' \geq 2$ where D_j'', E'' are images of D_j', E'. Now D_j'' will not be contracted. Hence $D_j'^2 = -1$ which implies $D_j^2 = -2$ and further $D_j \cdot (E + K_{\overline{S}}) = 0$. If C_i is contracted then $C_i^2 = -1$ by (ii) and $C_i \cdot E = 1$ so $C_i \cdot (E + K_{\overline{S}}) = 0$. Now (iii) follows from (i). □

2.2.5. By Lemma 1.6(a) we have $\kappa(K_{\overline{N}} + E_0) \geq 0$. Since every exceptional curve in $\overline{\mathbb{N}}$ meets E_0 at least twice, the pair $(\overline{\mathbb{N}}, E_0)$ is relatively minimal. Hence $K_{\overline{N}} + E_0 \equiv (K_{\overline{N}} + E_0)^+ + 2E_0/\gamma_0$ is the Zariski – Fujita decomposition of $K_{\overline{N}} + E_0$; $\gamma_0 = -E_0^2$. By Lemma 1.6(b) $\gamma_0 \geq 4$. We have

$$2 + K_{\overline{N}} \cdot (K_{\overline{N}} + E_0) = (K_{\overline{N}} + E_0)^2 = ((K_{\overline{N}} + E_0)^+)^2 - \frac{4}{\gamma_0} \geq -1.$$

Hence $K_{\overline{N}} \cdot (K_{\overline{N}} + E_0) \geq 1$. If $\kappa(K_{\overline{N}} + E_0) = 2$, i.e., if $((K_{\overline{N}} + E_0)^+)^2 > 0$ or if $\gamma_0 > 4$ then $(K_{\overline{N}} + E_0)^2 > -1$, i.e., $K_{\overline{N}} \cdot (K_{\overline{N}} + E_0) \geq 2$. Notice that $K_{\overline{N}} \cdot (K_{\overline{N}} + E_0) = K_{\overline{S}} \cdot (K_{\overline{S}} + E)$. Thus, in view of 1.9

2.2.5.1. *$K_{\overline{S}} \cdot D = 2 - \varepsilon - K_{\overline{S}} \cdot (K_{\overline{S}} + E) \leq 1 - \varepsilon$. Moreover $K_{\overline{S}} \cdot D \leq -\varepsilon$ if $\kappa(K_{\overline{N}} + E_0) = 2$.*

2.2.5.2. *If $C_1 \neq C_2$ and $C_i^2 = -1$ then C_i is contracted by ψ.*

PROOF. $C_1 \neq C_2$ implies that $C_i \cdot E = 1$. Suppose that C_i is not contracted by ψ. Then, by 2.2.3, $\psi(C_i) \cdot (E_0 + 2K_{\overline{N}}) \leq C_i \cdot (E + 2K_{\overline{S}}) < 0$. We reach a contradiction by the Lemma 2.2.1. □

2.2.6. Assume that $\kappa(K_{\overline{N}} + E_0) = 2$. Then $K_{\overline{N}} \cdot (K_{\overline{N}} + E_0) \geq 2$. Since $K_{\overline{S}} \cdot D \geq -2$ by Lemma 2.2.2 we obtain $K_{\overline{S}} \cdot (K_{\overline{S}} + E) = 2 - \varepsilon - K_{\overline{S}} \cdot D \leq 4$.

Suppose that $K_{\overline{S}} \cdot (K_{\overline{S}} + E) \geq 3$. Then $K_{\overline{S}} \cdot D \leq -1 - \varepsilon \leq -1$. We have three cases

(1) $C_1^2 = C_2^2 = -1$, $C_1 \neq C_2$ and $D_i^2 = -2$ if $D_i \neq C_j$ or
(2) $C_1^2 = -1$, $D_i^2 = -2$ for every $D_i \neq C_1$ or
(3) $C_1^2 = C_2^2 = -1$, $C_1 \neq C_2$, there exists $D_0 \subset D$ such that $D_0^2 = -3$ and $D_i^2 = -2$ for every $D_i \neq D_0, C_1, C_2$.

Consider (1). Since both C_1, C_2 get contracted by Lemma 2.2.4(ii) they are disjoint. C_j is not a tip of D since otherwise $D+E$ is not NC-minimal. The adjacent components of C_j are (-2)-curves. We get a $\mathbb{P}^1$-ruling F of $\overline{S}$ with $E \cdot F = 2$; contradiction by Lemma 1.6(d).

Consider (2). Again C_1 is not a tip. We get a $\mathbb{P}^1$-ruling F with $E \cdot F \leq 3$; contradiction.

Consider (3). Let $E = \sum_{i \neq 0} d_i D_i + c_1 C_1 + c_2 C_2 + d_0 D_0$ in $\mathrm{Pic}(\overline{S})$. Suppose that $d_0 = 0$. By multiplying by E and $K_{\overline{S}}$ we obtain $E^2 = c_1 + c_2$, $K_{\overline{S}} \cdot E = -c_1 - c_2$. From this $-2 = E^2 + K_{\overline{S}} \cdot E = 0$; contradiction. Therefore $d_0 \neq 0$ which implies that $\mathrm{Pic}(\overline{S})$ is numerically generated by E, C_1, C_2 and components D_i for $i \neq 0$. Hence $\mathrm{Pic}(\overline{N})$ is numerically generated by E_0 and $\psi(D_i)$, $i \neq 0$. By Lemma 2.2.4(i) $\psi(D_i) \cdot (K_{\overline{N}} + E_0) = 0$ for $i \neq 0$. Then $\psi(D_i) \cdot (K_{\overline{N}} + E_0)^+ = 0$. Also $E_0 \cdot (K_{\overline{N}} + E_0)^+ = 0$. We obtain that $(K_{\overline{N}} + E_0)^+ \equiv 0$, i.e., $\kappa(K_{\overline{N}} + E_0) = 0$; contradiction.

Suppose that $K_{\overline{S}} \cdot (K_{\overline{S}} + E) = 2$. By Proposition 1.5(c) $|E_0 + 2K_{\overline{N}}| \neq \varnothing$. Hence, by Lemma 2.2.1, $(E_0 + 2K_{\overline{N}})^2 = E_0^2 + 4K_{\overline{N}} \cdot E_0 + 4K_{\overline{N}}^2 \geq 0$. Since $K_{\overline{N}} \cdot (K_{\overline{N}} + E_0) = 2$ we obtain $K_{\overline{N}}^2 \geq -4$. Since $(E_0 + 2K_{\overline{N}})(K_{\overline{N}} + T) \leq 0$ we get $(E_0 + 2K_{\overline{N}}) \cdot T \leq -(E_0 + 2K_{\overline{N}}) \cdot K_{\overline{N}} = -2 - K_{\overline{N}}^2 \leq 2$.

Suppose that $(E_0 + 2K_{\overline{N}}) \cdot T = 0$. Then $(E_0 + 2K_{\overline{N}}) \cdot T_i = 0$ for any component $T_i \subset T$. Since T_i generate $\mathrm{Pic}(\overline{N})$ we get that $E_0 + 2K_{\overline{N}} \sim 0$. Then $K_{\overline{N}} \cdot E_0 = -(E_0 + K_{\overline{N}}) \cdot E_0 = 2$. We get $K_{\overline{N}} \cdot E_0 + 2K_{\overline{N}}^2 = 0$ so $K_{\overline{N}}^2 = -1$. But now $K_{\overline{N}} \cdot (E_0 + K_{\overline{N}}) = 1$; contradiction.

Suppose that $(E_0 + 2K_{\overline{N}}) \cdot T = 1$. There exists $T_0 \subset T$ such that $(E_0 + 2K_{\overline{N}}) \cdot T_0 = 1$ and $(E_0 + 2K_{\overline{N}}) \cdot T_i = 0$ for $i \neq 0$. We write $E_0 \sim \sum_{i \neq 0} t_i T_i + t_0 T_0$. By multiplying by $E_0 + 2K_{\overline{N}}$ we get $E_0 \cdot (E_0 + 2K_{\overline{N}}) = t_0$ thus $t_0 \geq 0$. On the other hand, multiplying by E_0, we get $0 > E_0^2 \geq t_0 E_0 \cdot T_0$. Since $(E_0 + 2K_{\overline{N}}) \cdot T_0 = 1$ $E_0 \cdot T_0 > 0$. We obtain $0 > t_0$; contradiction.

Suppose that $(E_0 + 2K_{\overline{N}}) \cdot T = 2$. Then $(E_0 + 2K_{\overline{N}})^2 = 0$ and $K_{\overline{N}}^2 = -4$, $K_{\overline{N}} \cdot E_0 = 6$. The quantity $K \cdot D$ (resp. $E \cdot D$) decreases by 1 (resp. increases by 1) under each subdivisional contraction in ψ. Let s denotes the number of subdivisional contractions in ψ. Then $E_0 \cdot T = 2 + s$ and $K_{\overline{N}} \cdot T = K_{\overline{S}} \cdot D - s = 2 - \varepsilon - K_{\overline{S}} \cdot (K_{\overline{S}} + E) - s = -\varepsilon - s$. Now $(E_0 + 2K_{\overline{N}}) \cdot T = 2 = 2 + s + 2(-\varepsilon - s)$ which gives $s + 2\varepsilon = 0$. So $\varepsilon = 0$ and $s = 0$. It follows that $\psi = \mathrm{id}$„ i.e., $T = D$. Then $K_{\overline{S}} \cdot D = 2 - K_{\overline{S}} \cdot (K_{\overline{S}} + E) = 0$. Suppose that $C_1 \neq C_2$. Then $C_i^2 \neq -1$

since $s=0$. We obtain that D consists of (-2)-curves but then $K_{\overline{S}} \sim 0$ since components of D form a basis of $\mathrm{Pic}(\overline{S})$. So $C_1 = C_2 = C$. By the argument above D cannot consist of (-2)-curves therefore $C^2 = -1$ and there exists a component $D_0 \subset D$ such that $D_0^2 = -3$ and $D_i^2 = -2$ for every $D_i \neq D_0, C$. The divisor $D+E$ has at most 2 tips otherwise we have contradiction with Corollary 1.13, (use 1.3.1). Hence either D is a chain or D is a fork and C is a tip of D. If D is a chain then the intersection matrix of $D-C+E$ is negative definite and we reach contradiction by Lemma 1.14. We may assume that D is a fork. Let Z_1, Z_2, Z_3 be the maximal twigs of D and let H be the branching component of D. We may assume that C is the tip of Z_1. Consider the surface $X = \overline{S} \setminus (D - C + E)$. We claim that the pair $(\overline{S}, D-C+E)$ is relatively minimal. Indeed, if not then there exist a (-1)-curve L such that $L \cdot E \leq 1$ and we have contradiction with Lemma 2.2.1. Also $\bar{\kappa}(X) = 2$ since $X \subset \overline{S} \setminus E$ and $\bar{\kappa}(\overline{S} \setminus E) = 2$. $Z_1 \neq C$, i.e., $D-C+E$ is a fork since otherwise $D-C$ is an admissible chain and the intersection matrix of $D-C+E$ is negative definite; contradiction by Lemma 1.14. Also by Lemma 1.14 $D-C+E$ is not negative definite. By the Kobayashi inequality $P^2 \leq 3\frac{1}{-E^2} = \frac{3}{8}$ where P denotes the positive part in the Zariski decomposition of the divisor $K_{\overline{S}} + (D-C+E)$. We have $K_{\overline{S}} \cdot (K_{\overline{S}} + D - C + E) = K_{\overline{S}} \cdot (K_{\overline{S}} + D + E) + 1 = 3$ so $\left(K_{\overline{S}} + (D-C+E)\right)^2 = -1$. We obtain

$$-1 = \left(K_{\overline{S}} + (D-C+E)\right)^2 = P^2 + \left(\mathrm{Bk}(D-C)\right)^2 - \frac{4}{-E^2}.$$

From this

$$e(Z_1 - C) + e(Z_2) + e(Z_3) \leq \tfrac{7}{8}.$$

But, by 1.3.1, $e(Z_1 - C) \geq \frac{1}{3}$, $e(Z_i) \geq \frac{1}{3}$, $i = 2, 3$. We reach a contradiction.

2.2.7. . We may assume $\kappa(K_{\overline{N}} + E_0) \leq 1$. By 2.2.5 $\gamma_0 = -4$. We have $\left((K_{\overline{N}} + E_0)^+\right)^2 = 0$. By 2.2.5 $K_{\overline{N}} \cdot (K_{\overline{N}} + E_0) = 1$ therefore $K_{\overline{S}} \cdot (K_{\overline{S}} + E) = 1$. 2.2.5.1 gives $K_{\overline{S}} \cdot D + \varepsilon = 1$.

2.2.8. Lemma. $E_0 + 2K_{\overline{N}} \sim 0$.

PROOF. By Proposition 1.5 $K_{\overline{N}}^2 = -1$. Hence $(E_0 + 2K_{\overline{N}}) \cdot K_{\overline{N}} = 0$. It follows from the assumption in Proposition 2.2 that $(E_0 + 2K_{\overline{N}}) \cdot T \leq 0$. In view of Lemma 2.2.1 we have $(E_0 + 2K_{\overline{N}}) \cdot T_i = 0$ for every component T_i of T. Hence $E_0 + 2K_{\overline{N}}$ is numerically equivalent to 0. The statement follows. □

2.2.9. Corollary. *$T_j \cdot E_0 = -2K_{\overline{N}} \cdot T_j$ is even for any component T_j of T. $K_{\overline{N}} \cdot T_j < 0$ iff $T_j \cdot E_0 > 0$*

2.2.10. (i) $K_{\overline{S}} \cdot D = 1 - \varepsilon$.
(ii) $K_{\overline{N}} \cdot T = \varepsilon - 3$.
(iii) $\varepsilon \leq 3$.

PROOF. (i) follows from the fact that $K_{\overline{S}} \cdot (K_{\overline{S}} + E) = K_{\overline{N}} \cdot (K_{\overline{N}} + E_0) = 1$ and from Corollary 1.11.

(ii) In view of Lemma 2.2.4(iii), $K_{\overline{N}} \cdot T = -(K_{\overline{N}} + E_0) \cdot T = -(K_{\overline{S}} + E) \cdot D = -K_{\overline{S}} \cdot D - 2 = -(1 - \varepsilon) - 2 = \varepsilon - 3$.

(iii) follows from (i) and from Lemma 2.2.2. □

Let $T_1, \ldots, T_l$ be the all components of T such that $T_i \cdot E_0 > 0$. Let $T_i = \psi(D_i)$ for $i = 1, \ldots, l$.

Assume $\varepsilon = 3$. Then $K_{\overline{S}} \cdot D = -2$ by 2.2.10. Hence $C_1 \neq C_2$, $C_1^2 = C_2^2 = -1$ and $D_j = -2$ for any $D_j \neq C_1, C_2$. Then there exists a $\mathbb{P}^1$-ruling F such that $E \cdot F \leq 2$; contradiction by Lemma 1.6(d).

Assume $\varepsilon = 2$. Then $K_{\overline{N}} \cdot T = -1$ by 2.2.10. In view of Corollary 2.2.9 $l = 1$ and $T_1^2 = -1$, $E_0 \cdot T_1 = 2$. Since $E_0 \setminus (E_0 \cap T) \simeq \mathbb{C}^*$, E_0 meets T_1 in two different points transversally. It follows that $\psi = \mathrm{id}$, i.e., $\overline{N} = \overline{S}$ otherwise D is obtained from T by a sequence of blowing ups over $E_0 \cap T_1$ but then $D + E$ is not NC-minimal. Hence $C_1 = C_2 = C$. Suppose that C is not tip of D. Then after successive contraction of C we obtain an NC-minimal boundary divisor of $\mathbb{C}^2$ which is not a chain. It is a contradiction with the Morrow's theorem, [12]. So C is a tip of D. Let $C + D_1 + \cdots + D_s$ be the maximal twig of D containing C. Let $D_s \cdot D_{s+1} = 1$. Let D_{s+2}, D_{s+3} be two adjacent components of D_{s+1}. Contract the twig and D_{s+1}, let T' be the image of D. Then D_{s+2}, D_{s+3} become (-1)-curves. T' satisfies assumptions of Lemma 1.17. It follows that D is a fork with D_{s+1} as the branching component. One maximal twig of D is a single (-2)-curve, the second consists of two (-2)-curves and the third maximal twig is $C + D_1 + \cdots + D_s$. $s = 6$ since $b_2(\overline{S}) = 10 - K_{\overline{S}}^2 = 11 = 5 + s$.

Let $D' = D - C$. Consider the surface $Y = \overline{S} \setminus D'$. We have $\pi_1(Y) = 0$ since Y contains $\mathbb{C}^2$. Arguing as in the proof of 2.9 in [8] we get $|d(D')| = d(E)$; this is contradiction since $d(D') = -1$ by the formula 1.2(i).

Assume $\varepsilon = 1$. Then $K_{\overline{N}} \cdot T = -2$ by 2.2.10. Suppose that $l = 1$. Then $K_{\overline{N}} \cdot T_1 = -2$, i.e., $T_1^2 = 0$. By Lemma 1.16 $\#T \leq 2$ hence $\overline{\mathbb{N}}$ is isomorphic to $\mathbb{P}^2$ or to a Hirzebruch surface. Since $E_0^2 = -4$ $\overline{N}$ is isomorphic to the Hirzebruch surface F_4 but the complement of the negative curve E_0 in F_4 has negative Kodaira dimension; contradiction. Hence $l = 2$ and $T_i^2 = -1$, $E_0 \cdot T_i = 2$ for $i = 1, 2$. Suppose $T_1 \cdot T_2 > 0$. By Lemma 1.17 $\#T \leq 3$. By the Noether formula $-1 = K_{\overline{N}}^2 = 10 - b_2(\overline{\mathbb{N}}) = 10 - \#T \geq 7$; contradiction. Hence T_1 and T_2 are disjoint. Let $Z = Z_1 + Z_2 + \cdots + Z_r$, where $Z_1 = T_1$, $Z_r = T_2$ be the chain connecting T_1 and T_2 in T. Let $T - Z = Z^{(1)} \cdots \cup Z^{(s)}$ be the decomposition into connected components. It follows from Lemma 1.15 that $Z^{(i)}$ is a chain and that Z meets $Z^{(i)}$ in a tip, $i = 1, \ldots, s$. In the process ψ^{-1} we blow up twice over $T_i \cap E_0$, $i = 1, 2$. Thus $D + E$ has $s + 2$ tips, among them there are two (-2)-tips created by ψ^{-1}. Let $T_i = \psi(D_i)$, $i = 1, 2$. The proper transforms of curves in $\overline{N}$ which are not affected by ψ^{-1} we denote by the same letter. All tips of $D + E$ are (-2)-curves. By Corollary 1.13 $s + 2 \leq 4$, i.e., $s \leq 2$. Suppose that $s = 2$. Then $Z^{(1)}$ and $Z^{(2)}$ are single (-2)-curves otherwise $\sum e_i$ for the divisor $D + E$ is greater than 2 which is contradiction with Corollary 1.13. Consider the surface $Y = \overline{S} \setminus (D + E - C_1)$. The surface Y is bounded by the divisor $D + E - C_1 = R_1 + R_2$ where R_2 is the (-2)-tip of D meeting C_1. We have $K_{\overline{S}} \cdot (K_{\overline{S}} + R_1 + R_2) = K_{\overline{S}} \cdot (K_{\overline{S}} + D + E) + 1 = 2 - \varepsilon + 1 = 2$. Hence $(K_{\overline{S}} + R_1 + R_2)^2 = -2$. We claim that the pair $(\overline{S}, R_1 + R_2)$ is relatively minimal. Indeed if it is not then there exists a (-1)-curve $L \subset \overline{S}$ such that $L \cdot R_i \leq 1$, $i = 1, 2$. It follows that $L \neq C_1$ and $L \neq C_2$. So L is not contracted by ψ. We have $L \cdot (E + 2K_{\overline{S}}) < 0$, by 2.2.3 $\psi(L)(E_0 + 2K_{\overline{N}}) < 0$ and we reach contradiction with Lemma 2.2.1. $\kappa(K_{\overline{S}} + R_1 + R_2) \geq 0$ otherwise $\overline{\kappa}(\overline{S} \setminus E) = -\infty$. The connected component R_2 contracts to a quotient singular point with local fundamental group isomorphic to $\mathbb{Z}_2$. Consider the Zariski decomposition $K_{\overline{S}} + R_1 + R_2 = P + N$ where P, N are the positive and the negative parts respectively. Since $(\overline{S}, R_1 + R_2)$

is relatively minimal we have $N = \text{Bk}(R_1 + R_2)$. We have

$$-2 = (K_{\overline{S}} + R_1 + R_2)^2 = P^2 + \big(\text{Bk}(R_1)\big)^2 + \big(\text{Bk}(R_2)\big)^2.$$

If $\bar\kappa(Y) = 2$ the Kobayashi inequality gives $P^2 \le \frac{3}{2}$ since $\chi(Y) = 0$ and R_1 does not contract to a quotient singularity (it follows from Lemma 1.14). If $\bar\kappa(Y) = 0, 1$ then the same inequality holds since then $P^2 = 0$. R_1 has three (-2)-tips, namely $Z^{(1)}, Z^{(2)}$ and the one created by ψ^{-1} over $E_0 \cap T_2$. E is also a tip of R_1, $E^2 = -8$. Hence $-\big(\text{Bk}(R_1)\big)^2 \ge \frac{3}{2} + \frac{1}{8}$ in view of 1.3.1; also $\text{Bk}(R_2) = R_2$. We get $-2 \le \frac{3}{2} - \frac{3}{2} - \frac{1}{8} - 2$; contradiction.

Now suppose that $s = 1$. We consider the surface $M = \overline{S} \setminus (D + E - C_1 - C_2)$. We write $D + E - C_1 - C_2 = R_1 + R_2 + R_3 + E$ where R_1 is the connected component containing D_1, D_2; R_2 and R_3 are the tips of D meeting C_1 and C_2 respectively. As above in case of the surface Y we prove that the pair $(\overline{S}, R_1 + R_2 + R_3 + E)$ is relatively minimal and that $\bar\kappa(M) \ge 0$.

Suppose that $Z^{(1)}$ does not meet T_1, T_2. Let $Z^{(1)} \cdot Z_j = 1$. We may contract $Z_1, \dots, Z_{j-1}$ and $Z_{j+1}, \dots, Z_r$ to points; the determinant of T doesn't change. After this T becomes a chain $Z_j + Z^{(1)}$, now $Z_j^2 = 0$. The determinant equals -1, it easily follows that $Z^{(1)}$ must be a single (-2)-curve. We have $r + 1 = b_2(\overline{N})$. The Noether formula, since $K_{\overline{N}}^2 = -1$, gives $b_2(\overline{N}) = 11$. Hence $r = 10$. R_1 is a fork with branching component Z_j. The twig containing D_1 has $j - 1$ components, the twig containing D_2 has $r - j$ components.

R_1 is not a contractible fork, i.e., it does not contract to a quotient singular point. Consider the Zariski decomposition $K_{\overline{S}} + R_1 + R_2 + R_3 + E = P + \text{Bk}(R_1 + R_2 + R_3 + E)$. Since $\chi(M) = -1$, the Kobayashi inequality gives $P^2 \le 3(\frac{1}{2} + \frac{1}{2} + \frac{1}{8} - 1) = \frac{3}{8}$. We have $\big(\text{Bk}(R_2)\big)^2 = \big(\text{Bk}(R_3)\big)^2 = -2$, $\big(\text{Bk}(E)\big)^2 = -\frac{1}{2}$, $\big(\text{Bk}(R_1)\big)^2 = -(j-1)/(2j-1) - (r-j)/(2(r-j)+1) - \frac{1}{2}$. We have $(K_{\overline{S}} + R_1 + R_2 + R_3 + E)^2 = -5$. Thus we obtain

$$-5 = P^2 - 4 - \frac{1}{2} - \frac{j-1}{2j-1} - \frac{r-j}{2(r-j)+1} - \frac{1}{2}.$$

Since $P^2 \le \frac{3}{8}$ we obtain

$$\frac{j-1}{2j-1} + \frac{r-j}{2(r-j)+1} \le \frac{3}{8}.$$

It leads to contradiction since $1 < j < r$.

Suppose that $Z^{(1)}$ meets T_1. Then R_1 is a chain. We compute $d(R_1) = 61$. All connected components of $D + E - (C_1 + C_2)$ contract to quotient singularities. The Kobayashi inequality gives $P^2 \le 3(1/61 + \frac{1}{2} + \frac{1}{2} + \frac{1}{8} - 1) = 207/488$. We find $\big(\text{Bk}(R_1)\big)^2 = -71/61$. Hence $-5 = (K_{\overline{S}} + R_1 + R_2 + R_3 + E)^2 \le 207/488 - 4 - \frac{1}{2} - 71/61$. This is contradiction.

If $s = 0$ then T and therefore D contracts to a 0-curve. It is impossible.

Assume $\varepsilon = 0$. Then $K_{\overline{N}} \cdot T = -3$ by 2.2.10. The case $l = 1$ we rule out as above. Suppose that $l = 2$. As above we prove that T_1 and T_2 are disjoint. We may assume that $K_{\overline{N}} \cdot T_1 = -1$, $K_{\overline{N}} \cdot T_2 = -2$. Then $E_0 \cdot T_1 = 2$, $E_0 \cdot T_2 = 4$. The process ψ^{-1} consists of blowing up twice over $E_0 \cap T_1$ and four times over $E_0 \cap T_2$. Hence $D + E$ has at least one (-2)-tip and a (-2)-twig consisting of three curves. Hence $\sum e_i$ for the divisor $D + E$ is greater than $\frac{1}{2} + \frac{3}{4} > 1$; contradiction by Corollary 1.13. We may assume that $l = 3$. We have $K_{\overline{N}} \cdot T_i = -1$, $E_0 \cdot T_i = 2$, $i = 1, 2, 3$. Since

$E_0 \cap T$ consists of two points there must be $T_1 \cdot T_2 = 1 = T_2 \cdot T_3$ and $E_0 \cap T$ consists of intersection points $T_1 \cap T_2$ and $T_2 \cap T_3$. So E_0 is twice tangent to T_1 at $T_1 \cap T_2$ and it is twice tangent to T_3 at $T_2 \cap T_3$. Let $Z = T_1 \cup T_2 \cup T_3$ and let, as above, $T - Z = Z^{(1)} \cdots \cup Z^{(s)}$ be the decomposition into connected components. We have $s \leq 2$ otherwise $D + E$ has at least three (-2)-tips so $\sum e_i > 1$; contradiction by Corollary 1.13. In order to recover D from T we blow up twice over $T_1 \cap T_2$ and twice over $T_2 \cap T_3$. The reduced inverse image of Z is a chain with two (-3)-tips. By Lemma 1.15 every $Z^{(i)}$ is a chain and Z meets $Z^{(i)}$ in a tip of $Z^{(i)}$. Suppose that $s = 2$. If T_1 or T_3 is disjoint from $Z^{(1)} + Z^{(2)}$ then $D + E$ has two (-2)-tips and at least one (-3)-tip; D_1 or D_3. Then $\sum e_i > 1$; contradiction in view of Corollary 1.13. So we may assume that $Z^{(1)} \cdot T_1 = 1$ and $Z^{(2)} \cdot T_2 = 1$. $D + E$ has two maximal twigs and $e_1 > \frac{1}{2}$, $e_2 > \frac{1}{2}$; contradiction by Corollary 1.13. Suppose that $s = 1$. If $Z^{(1)}$ meets T_2 then $D + E$ has one (-2)-tip and two (-3)-tips. Then $\sum e_i > 1$; contradiction. We may assume that $Z^{(1)}$ meets T_1. Then $D + E$ has one twig $Z^{(1)} + D_1$ and one tip D_3. Let t denotes the number of components in $Z^{(1)}$. Then $\sum e_i = \frac{2t+1}{2t+3} + \frac{1}{3}$. By Corollary 1.13 $\sum e_i \leq 1$, we obtain $2t \leq 3$. However, then $t = 1$ and so $\#T = 4 = \rho(\overline{\mathbb{N}})$, which contradicts $K_{\overline{N}}^2 = -1$. We may assume that $s = 0$. Then $b_2(\overline{N}) = \#T = 3$. But $K_{\overline{N}}^2 = -1$, we reach contradiction with the Noether formula. □

2.3. Proposition. *Let t denotes the number of sprouting contractions in ψ, see 2.1. Then*

$$5 + t \geq 2\varepsilon + K_{\overline{S}} \cdot E.$$

PROOF. We keep notation of the proof of Proposition 2.2. ψ is a sequence of blowing downs. A subdivisional blowing down does not change the quantities $K \cdot (K + D)$ and $E \cdot (K + D)$. Under sprouting blowing down $K \cdot (K + D)$ increases by 1 and $E \cdot (K + D)$ decreases by 1. Here, by the abuse of notation, D denotes the image of $D \subset \overline{S}$ at some stage of ψ and K denotes the canonical divisor on the image of $\overline{S}$. Hence $(E_0 + 2K_{\overline{N}}) \cdot (K_{\overline{N}} + T) = (E + 2K_{\overline{S}}) \cdot (K_{\overline{S}} + D) + t = 6 - 2\varepsilon - K_{\overline{S}} \cdot E + t$. Suppose that $5 + t < 2\varepsilon + K_{\overline{S}} \cdot E$. Then $(E_0 + 2K_{\overline{N}}) \cdot (K_{\overline{N}} + T) \leq 0$. Let A be a (-1)-curve as in 2.1. Then A is not contained in T since T is 2-reduced w.r.t. E_0. $A \cdot T \geq 2$ otherwise the proper transform of A in $\overline{S}$ is a good asymptote of $\overline{U}$. This implies that $|A + K_{\overline{N}} + T| \neq \varnothing$. Let m be the greatest integer such that $|A + m(K_{\overline{N}} + T)| \neq \varnothing$. We write $A + m(K_{\overline{N}} + T) = \sum A_i$. By the choice of m $|K_{\overline{N}} + A_i + T| = \varnothing$ for every i. Hence $A_i \simeq \mathbb{P}^1$ and $A_i \cdot T \leq 1$. Since, by Lemma 1.6, $\overline{N}$ is not isomorphic to a Hirzebruch surface or to $\mathbb{P}^2$ we may assume that $A_i^2 < 0$ for every i. Write

$$A + m(K_{\overline{N}} + T) = \sum_{L_i^2 = -1} L_i + \sum_{A_j^2 \leq -2} A_j.$$

We take intersection of both sides with $E_0 + 2K_{\overline{N}}$. Since $A \cdot (E_0 + 2K_{\overline{N}}) \leq -1$ we get

$$-1 \geq \sum L_i \cdot (E_0 + 2K_{\overline{N}}) + \sum A_j \cdot (E_0 + 2K_{\overline{N}}).$$

If $A_j = E_0$ then $A_j \cdot (E_0 + 2K_{\overline{N}}) = -2 + K_{\overline{N}} \cdot A_j \geq 0$ since $E_0^2 \leq -4$ by Lemma 1.6(b). If $A_j \neq E_0$ then again $A_j \cdot (E_0 + 2K_{\overline{N}}) \geq 0$. It follows that there exists L_{i_0} such that $L_{i_0} \cdot (E_0 + 2K_{\overline{N}}) < 0$, i.e., $L_{i_0} \cdot E_0 \leq 1$. By the definition of 2-reduction

$L_{i_0} \not\subseteq T$. Now L, the proper transform of L_{i_0} in $\overline{S}$, is a good asymptote of $\overline{U}$. This is contradiction. □

2.4. Let $\lambda, \tilde{\lambda}$ be the branches of $\overline{U}$ at L_∞. The resolution process Φ', see Definition 0.3, can be described in terms of Hamburger–Noether (HN-) pairs. For definition in our context and basic properties of HN-pairs we refer to [1, 1.12]; see also [7, Appendix] or [13]. Let $\binom{c_1}{p_1}, \dots, \binom{c_h}{p_h}$ (resp. $\binom{\tilde{c}_1}{\tilde{p}_1}, \dots, \binom{\tilde{c}_{\tilde{h}}}{\tilde{p}_{\tilde{h}}}$) be the sequence of HN-pairs of λ (resp. $\tilde{\lambda}$). We have $c_1 = \lambda \cdot L_\infty$, $\tilde{c}_1 = \tilde{\lambda} \cdot L_\infty$. Let $\mu_1, \mu_2, \dots$ (resp. $\tilde{\mu}_1, \tilde{\mu}_2, \dots$) be the sequence of multiplicities of all singular points of $\overline{U}$ infinitely near $q = \lambda \cap L_\infty$ (resp. $\tilde{q} = \tilde{\lambda} \cap L_\infty$). Then

$$\sum_{i \geq 1} \mu_i = c_1 + p_1 + p_2 + \cdots + p_h - 1. \tag{2.1}$$

$$\sum_{i \geq 1} \mu_i^2 = c_1 p_1 + c_2 p_2 + \cdots + c_h p_h. \tag{2.2}$$

$$\sum_{i \geq 1} \tilde{\mu}_i = \tilde{c}_1 + \tilde{p}_1 + \tilde{p}_2 + \cdots + \tilde{p}_{\tilde{h}} - 1. \tag{2.3}$$

$$\sum_{i \geq 1} \tilde{\mu}_i^2 = \tilde{c}_1 \tilde{p}_1 + \tilde{c}_2 \tilde{p}_2 + \cdots + \tilde{c}_{\tilde{h}} \tilde{p}_{\tilde{h}}. \tag{2.4}$$

Let $\gamma' = -E'^2$. We call a branch *simple* if it is transversal to L_∞.

2.5. Definition. A contraction of a divisor T in D' (resp. D) or in an image of D' (resp. D) is called *a smooth contraction* if the image of E' is smooth and the image of D' is an SNC-divisor.

2.6. Definition. By C', $\widetilde{C}'$ we denote the components of D' such that $\lambda \cdot C' = 1$, $\tilde{\lambda} \cdot \widetilde{C}' = 1$. Here, by abuse of notation, λ, $\tilde{\lambda}$ denote the strict transforms in $\overline{S}'$ of the branches of $\overline{U}$ at infinity. Let $C = \Psi(C')$, $\widetilde{C} = \Psi(\widetilde{C}')$.

2.7. Lemma. *Suppose that C and $\widetilde{C}$ are (-1)-curves. Let D_1, D_2 (resp. $\widetilde{D}_1$, $\widetilde{D}_2$) be the adjacent component of C (resp. $\widetilde{C}$). Suppose that $D_1^2 = \widetilde{D}_1^2 = -2$ and that $C + D_1$ and $\widetilde{C} + \widetilde{D}_1$ are disjoint and that contractions of $C + D_1$ and $\widetilde{C}, \widetilde{D}_1$ are smooth. Let $g \colon \overline{S} \to \overline{Y}$ be the contraction of $C + D_1 + \widetilde{C} + \widetilde{D}_1$. Then*

(a) $\gamma = 8$ *or* 9, $\varepsilon = 0$.
(b) *Suppose also that $D_2 \neq \widetilde{D}_2$ and that $D_2^2 = \widetilde{D}_2^2 = -3$, $\gamma = 8$. Then $g(E) + 2K_{\overline{Y}} \sim 0$. In particular the intersection of any curve in $\overline{Y}$ with $g(E)$ is even.*

PROOF. (a) Let $E_0 = g(E)$. $E_0^2 = E^2 + 4$. Hence, by Lemma 1.6(b), $\gamma \geq 8$. From Proposition 2.3 it follows that $8 \leq \gamma \leq 9$ and $\varepsilon = 0$.

(b) We have $E_0 \cdot g(D_2) = E_0 \cdot g(\widetilde{D}_2) = 2$. We first show that $|E_0 + 3K_{\overline{Y}}| = \varnothing$. Suppose that $E_0 + 3K_{\overline{Y}} = \sum Z_j$. Since $g(D_2) \cdot (E_0 + 3K_{\overline{Y}}) = g(\widetilde{D}_2) \cdot (E_0 + 3K_{\overline{Y}}) = -1 < 0$ $g(D_2)$ and $g(\widetilde{D}_2)$ appear among Z_j. But then $E_0 \cdot (E_0 + 3K_{\overline{Y}}) = 2 \geq E_0 \cdot g(D_2) + E_0 \cdot g(\widetilde{D}_2) = 4$; contradiction.

Since $|E_0 + 3K_{\overline{Y}}| = \varnothing$, $-E_0 - 2K_{\overline{Y}} \geq 0$ by the Riemann–Roch Theorem. By Proposition 1.5(c) $E_0 + 2K_{\overline{Y}} \geq 0$. Hence $E_0 + 2K_{\overline{Y}} \sim 0$. □

3. The branches separate

3.1. Suppose that the branches λ, $\tilde{\lambda}$ are disjoint and none of them is simple. We write $D' = L'_\infty + F + \widetilde{F}$ where $F = \Phi^{-1}(\lambda \cap L_\infty)$, $\widetilde{F} = \Phi^{-1}(\tilde{\lambda} \cap L_\infty)$. Here F and $\widetilde{F}$ we regard as reduced divisors. Then $C' \subset F$, $\widetilde{C}' \subset \widetilde{F}$.

3.2. Lemma. *Suppose that the branches $\lambda, \tilde{\lambda}$ are disjoint and none of them is simple. Then*

(a) *$C', \widetilde{C}'$ and E' are not touched during the NC-minimalization $\Psi\colon \overline{S}' \to \overline{S}$. In particular $\gamma' = -E^2 = \gamma$.*

(b) *Ψ involves only subdivisional (with respect to D') contractions. Moreover none maximal twig of D' is touched by Ψ.*

PROOF. (a) Since none branch is simple E' is disjoint from L'_∞. Hence the minimalization process Ψ starts with contraction of L'_∞ (or $\Psi = \mathrm{id}$ if $L'^2_\infty < -1$.) If, say, C' is touched by Ψ at some stage then C' becomes a 0-curve and E' is then a 1-section of a $\mathbb{P}^1$-ruling. It contradicts Lemma 1.6(d). Therefore $C'$and $\widetilde{C}'$ are not touched. It follows that E' is not touched.

(b) There is a natural order of branching components of $D' + E'$ contained in F (resp. $\widetilde{F}$). Let T_1 (resp. $\widetilde{T}_1$) be the branching component of $D' + E'$ contained in F (resp. $\widetilde{F}$) nearest to L'_∞. T_1, $\widetilde{T}_1$ are connected in D' by a chain containing L'_∞. None component of that chain is a branching component of $D' + E'$. All contractions in Ψ take place inside this chain. The statement follows. □

3.3. Lemma. *The following holds*:

(3.1) $$\gamma' + 2(c_1 + \tilde{c}_1) = \sum_{i=1}^{h} p_i + \sum_{i=1}^{\tilde{h}} \tilde{p}_i;$$

(3.2) $$\gamma' + (c_1 + \tilde{c}_1)^2 = \sum_{i=1}^{h} p_i c_i + \sum_{i=1}^{\tilde{h}} \tilde{c}_i \tilde{p}_i.$$

PROOF. The degree of $\overline{U}$ equals $d = \overline{U} \cdot L_\infty = c_1 + \tilde{c}_1$. We have $K_{\mathbb{P}^2} \cdot \overline{U} = -3d = -3(c_1 + \tilde{c}_1)$, $K_{\overline{S}'} \cdot E' = -2 + \gamma'$. Since $K_{\overline{S}'} \cdot E' - K_{\mathbb{P}^2} \cdot \overline{U} = \sum_{i \geq 1} \mu_i + \sum_{i \geq 1} \tilde{\mu}_i$ the formula (a) follows from (2.1) and (2.3). $\overline{U}^2 - E'^2 = (c_1 + \tilde{c}_1)^2 + \gamma = \sum_{i \geq 1} \mu_i^2 + \sum_{i \geq 1} \tilde{\mu}_i^2$ hence (b) follows from (2.2),(2.4). □

3.4. Proposition. *Suppose that the branches of $\overline{U}$ are disjoint. Then one of them is simple.*

PROOF. Suppose opposite. By Proposition 2.3 and Lemma 3.2 we have

(3.3) $$7 + t \geq 2\varepsilon + \gamma.$$

One of the adjacent components of C' (resp. $\widetilde{C}'$) is a (-2)-curve since F (resp. $\widetilde{F}$) contracts to a point. After contracting C', $\widetilde{C}'$ and these two adjacent components the image of E' stays smooth. Let $f\colon \overline{S}' \to \overline{Y}$ be the contraction of these 4 curves. Since $f(E')^2 \leq -4$ by Lemma 1.6(b) $\gamma \geq 8$. It follows from (*) that $\varepsilon = 0$, $\gamma \leq 9$ and $t \geq 1$. Suppose that $t = 2$. This is equivalent to $p_h = 1, \tilde{p}_{\tilde{h}} = 1$. The twigs of $D' + E'$ produced by the pairs $\binom{c_h}{p_h}$, $\binom{\tilde{c}_{\tilde{h}}}{\tilde{p}_{\tilde{h}}}$ consist of (-2)-curves. Hence, in view of Lemma 3.2(b), $D + E$ has at least two maximal twigs composed of (-2)-curves.

If one of the twigs has more than one component then $\sum e_i > 1$; contradiction by Corollary 1.13. It follows that $c_h = \tilde{c}_{\tilde{h}} = 2$. Since c_h (resp. $\tilde{c}_{\tilde{h}}$) divides c_i, p_i (resp. $\tilde{c}_i, \tilde{p}_i$) for all $i < h$ (resp. $i < \tilde{h}$) it follows from Lemma 3.3(b) that $4 \mid \gamma$. Hence $\gamma = 8$. Corollary 1.13 implies also that $D' + E'$ has only the two maximal twigs. It implies that F and $\widetilde{F}$ are chains. This in turn implies that $c_i = p_i$ for $i = 1, \ldots, h-1$ and $c_1 = c_h = 2$. So the first $h-1$ pairs of λ and first $\tilde{h}-1$ pairs of $\tilde{\lambda}$ are all equal to $\binom{2}{2}$.

Now Lemma 3.3(a) gives $8+2(2+2) = 2(h-1)+1+2(\tilde{h}-1)+1$, i.e., $h+\tilde{h} = 9$. Lemma 3.3(b) gives $8 + (2+2)^2 = 4(h-1) + 2 + 4(\tilde{h}-1) + 2$, i.e., $h + \tilde{h} = 7$; contradiction.

Now assume that $t = 1$. (3.3) gives $\gamma' = 8$. We may assume that $p_h > 1$, $\tilde{p}_{\tilde{h}} = 1$. Let $f(E') = E_0$. We have $E_0^2 = -4$. It follows from Lemma 2.7(b) that the intersection of E_0 with any curve in $\overline{Y}$ is even. But, since $p_h > 1$ there exists a component in $f(F)$ which meets E_0 once; contradiction. □

3.5. Lemma. *The branches cannot be disjoint.*

PROOF. Suppose that the branches are disjoint. We may assume that the resolution tree D' has smallest possible number of maximal twigs among all embeddings of $\overline{U}$ with disjoint branches at infinity. By Proposition 3.4 we may assume that $\tilde{\lambda}$ is simple. Then $c_1 > p_1$ otherwise after the first blowing up over $\lambda \cap L_\infty$ the proper transform of L_∞ becomes a 0-curve which meets the proper transform of $\overline{U}$ transversally at one point; we get contradiction with Lemma 1.6(d). Let k be the number of contractions in $\Psi\colon \overline{S}' \to \overline{S}$. If $L_\infty'^2 \leq -2$ then $k = 0$. We have $-\gamma = E^2 = E'^2 + k$, i.e., $\gamma = \gamma' - k$. Since $\deg(\overline{U}) = c_1 + 1$ the formulas (3.1) and (3.2) take the form

$$\gamma' + 2c_1 + 2 = p_1 + p_2 + \cdots + p_h \tag{3.4}$$

$$\gamma' + (c_1+1)^2 = p_1c_1 + p_2c_2 + \cdots + p_hc_h. \tag{3.5}$$

Let $c_1 = sc_2, p_1 = tc_2$. From (3.4) and (3.5) we obtain

$$\gamma' + 2 + (2s-t)c_2 = p_2 + \cdots + p_h \tag{3.6}$$

$$\gamma' + 1 + (s^2 - st)c_2^2 + 2sc_2 = p_2c_2 + \cdots + p_hc_h. \tag{3.7}$$

Multiply (3.6) by c_2 and subtract (3.7). We obtain

$$c_2(\gamma'+2) + c_2^2(2s - t - s^2 + st) - \gamma' - 1 - 2sc_2 = \sum_{i=3}^{h} p_i(c_2 - c_i). \tag{3.8}$$

Let r equals the number of pairs equal to $\binom{c_2}{c_2}$.

3.5.1. Suppose that $2s - t - s^2 + st \geq 0$. It implies $s \geq (s-1)(s-t)$. Suppose $s - t \geq 2$. Then $s \geq 2(s-1)$ which gives $s \leq 2$. Since $s - 2 \geq t \geq 1$ we reach contradiction. Hence $s - t = 1$, i.e., $s = t + 1$. In that case $k = s - 1$. Then $c_{r+2} = c_2$ and $c_i < c_2$ for $i \geq r+3$. Thus $c_i \leq \frac{c_2}{2}$ for $i \geq r+3$. Since $\gamma' = \gamma + k$ we get from (3.8) that

$$c_2(\gamma + s + 1) + c_2^2 - \gamma - s - 2sc_2 \geq \sum_{i=r+3}^{h} p_i \frac{c_2}{2}.$$

From this

$$\sum_{i=r+3}^{h} p_i \leq \frac{2}{c_2}(c_2(\gamma+s+1)+c_2^2-\gamma-s-2sc_2).$$

(3.6) takes form

$$\gamma+s+1+(s+1)c_2 = rc_2+p_{r+2}+\sum_{i=r+3} p_i.$$

We obtain

$$\gamma+s+1+(s+1)c_2 \leq rc_2+p_{r+2}+\frac{2}{c_2}(c_2(\gamma+s+1)+c_2^2-\gamma-s-2sc_2).$$

From this we obtain

$$c_2^2(s-r-1) \leq c_2(\gamma-3s+1+p_{r+2})-2s-2\gamma.$$

Since $p_{r+2} \leq c_2-1$

$$c_2^2(s-r-2) \leq c_2(\gamma-3s)-2s-2\gamma.$$

If $r \geq s-1$ then by performing $2s-2$ successive contractions starting with L'_∞ we pass to another embedding of $\overline{U}$ for which the resolution tree D' has smaller number of maximal twigs. Hence $r < s-1$. It follows that $\gamma-3s > 0$. Since $\gamma \leq 9$ by Proposition 2.3 we obtain $s=2$, $t=1$, $r=0$. (3.6) takes a form

(3.9) $$\gamma+3+3c_2 = p_2+\cdots+p_h$$

and (3.7) takes form

(3.10) $$\gamma+2+2c_2^2+4c_2 = p_2c_2+\cdots+p_hc_h.$$

Multiply (3.9) by $2c_2/3$ and subtract (3.10). We obtain

$$\frac{2}{3}c_2(\gamma+3)-\gamma-2-4c_2 = -\frac{p_2c_2}{3}+\sum_{i=3} p_i\left(\frac{2}{3}c_2-c_i\right).$$

Since $c_i \leq c_2/2$ for $i>2$ we get

$$\frac{2}{3}c_2(\gamma+3)-\gamma-2-4c_2+\frac{p_2c_2}{3} \geq \frac{c_2}{6}\sum_{i=3} p_i.$$

From this

$$\frac{2}{3}(\gamma+3)-4+\frac{p_2}{3} > \frac{1}{6}\sum_{i=3} p_i$$

and

$$\sum_{i=3} p_i < 4\gamma-12+2p_2.$$

Now from (3.9) we get

$$\gamma+3+3c_2 < 3p_2+4\gamma-12$$

and

$$c_2-p_2 < \gamma-5.$$

From (3.9) it follows that c_h divides $p_h-\gamma-3$. From (3.10) it follows that c_h divides $\gamma+2$. Hence $c_h \mid p_h-1$. So $p_h=1$. $h>1$ otherwise λ is smooth and $\lambda\cdot L_\infty=2$ so $\deg\overline{U}=3$ and $E^2>0$; contradiction by Lemma 1.6(b). It follows that $D+E$ has at least two (-2)-tips; one is produced by the pair $\binom{c_1}{p_1}$, the second one by the pair $\binom{c_h}{p_h}$. Suppose that $\varepsilon=0$. Then there are no other tips by Corollary 1.13.

Also $c_h = 2$. Since $c_2 > p_2$ it follows that $h = 2$ so $c_2 = 2$. From (3.9) we get $\gamma + 9 = 1$; contradiction. Hence $\varepsilon \geq 1$. By Proposition 2.3 $\gamma \leq 6$ since $t = 1$ ($p_h = 1$). We have $c_2 - p_2 \leq \gamma - 5 \leq 1$. Hence $\gamma = 6$ and $c_2 = p_2 + 1$. It follows that $h = 2$. Now (3.9) gives $\gamma + 4 + 2c_2 = 0$; contradiction.

3.5.2. Suppose that $2s - t - s^2 + st = s - (s-t)(s-1) < 0$. It implies $s - t \geq 2$. From (3.8) we obtain

$$c_2(\gamma + k + 2) > c_2^2(-2s + t + s^2 - st). \tag{3.11}$$

The upper half chain produced by $\binom{c_1}{p_1}$ starts with $k-1$ (-2)-curves and contains at least one ≤ -3-curve. Hence the determinant of it, which is equal to t, is not less than $2k+1$. Thus $2k + 1 \leq t \leq s - 2$. It gives $k \leq (s-3)/2$. Thus

$$c_2\left(\gamma + \frac{s+1}{2}\right) > c_2^2(-2s + t + s^2 - st). \tag{3.12}$$

Suppose that $c_2 = 2$. (3.6) takes the form

$$\gamma + k + 2 + 2(2s - t) = 2r + 1 \tag{3.13}$$

and (3.7) takes the form

$$\gamma + k + 1 + 4(s^2 - st) + 4s = 4r + 2. \tag{3.14}$$

Subtract (3.13) from (3.14). We get

$$2(s^2 - st) + t = r + 1.$$

The number of sprouting (with respect to D') contractions in $\Phi\colon \overline{S}' \to \mathbb{P}^2$ equals h. Each sprouting contraction increases the quantity $K\cdot(K+D')$ by 1, subdivisional contractions don't change it. Φ contracts D' to L_∞. Hence $h = K_{\mathbb{P}^2}\cdot(K_{\mathbb{P}^2} + L_\infty) - K_{\overline{S}'}\cdot(K_{\overline{S}'} + D') = 6 - K_{\overline{S}'}\cdot(K_{\overline{S}'} + D')$. $\Psi\colon \overline{S}' \to \overline{S}$ involves k contractions, all sprouting with respect to D'. Hence $K_{\overline{S}}\cdot(K_{\overline{S}} + D) = K_{\overline{S}'}\cdot(K_{\overline{S}'} + D') + k$. We get $K_{\overline{S}}\cdot(K_{\overline{S}} + D) = 6 - h + k$. On the other hand $K_{\overline{S}}\cdot(K_{\overline{S}} + D) = 2 - \varepsilon - K_{\overline{S}}\cdot E = 4 - \varepsilon - \gamma$. Hence $h = 2 + k + \varepsilon + \gamma$. Since $h = r + 2$ we get $r = k + \varepsilon + \gamma$. By Proposition 2.3 $\varepsilon + \gamma \leq 8$ ($t \leq 1$). We get $r = k + \varepsilon + \gamma \leq k + 8 \leq (s-3)/2 + 8$. Thus

$$2s(s-t) + t = r + 1 \leq \frac{s-3}{2} + 9,$$

i.e.,

$$4s(s-t) + 2t \leq s + 15.$$

From this $7s + 2t \leq 15$; contradiction.

Suppose that $-2s + t + s^2 - st \geq (s+5)/2$. (3.12) gives

$$8 \geq \gamma > (c_2 - 1)\frac{s+5}{2} + 2.$$

Since $s \geq t + 2$, $s \geq 3$. Thus $6 > (c_2 - 1)4$. It gives $c_2 = 2$. It is impossible by 3.5.2.

Suppose that $-2s + t + s^2 - st < (s+5)/2$. It follows that $-5s + 2t + 2s(s-t) < 5$. If $s - t \geq 3$ then $s + 2t < 5$; contradiction. Hence $s - t = 2$. The upper half chain produced by $\binom{c_1}{p_1}$ consists of $k-1$ (-2)-curves and one (-3)-curve. Thus $t = 2k+1$. Now (3.11) gives $\gamma + k + 2 > tc_2$ and $\gamma + 2 > (2k+1)c_2 - k$. By 3.5.2 $c_2 \geq 3$. It follows that $k \leq 1$. We have two cases

$$k = 0, t = 1, s = 3 \quad \text{or} \quad k = 1, t = 3, s = 5.$$

Consider the first case. (3.8) takes form $c_2(\gamma - 4) - \gamma - 1 = c_2^2 + \sum_{i=3} p_i(c_2 - c_i)$. This implies that $c_2 < \gamma - 4 \leq 4$. In view of 3.5.2 $c_2 = 3$. Now (3.6) takes form

$\gamma + 17 = 3r + p_h$ and (3.7) takes form $\gamma + 73 = 9r + 3p_h$. We obtain $\gamma = 11$; contradiction.

Consider the second case. From (3.8) we obtain

$$c_2(\gamma+3) - 3c_2^2 - \gamma - 2 - 10c_2 \geq 0.$$

This gives $\gamma - 7 > 3c_2$; contradiction. □

4. The branches together

In view of Lemma 3.5 we may assume that $\lambda \cap L_\infty = \tilde{\lambda} \cap L_\infty$ and that the branches cannot be separated by an automorphism of $\mathbb{C}^2$. We will also assume that the resolution tree D' has smallest possible number of maximal twigs, i.e., if $\sigma\colon \mathbb{C}^2 \to \mathbb{C}^2$ is an automorphism then the number of maximal twigs of the resolution tree of $\bar{\sigma}(U)$ is not less than the number of maximal twigs of D'. Let s denotes the number of common pairs of λ and $\tilde{\lambda}$. Hence $c_i/p_i = \tilde{c}_i/\tilde{p}_i$ for $i = 1, \ldots, s$ and the branches separate along the chain created by pairs $\binom{c_{s+1}}{p_{s+1}}$, $\binom{\tilde{c}_{s+1}}{\tilde{p}_{s+1}}$. The formulas (3.1) and (3.2) now take the form, see [7, Appendix].

4.1. Lemma.

$$\gamma' + 2d = \sum p_i + \sum \tilde{p}_i \tag{4.1}$$

$$\gamma' + d^2 = \sum_{i=1}^{s}(p_i + \tilde{p}_i)(c_i + \tilde{c}_i) + \sum_{i>s} p_i c_i + \sum_{i>s} \tilde{p}_i \tilde{c}_i + 2\min(\tilde{p}_{s+1}c_{s+1}, p_{s+1}\tilde{c}_{s+1}) \tag{4.2}$$

where $d = c_1 + \tilde{c}_1$.

4.2. Lemma. *Suppose that L'_∞, the proper transform of L_∞ in $\overline{S}'$, is not a (-1)-curve. Then the number of sprouting contractions in Φ equals $6 - K_{\overline{S}} \cdot (K_{\overline{S}} + D) = 2 + \varepsilon + \gamma$.*

4.3. Lemma. *We have $s = 0$.*

PROOF. We may assume that none branch is simple since otherwise they separate on the first blowing up. Let $c_1 - p_1 = \alpha c_2$, $\tilde{c}_1 - \tilde{p}_1 = \widetilde{\alpha}\tilde{c}_2$. Let $\alpha_0 = \min(\alpha, \widetilde{\alpha})$. Since $2\min(\tilde{p}_{s+1}c_{s+1}, p_{s+1}\tilde{c}_{s+1}) \leq \tilde{p}_{s+1}c_{s+1} + p_{s+1}\tilde{c}_{s+1}$, (4.2) gives

$$\gamma + d^2 \leq \sum_{i=1}^{s}(p_i + \tilde{p}_i)(c_i + \tilde{c}_i) + \sum_{i>s} p_i c_i + \sum_{i>s} \tilde{p}_i\tilde{c}_i + \tilde{p}_{s+1}c_{s+1} + p_{s+1}\tilde{c}_{s+1}.$$

and

$$d^2 + \gamma \leq (p_1 + \tilde{p}_1)d + (c_2 + \tilde{c}_2)(P + \widetilde{P})$$

where $P = \sum_{i=2} p_i$, $\widetilde{P} = \sum_{i=2} \tilde{p}_i$.

From this

$$d(d - p_1 - \tilde{p}_1) + \gamma \leq (c_2 + \tilde{c}_2)(P + \widetilde{P}). \tag{4.3}$$

Since $d - p_1 - \tilde{p}_1 = c_1 - p_1 + \tilde{c}_1 - \tilde{p}_1 \geq \alpha_0(c_2 + \tilde{c}_2)$ and since $\gamma \geq 1$ we get $\alpha_0 d < P + \widetilde{P}$.

Suppose that $s \geq 1$. Then $\alpha = \widetilde{\alpha} = \alpha_0$. Also $\alpha_0 \neq 0$, i.e., $c_1 \neq p_1$ since otherwise after first blowing up L_∞ becomes a 0-curve disjoint from E and then $\bar{\kappa}(\overline{S}\setminus E) = -\infty$. It is easy to see that E' is not touched by $\Psi\colon \overline{S}' \to \overline{S}$ hence $\gamma' = \gamma$.

We will use so called *elementary transformations* of $\mathbb{C}^2$. Let $p \in L_\infty$. Consider a sequence of blowing ups over p given by HN-pairs $\binom{s+1}{s}$ and then s pairs equal to $\binom{1}{1}$.

After all these blowing ups we contract successively the proper transform of L_∞. The composition of all these blowing ups and downs is a Cremona transformation which induces an isomorphism of $\mathbb{C}^2$. We call it an elementary transformation.

Suppose that $\alpha = 1$. Consider the elementary transformation of $\mathbb{C}^2$ which follows the pair $\binom{c_1}{p_1}$ and then it follows the branches during the first blow up in pair $\binom{c_2}{p_2}$. The remaining blow ups we choose arbitrarily. That transformation defines a new embedding of U for which the resolution tree has smaller number of twigs than D' has. By our assumption on the minimality of D' it is impossible. Hence $\alpha \geq 2$.

By Proposition 2.3 $\gamma \leq 9$. Suppose that $\alpha \geq 3$. We have $3d < P + \widetilde{P}$. From (4.1) we obtain $2d = P + \widetilde{P} + p_1 + \tilde{p}_1 - \gamma$. We get that $d + p_1 + \tilde{p}_1 < \gamma \leq 9$. Thus $d < 9 - p_1 - \tilde{p}_1 \leq 7$. But $d = c_1 + \tilde{c}_1 \geq 2(\alpha + 1) \geq 8$; contradiction.

Hence $\alpha = 2$. We have $d - p_1 - \tilde{p}_1 = 2(c_2 + \tilde{c}_2)$. Substitute this to (4.3). We obtain

$$2d(c_2 + \tilde{c}_2) + \gamma \leq (c_2 + \tilde{c}_2)(P + \widetilde{P}).$$

and

$$(c_2 + \tilde{c}_2)(2d - P - \widetilde{P}) + \gamma \leq 0. \tag{4.4}$$

By (4.1) $2d - P - \widetilde{P} = p_1 + \tilde{p}_1 - \gamma$. Let $p_1 = lc_2$, $\tilde{p}_1 = l\tilde{c}_2$. Then

$$(c_2 + \tilde{c}_2)(l(c_2 + \tilde{c}_2) - \gamma) + \gamma \leq 0.$$

From this $l(c_2 + \tilde{c}_2) < \gamma$. Thus $l(c_2 + \tilde{c}_2) \leq 8$ which implies $l \leq 4$. Let $c_1 = kc_2$. Then $k - l = 2$. Since $GCD(k, l) = 1$ l is odd. Suppose that $l = 3$. Then $c_2 + \tilde{c}_2 = 2$ and we have $2(6 - \gamma) + \gamma \leq 0$ which gives $\gamma \geq 12$; contradiction. Therefore $l = 1$ and $k = 3$.

It follows from (4.4) that $2d - P - \widetilde{P} < 0$. Put $\delta = \gamma - p_1 - \tilde{p}_1$. In view of (4.1) $\delta = P + \widetilde{P} - 2d > 0$. (4.4) gives $\gamma \leq \delta(c_2 + \tilde{c}_2)$. It follows that $\delta \geq 2$ otherwise $\gamma \leq c_2 + \tilde{c}_2 \leq p_1 + \tilde{p}_1$ which implies $\delta \leq 0$. Therefore

$$c_2 + \tilde{c}_2 = p_1 + \tilde{p}_1 \leq \gamma - 2.$$

The proper transform of L_∞ in $\overline{S}$ is a (-2)-curve and it is a tip of D. The lower subchain produced by the pair $\binom{c_1}{p_1}$ consists of two (-2)-curves. Therefore the contribution of both twigs into $\sum e_i$ is $\frac{1}{2} + \frac{2}{3} > 1$. In view of Corollary 1.13 $\varepsilon \geq 1$. By Proposition 2.3 $\gamma \leq 7$. If $\gamma = 7$ then $\epsilon = 1$ and $t = 2$ by Proposition 2.3 so D has at least two (-2)-tips contained in the twigs produced by the pair $\binom{c_h}{p_h}$, $\binom{\tilde{c}_{\tilde{h}}}{\tilde{p}_{\tilde{h}}}$; these tips are not contained in the twigs produced by $\binom{c_1}{p_1}$. We get that $\sum e_i > 2$; contradiction by Corollary 1.13.

Hence we have $\gamma \leq 6$ so $c_2 + \tilde{c}_2 \leq 4$.

Suppose that $c_2 = \tilde{c}_2 = 2$. Let a be the number of common pairs of type $\binom{2}{2}$. Thus $s \geq a + 1$. If $s \geq a + 2$ then $s = a + 2$ and the last common pair is $\binom{2}{1}$. We have then $h = \tilde{h} = s + 1$ and $\binom{c_h}{p_h} = \binom{\tilde{c}_h}{\tilde{p}_{\tilde{h}}} = \binom{1}{1}$. The branches then both meet the last (-1)-curve transversally but at two different points. Also $t = 0$ so Proposition 2.3 gives $\gamma \leq 5$. (4.1) gives $\gamma + 16 = 4a$ so $\gamma = 4$ and $a = 5$, $h = 8$. We find $K_{\overline{S}} \cdot (K_{\overline{S}} + D) = K_{\mathbb{P}^2} \cdot (K_{\mathbb{P}^2} + L_\infty) - h = 6 - 8 = -2$. On the other hand $K_{\overline{S}} \cdot (K_{\overline{S}} + D) = 2 - \varepsilon - K_{\overline{S}} \cdot E = 2 - 1 - 2 = -1$.

Hence $s = a + 1$. Also either $\binom{c_{s+1}}{p_{s+1}} = \binom{\tilde{c}_{s+1}}{\tilde{p}_{s+1}} = \binom{2}{2}$ or $\binom{c_{s+1}}{p_{s+1}} = \binom{\tilde{c}_{s+1}}{\tilde{p}_{s+1}} = \binom{2}{1}$ or $\binom{c_{s+1}}{p_{s+1}} = \binom{2}{1}$, $\binom{\tilde{c}_{s+1}}{\tilde{p}_{s+1}} = \binom{2}{2}$. Hence $m = \min(c_{s+1}\tilde{p}_{s+1}, \tilde{c}_{s+1}p_{s+1}) = 2$ or 4. We have also $d = c_1 + \tilde{c}_1 = 3(c_2 + \tilde{c}_2) = 12$. Now (4.2) gives

$$\gamma + 144 = 48 + 16a + 2m + \sum_{i=s+1}^{h-1} p_i c_i + 2 + \sum_{i=s+1}^{\tilde{h}-1} \tilde{p}_i \tilde{c}_i + 2.$$

The right-hand side is divisible by 4, hence $\gamma = 4$. Hence no smooth contraction is allowed in D. It follows that only one case possible: $h = \tilde{h} = s + 1$ and $\binom{c_h}{p_h} = \binom{\tilde{c}_h}{\tilde{p}_{\tilde{h}}} = \binom{2}{1}$. Now (4.1) gives $4 + 2d = 2(2 + 2a + 1)$, i.e., $22 = 4a$; contradiction.

Suppose now that $c_2 = 1$. Then $\tilde{c}_2 \le 3$. Let a be the number of common pairs of type $\binom{\tilde{c}_2}{\tilde{c}_2}$. The branches get separated on the $(a+2)$th pair, i.e., $s = 1 + a$. Write $\tilde{h} = 1 + a + b + 1$ where $a + b$ is the number of pairs equal to $\binom{\tilde{c}_2}{\tilde{c}_2}$. We have $m = \min(c_{s+1}\tilde{p}_{s+1}, \tilde{c}_{s+1}p_{s+1}) = \tilde{p}_{s+1}$. The formulas (4.1) and (4.2) take the form

(4.5) $\quad \gamma + 2(3 + 3\tilde{c}_2) = s + 1 + (s + b)\tilde{c}_2 + \tilde{p}_{\tilde{h}}$

(4.6) $\quad \gamma + (3 + 3\tilde{c}_2)^2 = (3 + 3\tilde{c}_2)(1 + \tilde{c}_2) + (s - 1)(1 + \tilde{c}_2)^2 + 2\tilde{p}_{s+1} + 1 + b\tilde{c}_2^2 + \tilde{c}_2\tilde{p}_{\tilde{h}}.$

Suppose that $\tilde{c}_2 = 3$. Then $\gamma = 6$. By Proposition 2.3 $t \ge 1$ which implies that $\tilde{c}_{s+1} = \tilde{p}_{s+1} = 3$ and $\tilde{p}_{\tilde{h}} = 1$. (4.5) and (4.6) give now $6 + 24 = s + 1 + 3(s + b) + 1$, i.e., $28 = 4s + 3b$ and $6 + 144 = 48 + (s - 1)16 + 9b + 10$, i.e., $108 = 16s + 9b$. The system of equations has no integer solutions.

Suppose that $\tilde{c}_2 = 2$. (4.5) and (4.6) give $\gamma + 18 = s + 1 + (s + b)2 + 1$, i.e., $\gamma + 16 = 3s + 2b$ and $\gamma + 81 = 27 + (s-1)9 + 4b + 3 + 2\tilde{p}_{s+1}$, i.e., $\gamma + 60 = 9s + 4b + 2\tilde{p}_{s+1}$. We have $5 \le \gamma \le 6$ and $\tilde{p}_{s+1} \le 2$. We find two solutions:

(i) $\gamma = 5$, $s = 7$, $b = 0$, $\tilde{p}_{s+1} = 1$
(ii) $\gamma = 6$, $s = 6$, $b = 2$, $\tilde{p}_{s+1} = 2$.

In case (i) we may contract once inside D and in case (ii) twice to keep E smooth. Let E_0 be the image of E. In both cases $E_0^2 = -4$. Let $f\colon \overline{S} \to \overline{X}$ be the morphism. By Lemma 2.7(b) $E_0 + 2K_{\overline{X}} \equiv 0$. But in both cases there exists a component $B \subset f(F)$ such that $B \cdot E_0 = 1$. We reach contradiction.

Suppose that $\tilde{c}_2 = 1$. Then $d = 6$. The formulas give $\gamma + 10 = 2s + b$ and $\gamma + 24 = 4s + b$. We get solution

(iii) $\gamma = 4$, $s = 7$.

This case was already eliminated, see the proof of 2.2.10, the case $\varepsilon = 2$. □

4.4. By Lemma 4.3 $s = 0$. Suppose that the branches stay together after the first blowing up, i.e., they both are tangent to L_∞. In particular none branch is simple. Let, as in the proof of Lemma 4.3, $c_1 - p_1 = \alpha c_2$, $\tilde{c}_1 - \tilde{p}_1 = \tilde{\alpha}\tilde{c}_2$. We will show that the case can be reduced to the case $\alpha = \tilde{\alpha} = 1$ and $c_1/p_1 = \tilde{c}_1/\tilde{p}_1 = (u+1)/u$.

Assume that $\gamma - p_1 - \tilde{p} \le 0$.

From (4.1) we obtain

$$c_1 + \tilde{c}_1 > \frac{1}{2}\Big(\sum_{i=2} p_i + \sum_{i=2} \tilde{p}_i\Big). \tag{4.7}$$

Since $2\min(c_1\tilde{p}_1, \tilde{c}_1 p_1) \le c_1\tilde{p}_1 + \tilde{c}_1 p_1$ the formula (4.2) implies

$$(c_1 + \tilde{c}_1)^2 < (c_1 + \tilde{c}_1)(p_1 + \tilde{p}_1) + \sum_{i=2} p_i c_i + \sum_{i=2} \tilde{p}_i \tilde{c}_i$$

and

$$(c_1 + \tilde{c}_1)(c_1 + \tilde{c}_1 - p_1 - \tilde{p}_1) < \sum_{i=2} p_i c_i + \sum_{i=2} \tilde{p}_i \tilde{c}_i.$$

From this and (4.7) we get

$$\frac{1}{2}(\alpha c_2 + \widetilde{\alpha}\tilde{c}_2)(\sum_{i=2} p_i + \sum_{i=2} \tilde{p}_i) < \sum_{i=2} p_i c_i + \sum_{i=2} \tilde{p}_i \tilde{c}_i.$$

If $\alpha \geq 2$ and $\widetilde{\alpha} \geq 2$ we get contradiction. Assume that $\widetilde{\alpha} = 1$. Let $\tilde{p}_1 = u\tilde{c}_2$. Then $\tilde{c}_1 = (u+1)\tilde{c}_2$. Let T_1 denotes the (-1)-curve produced by the pair $\binom{\tilde{c}_1}{\tilde{p}_1}$. Let T^u denotes the upper chain produced by $\binom{\tilde{c}_1}{\tilde{p}_1}$ (i.e. the chain which contains the proper transform of L_∞) and let T^l denotes the lower chain (component). If λ separates from $\tilde{\lambda}$ on $T_1 + T^l$, i.e., if $p_1 \geq u(c_1 - p_1)$ then complete the pair $\binom{\tilde{c}_1}{\tilde{p}_1}$ to an elementary transformation by blowing up u times over $\tilde{\lambda} \cap T_1$. We get new embedding of U with smaller resolution tree. It is impossible by the assumption. Hence the branches separate on T^u. Now the elementary transformation described above gives new embedding with $\alpha = \widetilde{\alpha} = 1$.

Assume that $\gamma - p_1 - \tilde{p}_1 > 0$. Let $\alpha_0 = \min(\alpha, \widetilde{\alpha})$. If $\alpha_0 = 1$ we may reduce to a case with $\alpha = \widetilde{\alpha} = 1$ as above. So $\alpha_0 \geq 2$.

By Proposition 2.3 $\gamma \leq 9$. Suppose that $\gamma = 9$. Then $\varepsilon = 0$ and $t = 2$, by Proposition 2.3. It follows that the 2-reduction $\psi\colon \bar{\mathbb{S}} \to \bar{\mathbb{N}}$, see 2.1, involves two sprouting contractions. In particular D has at least two (-2)-tips. It follows also that D is not a chain. Hence $\sum e_i > 1$; contradiction by Corollary 1.13. Hence $\gamma \leq 8$.

We write (4.2) as

$$\gamma + \alpha c_1 c_2 + \widetilde{\alpha}\tilde{c}_1\tilde{c}_2 + 2c_1\tilde{c}_1 = \sum_{i=2} p_i c_i + \sum_{i=2} \tilde{p}_i \tilde{c}_i + 2\min(p_1\tilde{c}_1, \tilde{p}_1 c_1).$$

From this

$$\begin{aligned}\gamma + 2\max(p_1\tilde{c}_1, \tilde{p}_1 c_1) &+ \alpha c_1 c_2 + \widetilde{\alpha}\tilde{c}_1\tilde{c}_2 + 2c_1\tilde{c}_1 \\ &= \sum_{i=2} p_i c_i + \sum_{i=2} \tilde{p}_i \tilde{c}_i + 2\min(p_1\tilde{c}_1, \tilde{p}_1 c_1) + 2\max(p_1\tilde{c}_1, \tilde{p}_1 c_1)\end{aligned}$$

and

$$\gamma + 2\max(p_1\tilde{c}_1, \tilde{p}_1 c_1) + \alpha c_1 c_2 + \widetilde{\alpha}\tilde{c}_1\tilde{c}_2 + \alpha\tilde{c}_1 c_2 + \widetilde{\alpha} c_1 \tilde{c}_2 = \sum_{i=2} p_i c_i + \sum_{i=2} \tilde{p}_i \tilde{c}_i + p_1\tilde{c}_1 + \tilde{p}_1 c_1.$$

Let $\beta = 2\max(p_1\tilde{c}_1, \tilde{p}_1 c_1) - (p_1\tilde{c}_1 + \tilde{p}_1 c_1) = \max(p_1\tilde{c}_1, \tilde{p}_1 c_1) - \min(p_1\tilde{c}_1, \tilde{p}_1 c_1)$. We get

$$\gamma + \beta + (c_1 + \tilde{c}_1)(\alpha c_2 + \widetilde{\alpha}\tilde{c}_2) = \sum_{i=2} p_i c_i + \sum_{i=2} \tilde{p}_i \tilde{c}_i. \tag{4.8}$$

From (4.1) we get

$$\gamma + c_1 + \tilde{c}_1 + \alpha c_2 + \widetilde{\alpha}\tilde{c}_2 = \sum_{i=2} p_i + \sum_{i=2} \tilde{p}_i. \tag{4.9}$$

We may assume by symmetry that $c_2 \geq \tilde{c}_2$. Multiply (4.9) by c_2 and subtract (4.8). We obtain

$$\gamma c_2 + (c_1 + \tilde{c}_1)c_2 + c_2(\alpha c_2 + \widetilde{\alpha}\tilde{c}_2) \geq \gamma + \beta + (c_1 + \tilde{c}_1)(\alpha c_2 + \widetilde{\alpha}\tilde{c}_2).$$

So

$$\gamma(c_2 - 1) + (c_1 + \tilde{c}_1)c_2 \geq \beta + (\alpha c_2 + \widetilde{\alpha}\tilde{c}_2)(c_1 + \tilde{c}_1 - c_2)$$

and

$$\gamma c_2 + (c_1 + \tilde{c}_1)c_2 > (\alpha c_2 + \widetilde{\alpha}\tilde{c}_2)(c_1 + \tilde{c}_1 - c_2) \geq 2(c_2 + \tilde{c}_2)(c_1 + \tilde{c}_1 - c_2) > 2c_2(c_1 + \tilde{c}_1 - c_2).$$

We obtain

$$\gamma + c_1 + \tilde{c}_1 > 2(c_1 + \tilde{c}_1 - c_2).$$

It follows that

$$\gamma > c_1 + \tilde{c}_1 - 2c_2. \tag{4.10}$$

Since $\alpha \geq 2, \widetilde{\alpha} \geq 2$, $c_1 \geq 3c_2$ and $\tilde{c}_1 \geq 3\tilde{c}_2$. Hence $\gamma > c_2 + 3\tilde{c}_2 \geq 4\tilde{c}_2$. Since $\gamma \leq 8$ we obtain $7 \geq 4\tilde{c}_2$ which implies $\tilde{c}_2 = 1$. In particular $\tilde{h} = 1$. Now (4.8) takes the form

$$\gamma + \beta + (c_1 + \tilde{c}_1)(\alpha c_2 + \widetilde{\alpha}) = \sum_{i=2} p_i c_i.$$

Suppose that $\alpha \geq 3$. Then $c_1 = p_1 + \alpha c_2 \geq 4c_2$. Also $\alpha c_1 c_2 < \sum_{i=2} p_i c_i$. We have $\alpha c_1 c_2 \geq 12c_2^2$ and $\sum_{i=2} p_i c_i \leq (h-1)c_2^2 \leq 9c_2^2$ since $h - 1 = 1 + \varepsilon + \gamma \leq 9$ by Lemma 4.2 and Proposition 2.3. We reach contradiction. Hence $\alpha = 2$. If $c_1 \geq 4c_2$ then $c_1 \geq 5c_2$ (since $\alpha = 2$), $\alpha c_1 c_2 \geq 10c_2^2$ and we reach contradiction as above. So $c_1 = 3c_2$ and $p_1 = c_2$.

From (4.10) we get $\gamma > c_2 + \tilde{c}_1$. Since $h = \varepsilon + \gamma + 2 - 1 > 2$ by Lemma 4.2, $c_2 \geq 2$. So $\gamma - 3 \geq \tilde{c}_1$. Also $\gamma - 3 > c_2$ since $\tilde{c}_1 \geq 3$. It follows that $c_2 \leq 4$. It follows that $c_h = c_2$ or $c_h = \frac{1}{2}c_2$.

Suppose that $c_h = c_2$. Then (4.8) takes the form

$$\gamma + \beta + (3c_2 + \tilde{c}_1)(2c_2 + \widetilde{\alpha}) = (h-2)c_2^2 + c_2 p_h.$$

And (4.9) takes the form

$$\gamma + 3c_2 + \tilde{c}_1 + 2c_2 + \widetilde{\alpha} = (h-2)c_2 + p_h.$$

From this

$$c_2(\gamma + 5c_2 + \tilde{c}_1 + \widetilde{\alpha}) = \gamma + \beta + (3c_2 + \tilde{c}_1)(2c_2 + \widetilde{\alpha}).$$

It gives

$$c_2\gamma = \gamma + \beta + c_2^2 + c_2\tilde{c}_1 + 2c_2\widetilde{\alpha} + \tilde{c}_1\widetilde{\alpha}.$$

We obtain

$$\gamma > c_2 + \tilde{c}_1 + 2\widetilde{\alpha} \geq c_2 + 3 + 4.$$

This implies $c_2 < \gamma - 7 \leq 1$; a contradiction. Suppose that $c_h = \frac{1}{2}c_2$. Then $c_2 = 4$, $c_h = 2$. By (4.10) $\gamma > c_2 + \tilde{c}_1$ so $\tilde{c}_1 \leq \gamma - 5$. It follows that $\gamma = 8$, $\tilde{c}_1 = 3$, $\tilde{p}_1 = 1$. Since $c_1/p_1 = \tilde{c}_1/\tilde{p}_1$ the branches separate on the curve produced by the pair $\binom{c_1}{p_1}$. The divisor $D+E$ has at least three tips, at least two of them are (-2)-curves hence $\sum e_i > 1$. $\gamma = 8$ implies $\varepsilon = 0$ by Proposition 2.3. Now we reach contradiction with Corollary 1.13.

4.5. We consider here the case $s = 0$, $c_1 = (u+1)c_2$, $p_1 = uc_2$, $\tilde{c}_1 = (u+1)\tilde{c}_2$, $\tilde{p}_1 = u\tilde{c}_2$. We will prove that this case does not occur. Suppose opposite. Notice that $c_2 > 1$ and $\tilde{c}_2 > 1$ otherwise by suitable elementary transformation we can separate the branches. Let H' denotes the (-1)-curve produced by the pair $\binom{c_1}{p_1}$ and let $H = \Psi(H')$. Ψ involves u successive contractions of L'_∞. H' is not contracted by Ψ. Let F (resp. $\widetilde{F}$) denotes the part of D' produced by pairs $\binom{c_2}{p_2}, \ldots, \binom{c_h}{p_h}$ (resp. $\binom{\tilde{c}_2}{\tilde{p}_2}, \ldots, \binom{\tilde{c}_{\tilde{h}}}{\tilde{p}_{\tilde{h}}}$). Let C (resp. $\widetilde{C}$) be the unique (-1)-curve in F (resp. $\widetilde{F}$)

One of the adjacent components of C (resp. $\widetilde{C}$) is a (-2)-curve since F (resp. $\widetilde{F}$) contracts to a point. After contracting C, $\widetilde{C}$ and these two adjacent components the image of E stays smooth. Let $f\colon \overline{S} \to \overline{Y}$ be the contraction of these 4 curves. Since $f(E)^2 \leq -4$ by Lemma 1.6(b) $\gamma \geq 8$. It follows from Proposition 2.3 that $\varepsilon = 0$, $\gamma \leq 9$ and $t \geq 1$. Suppose that $t = 2$. Then the twigs of $D + E$ produced by the pairs $\binom{c_h}{p_h}$, $\binom{\tilde{c}_{\tilde{h}}}{\tilde{p}_{\tilde{h}}}$ consist of (-2)-curves. Hence $D + E$ has at least two maximal twigs composed of (-2)-curves. $D + E$ has at least one more twig. This is the tip produced by the pair $\binom{c_1}{p_1}$, the tip meets H. We obtain that $\sum e_i > 1$; contradiction by Corollary 1.13.

Now assume that $t = 1$. Proposition 2.3 gives $\varepsilon = 0$ and $\gamma = 8$. We may assume that $p_h > 1$, $\tilde{p}_{\tilde{h}} = 1$. Let $f(E') = E_0$. We have $E_0^2 = -4$.

Notice that $f(F)$ is not a point since f does not involve a sprouting contractions in F. We claim that that $f(\widetilde{F})$ is not a point. Suppose opposite. Then $\tilde{c}_2 = 2$, $\tilde{p}_2 = 1$, $\tilde{h} = 2$. The formulas (4.1) and (4.2) give

$$11 + 2c_1 = \sum p_i, \tag{4.11}$$

$$10 + c_1^2 + 4c_1 = \sum p_i c_i. \tag{4.12}$$

It follows from (4.12) that c_h divides 10. Since $p_h > 1$, $c_h = 5$ or $c_h = 10$. (4.11) implies that c_h divides $11 - p_h$. If $c_h = 5$ then p_h must be therefore 1; contradiction. If $c_h = 10$ then 10 divides $1 - p_h$ which implies that $p_h = 1$; again contradiction.

The assumptions of Lemma 2.7(b) are fulfilled. Hence an intersection of E_0 with every curve in $\overline{Y}$ is even. But, since $p_h > 1$, there exists a component of $f(F)$ which meets E_0 once. We reach contradiction.

4.6. Suppose that the branch λ is tangent to L_∞ and $\tilde{\lambda}$ is not. This is the last case to consider. We change slightly our usual labelling. The pairs of $\tilde{\lambda}$ we label now: $\binom{\tilde{c}_1}{\tilde{c}_1}, \binom{\tilde{c}_1}{\tilde{p}_1}, \binom{\tilde{c}_2}{\tilde{p}_2}, \ldots$. Consider the line in $\mathbb{P}^2$ which passes through $\tilde{\lambda} \cap L_\infty$ and which is tangent to $\tilde{\lambda}$. Let L be the proper transform of that line after the first blowing up. Then $\tilde{\lambda} \cdot L = \tilde{p}_1$. At this stage $L_\infty^2 = 0$ and the intersection of λ with L_∞ equals $c_1 - p_1$. Since $L \sim L_\infty$ we obtain that $c_1 - p_1 \geq \tilde{p}_1 + 2$ otherwise L is a good asymptote of $\overline{U}$.

Let $c_1 - p_1 = \alpha c_2$. If $\alpha = 1$ then by suitable elementary transformation we get a new embedding of $\overline{U}$ with smaller resolution tree. Hence $\alpha \geq 2$.

The formulas (4.1) and (4.2) take the form

$$\gamma + 2c_1 + \tilde{c}_1 = \sum_{i=1} p_i + \sum_{i=1} \tilde{p}_i \tag{4.13}$$

$$\gamma + c_1^2 + 2c_1\tilde{c}_1 = \sum_{i=1} p_i c_i + \sum_{i=1} \tilde{p}_i \tilde{c}_i + 2p_1\tilde{c}_1. \tag{4.14}$$

From this

$$\gamma + c_1^2 + 2\alpha c_2\tilde{c}_1 = \sum_{i=1} p_i c_i + \sum_{i=1} \tilde{p}_i\tilde{c}_i.$$

and

$$\gamma + 2\alpha c_2\tilde{c}_1 + \alpha c_2 c_1 = \sum_{i=2} p_i c_i + \sum_{i=2} \tilde{p}_i\tilde{c}_i + \tilde{p}_1\tilde{c}_1.$$

Since $\tilde{p}_1 \leq \alpha c_2 - 2$

$$\gamma + \alpha c_2 c_1 + \alpha c_2\tilde{c}_1 + 2\tilde{c}_1 \leq \sum_{i=2} p_i c_i + \sum_{i=2} \tilde{p}_i\tilde{c}_i.$$

(4.1) takes the form

$$\gamma + 2c_1 + \tilde{c}_1 - p_1 - \tilde{p}_1 = \sum_{i=2} p_i + \sum_{i=2} \tilde{p}_i$$

and

$$\gamma + 2(c_1 + \tilde{c}_1) - \tilde{c}_1 - p_1 - \tilde{p}_1 = \sum_{i=2} p_i + \sum_{i=2} \tilde{p}_i.$$

Let F be the part of D' produced by pairs $\binom{c_1-p_1}{p_1}, \binom{c_2}{p_2}, \dots, \binom{c_h}{p_h}$ and let $\widetilde{F}$ be the part of D' produced by pairs $\binom{\tilde{c}_1}{\tilde{p}_1}, \dots, \binom{\tilde{c}_h}{\tilde{p}_h}$. Let T_1 be the proper transform in D' of the (-1)-curve produced by the pair $\binom{c_1-p_1}{p_1}$. Hence T_1 is the first, counting from L'_∞, branching component of $D' + E'$. Let R' be the maximal twig of $D' + E'$ which meets T_1. The tip of R' is L'_∞. Let $R = \Psi(R')$.

4.6.1. Lemma. *T_1 and therefore E' is not touched by Ψ. In particular $\gamma = \gamma'$.*

PROOF. Suppose T_1 is touched. Then $L'^2_\infty = -1$ and R' consists of (-2)-curves. This means that $c_1/c_2 - p_1/c_2 = 1$, i.e., $\alpha = 1$; contradiction. □

4.6.2. Suppose that $\tilde{c}_1 > 1$ and that $h > 1$. Then the assumptions of Lemma 2.7 are fulfilled hence $\varepsilon = 0$ and $\gamma = 8, 9$. By Proposition 2.3 $t \geq 1$. If $t = 2$ then D has at least two (-2)-tips contained in the twigs produced by pairs $\binom{c_h}{p_h}$, $\binom{\tilde{c}_{\tilde{h}}}{\tilde{p}_{\tilde{h}}}$. D has one more tip, the one contained in R. We reach contradiction with Corollary 1.13. Hence $t = 1$, i.e., $p_h > 1$ or $\tilde{p}_{\tilde{h}} > 1$. Suppose that $\tilde{h} > 1$ or $\tilde{h} = 1$ and $\tilde{c}_1 > 2$. Again we use Lemma 2.7. Since $t < 2$ there exists a component of D which meets E once. We reach contradiction with Lemma 2.7(b). Thus $\tilde{h} = 1$ and $\tilde{c}_1 = 2$. Then $\tilde{p}_1 = 1$ so $p_h > 1$. Since $t = 1$ $\gamma = 8$ by Proposition 2.3. The formulas (4.13) and (4.14) take the form

$$9 + 2c_1 = \sum_{i=1}^{h} p_i$$

$$6 + c_1^2 + 4c_1 = \sum_{i=1}^{h} p_i c_i + 4p_1.$$

The first formula gives $c_h \mid 9 - p_h$. The second one gives $c_h \mid 6$. Since $p_h > 1$ $c_h \neq 2$. If $c_h = 3$ then $c_h \mid p_h$; contradiction. If $c_h = 6$ then $p_h = 5$ since $\gcd(c_h, p_h) = 1$. We obtain $6 \mid 9 - 5$; contradiction.

4.6.3. Suppose that $\tilde{c}_1 > 1$ and $h = 1$. Since $c_1 - p_1 \geq \tilde{p}_1 + 2$ we have $c_1 - p_1 \geq 3$. It follows in particular that F has at least 3 components. $C' = T_1$ so C' is not touched by Ψ in view of Lemma 4.6.1. Also the adjacent (-2)-component of C' is not touched by Ψ since otherwise it would get contracted so C' would be touched. Hence we have at least two smooth contractions in $\Psi(F)$.

Suppose that $\tilde{c}_1 > 2$ or $\tilde{h} > 1$. Then we have at least two smooth contractions in $\Psi(\widetilde{F})$. By Lemma 2.7(a) $\varepsilon = 0$ and $\gamma = 8, 9$. Suppose that $t = 2$. Then D has two (-2)-tips and moreover they are the only tips of D and also they are the maximal twigs of D. Since $\alpha > 1$ the twig R' does not contract to a point. Therefore R is a single (-2)-curve. Let B be the proper transform of R in D'. Since $R' - B$ contracts to a point starting from contraction of L'_∞ $B^2 = -3$. Let A be the second adjacent component of C' in D'. Then $A^2 = -2$. Now $A+2C+R$ defines a $\mathbb{P}^1$-ruling of $\overline{S}$ and E is a 2-section. We reach contradiction in view of Lemma 1.6(d). Now suppose $t = 1$. By Proposition 2.3 $\gamma = 8$. Let D_1 be the (-2)-adjacent component of C' in D'. Suppose that $D_1 \subset R'$. Then D_2, the second adjacent component of C' in D', is a (-3)-curve and is not touched by Ψ. Let D_0 be the adjacent component of D_1 different from C'. Then D_0 is not contracted by Ψ since otherwise D_1 would be contracted. After contraction of $C' + D_1$ E meets D_0 once; we reach contradiction with Lemma 2.7(b). Now suppose that $D_1 \not\subset R'$. Let D_2 be as above. Notice that we have exactly two smooth contractions in $\Psi(F)$. It follows that $D_2^2 = -3$. It follows that D_2 is not touched by Ψ since otherwise R is a single (-2)-curve and then $D_1 + 2C' + R$ defines a $\mathbb{P}^1$-ruling for which E is a 2-section; contradiction by Lemma 1.6(d). Let D_0 be the adjacent component of D_1 different from C'. After contraction of $C' + D_1$ E meets D_0 once; contradiction by Lemma 2.7(b).

We may assume that $\tilde{h} = 1$ and $\tilde{c}_1 = 2$. This we rule out , using the formulas (4.13) and (4.14), exactly as in 4.6.2.

4.6.4. Suppose that $\tilde{c}_1 = 1$. Then $\tilde{h} = 1$ and the formulas (4.13) and (4.14) take the form

$$1 + \gamma + 2c_1 = \sum_{i=1}^{h} p_i \tag{4.15}$$

$$\gamma + c_1^2 + 2c_1 = \sum_{i=1}^{h} c_i p_i + 2p_1 \tag{4.16}$$

or

$$1 + \gamma + 2c_1 - p_1 = \sum_{i=2}^{h} p_i$$

$$\gamma + c_1^2 + 2c_1 - c_1 p_1 - 2p_1 = \sum_{2}^{h} c_i p_i.$$

We multiply the first equation by c_2 and subtract the second one. Since $c_i \leq c_2$ for $i \geq 2$ we obtain

$$c_2(\gamma + 1) + 2c_1 c_2 - c_2 p_1 - \gamma - c_1^2 - 2c_1 + c_1 p_1 + 2p_1 \geq 0.$$

Let $c_1 = kc_2, p_1 = lc_2$. We get

$$\gamma + 1 > lc_2 + k^2 c_2 - 2kc_2 - klc_2 + 2(k - l). \tag{4.17}$$

$$\gamma \geq \big((k-2)(k-l) - l\big)c_2 + 2(k - l). \tag{4.18}$$

Notice that $k-l=\alpha\geq 2$. It follows that $k\geq l+2\geq 3$. We have also $\gamma\geq (2(k-2)-l)c_2$. If $(k-2)(k-l)-l\leq 0$ then $2(k-2)-l\leq 0$ but $2(k-2)-l=k-l+k-4\geq 2-1=1$. Hence $(k-2)(k-l)-l\geq 1$. Suppose that $(k-2)(k-l)-l=2$. Then $2(k-2)-l\leq 2$ which gives $2k-l\leq 6$. So $k\leq 6-(k-l)\leq 4$. Since $k-l\geq 2$ and $\gcd(k,l)=1$ $l=1$. Thus $(k-2)(k-1)=3$ which is impossible. Therefore $(k-2)(k-l)-l=1$ or $(k-2)(k-l)-l\geq 3$.

Suppose that $h>1$. Suppose $(k-2)(k-l)-l\geq 3$. From (4.18) we obtain $c_2\leq(\gamma-4)/3$. Hence $c_2=1$ since $\gamma\leq 9$ by Proposition 2.3. It means that $h=1$; contradiction. Suppose that $(k-2)(k-l)-l=1$. Then $k=3, l=1$. By Proposition 2.3, since $t\leq 1$, we have $\gamma\leq 8$. Now (4.18) gives $c_2\leq 4$ and $c_2\leq\gamma-4$. From Proposition 2.3 $\varepsilon=0$. If $c_2=4$ then $\gamma=8$, $c_1=12, p_1=4$ but then we get contradiction with (4.17). So $c_2\leq 3$. This implies that $c_i=c_2$ for $i\geq 2$. From (4.16) we get that $c_2\mid\gamma$. If $c_2=3$ then $7\leq\gamma\leq 8$ but $c_2\mid\gamma$. Hence $c_2=2$ since $c_2=1$ implies that $h=1$. Suppose that $c_2=2$. Then $p_h=1$ and (4.15) gives $h=11$. $(L'_\infty)^2=-2$ hence $D'=D$ and we have $K_{\overline{S}}\cdot(K_{\overline{S}}+D)=K_{\mathbb{P}^2}\cdot(K_{\mathbb{P}^2}+L_\infty)-h=6-11=-5$. On the other hand, by 1.9, $K_{\overline{S}}\cdot(K_{\overline{S}}+D)=2-\varepsilon-K_{\overline{S}}\cdot E=-4$; contradiction.

We may assume that $h=1$. Then $K_{\overline{S}'}\cdot(K_{\overline{S}'}+D')=K_{\mathbb{P}^2}\cdot(K_{\mathbb{P}^2}+L_\infty)-1=5$. Thus $2-\varepsilon-K_{\overline{S}}\cdot E=K_{\overline{S}}\cdot(K_{\overline{S}}+D)\geq K_{\mathbb{P}^2}\cdot(K_{\mathbb{P}^2}+L_\infty)=5$. It gives $K_{\overline{S}}\cdot E\leq -3-\varepsilon\leq -3$; contradiction by Lemma 1.6(b).

References

1. P. Cassou-Noguès, M. Koras, and P. Russell, *Closed embeddings of $\mathbb{C}^*$ in $\mathbb{C}^2$*. I, J. Algebra **322** (2009), no. 9, 2950–3002.
2. T. Fujita, *On the topology of noncomplete algebraic surfaces*, J. Fac. Sci. Univ. Tokyo Sect. IA Math. **29** (1982), no. 3, 503–566.
3. ———, *On Zariski problem*, Proc. Japan Acad. Ser. A Math. Sci. **55** (1979), no. 3, 106–110.
4. S. Iitaka, *On logarithmic Kodaira dimension of algebraic varieties*, Complex Analysis and Algebraic Geometry (W. L. Baily Jr. and T. Shioda, eds.), Iwanami Shoten, Tokyo, 1977, pp. 175–189.
5. Y. Kawamata, *On the classification of noncomplete algebraic surfaces*, Algebraic Geometry (Copenhagen, 1978) (K. Lønsted, ed.), Lecture Notes in Math., vol. 732, Springer, Berlin, 1979, pp. 215–232.
6. R. Kobayashi, *Uniformization of complex surfaces*, Kähler Metric and Moduli Spaces (T. Ochiai, ed.), Adv. Stud. Pure Math., vol. 18, Academic Press, Boston, MA, 1990, pp. 313–394.
7. M. Koras and P. Russell, *$\mathbf{C}^*$-actions on $\mathbf{C}^3$: the smooth locus of the quotient is not of hyperbolic type*, J. Algebraic Geom. **8** (1999), no. 4, 603–694.
8. ———, *Contractible affine surfaces with quotient singularities*, Transform. Groups **12** (2007), no. 2, 293–340.
9. M. Miyanishi, *Open algebraic surfaces*, CRM Monogr. Ser., vol. 12, Amer. Math. Soc., Providence, RI, 2001.
10. Y. Miyaoka, *The maximal number of quotient singularities on surfaces with given numerical invariants*, Math. Ann. **268** (1984), no. 2, 159–171.
11. N. Mohan Kumar and M. P. Murthy, *Curves with negative self-intersection on rational surfaces*, J. Math. Kyoto Univ. **22** (1982/83), no. 4, 767–777.
12. J. A. Morrow, *Minimal normal compactifications of $\mathbf{C}^2$*, Rice Univ. Studies **59** (1973), no. 1, 97–112.
13. P. Russell, *Hamburger-Noether expansions and approximate roots of polynomials*, Manuscripta Math. **31** (1980), no. 1-3, 25–95.
14. ———, *On affine-ruled rational surfaces*, Math. Ann. **255** (1981), no. 3, 287–302.

INSTITUTE OF MATHEMATICS, FACULTY OF MATHEMATICS, INFORMATICS AND MECHANICS, UNIVERSITY OF WARSAW, BANACHA 2, 02-097 WARSZAWA, POLAND
E-mail address: koras@mimuw.edu.pl

Centre de Recherches Mathématiques
CRM Proceedings and Lecture Notes
Volume **54**, 2011

Initial Algebras and the Jung – van der Kulk Theorem

Shigeru Kuroda

Dedicated to Professor Peter Russell on his seventieth birthday

ABSTRACT. An initial algebra is an algebra which is obtained from a subalgebra of some ambient algebra analogously to the initial ideal of an ideal in Gröbner basis theory.

To find hidden properties of subalgebras, it is sometimes useful to study their initial algebras. In general, however, it is not easy to determine the structure of the initial algebra of a given subalgebra.

In this paper, we determine the structure of the initial algebra of a certain subalgebra of a field. As an application of the main theorem, we reconstruct the proof of the Jung – van der Kulk theorem on polynomial automorphisms given by Makar-Limanov which was modified by Dicks and by Cohn.

1. Introduction: Initial ideals and initial algebras

In this section, we introduce initial algebras as a direct analogue of initial ideals which are common in Gröbner basis theory, and discuss the difference between initial ideals and initial algebras. We redefine the initial algebra in a more general setting in Section 2, where the main theorem is proved.

Throughout this paper, let k denote a field of an arbitrary characteristic. For $n \in \mathbb{N}$, let $k[\mathbf{x}] = k[x_1, \ldots, x_n]$ and $k[\mathbf{x}, \mathbf{x}^{-1}] = k[\mathbf{x}][x_1^{-1}, \ldots, x_n^{-1}]$ respectively be the polynomial ring and the Laurent polynomial ring in n variables over k. Let Ω be the set of total orders on $\mathbb{Z}^n$ such that $a \preceq b$ implies $a + c \preceq b + c$ for each $a, b, c \in \mathbb{Z}^n$. An element of Ω is called a *monomial order* if the zero vector is the least element of $(\mathbb{Z}_{\geq 0})^n$. Here, $\mathbb{Z}_{\geq 0}$ denotes the set of nonnegative integers. Assume that $f = \sum_{a \in \mathbb{Z}^n} \alpha_a \mathbf{x}^a$ is an element of $k[\mathbf{x}, \mathbf{x}^{-1}] \setminus \{0\}$, where $\alpha_a \in k$ and $\mathbf{x}^a = x_1^{a_1} \cdots x_n^{a_n}$ for each $a = (a_1, \ldots, a_n) \in \mathbb{Z}^n$. Then, we define $\mathrm{in}_{\preceq}(f) := \alpha_b \mathbf{x}^b$ for each $\preceq \in \Omega$, where $b = \max_{\preceq}\{a \in \mathbb{Z}^n \mid \alpha_a \neq 0\}$. For $\preceq \in \Omega$ and a k-vector subspace K of $k[\mathbf{x}, \mathbf{x}^{-1}]$, we define the *initial vector space* $\mathrm{in}_{\preceq}(K)$ to be the k-vector subspace of $k[\mathbf{x}, \mathbf{x}^{-1}]$ generated by $\mathrm{in}_{\preceq}(f)$ for $f \in K \setminus \{0\}$. If I is an ideal of $k[\mathbf{x}]$, then $\mathrm{in}_{\preceq}(I)$ is

2010 *Mathematics Subject Classification.* Primary 14R10; Secondary 13F20, 13F25.

Key words and phrases. Initial algebra, polynomial automorphism, tame generators problem.

Partly supported by the Grant-in-Aid for Young Scientists (B) 21740026, The Ministry of Education, Culture, Sports, Science and Technology, Japan.

This is the final form of the paper.

an ideal of $k[\mathbf{x}]$, since $\mathrm{in}_{\preceq}(fg) = \mathrm{in}_{\preceq}(f)\,\mathrm{in}_{\preceq}(g)$ holds for each $f, g \in k[\mathbf{x}, \mathbf{x}^{-1}] \setminus \{0\}$. The ideal $\mathrm{in}_{\preceq}(I)$ is called the *initial ideal* of I. Similarly, if A is a k-subalgebra of $k[\mathbf{x}, \mathbf{x}^{-1}]$, then $\mathrm{in}_{\preceq}(A)$ is a k-subalgebra of $k[\mathbf{x}, \mathbf{x}^{-1}]$, which is called the *initial algebra* of A.

In general, the initial ideal $\mathrm{in}_{\preceq}(I)$ is not generated by $\mathrm{in}_{\preceq}(f_1), \dots, \mathrm{in}_{\preceq}(f_r)$ even if the ideal I is generated by $f_1, \dots, f_r$. However, $\mathrm{in}_{\preceq}(I)$ is always finitely generated as an ideal of $k[\mathbf{x}]$, because $k[\mathbf{x}]$ is a Noetherian ring. It is possible that $\mathrm{in}_{\preceq}(I) = \mathrm{in}_{\preceq'}(I)$ holds for distinct $\preceq, \preceq \in \Omega$. In fact, for a fixed ideal I, there exist only finitely many distinct initial ideals $\mathrm{in}_{\preceq}(I)$ for monomial orders $\preceq$ of $\mathbb{Z}^n$ due to Schwartz [13], although there exist infinitely many monomial orders when $n \geq 2$. On the other hand, the initial algebra $\mathrm{in}_{\preceq}(A)$ is not always finitely generated over k even if the k-subalgebra A is finitely generated. Moreover, some finitely generated k-subalgebras have infinitely many distinct initial algebras. We have no general criterion to decide whether $\mathrm{in}_{\preceq}(A)$ is finitely generated over k, and no effective way to determine the number of $\mathrm{in}_{\preceq}(A)$'s. The initial algebras are thus quite difficult to study. We mention that Robbiano – Sweedler [12] and Kapur – Madlener [8] proposed the notion of a Subalgebra Analogue to the Gröbner Bases for Ideals (*SAGBI* basis) in the context of computer science in 1980s, where finite generation of initial algebras crucially relates to termination of computation.

For example, let S be an additive subsemigroup of $\mathbb{Z}^n$ with zero, and G a finite group acting faithfully on the semigroup S. Then, the action of G on S naturally extends to that on the k-subalgebra $k[S] := k[\{\mathbf{x}^a \mid a \in S\}]$ of $k[\mathbf{x}]$. Consider the initial algebras of the invariant ring $k[S]^G$. We call $\sigma \in G$ a *reflection* if the $\mathbb{R}$-vector subspace of $\mathbb{R}^n$ generated by $a \in S$ with $\sigma(a) = a$ is of codimension one in the $\mathbb{R}$-vector subspace of $\mathbb{R}^n$ generated by S. We call G a *reflection group* if G is generated by reflections in G.

The following theorems can be proved by the technique exactly the same as that used in [5].

Theorem 1.1 ([7, Theorem 3.2]). *The following assertions hold for each* $\preceq \in \Omega$:

(i) *Assume that the semigroup S is finitely generated. Then,* $\mathrm{in}_{\preceq}(k[S]^G)$ *is finitely generated over k if and only if G is a reflection group.*
(ii) *If the semigroup S is not finitely generated, then* $\mathrm{in}_{\preceq}(k[S]^G)$ *is not finitely generated over k for every G.*

Theorem 1.2 ([7, Theorem 3.3]). *The cardinality of* $\{\mathrm{in}_{\preceq}(k[S]^G) \mid \preceq \in \Omega\}$ *is equal to the cardinality of G if G is a reflection group, while uncountable otherwise.*

The assertions of the above theorems for nonreflection groups are obtained by an analysis of the topological properties of the subsets

$$U(\preceq) := \{\preceq' \in \Omega \mid \mathrm{in}_{\preceq'}(k[S]^G) = \mathrm{in}_{\preceq}(k[S]^G)\} \quad (\preceq \in \Omega)$$

of Ω as follows. Here, we define a certain structure of metric space on Ω, for which Ω is compact. It holds that $U(\preceq)$ is an open subset of Ω if the k-algebra $\mathrm{in}_{\preceq}(k[S]^G)$ is finitely generated (cf. [7, Lemma 1.6]). When G is a nonreflection group, it is shown that $U(\preceq)$ is a nowhere dense subset of Ω, and hence is not open for every $\preceq \in \Omega$. Thus, it follows that $\mathrm{in}_{\preceq}(k[S]^G)$ is not finitely generated for every $\preceq \in \Omega$. Moreover, Baire's category theorem implies that $\{U(\preceq) \mid \preceq \in \Omega\}$ is an uncountable set, since Ω is a compact metric space. The cardinality of $\{\mathrm{in}_{\preceq}(k[S]^G) \mid \preceq \in \Omega\}$

is equal to that of $\{U(\preceq) \mid \preceq \in \Omega\}$, and therefore is uncountable. For the same reason, $\{\mathrm{in}_{\preceq}(k[S]^G) \mid \preceq \in \Omega'\}$ is an uncountable set if Ω' is a nonempty complete subset of Ω such that $\Omega' \cap U(\preceq)$ is nowhere dense in Ω' for each $\preceq \in \Omega'$.

Let $\mathbf{S}$ be the set of $\mathbf{w} = (w_1, \ldots, w_n) \in \mathbb{R}^n$ such that $w_1, \ldots, w_n$ are linearly independent over $\mathbb{Q}$ and $\sum_{i=1}^n w_i^2 = 1$. We remark that, for each $\mathbf{w} \in \mathbf{S}$, an element $\preceq_{\mathbf{w}}$ of Ω is defined by $a \preceq_{\mathbf{w}} b$ if $a \cdot \mathbf{w} \leq b \cdot \mathbf{w}$ for each $a, b \in \mathbb{Z}^n$. Then, $\iota \colon \mathbf{S} \ni \mathbf{w} \mapsto \preceq_{\mathbf{w}} \in \Omega$ is an injective continuous map such that $\iota(\mathbf{S})$ is dense in Ω, where we regard $\mathbf{S}$ as a subspace of the Euclidean space $\mathbb{R}^n$ (cf. [7, Theorem 1.1]). What is actually proved is that $\iota^{-1}(U(\preceq_{\mathbf{w}}))$ is nowhere dense in $\mathbf{S}$ for each $\mathbf{w} \in \mathbf{S}$, from which it follows that $U(\preceq)$ is nowhere dense in Ω for each $\preceq \in \Omega$. The set Ω_0 of monomial orders of $\mathbb{Z}^n$ is equal to the closure of $\iota(\mathbf{S} \cap (\mathbb{R}_{\geq 0})^n)$ in Ω. Hence, Ω_0 is compact since so is Ω. Moreover, it follows that $\Omega_0 \cap U(\preceq)$ is nowhere dense in Ω_0 for each $\preceq \in \Omega_0$. Therefore, $\{\mathrm{in}_{\preceq}(k[S]^G) \mid \preceq \in \Omega_0\}$ is also an uncountable set if G is a nonreflection group.

Göbel [3] first showed the special case of Theorem 1.1 where $\preceq$ is the lexicographic order and G acts on $S = (\mathbb{Z}_{\geq 0})^n$ by permutation of coordinates. Following Göbel, the author [5], [7] showed the special case of the above theorems where G acts on $S = (\mathbb{Z}_{\geq 0})^n$ and $S = \mathbb{Z}^n$ by permutation of coordinates, which were generalized in [6] as in the above theorems. Results similar to Theorem 1.1 are also obtained by Reichstein [11] and Thiéry–Thomassé [15] independently.

The author would like to thank the referee for helpful comments.

2. A structure theorem for initial algebra

We define the initial algebra in a more general setting. Let L be an extension field of k, Γ a totally ordered abelian group, and $\delta \colon L \to \Gamma \cup \{-\infty\}$ a *degree function* over k, i.e., a map satisfying

(i) $\delta(k^\times) = \{0\}$;
(ii) $\delta(a+b) \leq \max\{\delta(a), \delta(b)\}$ for each $a, b \in L$;
(iii) $\delta(ab) = \delta(a) + \delta(b)$ for each $a, b \in L$;
(iv) $\delta(a) = -\infty$ if and only if $a = 0$,

where $-\infty$ is a symbol which is less than any element of Γ. We note that the notion of degree functions on L over k is equivalent to the notion of additive valuations of L over k. Actually, if δ is a degree function, then $L \ni a \mapsto -\delta(a) \in \Gamma \cup \{\infty\}$ is an additive valuation. Similarly, we obtain a degree function from an additive valuation.

For each $\gamma \in \Gamma$, the subsets $\{a \in L \mid \delta(a) \leq \gamma\}$ and $\{a \in L \mid \delta(a) < \gamma\}$ of L form k-vector subspaces of L. Actually, these subsets are closed under addition by (ii) and (iv), and under scalar multiplication by (i), (iii) and (iv). We define L_γ to be the quotient of the k-vector space $\{a \in L \mid \delta(a) \leq \gamma\}$ by the subspace $\{a \in L \mid \delta(a) < \gamma\}$. Then, $\mathrm{gr}\, L := \bigoplus_{\gamma \in \Gamma} L_\gamma$ naturally becomes a graded k-algebra. For each $a \in L \setminus \{0\}$, we define the *initial form* a^δ of a to be the image of a in $L_{\delta(a)}$. Then, $(ab)^\delta = a^\delta b^\delta$ and $(a/b)^\delta = a^\delta / b^\delta$ hold for each $a, b \in L \setminus \{0\}$ by (iii). For each nonzero elements $f \in L_\lambda$ and $g \in L_\mu$, we have $f = a^\delta$ and $g = b^\delta$ for some $a, b \in L \setminus \{0\}$ with $\delta(a) = \lambda$ and $\delta(b) = \mu$. Hence, $fg = a^\delta b^\delta = (ab)^\delta$ is a nonzero element of $L_{\lambda+\mu}$. This shows that $\mathrm{gr}\, L$ is an integral domain. Observe that $\delta(a+b) = \max\{\delta(a), \delta(b)\}$ holds for $a, b \in L$ with $\delta(a) \neq \delta(b)$. In fact, supposing that $\delta(b)$ is greater than $\delta(a)$ and $\delta(a+b)$, we have $\delta((a+b)-a) \leq \max\{\delta(a+b), \delta(a)\}$

by (ii), a contradiction. It holds that $\delta(a+b) < \max\{\delta(a), \delta(b)\}$ if and only if $a^\delta + b^\delta = 0$ for $a, b \in L \setminus \{0\}$.

Now, for a k-subalgebra A of L, we define the *initial algebra* A^δ of A to be the k-vector subspace of $\operatorname{gr} L$ generated by a^δ for $a \in A \setminus \{0\}$, which forms a k-subalgebra of $\operatorname{gr} L$. We show that

$$(1) \qquad k[f_1, \dots, f_r]^\delta = k[f_1^\delta, \dots, f_r^\delta]$$

holds for $f_1, \dots, f_r \in L \setminus \{0\}$ if $f_1^\delta, \dots, f_r^\delta$ are algebraically independent over k. Obviously, the right-hand side is contained in the left-hand side. For the converse, take any

$$\phi = \sum_{i_1, \dots, i_r} \alpha_{i_1, \dots, i_r} f_1^{i_1} \cdots f_r^{i_r} \in k[f_1, \dots, f_r] \setminus \{0\}$$

with $\alpha_{i_1, \dots, i_r} \in k$, and let μ be the maximum among $\delta(f_1^{i_1} \cdots f_r^{i_r})$ for $i_1, \dots, i_r$ with $\alpha_{i_1, \dots, i_r} \neq 0$, and ϕ_0 the sum of $\alpha_{i_1, \dots, i_r} (f_1^\delta)^{i_1} \cdots (f_r^\delta)^{i_r}$ for $i_1, \dots, i_r$ with $\delta(f_1^{i_1} \cdots f_r^{i_r}) = \mu$. Then, ϕ_0 is not zero by the assumption that $f_1^\delta, \dots, f_r^\delta$ are algebraically independent over k. By the definition of ϕ_0, it follows that $\phi^\delta = \phi_0$, which belongs to the right-hand side of (1). Thus, the left-hand side of (1) is contained in the right-hand side if $f_1^\delta, \dots, f_r^\delta$ are algebraically independent over k.

When $L = k(\mathbf{x})$ is the field of fractions of $k[\mathbf{x}]$, the following degree function $\deg_{\mathbf{w}} \colon k(\mathbf{x}) \to \Gamma \cup \{-\infty\}$ defined by an n-tuple $\mathbf{w} = (w_1, \dots, w_n)$ of elements of Γ is important: Define $\deg_{\mathbf{w}} x_1^{a_1} \cdots x_n^{a_n} := \sum_{i=1}^n a_i w_i$ for $a_1, \dots, a_n \in \mathbb{Z}$. For $f = \sum_{a \in \mathbb{Z}^n} \alpha_a \mathbf{x}^a \in k[\mathbf{x}, \mathbf{x}^{-1}]$, define $\deg_{\mathbf{w}} f$ to be the maximum among $\deg_{\mathbf{w}} \mathbf{x}^a$ for $a \in \mathbb{Z}^n$ with $\alpha_a \neq 0$. If $f = g/h \in k(\mathbf{x})$ with $g, h \in k[\mathbf{x}]$ and $h \neq 0$, then define $\deg_{\mathbf{w}} f := \deg_{\mathbf{w}} g - \deg_{\mathbf{w}} h$, which does not depend on the choice of g and h. Then, $\deg_{\mathbf{w}}$ is a degree function of $k(\mathbf{x})$. We remark that $\operatorname{gr}_{\mathbf{w}} k(\mathbf{x}) := \operatorname{gr} k(\mathbf{x})$ is identified with the localization of $k[\mathbf{x}]$ by the set of nonzero $\mathbf{w}$-homogeneous elements of $k[\mathbf{x}]$. Here, we say that $f = \sum_{a \in \mathbb{Z}^n} \alpha_a \mathbf{x}^a \in k[\mathbf{x}, \mathbf{x}^{-1}]$ is $\mathbf{w}$-*homogeneous* if $\deg_{\mathbf{w}} \mathbf{x}^a = \deg_{\mathbf{w}} f$ for every $a \in \mathbb{Z}^n$ with $\alpha_a \neq 0$. In fact, $L_\gamma \setminus \{0\}$ is identified with the set of g/h for $\mathbf{w}$-homogeneous elements $g, h \in k[\mathbf{x}] \setminus \{0\}$ with $\deg_{\mathbf{w}} g - \deg_{\mathbf{w}} h = \gamma$ for each γ. With this identification, $f^{\mathbf{w}} := f^{\deg_{\mathbf{w}}}$ is regarded as the highest $\mathbf{w}$-homogeneous part of f, i.e., the sum of $\alpha_a \mathbf{x}^a$ for $a \in \mathbb{Z}^n$ with $\deg_{\mathbf{w}} \mathbf{x}^a = \deg_{\mathbf{w}} f$, if $f = \sum_{a \in \mathbb{Z}^n} \alpha_a \mathbf{x}^a$, and as $g^{\mathbf{w}}/h^{\mathbf{w}}$ if $f = g/h$ with $g, h \in k[\mathbf{x}]$ and $h \neq 0$. We denote the initial algebra of A for the degree function $\deg_{\mathbf{w}}$ by $A^{\mathbf{w}}$ instead of $A^{\deg_{\mathbf{w}}}$.

Note that each $\preceq \in \Omega$ gives the structure of a totally ordered additive group on $\mathbb{Z}^n$. Let $\mathbf{w} = (\mathbf{e}_1, \dots, \mathbf{e}_n) \in (\mathbb{Z}^n)^n$, where $\mathbf{e}_1, \dots, \mathbf{e}_n$ are the coordinate unit vectors of $\mathbb{R}^n$. Then, with the above identification, $\operatorname{gr}_{\mathbf{w}} k(\mathbf{x})$ is regarded as $k[\mathbf{x}, \mathbf{x}^{-1}]$. Moreover, we can regard $f^{\mathbf{w}} = \operatorname{in}_\preceq(f)$ for each $f \in k[\mathbf{x}, \mathbf{x}^{-1}] \setminus \{0\}$, and $A^{\mathbf{w}} = \operatorname{in}_\preceq(A)$ for each k-subalgebra A of $k[\mathbf{x}, \mathbf{x}^{-1}]$.

Now, let M be a subfield of L containing k, and g an element of L. Then, $M[g]$ is clearly a k-subalgebra of L. We consider the structure of the initial algebra $M[g]^\delta$ of $M[g]$ in the case where $Q(M[g]^\delta)$ is a transcendental extension of $Q(M^\delta)$. Here, $Q(A)$ denotes the field of fractions of A for each integral domain A. In this case, we can find a polynomial $\phi_0(t) \in M[t]$ such that $\phi_0(g)^\delta$ is transcendental over $Q(M^\delta)$. Among such polynomials, choose ϕ_0 so that $d := \deg_t \phi_0$ is minimal, and define $q := \phi_0(g)^\delta$ and

$$(2) \qquad \mathcal{A} := k[\{\psi(g)^\delta \mid \psi \in M[t] \setminus \{0\}, \deg_t \psi < d\}].$$

By the minimality of d, every element of $\mathcal{A}$ is algebraic over $Q(M^\delta)$. Hence, q is transcendental over $Q(\mathcal{A})$, since q is transcendental over $Q(M^\delta)$. Thus, $\mathcal{A}[q]$ is a polynomial ring in q over $\mathcal{A}$.

Here is the main theorem.

Theorem 2.1. *Let $k \subset M \subset L$ be extensions of fields, and δ a degree function on L over k. If $Q(M[g]^\delta)$ is a transcendental extension of $Q(M^\delta)$ for $g \in L$, then it holds that $M[g]^\delta = \mathcal{A}[q]$.*

PROOF. Clearly, $\mathcal{A}[q]$ is contained in $M[g]^\delta$, since so are $\mathcal{A}$ and q by definition. We show the reverse inclusion by contradiction. Supposing the contrary, we find $\psi \in M[t]$ such that $\psi(g)^\delta$ does not belong to $\mathcal{A}[q]$. Among such polynomials, choose ψ so that $e := \deg_t \psi$ is minimal. Then, e is at least d, for otherwise $\psi(g)^\delta$ would belong to $\mathcal{A}$. Let θ_0 and θ_1 be the quotient and remainder of ψ divided by ϕ_0, respectively. Then, since $\deg_t \theta_0 = e - d < e$, it follows that $\theta_0(g)^\delta$ belongs to $\mathcal{A}[q]$ by the minimality of e. Since $\deg_t \theta_1 < d$, it follows that $\theta_1(g)^\delta$ belongs to $\mathcal{A}$ by the definition of $\mathcal{A}$. From the equation $\psi = \phi_0\theta_0 + \theta_1$, we get

$$\psi(g) = \phi_0(g)\theta_0(g) + \theta_1(g) \tag{3}$$

by substituting g for t.

First, assume that $\delta(\psi(g)) < \delta(\theta_1(g))$. Then, we have $(\theta_0(g)\phi_0(g))^\delta + \theta_1(g)^\delta = 0$ by (3), for otherwise

$$\delta(\psi(g)) = \delta(\phi_0(g)\theta_0(g) + \theta_1(g)) = \max\{\delta(\phi_0(g)\theta_0(g)), \delta(\theta_1(g))\} \geq \delta(\theta_1(g)),$$

a contradiction to the assumption. Hence, $(\theta_0(g)\phi_0(g))^\delta$ is equal to $-\theta_1(g)^\delta$, which belongs to $\mathcal{A}$ as mentioned. On the other hand, we have

$$(\theta_0(g)\phi_0(g))^\delta = \theta_0(g)^\delta\phi_0(g)^\delta = \theta_0(g)^\delta q,$$

which belongs to $q\mathcal{A}[q]$ since $\theta_0(g)^\delta$ belongs to $\mathcal{A}[q]$ as mentioned. Because q is transcendental over $Q(\mathcal{A})$, this implies that $(\theta_0(g)\phi_0(g))^\delta$ does not belong to $\mathcal{A}$, a contradiction.

Next, assume that $\delta(\psi(g)) \geq \delta(\theta_1(g))$. Then, we get $\delta(\psi(g)) \geq \delta(\theta_0(g)\phi_0(g))$ by (3). In fact, otherwise we have $\delta(\theta_0(g)\phi_0(g)) > \delta(\psi(g)) \geq \delta(\theta_1(g))$, implying that the δ-degree of the right-hand side of (3) is greater than that of the left-hand side. Similarly, we have either $\delta(\psi(g)) = \delta(\theta_0(g)\phi_0(g))$ or $\delta(\psi(g)) = \delta(\theta_1(g))$, for otherwise the δ-degree of the left-hand side of (3) is greater than that of the right-hand side. It is easy to check that $\psi(g)^\delta$ is equal to

(i) $(\theta_0(g)\phi_0(g))^\delta$ if $\delta(\psi(g)) = \delta(\theta_0(g)\phi_0(g))$ and $\delta(\psi(g)) > \delta(\theta_1(g))$;
(ii) $\theta_1(g)^\delta$ if $\delta(\psi(g)) > \delta(\theta_0(g)\phi_0(g))$ and $\delta(\psi(g)) = \delta(\theta_1(g))$;
(iii) $(\theta_0(g)\phi_0(g))^\delta + \theta_1(g)^\delta$ if $\delta(\psi(g)) = \delta(\theta_0(g)\phi_0(g)) = \delta(\theta_1(g))$,

which all belong to $\mathcal{A}[q]$, since so do $\phi_0(g)^\delta = q$ and $\theta_i(g)^\delta$ for $i = 0, 1$. This contradicts the supposition that $\psi(g)^\delta$ does not belong to $\mathcal{A}[q]$.

Therefore, $\mathcal{A}[q]$ is contained in $M[g]^\delta$, and thereby proving $M[g]^\delta = \mathcal{A}[q]$. □

Let us look into the structure of $\mathcal{A}$ in detail. Set

$$\mathcal{A}_i := k[\{\psi(g)^\delta \mid \psi \in M[t] \setminus \{0\}, \deg_t \psi < i\}] \tag{4}$$

for $i = 1, \ldots, d$. Then, we have $M^\delta = \mathcal{A}_1 \subset \mathcal{A}_2 \subset \cdots \subset \mathcal{A}_d = \mathcal{A}$. Moreover, $\mathcal{A}_{i+1}$ is generated by $\psi(g)^\delta$ for $\psi \in M[t]$ with $\deg_t \psi = i$ over $\mathcal{A}_i$ for each i. More precisely, we have the following lemma.

Lemma 2.2. *If $\mathcal{A}_{i+1} \neq \mathcal{A}_i$ for $i \in \{1, \ldots, d-1\}$, then $\mathcal{A}_{i+1} = \mathcal{A}_i[\psi(g)^\delta]$ holds for every $\psi \in M[t]$ such that $\deg_t \psi = i$ and $\psi(g)^\delta$ does not belong to $\mathcal{A}_i$. Hence, $\mathcal{A}$ is generated by at most $d-1$ elements of $\bigcup_{\gamma \in \Gamma} L_\gamma$ over M^δ.*

PROOF. Let S_i be the set of $\psi(g)^\delta$ for $\psi \in M[t]$ with $\deg_t \psi = i$. Then, $\mathcal{A}_{i+1}$ is generated by S_i over $\mathcal{A}_i$. We show that, for each $h_1, h_2 \in S_i \setminus \mathcal{A}_i$, there exist $c \in M \setminus \{0\}$ and $h' \in \mathcal{A}_i$ such that $h_1 = c^\delta h_2 + h'$. Then, it follows that $\mathcal{A}_{i+1} = \mathcal{A}_i[h]$ for every $h \in S_i \setminus \mathcal{A}_i$. Let $\psi_l \in M[t]$ be such that $\deg_t \psi_l = i$ and $\psi_l(g)^\delta = h_l$ for $l = 1, 2$, and set $\psi := \psi_1 - (c_1/c_2)\psi_2$, where $c_l \in M\setminus\{0\}$ is the leading coefficient of ψ_l for $l = 1, 2$. Then, $\deg_t \psi$ is less than i, and so $\psi(g)^\delta = (\psi_1(g) - (c_1/c_2)\psi_2(g))^\delta$ belongs to $\mathcal{A}_i$. Since $\psi_1(g)^\delta = h_1$ and $((c_1/c_2)\psi_2(g))^\delta = (c_1/c_2)^\delta h_2$ do not belong to $\mathcal{A}_i$, it follows that $\delta(\psi_1(g)) = \delta((c_1/c_2)\psi_2(g))$. Hence, we get $(\psi_1(g) - (c_1/c_2)\psi_2(g))^\delta = h_1 - (c_1/c_2)^\delta h_2$, and therefore $h_1 = (c_1/c_2)^\delta h_2 + \psi(g)^\delta$. This proves the first assertion. The last assertion is immediate. □

We call an element p of a k-domain A a *proper power* in A if $p = \alpha q^m$ for some $\alpha \in k^\times$, $q \in A \setminus k$ and $m \geq 2$. Note that, when A is a unique factorization domain with $A^\times = k^\times$, each $h \in A \setminus \{0\}$ can be expressed as $h = \alpha p^m$, where $\alpha \in k^\times$, $m \in \mathbb{Z}_{\geq 0}$, and $p \in A$ which is not a proper power in A. We say that $\lambda, \mu \in \Gamma$ are *linearly dependent* if λ and μ are linearly dependent over $\mathbb{Z}$ as elements of the $\mathbb{Z}$-module Γ.

Lemma 2.3. *Assume that k is algebraically closed in $Q(\operatorname{gr} L)$. Let $h \in L_\lambda$ and $p \in L_\mu$ be such that h is algebraic over $k(p)$, where $\lambda, \mu \in \Gamma$ with $\mu \neq 0$. Then, λ and μ are linearly dependent. Moreover, the following assertions hold:*

(i) *If $\lambda = 0$, then h belongs to k.*
(ii) *If $\lambda \neq 0$, then there exist $\alpha \in k^\times$ and $l, m \in \mathbb{Z}$ with $l \neq 0$, $m \geq 1$ and $\gcd(|l|, m) = 1$ such that $h^m = \alpha p^l$.*

In particular, if p is not a proper power in $\operatorname{gr} L$, then h belongs to $k[p, p^{-1}]$.

PROOF. By assumption, h is a root of a monic polynomial over $k(p)$, and hence is a root of a nonzero polynomial over $k[p]$. Write $\sum_{i,j} \alpha_{i,j} h^i p^j = 0$, where $\alpha_{i,j} \in k$ for each (i, j) such that $\alpha_{i_0,j_0} \neq 0$ for some (i_0, j_0). For each $\gamma \in \Gamma$, the sum of $\alpha_{i,j} h^i p^j$ for (i, j) with $i\lambda + j\mu = \gamma$ must be zero, since $\sum_{i,j} \alpha_{i,j} h^i p^j$ is an element of the graded ring $\operatorname{gr} L$, and p and h belong to L_λ and L_μ, respectively. Hence, without loss of generality, we may assume that the sum in $\sum_{i,j} \alpha_{i,j} h^i p^j = 0$ is taken over (i, j) with $i\lambda + j\mu = \gamma$, where $\gamma := i_0\lambda + j_0\mu$.

If $\lambda = 0$, then we have $\sum_{i,j} \alpha_{i,j} h^i p^j = p^{j_0} \sum_i \alpha_{i,j_0} h^i$. Hence, $\sum_i \alpha_{i,j_0} h^i = 0$, and so h is algebraic over k. Since k is algebraically closed in $Q(\operatorname{gr} L)$ by assumption, it follows that h belongs to k. Trivially, $\lambda = 0$ and μ are linearly dependent, and h belongs to $k[p, p^{-1}]$.

Assume that $\lambda \neq 0$. Since $\sum_{i,j} \alpha_{i,j} h^i p^j = 0$, there exists $(i_1, j_1) \neq (i_0, j_0)$ such that $\alpha_{i_1,j_1} \neq 0$. Then, we have $(i_1 - i_0)\mu + (j_1 - j_0)\lambda = 0$, so λ and μ are linearly dependent. Since λ and μ are nonzero, it follows that $i_1 - i_0$ and $j_1 - j_0$ are nonzero. So, we can find $l, m \in \mathbb{Z}$ with $l \neq 0$, $m \geq 1$ and $\gcd(|l|, m) = 1$ such that $l\mu = m\lambda$. Then, $h_1 := h^m/p^l$ belongs to L_0. Moreover, h_1 is algebraic over $k(p)$, since so is h by assumption. Thus, thanks to (i), we get $h_1 = \alpha$ for some $\alpha \in k^\times$, and therefore $h^l = \alpha p^m$. If p is not a proper power in $\operatorname{gr} L$, then $l = 1$ or $l = -1$, and so h belongs to $k[p, p^{-1}]$. □

We note that k is algebraically closed in $Q(\operatorname{gr}_{\mathbf{w}} k(\mathbf{x}))$ for any Γ and $\mathbf{w} \in \Gamma^n$, since $Q(\operatorname{gr}_{\mathbf{w}} k(\mathbf{x}))$ is identified with a k-subalgebra of $k(\mathbf{x})$. In general, however, k is not necessarily algebraically closed in $Q(\operatorname{gr} L)$ even if k is algebraically closed in L. For instance, let k' be a field, and consider the subfields $k := k'(x_1^2)$ and $L := k(x_1+x_2)$ of $k'(x_1, x_2)$. We define the degree function δ on L as the restriction of $\deg_{\mathbf{w}} \colon k'(x_1, x_2) \to \mathbb{Z} \cup \{-\infty\}$ for $\mathbf{w} = (0, -1)$. Then, k is algebraically closed in L, since L is a rational function field in one variable over k. On the other hand, $h := (x_1 + x_2)^\delta \in L_0$ does not belong to $k \subset L_0$, since $\delta(x_1 + x_2 + f) = 0$ for each $f \in k$. However, $h^2 - x_1^2 = 0$ holds in $\operatorname{gr} L$ since $\delta((x_1 + x_2)^2 - x_1^2) < 0$. Hence, h is algebraic over k. Therefore, k is not algebraically closed in $Q(\operatorname{gr} L)$.

To end this section, we consider the case where $M = k(f)$ for some $f \in L$ such that $\delta(f) \neq 0$. Observe that

$$p^\delta = \begin{cases} \alpha_m (f^\delta)^m & \text{if } \delta(f) > 0 \\ \alpha_l (f^\delta)^l & \text{if } \delta(f) < 0 \end{cases}$$

for each $p = \sum_{i=l}^m \alpha_i f^i \in k[f] \setminus \{0\}$, where $\alpha_l, \ldots, \alpha_m \in k$ with $l \leq m$, $\alpha_l \neq 0$ and $\alpha_m \neq 0$. Hence, q^δ belongs to $k[f^\delta, (f^\delta)^{-1}]$ for every $q \in k(f) \setminus \{0\}$. Since f^δ and $(f^\delta)^{-1}$ belong to $k(f)^\delta$, we get $k(f)^\delta = k[f^\delta, (f^\delta)^{-1}]$, and therefore $Q(k(f)^\delta) = k(f^\delta)$.

The statement (i) of the following proposition will be used in Section 3 to derive the Jung – van der Kulk theorem. This is a consequence of only Theorem 2.1 and Lemma 2.3.

Proposition 2.4. *Assume that k is algebraically closed in $Q(\operatorname{gr} L)$. Let $f, g \in L \setminus \{0\}$ be such that $\delta(f) \neq 0$ and $Q(k[f, g]^\delta)$ is a transcendental extension of $k(f^\delta)$. Then, there exists $h \in k[f, g] \setminus \{0\}$ such that the following assertions hold for $q := h^\delta$:*

(i) *Assume that $p \in \operatorname{gr} L$ is such that $f^\delta = \alpha p^l$ for some $\alpha \in k^\times$ and $l \in \mathbb{Z}$, and is not a proper power in $\operatorname{gr} L$. Then, $k[f, g]^\delta$ is contained in $k[p, p^{-1}, q]$.*

(ii) *There exists $p \in L_\mu \setminus \{0\}$ for some $\mu \in \Gamma \setminus \{0\}$ such that $k[f, g]^\delta$ is contained in $k[p, p^{-1}, q]$ and $f^\delta = \alpha p^l$ for some $\alpha \in k^\times$ and $l \in \mathbb{Z}$ with $l \geq 1$.*

PROOF. Since $Q(k[f, g]^\delta)$ is a transcendental extension of $k(f^\delta)$ by assumption, so is $Q(k(f)[g]^\delta)$. Hence, we can find $\phi \in k(f)[t]$ such that $\phi(g)^\delta$ is transcendental over $k(f^\delta)$. Among such polynomials, choose ϕ_0 so that $d := \deg_t \phi_0$ is minimal, and take $c \in k[f] \setminus \{0\}$ such that $c\phi_0$ belongs to $k[f][t]$. Then, $h := c\phi_0(g)$ belongs to $k[f, g]$, $\deg_t(c\phi_0)$ is equal to d, and $q := h^\delta = c^\delta \phi_0(g)^\delta$ is transcendental over $k(f^\delta)$. Applying Theorem 2.1 with $M = k(f)$, we get $k(f)[g]^\delta = \mathcal{A}[q]$, where $\mathcal{A}$ is a k-subalgebra of $\operatorname{gr} L$ defined as in (2) for $M = k(f)$. Since $k[f, g]$ is contained in $k(f)[g]$, it follows that $k[f, g]^\delta$ is contained in $k(f)[g]^\delta$, and therefore contained in $\mathcal{A}[q]$.

To show (i), it suffices to check that $\mathcal{A}$ is contained in $k[p, p^{-1}]$. Since the k-algebra $\mathcal{A}$ is generated by $\psi(g)^\delta$ for $\psi \in k(f)[t] \setminus \{0\}$ with $\deg_t \psi < d$, we verify that each $\psi(g)^\delta$ belongs to $k[p, p^{-1}]$ using the last statement of Lemma 2.3. Note that k is algebraically closed in $Q(\operatorname{gr} L)$ by assumption, and $\psi(g)^\delta$ is an element of $L_\lambda \setminus \{0\}$ which is algebraic over $k(f^\delta)$ since $\deg_t \psi < d$, where $\lambda = \delta(\psi(g))$. Since $f^\delta = \alpha p^l$ by assumption, $k(f^\delta)$ is an algebraic extension of $k(p)$. Hence, $\psi(g)^\delta$ is algebraic over $k(p)$. By assumption, p is not a proper power in $\operatorname{gr} L$, and is an element of $L_\mu \setminus \{0\}$, where $\mu \in \Gamma$ is such that $l\mu = \delta(f)$. Since $\delta(f) \neq 0$ by assumption, we get

$\mu \neq 0$. Thus, by the last statement of Lemma 2.3, we conclude that $\psi(g)^\delta$ belongs to $k[p, p^{-1}]$. Therefore, $\mathcal{A}$ is contained in $k[p, p^{-1}]$, proving (i).

Similarly, to show (ii), it suffices to find $p \in L_\mu \setminus \{0\}$ for some $\mu \in \Gamma \setminus \{0\}$ such that $\mathcal{A}$ is contained in $k[p, p^{-1}]$, and $f^\delta = \alpha p^l$ for some $\alpha \in k^\times$ and $l \in \mathbb{Z}$ with $l \geq 1$. By Lemma 2.2, there exists $p_i \in L_{\gamma_i}$ for some $\gamma_i \in \Gamma$ for $i = 1, \ldots, d-1$ such that $\mathcal{A}$ is generated by $p_1, \ldots, p_{d-1}$ over $k(f)^\delta = k[f^\delta, (f^\delta)^{-1}]$. Since every element of $\mathcal{A}$ is algebraic over $k(f^\delta)$, it follows that p_i and f^δ are algebraically dependent over k for each i. Hence, γ_i and $\delta(f)$ are linearly dependent by Lemma 2.3, since $\delta(f) \neq 0$ by assumption. Let Γ_1 be the $\mathbb{Z}$-submodule of Γ generated by $\delta(f)$ and γ_i for $i = 1, \ldots, d-1$. Then, the torsion part of Γ_1 is trivial, for otherwise we have $0 < \eta < t\eta = 0$ for some $\eta \in \Gamma_1$ and $t \geq 2$, a contradiction. Hence, we conclude that Γ_1 is a free $\mathbb{Z}$-module of rank one. Let $\lambda \neq 0$ be a basis Γ_1. Replacing λ with $-\lambda$ if necessary, we may assume that $\delta(f) = l\lambda$ for some $l \geq 1$. Write $\lambda = m\delta(f) + \sum_i m_i \gamma_i$, where $m, m_i \in \mathbb{Z}$ for each i, and define $p := (f^\delta)^m \prod_i p_i^{m_i}$. Then, p is a nonzero element of L_λ which is algebraic over $k(f^\delta)$. Hence, f^δ / p^l is a nonzero element of L_0 which is algebraic over $k(f^\delta)$. By Lemma 2.3 (i), it follows that $f^\delta / p^l = \alpha$ for some $\alpha \in k^\times$, and so $f^\delta = \alpha p^l$. For each i, take $l_i \in \mathbb{Z}$ such that $\gamma_i = l_i \lambda$. Then, we have $p_i = \alpha_i p^{l_i}$ for some $\alpha_i \in k^\times$ similarly. Thus, f^δ, $(f^\delta)^{-1}$ and $p_1, \ldots, p_{d-1}$ are contained in $k[p, p^{-1}]$. Therefore, $\mathcal{A}$ is contained in $k[p, p^{-1}]$, proving (ii). □

3. The Jung – van der Kulk theorem

The study of the automorphism group $\mathrm{Aut}_k\, k[\mathbf{x}]$ of $k[\mathbf{x}]$ over k is of great interest in affine algebraic geometry. In this section, we study the structure of this group using the results in the preceding section.

Let F be an endomorphism of $k[\mathbf{x}]$ over k. Since F is a substitution map $x_i \mapsto f_i$ for $i = 1, \ldots, n$, we identify F with the n-tuple $(f_1, \ldots, f_n)$ of polynomials, where $f_i := F(x_i)$ for each i. Then, F is an automorphism of $k[\mathbf{x}]$ over k if and only if $k[\mathbf{x}]$ is generated by $f_1, \ldots, f_n$ over k. For $\mathbf{w} \in \Gamma^n$, consider the degree function $\deg_{\mathbf{w}}$ on $k(\mathbf{x})$. We note that, if $f_1^{\mathbf{w}}, \ldots, f_n^{\mathbf{w}}$ are algebraically independent over k for $F \in \mathrm{Aut}_k\, k[\mathbf{x}]$, then $F^{\mathbf{w}} := (f_1^{\mathbf{w}}, \ldots, f_n^{\mathbf{w}})$ also belongs to $\mathrm{Aut}_k\, k[\mathbf{x}]$, since

$$k[f_1^{\mathbf{w}}, \ldots, f_n^{\mathbf{w}}] = k[f_1, \ldots, f_n]^{\mathbf{w}} = k[\mathbf{x}]^{\mathbf{w}} = k[\mathbf{x}]$$

by (1). For $F \in \mathrm{Aut}_k\, k[\mathbf{x}]$, we define $\deg_{\mathbf{w}} F := \sum_{i=1}^n \deg_{\mathbf{w}} f_i$. When $\Gamma = \mathbb{Z}$ and $\mathbf{w}_0 := (1, \ldots, 1)$, we simply denote "$\deg_{\mathbf{w}_0}$" by "deg". Then, since $\deg f_i \geq 1$ for each i, we have $\deg F \geq n$ for every $F \in \mathrm{Aut}_k\, k[\mathbf{x}]$. We say that $F \in \mathrm{Aut}_k\, k[\mathbf{x}]$ is *affine* if $\deg F = n$, in which case there exist $(a_{i,j})_{i,j} \in GL_n(k)$ and $(b_i)_i \in k^n$ such that $f_i = \sum_{j=1}^n a_{i,j} x_j + b_i$ for each i. We remark that $f_1^{\mathbf{w}_0}, \ldots, f_n^{\mathbf{w}_0}$ are algebraically dependent over k if F is not affine. Actually, the k-algebra $k[\mathbf{x}]$ cannot be generated by the $\mathbf{w}_0$-homogeneous elements $f_1^{\mathbf{w}_0}, \ldots, f_n^{\mathbf{w}_0}$ if $\deg f_i > 1$ for some i. We say that $F \in \mathrm{Aut}_k\, k[\mathbf{x}]$ is *elementary* if there exists $i \in \{1, \ldots, n\}$ such that $F(x_l) = x_l$ for $l \neq i$, and $F(x_i) = x_i + g$ for some $g \in k[\mathbf{x}]$. Then, g necessarily belongs to $k[\{x_l \mid l \in S_i\}]$, where $S_i := \{1, \ldots, n\} \setminus \{i\}$. We say that $F \in \mathrm{Aut}_k\, k[\mathbf{x}]$ admits an *elementary reduction* if there exists $i \in \{1, \ldots, n\}$ such that $f_i^{\mathbf{w}_0}$ belongs to $k[\{f_l \mid l \in S_i\}]^{\mathbf{w}_0}$. If this is the case, then $\deg F \circ E < \deg F$ holds for the elementary automorphism E defined by $E(x_l) = x_l$ for $l \neq i$, and $E(x_i) = x_i - g$ for $g \in k[\{f_l \mid l \in S_i\}]$ with $g^{\mathbf{w}_0} = f^{\mathbf{w}_0}$.

Now, let $\mathrm{T}_k\, k[\mathbf{x}]$ denote the subgroup of $\mathrm{Aut}_k\, k[\mathbf{x}]$ generated by all the affine automorphisms and elementary automorphisms of $k[\mathbf{x}]$. This subgroup is called the *tame subgroup*. One of the major problems in affine algebraic geometry is the *Tame Generators Problem* which asks whether $\mathrm{Aut}_k\, k[\mathbf{x}] = \mathrm{T}_k\, k[\mathbf{x}]$ holds true.

Note that $\mathrm{Aut}_k\, k[\mathbf{x}] = \mathrm{T}_k\, k[\mathbf{x}]$ is true if every $F \in \mathrm{Aut}_k\, k[\mathbf{x}]$ with $\deg F > n$ admits an elementary reduction. In fact, if this is the case, for each $F \in \mathrm{Aut}_k\, k[\mathbf{x}]$ with $\deg F > n$, there exist finitely many elementary automorphisms $E_1, \ldots, E_r$ such that

$$\deg F > \deg F \circ E_1 > \cdots > \deg F \circ E_1 \circ \cdots \circ E_r = n.$$

Hence, F belongs to $\mathrm{T}_k\, k[\mathbf{x}]$. Recall that, if $\deg F > n$, then $f_1^{\mathbf{w}_0}, \ldots, f_n^{\mathbf{w}_0}$ are algebraically dependent over k. When $n = 2$, this implies further that $(f_1^{\mathbf{w}_0})^{l_2} = \alpha(f_2^{\mathbf{w}_0})^{l_1}$ for some $\alpha \in k^\times$ in view of Lemma 2.3. Here, l_1 and l_2 are mutually prime positive integers such that $l_2 \deg f_1 = l_1 \deg f_2$. Thus, if $l_i = 1$ for some i, then $f_i^{\mathbf{w}_0}$ belongs to $k[f_j^{\mathbf{w}_0}] = k[f_j]^{\mathbf{w}_0}$ for $j \neq i$, and so F admits an elementary reduction. Therefore, if either $\deg f_1 \mid \deg f_2$ or $\deg f_2 \mid \deg f_1$ holds for every $F \in \mathrm{Aut}_k\, k[x_1, x_2]$ with $\deg F > 2$, then $\mathrm{Aut}_k\, k[x_1, x_2] = \mathrm{T}_k\, k[x_1, x_2]$ holds true. This was shown by Jung [4] in 1942 when k is of characteristic zero, and by van der Kulk [16] in 1958 for general k.

Theorem 3.1 (Jung – van der Kulk). *For every $F \in \mathrm{Aut}_k\, k[x_1, x_2]$ with $\deg F > 2$, we have $\deg f_1 \mid \deg f_2$ or $\deg f_2 \mid \deg f_1$. In particular, $\mathrm{Aut}_k\, k[x_1, x_2] = \mathrm{T}_k\, k[x_1, x_2]$ holds true.*

When $n \geq 3$, some elements of $\mathrm{Aut}_k\, k[\mathbf{x}]$ admit no elementary reductions. In fact, Shestakov-Umirbaev [14] recently showed that $\mathrm{Aut}_k\, k[\mathbf{x}] \neq \mathrm{T}_k\, k[\mathbf{x}]$ when $n = 3$ and k is of characteristic zero. The Tame Generators Problem is open when $n \geq 4$, or $n = 3$ and k is of positive characteristic.

We mention that there are several different proofs of Theorem 3.1 in the case where k is a field of characteristic zero. However, there are few proofs for the general case. In addition, for any n, no counterexample to the Tame Generators Problem is found when k is of positive characteristic. Therefore, it is of importance to develop a characteristic-free technique to study $\mathrm{Aut}_k\, k[\mathbf{x}]$. For this purpose, we revisit the Jung – van der Kulk theorem from the viewpoint of initial algebras. In what follows, using the results on initial algebras, we reconstruct the proof of the Jung – van der Kulk theorem given by Makar-Limanov [9] which was modified by Dicks [2] and by Cohn [1].

Now, let $F = (f_1, \ldots, f_n)$ be an element of $\mathrm{Aut}_k\, k[\mathbf{x}]$ whose inverse is $G = (g_1, \ldots, g_n)$, and let $\mathbf{v} = (v_1, \ldots, v_n) \in \Gamma^n$ be such that $v_i \geq 0$ for $i = 1, \ldots, n$. We define an element $\mathbf{w} = (w_1, \ldots, w_n)$ of Γ^n by $w_i = \deg_{\mathbf{v}} f_i$ for each i. Then, it holds that

$$\deg_{\mathbf{v}} F(x_1^{a_1} \cdots x_n^{a_n}) = \deg_{\mathbf{v}} f_1^{a_1} \cdots f_n^{a_n} = \sum_{i=1}^{n} a_i w_i = \deg_{\mathbf{w}} x_1^{a_1} \cdots x_n^{a_n} \tag{5}$$

for each $a_1, \ldots, a_n \in \mathbb{Z}_{\geq 0}$.

With this notation, the following lemma holds.

Lemma 3.2. *If $\deg_{\mathbf{v}} F > \sum_{i=1}^{n} v_i$, then there exists $i \in \{1, \ldots, n\}$ such that $g_i^{\mathbf{w}}$ is not a monomial, i.e., at least two monomials appear in $g_i^{\mathbf{w}}$.*

PROOF. Suppose to the contrary that $g_1^{\mathbf{w}}, \ldots, g_n^{\mathbf{w}}$ are all monomials. Then, in view of (5), we get

$$\deg_{\mathbf{v}} F(\mathbf{x}^a) = \deg_{\mathbf{w}} \mathbf{x}^a < \deg_{\mathbf{w}} g_i^{\mathbf{w}} = \deg_{\mathbf{v}} F(g_i^{\mathbf{w}})$$

for every monomial $\mathbf{x}^a$ appearing in $g_i - g_i^{\mathbf{w}}$ for each i. Hence, $\deg_{\mathbf{v}} F(g_i^{\mathbf{w}})$ is greater than $\deg_{\mathbf{v}} F(g_i - g_i^{\mathbf{w}})$, and so

$$F(g_i)^{\mathbf{v}} = (F(g_i^{\mathbf{w}}) + F(g_i - g_i^{\mathbf{w}}))^{\mathbf{v}} = F(g_i^{\mathbf{w}})^{\mathbf{v}}.$$

Since G is the inverse of F, we have $F(g_i) = x_i$. Hence,

$$x_i = x_i^{\mathbf{v}} = F(g_i)^{\mathbf{v}} = F(g_i^{\mathbf{w}})^{\mathbf{v}} = F(\alpha_i x_1^{a_1^i} \cdots x_n^{a_n^i})^{\mathbf{v}} = \alpha_i (f_1^{\mathbf{v}})^{a_1^i} \cdots (f_n^{\mathbf{v}})^{a_n^i}, \tag{6}$$

where $\alpha_i \in k^{\times}$ and $a_1^i, \ldots, a_n^i \in \mathbb{Z}_{\geq 0}$ are such that $g_i^{\mathbf{w}} = \alpha_i x_1^{a_1^i} \cdots x_n^{a_n^i}$. Observe that $f_l^{\mathbf{v}}$ does not belong to k for every l, since neither does f_l, and $v_1, \ldots, v_n$ are nonnegative by assumption. Hence, by (6), there exists $\sigma(i) \in \{1, \ldots, n\}$ such that $a_j^i = 0$ for $j \neq \sigma(i)$, $a_{\sigma(i)}^i = 1$ and $f_{\sigma(i)}^{\mathbf{v}} = \alpha_i^{-1} x_i$. Then, the map $\sigma \colon \{1, \ldots, n\} \to \{1, \ldots, n\}$ is bijective, since $F \circ G = (x_1, \ldots, x_n)$. Thus, we have

$$\deg_{\mathbf{v}} F = \sum_{i=1}^{n} \deg_{\mathbf{v}} f_i = \sum_{i=1}^{n} \deg_{\mathbf{v}} f_{\sigma(i)} = \sum_{i=1}^{n} \deg_{\mathbf{v}} \alpha_i^{-1} x_i = \sum_{i=1}^{n} v_i,$$

a contradiction to the assumption that $\deg_{\mathbf{v}} F > \sum_{i=1}^{n} v_i$. Therefore, $g_i^{\mathbf{w}}$ is not a monomial for some i. □

The following lemma is elementary.

Lemma 3.3. *Assume that $n = 2$. Let $\mathbf{w} = (w_1, w_2) \in \mathbb{Z}^2$ be such that $w_i > 0$ for $i = 1, 2$, and let p and q be $\mathbf{w}$-homogeneous elements of $k[x_1, x_2] \setminus \{0\}$ such that p is not a monomial. If $k[p, p^{-1}, q]$ contains x_1 and x_2, then we have $w_1 \mid w_2$ or $w_2 \mid w_1$.*

PROOF. We may assume that w_1 and w_2 are mutually prime by replacing w_i with $w_i / \gcd(w_1, w_2)$ if necessary. Since p and q are $\mathbf{w}$-homogeneous elements of $k[x_1, x_2]$, we may write

$$p = x_1^{a_1} x_2^{a_2} \sum_{i \geq 0} \alpha_i (x_1^{w_2} x_2^{-w_1})^i, \quad q = x_1^{b_1} x_2^{b_2} \sum_{i \geq 0} \beta_i (x_1^{w_2} x_2^{-w_1})^i, \tag{7}$$

where $a_1, a_2, b_1, b_2 \in \mathbb{Z}_{\geq 0}$, and $\alpha_i, \beta_i \in k$ for each i such that $\alpha_0 \neq 0$ and $\beta_0 \neq 0$.

First, assume that (a_1, a_2) and (b_1, b_2) are linearly dependent over $\mathbb{Z}$. Then, the $\mathbb{Z}$-submodule M of $\mathbb{Z}^2$ generated by (a_1, a_2) and (b_1, b_2) is a free $\mathbb{Z}$-module of rank one. Let (c_1, c_2) be a basis of M, and set $w := c_1 w_1 + c_2 w_2$. Since $(a_1, a_2) = l(c_1, c_2)$ and $(b_1, b_2) = m(c_1, c_2)$ for some $l, m \in \mathbb{Z}$, it follows that $\deg_{\mathbf{w}} p = a_1 w_1 + a_2 w_2 = lw$ and $\deg_{\mathbf{w}} q = b_1 w_1 + b_2 w_2 = mw$ belong to $w\mathbb{Z}$. Moreover, we may assume that c_i is nonnegative for $i = 1, 2$, since so are a_i and b_i. By the assumption that x_i belongs to $k[p, p^{-1}, q]$, we may write $x_i = p^{s_i} f_i$, where $s_i \in \mathbb{Z}$ and $f_i \in k[p, q]$. Because p and q are $\mathbf{w}$-homogeneous, $\deg_{\mathbf{w}} f_i$ is a linear combination of $\deg_{\mathbf{w}} p$ and $\deg_{\mathbf{w}} q$ over $\mathbb{Z}$, and hence so is $\deg_{\mathbf{w}} x_i = w_i$. Thus, w_1 and w_2 belong to $w\mathbb{Z}$. Since $w = c_1 w_1 + c_2 w_2$, and c_1 and c_2 are nonnegative, it holds that $(c_1, c_2) = (1, 0)$ and $w_1 \mid w_2$, or $(c_1, c_2) = (0, 1)$ and $w_2 \mid w_1$. Therefore, the assertion is true.

Next, assume that (a_1, a_2) and (b_1, b_2) are linearly independent over $\mathbb{Z}$. Then, $p^{\mathbf{v}} = \alpha_0 x_1^{a_1} x_2^{a_2}$ and $q^{\mathbf{v}} = \beta_0 x_1^{b_1} x_2^{b_2}$ are algebraically independent over k, where

$\mathbf{v} = (-1, 1)$. Hence, we have $k[p, q]^{\mathbf{v}} = k[p^{\mathbf{v}}, q^{\mathbf{v}}]$ by (1). We show that

$$k[p, p^{-1}, q]^{\mathbf{v}} = k[p^{\mathbf{v}}, (p^{\mathbf{v}})^{-1}, q^{\mathbf{v}}].$$

It is easy to check that the right-hand side is contained in the left-hand side. For the reverse inclusion, take any $f \in k[p, p^{-1}, q] \setminus \{0\}$. Then, we may write $f = gp^l$, where $l \in \mathbb{Z}$ and $g \in k[p, q]$. Hence, $f^{\mathbf{v}} = g^{\mathbf{v}}(p^{\mathbf{v}})^l$ belongs to $k[p, q]^{\mathbf{v}}[(p^{\mathbf{v}})^{-1}]$. Since $k[p, q]^{\mathbf{v}} = k[p^{\mathbf{v}}, q^{\mathbf{v}}]$, it follows that $f^{\mathbf{v}}$ belongs to the right-hand side. This proves the above equality. Since x_i belongs to $k[p, p^{-1}, q]$ by assumption, it follows that $x_i = x_i^{\mathbf{v}}$ belongs to $k[p^{\mathbf{v}}, (p^{\mathbf{v}})^{-1}, q^{\mathbf{v}}]$ for $i = 1, 2$. Because x_i, $p^{\mathbf{v}}$ and $q^{\mathbf{v}}$ are monomials, we can express

$$x_i = (p^{\mathbf{v}})^{t_i}(q^{\mathbf{v}})^{u_i} = \alpha_0\beta_0(x_1^{a_1}x_2^{a_2})^{t_i}(x_1^{b_1}x_2^{b_2})^{u_i}$$

for $i = 1, 2$, where $t_i \in \mathbb{Z}$ and $u_i \in \mathbb{Z}_{\geq 0}$. Then, it follows that

$$\begin{pmatrix} t_1 & u_1 \\ t_2 & u_2 \end{pmatrix}\begin{pmatrix} a_1 & a_2 \\ b_1 & b_2 \end{pmatrix} = \begin{pmatrix} 1 & 0 \\ 0 & 1 \end{pmatrix}, \quad \begin{pmatrix} a_1 & a_2 \\ b_1 & b_2 \end{pmatrix}\begin{pmatrix} t_1 & u_1 \\ t_2 & u_2 \end{pmatrix} = \begin{pmatrix} 1 & 0 \\ 0 & 1 \end{pmatrix},$$

where the second equality is due to the first equality. Hence, we get

$$(8) \qquad t_1a_1 + u_1b_1 = 1, \quad t_1a_2 + u_1b_2 = 0, \quad a_1t_1 + a_2t_2 = 1, \quad a_1u_1 + a_2u_2 = 0.$$

By the assumption that p is not a monomial, we have $\alpha_{i_1} \neq 0$ for some $i_1 \geq 1$ in the expression (7). Since p is an element of $k[x_1, x_2]$, it follows that $a_2 - i_1w_1 \geq 0$. Hence, $a_2 \geq i_1w_1 \geq w_1 \geq 1$. Let us demonstrate that $a_1 = 0$. Then, we get $a_2t_2 = a_1t_1 + a_2t_2 = 1$ by the third equality in (8). This implies that $|a_2| = 1$, since a_2 and t_2 are integers. Then, from the preceding inequality, we obtain that $a_2 = w_1 = 1$, and thereby conclude that $w_1 \mid w_2$. Suppose to the contrary that $a_1 \neq 0$. Then, u_1 must be zero by the last equality in (8), in which $a_i \geq 0$ and $u_i \geq 0$ for $i = 1, 2$. Then, $t_1a_1 = t_1a_1 + u_1b_1 = 1$ by the first equality in (8), so $t_1 \neq 0$. Since $t_1 \neq 0$ and $u_1 = 0$, the second equality in (8) yields $a_2 = 0$. This contradicts that $a_2 \geq 1$. Therefore, we have $a_1 = 0$, and proved that $w_1 \mid w_2$. □

Now, let us derive the Jung – van der Kulk theorem from Proposition 2.4 (i) and Lemmas 3.2 and 3.3. Take any $F = (f_1, f_2) \in \operatorname{Aut}_k k[x_1, x_2]$ with $\deg F > 2$ whose inverse is $G = (g_1, g_2)$, and set $w_i := \deg f_i > 0$ for $i = 1, 2$. Since $\deg F > 2$, we know by Lemma 3.2 that $g_i^{\mathbf{w}}$ is not a monomial for some $i \in \{1, 2\}$, where $\mathbf{w} = (w_1, w_2)$. Set $f = g_i$ and $g = g_j$ for $j \in \{1, 2\}$ with $j \neq i$. Since $k[x_1, x_2]$ is a unique factorization domain with $k[x_1, x_2]^{\times} = k^{\times}$, we can write $f^{\mathbf{w}} = \alpha p^l$, where $\alpha \in k^{\times}$, $l \in \mathbb{Z}_{\geq 0}$, and $p \in k[x_1, x_2]$ is not a proper power in $k[x_1, x_2]$. Note that p is not a monomial, since neither is $f^{\mathbf{w}}$. Let us check that the assumptions of Proposition 2.4 are fulfilled for $L = k(x_1, x_2)$ and $\delta = \deg_{\mathbf{w}}$. As mentioned after Lemma 2.3, k is algebraically closed in $Q(\operatorname{gr}_{\mathbf{w}} L)$. Moreover, $\delta(f) = \deg_{\mathbf{w}} g_i$ is not zero, since $w_i > 0$ for $i = 1, 2$. Since G is an automorphism, we have $k[f, g] = k[x_1, x_2]$, and so $Q(k[f, g]^{\delta}) = k(x_1, x_2)$ is a transcendental extension of $k(f^{\delta})$. Thus, the assumptions of Proposition 2.4 are fulfilled. Therefore, by the statement (i) of this proposition, we know that $k[f, g]^{\delta}$ is contained in $k[p, p^{-1}, h^{\mathbf{w}}]$ for some $h \in k[f, g] \setminus \{0\}$. Since $k[f, g]^{\delta} = k[x_1, x_2]$, it follows that x_1 and x_2 belong to $k[p, p^{-1}, h^{\mathbf{w}}]$, in which p is not a monomial as mentioned. Therefore, we get $w_1 \mid w_2$ or $w_2 \mid w_1$ by Lemma 3.3. This proves the Jung – van der Kulk theorem.

Remark 3.4. The proof of the Jung – van der Kulk theorem which we give here follows a line similar to Cohn [1, Section 6.8]. Proposition 2.4 (i) and Lemma 3.3 correspond to Lemmas 8.3 and 8.4 in [1], respectively. The major differences are as follows. Lemma 8.3 in [1] says that, for $\mathbf{w} \in \mathbb{Z}^2 \setminus \{0\}$ and $f, g \in k(x_1, x_2)^\times$ with $\deg_{\mathbf{w}} f \neq 0$, there exist $p, q \in \{h^{\mathbf{w}} \mid h \in k(x_1, x_2)^\times\}$ such that $k[f, g]^{\mathbf{w}} \subset k[p, p^{-1}, q]$ and $f^{\mathbf{w}} = \lambda p^a$ for some $a \geq 1$ and $\lambda \in k^\times$. The idea of the proof of this lemma essentially differs from that of Theorem 2.1 from which Proposition 2.4 is derived. In fact, in the argument of the proof of the lemma of Cohn, it is not clear whether q can be taken from $k[x_1, x_2]$ if f and g are elements of $k[x_1, x_2]$. On the other hand, Proposition 2.4 states that q can be taken from $k[\mathbf{x}]$ if f and g are elements of $k[\mathbf{x}]$. This makes the proof of Lemma 3.3 rather simpler in the case where (a_1, a_2) and (b_1, b_2) are linearly dependent over $\mathbb{Z}$.

References

1. P. M. Cohn, *Free rings and their relations*, 2nd ed., London Math. Soc. Monogr., vol. 19, Academic Press, London, 1985.
2. W. Dicks, *Automorphisms of the polynomial ring in two variables*, Publ. Sec. Mat. Univ. Autònoma Barcelona **27** (1983), no. 1, 155–162.
3. M. Göbel, *A constructive description of SAGBI bases for polynomial invariants of permutation groups*, J. Symbolic Comput. **26** (1998), no. 3, 261–272.
4. H. W. E. Jung, *Über ganze birationale Transformationen der Ebene*, J. Reine Angew. Math. **184** (1942), 161–174.
5. S. Kuroda, *The infiniteness of the SAGBI bases for certain invariant rings*, Master's Thesis, Tohoku University, 2000.
6. ———, *The infiniteness of the SAGBI* (*subalgebra analogue to Gröbner bases for ideals*) *bases for certain invariant rings*, Sūrikaisekikenkyūsho Kōkyūroku **1175** (2000), 51–62 (Japanese).
7. ———, *The infiniteness of the SAGBI bases for certain invariant rings*, Osaka J. Math. **39** (2002), no. 3, 665–680.
8. D. Kapur and K. Madlener, *A completion procedure for computing a canonical basis for a k-subalgebra*, Computers and Mathematics (Cambridge, MA, 1989) (E. Kaltofen and S. M. Watt, eds.), Springer, New York, 1989, pp. 1–11.
9. L. Makar-Limanov, *On automorphisms of certain algebras*, Ph.D. Thesis, Moscow State University, 1970 (Russian).
10. M. Nagata, *On automorphism group of* $k[x, y]$, Lectures in Math., vol. 5, Kinokuniya, Tokyo, 1972.
11. Z. Reichstein, *SAGBI bases in rings of multiplicative invariants*, Comment. Math. Helv. **78** (2003), no. 1, 185–202.
12. L. Robbiano and M. Sweedler, *Subalgebra bases*, Commutative Algebra (Salvador, 1988) (W. Bruns and A. Simis, eds.), Lecture Notes in Math., vol. 1430, Springer, Berlin, 1990, pp. 61–87.
13. N. Schwartz, *Stability of Gröbner bases*, J. Pure Appl. Algebra **53** (1988), no. 1-2, 171–186.
14. I. P. Shestakov and U. U. Umirbaev, *The tame and the wild automorphisms of polynomial rings in three variables*, J. Amer. Math. Soc. **17** (2004), no. 1, 197–227 (electronic).
15. N. M. Thiéry and S. Thomassé, *Convex cones and SAGBI bases of permutation invariants*, Invariant Theory in All Characteristics (Kingston, ON, 2002) (H. E. A. E. Campbell and D. L. Wehlau, eds.), CRM Proc. Lecture Notes, vol. 35, Amer. Math. Soc., Providence, RI, 2004, pp. 259–263.
16. W. van der Kulk, *On polynomial rings in two variables*, Nieuw Arch. Wiskunde (3) **1** (1953), 33–41.

Department of Mathematics and Information Sciences, Tokyo Metropolitan University, 1-1 Minami-Osawa, Hachioji, Tokyo 192-0397, Japan

E-mail address: kuroda@tmu.ac.jp

Centre de Recherches Mathématiques
CRM Proceedings and Lecture Notes
Volume **54**, 2011

Holomorphic Curves on Irregular Varieties of General Type Starting from Surfaces

Steven S.-Y. Lu

ABSTRACT. We obtain the finiteness of nonhyperbolic curves in a quasiprojective surface of general type with log irregularity two or more, validating Lang's conjecture concerning holomorphic curves in this context. Effective versions and generalizations to higher dimensions are discussed.

1. Introduction

The main aim of this paper is to resolve the conjecture of Lang on the behavior of curves in complex varieties of general type in the case of surfaces of log-general type of log irregularity two or more. This is done by giving a simple proof of the finiteness of the number algebraic nonhyperbolic curves in such a surface, alluded to in the paper [23], where a sharp effective version is obtained in the projective case. We will also give an overview of our earlier effective version and its generalization to the log case and to higher dimensions for which the surface case offer the basic model steps.

We will implicitly assume that a curve is reduced and irreducible and that all varieties involved are complex algebraic. Recall that given a nonsingular projective variety $\overline{X}$ over $\mathbb{C}$ and a normal crossing divisor D, $X = \overline{X} \setminus D$ is called of log-general type if its log-canonical divisor $\overline{K}_X := K_{\overline{X}} + D$ is of maximal Kodaira dimension. Both this log Kodaira dimension $\bar{\kappa}(X)$ and the log irregularity $\bar{q}(X) = h^0(\Omega_{\overline{X}}(\log D))$ are proper (more generally log) birational invariants of the quasiprojective variety X and hence make sense for singular varieties. Another such invariant of interest to us is given by the Albanese map $\mathrm{alb}_X \colon X \to \mathrm{Alb}(X)$, regular in the case where X is smooth, where $\mathrm{Alb}(X)$ is a semi-Abelian variety of dimension $\bar{q}(X)$ canonically associated to X. We say that X has maximal Albanese dimension if alb_X is generically finite. A possibly singular quasiprojective curve is said to be nonhyperbolic if it is dominated by a holomorphic map from $\mathbb{C}$, or equivalently, if it is not of log-general type. The logarithmic version of the conjecture of Lang mentioned above, see [18], has the following two forms.

2010 *Mathematics Subject Classification.* 14C17, 14E30, 14G25, 14H45, 14J29, 14K12, 32Q45, 32Q57.

Partially supported by an NSERC grant and UQÀM.

This is the final form of the paper.

Conjecture 1.1. *An algebraic variety of log-general type admits a proper Zariski closed subset that contains all nonhyperbolic curves.*

Conjecture 1.2. *An algebraic variety of log-general type admits a proper Zariski subset that contains all nontrivial holomorphic images of* $\mathbb{C}$.

In this article, we settle Conjecture 1.1 in the case of surfaces of log irregularity two by our main theorem:

Theorem 1.3. *Let* S *be a quasiprojective surface of log-general type with* $\bar{q}(S) \geq 2$. *Then* S *has only a finite number of nonhyperbolic algebraic curves.*

This theorem was known in the case where S is smooth and admits a proper finite map to a semi-abelian variety [36]. Conjecture 1.2 for surfaces as given in the theorem is then a corollary of this theorem and of the following main theorem of Lu – Winkelmann in [27].

Theorem 1.4. *Let* X *be a quasiprojective variety of log-general type with* $\bar{q}(X) \geq \dim X$. *Then* X *does not admit a Zariski dense holomorphic image of* $\mathbb{C}$.

The proof of Theorem 1.4 uses the main result of Noguchi – Winkelmann – Yamanoi in [37]. An overview of its proof, as well as a natural generalization, will be given below.

We will also indicate below our approach to the following effective strengthening of Theorem 1.3, see [24], that simplifies to some extent that given in [23].

Theorem 1.5. *Given a quasiprojective surface* S *equipped with a generically finite morphism to a semi-Abelian variety* J *and* $C \subset S$ *a nonhyperbolic curve. Let* $\overline{S}$ *be a compactification of* S *by adding to* S *a normal crossing divisor* B *and* $\overline{C}$ *the Zariski closure of* C *in* $\overline{S}$. *Let* P *be the positive part of the Zariski decomposition of the log-canonical divisor* $K_{\overline{S}} + B$. *Then*

$$P\overline{C} \leq 4P^2.$$

That this is a strengthening of Theorem 1.3 is well explained in the introduction of [23].

2. Curves in semi-Abelian surfaces

We recall that a semi-Abelian variety is a complex Abelian group J that admits a holomorphic exact sequence of groups

$$0 \to (\mathbb{C}^*)^k \to J \xrightarrow{\pi} A \to 0 \tag{2.1}$$

where A is an abelian variety and $k \geq 0$. The equivariant compactification $(\mathbb{C}^*)^k \subset (\mathbb{P}^1)^k$ gives rise to an equivariant compactification of the $(\mathbb{C}^*)^k$ bundle given by π to a $(\mathbb{P}^1)^k$ bundle $\bar{\pi}\colon \bar{J} \to A$ with $B_0 = \bar{J} \setminus J$ a divisor of simple normal crossings. One sees easily that the sheaf of logarithmic one-forms $\Omega_{\bar{J}}(\log B_0)$ is trivial over $\bar{J}$ and that the global logarithmic one forms correspond precisely to the linear one forms on the universal cover of J and that the same facts remain true for any smooth equivariant compactification $\tilde{J}$ of J, having the boundary $B = \tilde{J} \setminus J$ necessarily a divisor with simple normal crossings, see [34]. Another easy fact, which one can see for example by considering global logarithmic one forms, is that a nonhyperbolic curve in J is precisely a translate of a one-dimensional connected algebraic subgroup, being either an elliptic curve or $\mathbb{C}^*$. We will call such curves linear curves and also call their corresponding global logarithmic one forms on $\tilde{J}$ linear one forms.

We now focus on the case where J is a semi-Abelian surface. We observe that if B is the boundary divisor of a smooth equivariant compactification $\tilde{J}$ of J, then B has singularities only if $J = \mathbb{C}^* \times \mathbb{C}^*$, i.e., $\tilde{J}$ is a smooth toric surface. In this case, J has an infinite number of equivariant compactifications since any such can be obtained from one by blowing up a singular point of the boundary divisor, see for example [34]. In preparation for the proof of Theorem 1.3, we first give the following elementary lemmas.

Lemma 2.1. *Given a smooth equivariant compactification $\tilde{J}$ of J with boundary divisor B, let G be a component of B and G' the complementary divisor. Let C be a a noncompact nonhyperbolic curve in J and let $\overline{C}$ be its Zariski closure in $\tilde{J}$. Then $\overline{C}$ intersects $G^* = G \setminus G'$ if and only if C is a translate of the algebraic subgroup F of J leaving points of G^* fixed.*

Proof. Since G^* is an orbit, it is canonically the quotient of J by a subgroup F, necessarily one-dimensional. Thus G^* is a one-dimensional complex Lie group. Let π be the quotient map. We first observe that as C is noncompact, it is not an elliptic curve and therefore $\overline{C}$ is a rational curve and C is isomorphic to $\mathbb{C}^*$. Since C must be the translate of a connected algebraic subgroup in J, $\pi(C)$ must also be so in G^*. It follows that either $\pi(C)$ is a point or it is G^*. In the former case, C is a translate of F and $\overline{C}$ intersects G^*. In the latter case, C is not a translate of F and as π extends to a surjective regular map from $\overline{C}$ to G and G^* is a quotient of $\mathbb{C}^*$, we must have that $G^*(=\mathbb{C}^*)$ is $G(=\mathbb{P}^1)$ minus two points. Hence $\overline{C}$ cannot intersect G^* in this case because C, being nonhyperbolic, cannot differ from $\overline{C} = \mathbb{P}^1$ by more than the pre-images of the two distinct points $G \setminus G^*$. □

Lemma 2.2. *Let D be a nontrivial reduced divisor in J. Then there is a smooth equivariant compactification $\tilde{J}$ of J in which the Zariski closure of D does not intersect the singularities of the normal crossing divisor $B = \tilde{J} \setminus J$.*

Proof. This is done by repeatedly blowing up the singularities of the boundary divisor where the intersection occur. □

Definition 2.3. D is said to be in general position with respect to an equivariant compactification $\tilde{J}$ if it does not intersect the singularities of the boundary divisor.

We remark that this notion can be generalized to higher-dimensional semi-Abelian varieties and that the following lemma remains valid in that context (see [34]).

Lemma 2.4. *Let D be a nontrivial reduced divisor in J. Then there is a constant n such that the intersection multiplicity of D at any point of D with a nonhyperbolic curve not contained in D is less than n.*

Proof. Since such an intersection multiplicity for two distinct linear curves is one and since D has a finite number of components, we may assume that D is irreducible and nonlinear. Let $\tilde{J}$ be an equivariant compactification of J in which D is in general position. Let $\overline{D}$ be the Zariski closure of D and $r\colon \tilde{D} \to \tilde{J}$ the normalization of $\overline{D}$ followed by its inclusion to $\tilde{J}$. If the lemma is false, then there is a sequence of linear one forms $w_1, w_2, \ldots,$ such that $r^* w_n$ has an isolated zero

p_n whose vanishing order tends to infinity as n does so. By rescaling the w_n's and extracting subsequences, we may assume that $p = \lim p_n \in \tilde{D}$ and that $w = \lim w_n$ is a nontrivial linear (logarithmic) one form on $\bar{J}$. It follows that r^*w is identically zero as it has a zero of infinite order at p as a logarithmic one form on $\tilde{D}$. If p is not in the pre-image of D, this is only possible if w is a regular one form in a neighborhood of the image of p since a nonregular logarithmic one form germ must pull back to one of the same type on $\tilde{D}$. Hence w is a regular linear one form at p in all cases, forcing D to be itself linear, in contradiction with our assumption. We remark that w_n is chosen to be the linear one form that kills the tangent space of the linear curve C_n and $\mathrm{ord}_{p_n}(r^*w_n) = (C_n, D)_{p_n} - 1$, so that $(C_n, D)_{p_n} - 1$ is bounded by the degree of the log canonical bundle of $\widetilde{D}$. □

This lemma is true for higher-dimensional semi-Abelian varieties in the case where $\overline{D}$ is ample and for higher-dimensional abelian varieties without condition, see [34]. We remark that by identifying the order of vanishing of the w_n's above with the local multiplicity of the Gauss map of D, one can show that n is bounded by the degree of this Gauss map (see [23, 36]).

We quote the following well-known lemma of Kodaira that is basic to the theory of Kodaira dimension.

Lemma 2.5. *Let L and H be divisors on a smooth projective variety with H ample. Then L is of maximal Kodaira dimension if an only if there is a positive integer k such that the linear system $|kL - H|$ is nonempty.*

Since we will be using the Zariski decomposition of divisors on a surface, we recall its definition. Given an effective divisor K on a smooth surface, its Zariski decomposition is a maximal effective $\mathbb{Q}$-divisor P such that $N = K - P$ is $\mathbb{Q}$-effective and P is nef (i.e., $PC \geq 0$ for all curves C). One can show that the decomposition exists and is unique and that the Kodaira dimensions of P and K are the same, [40]. Also, N is supported on a contractible divisor and $PN_i = 0$ for every irreducible component N_i of N. The following lemma is evident from the definition of P and the fact that a contractible divisor has negative self-intersection. We will mean below, by a connected component of an effective divisor D, the part of D supported on a connected component of D_{red}.

Lemma 2.6. *Let P be the Zariski decomposition of an effective divisor $\overline{K}$ on a smooth surface and E a connected component of $\overline{K}$ that is contractible. Then P is also the Zariski decomposition of $\overline{K} - E$.*

3. Proof of the main theorem (Theorem 1.3)

Let $h = \mathrm{alb}_S$. We may assume without loss of generality that the quasiprojective surface S is nonsingular and that, after a translation in $\mathrm{Alb}(S)$, h maps a point in S to the identity element. We first note that $h(S)$ generates $\mathrm{Alb}(S)$ by the universal property of h, namely that any morphism of S into a semi-Abelian variety having the identity element as image of the same point factors uniquely through h by a morphism of semi-Abelian varieties. As we assumed that S is a surface and that the dimension of $\mathrm{Alb}(S)$ is $\bar{q} \geq 2$, the Zariski closure S' of $h(S)$ in $\mathrm{Alb}(S)$ is either a curve or a surface. In the former case, this curve must be nonlinear by the universal property and hence any nonhyperbolic curve in S must lie on a fibre of S. As the log-canonical divisor of S restricts to that of the generic fibre of h

by adjunction, the generic fibre is of log-general type since S is. Hence there are only a finite number of nonhyperbolic curves in this case. In the latter case, S' is a surface and its log Kodaira dimension $\bar{\kappa}(S')$ is $0, 1$ or 2. If $\bar{\kappa}(S')$ is two, then there are only a finite number of nonhyperbolic curves in S', either by the result of [25] via the Miyaoka – Yau inequality, as these curves are linear (hence smooth) curves, or by the main result of Kawamata in [15] generalized to the logarithmic case by Noguchi in [31] (see also [1]). Now there are only a finite number of vertical curves in S, those mapping to points by h. Also, each horizontal nonhyperbolic curve surjects to a linear curve in S' and the number of such curves is bounded by the degree of h above each linear curve in S'. Hence, there are only a finite number of nonhyperbolic curves in S in this case. If $\bar{\kappa}(S')$ is one, then, by considering the log Iitaka fibration of S', one sees that there is a one-dimensional algebraic subgroup of $\mathrm{Alb}(S)$ leaving S' invariant and that the quotient of S' by this action is a curve Z in a semi-Abelian variety. We can eliminate the case where Z is nonhyperbolic quite easily as otherwise Z would be semi-Abelian and hence so would S', being then an algebraic subgroup of $\mathrm{Alb}(S)$; another way to see the same is to apply Kawamata's structure theorem for sub-varieties of a semi-Abelian variety [16]. Hence, Z is hyperbolic and by composing h with this quotient, the same argument as in the case where S' is a curve applies to show the finiteness of the number of nonhyperbolic curves in S' in the case where $\bar{\kappa}(S')$ is one. Hence, we are left with the case where $\bar{\kappa}(S')$ is zero so that S' is a semi-Abelian surface by Kawamata's structure theorem in [16]. We have thus reduced our main theorem to the following proposition.

Proposition 3.1. *Let $h\colon S \to J$ be a generically finite morphism from a smooth quasiprojective surface S of log-general type to a semi-Abelian surface J. Then S has only a finite number of nonhyperbolic algebraic curves.*

PROOF. Let $\tilde{J}$ be an equivariant compactification of J and let $\bar{S}$ be a smooth compactification of S both with normal crossing boundary divisors. We may assume that h extends to a morphism $\bar{h} : \bar{S} \to \tilde{J}$ by the resolution of singularity theorem [5, 13]. Since $\bar{h}$ pulls back local log-forms to local log-forms, it pulls back the log-canonical sheaf of J on $\tilde{J}$ to a sub-sheaf of that of S on $\bar{S}$. Since $\mathcal{O}(\overline{K}_J)$ is the trivial sheaf $\mathcal{O}_{\tilde{J}}$, this gives rise to a section of $\mathcal{O}(\overline{K}_S)$ or equivalently to an effective divisor $\overline{K}$ on $\bar{S}$. A simple computation in local coordinates shows that $\overline{K}$ cannot have any horizontal components in the pre-image of the boundary divisor B of J and hence the divisorial part K' of the image in $\tilde{J}$ of $\overline{K}_{\mathrm{red}}$ cannot have any components in common with B. By blowing up the singularities of B, if any, we may assume that K' does not contain any of the singularities of B (see Lemma 2.2). Since there are only a finite number of vertical curves in S and of curves in the pre-image of K', we will only be interested in bounding the number of horizontal nonhyperbolic curves C in S that do not project to K'. It is sufficient by Kodaira's lemma to bound the intersection degree $P\overline{C}$, where $\overline{C}$ is the Zariski closure of C in $\bar{S}$ and P is the Zariski decomposition of $\overline{K}$. For then such curves lie in a finite number of families of curves in S, and so if there are an infinite number of them, then one can construct a one-dimensional connected family of such curves, contradicting the fact that the log Kodaira dimension of the total space $\tilde{S}$ of such a family, being a surface that dominates S, must be maximal as the log-canonical sheaf of S naturally pulls back to a sub-sheaf of that of a desingularization of $\tilde{S}$. Let K be $\overline{K}$ with isolated connected contractible components removed. Then P is also the Zariski decomposition of K by Lemma 2.6. Hence, to bound the degree of $\overline{C}$

with respect to P, it is sufficient to do so with respect to a component G of K that intersects $\overline{C}$ somewhere. We will consider the case where G is horizontal and the case where it is vertical separately starting with the latter below.

Note that as K' does not intersect B_{sing} by hypothesis, the vertical parts of $\overline{K}$ above B_{sing} are isolated and contractible (via Stein factorization). So K has no components above B_{sing}. Hence if G is a vertical component of K, then its image point Q in $\tilde{J}$ lies either in a complex one-dimensional orbit of J or in J. In the former case, the image of C in J is a translate C' of a $\mathbb{C}^*$ subgroup of J whose Zariski closure $\overline{C}'$ intersects this orbit transversally at Q. Hence $\overline{C}'$ is uniquely determined by G and $G\overline{C} \leq G(h^*\overline{C}')$ is bounded as there are only finitely many such G's in this case. In the latter case, choose any point p on the normalization $\widetilde{C}$ of C above an intersection point q of C and G. Let w be a linear one form on J that is nontrivial on C'. Then $w = df$ for a "linear" function f in a neighborhood of Q with $f(Q) = 0$ and G is contained in the divisor F defined by $f \circ h$ there. Since an algebraic morphism from a nonhyperbolic curve to itself is a local isomorphism, its differential is nowhere vanishing so that the intersection multiplicity at p of C and F is the same as that of C' and (f), which is one. It follows that through q, F is reduced and smooth, G is smooth and that every analytic component of C at q is transversal to G. Since the number of such components is bounded by the degree d of h, we get $G\overline{C} \leq d$ in this case.

Now let G be a horizontal component of K and G' the Zariski closure of its image in $\tilde{J}$. Since $G' \cap B$ is finite, there are finitely many nonhyperbolic curves in J whose compactifications touch $G' \cap B$ by Lemma 2.1. Hence we may ignore the finite number of nonhyperbolic curves in their pre-images in S. So doing, since also $\overline{C} \backslash C \subset h^{-1}(B)$ (as C is nonhyperbolic), we then have $G\overline{C} = GC$. Hence, to bound $G\overline{C}$, it is sufficient to bound $\#(G \cap C)$ and the multiplicity of the intersection at each point of $G \cap C$. We will do this separately below, starting with the former.

Since C' does not lie in G' by construction, the linear one form w whose kernel contains the tangent space to $C' \cap J$ pulls back to a nontrivial logarithmic one form $\widetilde{w}$ on the normalization $\widetilde{G}$ of G and $\widetilde{w}$ is regular on the pre-image $\widetilde{G}_0$ of $G \cap S$. Let $r \colon \widetilde{G} \to \overline{S}$ be the inclusion map of G followed by its normalization and let $p \in r^{-1}(C) \subset \widetilde{G}_0$. Since a nontrivial morphism from a nonhyperbolic curve to another is necessarily unramified and $r(p) \in G$ is necessarily a ramified point of h, we see that the image of dh at p is precisely the one-dimensional kernel of w at $h\big(r(p)\big)$ on J. It follows that p is a zero of $\widetilde{w}$. Since the number of zeros of $\widetilde{w}$ is bounded by the degree e of the log-cotangent bundle of $\widetilde{G}_0$, we see that $\#(G \cap C) \leq e$.

Now let $q \in G \cap C$. Let the pullbacks of the divisors C' and G' to $\overline{S}$ be denoted by the same. Let $C' = C'_{\mathrm{h}} + C'_{\mathrm{v}}$ be the decomposition of the divisor C' into its horizontal and vertical parts in $\overline{S}$. As $\overline{C}$ lies in C'_{h}, G lies in G' and $G'V = 0$ for all vertical divisors V, we have

$$(\overline{C}, G)_q \leq (\overline{C}, G')_q \leq (C'_{\mathrm{h}}, G')_q \leq (\deg h)(C', G')_{h(q)},$$

which has an upper bound independent of the nonhyperbolic curves C and its image $C' \cap J$ by Lemma 2.4.

This concludes the proof of our proposition. □

4. A brief overview of the effective version

We give a brief overview of the proof of the projective version of Theorem 1.5 as given in [23] with some simplifications and some indications in the quasiprojective case.

From our proof of Theorem 1.4 above, it is not difficult to see how one can reduce the proof of Theorem 1.5 to the case where J is a semi-Abelian surface. An important role is played in the above section by the use of the Zariski decomposition for the log-canonical divisor, via which one can reduce the problem to just intersections on S rather than on $\overline{S}$ as we have seen in the previous section. Then one can reduce further to just considering one component at a time. Here, the key to our sharp effective result is the consideration of the horizontal case which we explain below for simplicity for the case of projective S. Replacing the last paragraph of the previous section by it would already give a somewhat effective version to our theorem there.

Proposition 4.1. *Let S be a smooth complex projective surface, A a complex Abelian surface and $\alpha\colon S \to A$ a surjective morphism. We consider the following data on S:*

- $K = (\det d\alpha)$ *the canonical divisor of S determined by α*
- $D = K_{\mathrm{red}}$ *the reduced canonical divisor*
- C_0 *an elliptic curve in A considered as a reduced divisor*
- $\widehat{C} = \alpha^*(C_0)$ *the total transform of C_0*
- C' *the sum of nonelliptic horizontal components of $\widehat{C}$ on which $\widehat{C}$ has multiplicity one*
- $\overline{C} = \widehat{C} - C' - C''$ *where C'' is some vertical part of $\widehat{C}$, i.e., $C'' \leq \widehat{C}$ is α-exceptional.*

Let G be a horizontal component of D that is not a component of $\overline{C}$. Then we have

$$(G, \overline{C}) \leq 2(G, K + D) - n_G,$$

where n_G is the intersection number of $D - G$ with the smooth part of G.

Indication of proof. In view of the adjunction formula for a singular divisor and how it changes under blowing-ups of the divisor, the above inequality follows from the case where G is smooth. In this case, one obtains the following estimate at a point $x \in Q = G \cap \overline{C}_{\mathrm{red}}$ by considering the pullback γ to G of the linear one form w_0 that vanishes in the linear curve C_0.

$$(G, \overline{C})_x \leq (G, \widehat{C})_x = \mathrm{ord}_x(\gamma) + 1 \leq 2\,\mathrm{ord}_x(\gamma) - (G, D - G)_x. \tag{4.1}$$

The key to this comes from the fact that if $x \notin G \setminus (D - G)$, then either $\overline{C}$ is not reduced at x, in which case $(G, \widehat{C})_x \geq (G, \overline{C})_x \geq 2$ and so $\mathrm{ord}_x(\gamma) \geq 1$, or the components C of $\overline{C}$ through x are horizontal elliptic curves. In the latter case, since the tangent directions of these elliptic curves cannot lie in the kernel of $d\alpha$ as no elliptic curve can ramify over an elliptic curve, $d\alpha$ has rank one at x and maps the tangent "directions" of C at x, and therefore also T_xS, to $T_{\alpha(x)}C_0 = \ker w_0\big(\alpha(x)\big)$. Hence, in this case, $\gamma = \alpha^* w_0$ is identically zero on T_xS and so $\mathrm{ord}_x(\gamma) \geq 1$. It follows that in either case

$$(G, \overline{C})_x \leq (G, \widehat{C})_x = \mathrm{ord}_x(\gamma) + 1 \leq 2\,\mathrm{ord}_x(\gamma),$$

which is our desired estimate if $(G, D - G)_x = 0$. Hence if G is isolated in D, our proposition follows by summing over $x \in Q$ and noting that $n_G \leq (G, D - G)$:

$$\begin{aligned}(G, \overline{C}) \leq \sum_{x \in Q} (2 \operatorname{ord}_x(\gamma)) \leq \sum_{x \in G} (2 \operatorname{ord}_x(\gamma)) &= 2(G, K + G) \\ &= 2(G, K + D) - (G, D - G).\end{aligned}$$

If G is not isolated in D, the proposition comes from an easy consideration of the contribution of $(G, D - G)$. ☐

This is Proposition 3.2 in [23]. Observe that the theorem in the projective case would then follow if $G = K$. If $G \neq K$, we can still obtain $P'C \leq 2(P', K + D)$ for any effective nef $\mathbb{Q}$-divisor P' supported on the horizontal part of D by summing the obvious inequality for each component of P'. To get the general effective version, it is necessary to deal with vertical parts of D of negative self-intersection and what is shown in [23] is that they can be ignored for our estimate by absorbing their bad effects by the positive parts of K. That argument can be simplified quite a bit now and we present one simplification below (see [24]).

With the same setup as in Proposition 4.1, we will classify a curve as being vertical or horizontal according to whether it is α-exceptional or not. Let D_{h} be the horizontal part of D and let D_1 be the sum of all the components of D that have points in common with D_{h}. Then the vertical part of D_1 is given by $D_{1\mathrm{v}} = D_1 - D_{\mathrm{h}}$. The following is a simple consequence of the adjunction formula. Note that rational curves are vertical.

Lemma 4.2. *Let G be a component of D_1 such that $G(K + D_1) < 0$. Then G is a smooth rational curve intersecting the divisor $D_1 - G$ with multiplicity one and $G(K + D_1) = -1$.*

Our theorem in the projective case is as follows.

Theorem 4.3. *With the same setup and notations introduced, let C be a rational or an elliptic curve in S. If P is the positive part of the Zariski decomposition of K, then*

$$PC \leq 4P^2.$$

Proof. Since (-1)-curves are necessarily vertical, α will remain a morphism after their repeated blowing-downs. So we may assume that S is minimal, i.e., that $P = K$. If C is vertical, then $C \leq K$ and so $KC \leq K^2 \leq 4K^2$ by the nefness of K. As rational curves are necessarily vertical, it suffices to consider the case where C is a horizontal reduced elliptic curve for our proposition. In this case, $C_0 = \alpha(C)$ is an elliptic curve in A. We will also consider C_0 as a reduced divisor in A. Let C' be the sum of the nonelliptic horizontal components of $\widehat{C} = \alpha^* C_0$ on which $\widehat{C}$ has multiplicity one and let $\overline{C} = \alpha^* C_0 - C'$. Then $C \leq \overline{C} \leq \alpha^* C_0$ and so $KC \leq K\overline{C}$ as K is nef. Hence, it is sufficient to establish that

$$K\overline{C} \leq 4K^2.$$

Let D_1^- be the sum of the components of D_1 with strictly negative intersection with $K + D_1$. Then D_1^- is vertical by the above lemma. As K is nef and dominates both D_1 and $D_1^+ := D_1 - D_1^-$, the above inequality would follow from

$$K\overline{C} \leq 2K(K + D_1) - KD_1^- = K(K + D_1) + K(K + D_1^+). \tag{4.2}$$

This in turn would follow from

$$GC \leq 2G(K + D_1) - GD_1^- = G(K + D_1) + G(K + D_1^+) \tag{4.3}$$

for each components G of K. By considering the four possible types of components G of K below, we establish this last inequality for each of these types:

Type 1. G is horizontal but not a component of $\overline{C}$. Then G is not a component of $\widehat{C} := \alpha^* C_0$. By Proposition 4.1, we have

$$G\overline{C} \leq 2G(K + D) - n_G = 2G(K + D_1) - n_G \leq 2G(K + D_1) - GD_1^-$$

as each component of $D_1^- \leq D_1 - G \leq D - G$ intersects G with multiplicity at most one by Lemma 4.2 and hence necessarily at a smooth point of G if at all.

Type 2. G is a horizontal component of $\overline{C}$. As G is not a rational curve, it has nonnegative intersection with $K + D_1$ and with $K + D_1^+$. Hence,

$$G\overline{C} \leq G\widehat{C} = 0 \leq G(K + D_1) + G(K + D_1^+).$$

Type 3. G is a component of D_1^-. Then by Lemma 4.2, $G(K + D_1) = -1$ and $GD_1^+ = 1$. Hence,

$$G\overline{C} \leq G\widehat{C} = 0 \leq GK = G(K + D_1) + G(K + D_1^+)$$

Type 4. G is vertical but not a component of D_1^-. Then by construction, either G is a component of D_1^+ or it is not a component of D_1 at all. In the former case, G has nonnegative intersection with $K + D_1^+$ and with $K + D_1$ by the definition of D_1^+ and, in the latter case, this is so by the nefness of K. Hence,

$$G\overline{C} \leq G\widehat{C} = 0 \leq G(K + D_1) + G(K + D_1^+). \qquad \square$$

We remark that the only other general result on effective bounds on the canonical degrees of curves is due recently to Yoichi Miyaoka for curves on projective surfaces with positive Segre class in [29], which generalizes the same result for the case of nodal curves by him and the author in [25]. These results are obtained via the logarithmic Miyaoka – Yau inequality of Sakai in [38] and its orbifold generalizations, also used in [19] for similar purposes.

5. A tour on holomorphic curves and special varieties

We give an overview of the proof of Theorem 1.4 and its generalization as given in [27]. These are theorems about algebraic degeneracy of entire curves in the quasiprojective (or logarithmic) context.

Modulo the conjectures of Lang above for surfaces and outside certain birational class of K3 surfaces, a complete birational characterization (via the Kodaira dimension and the fundamental group) is given by Buzzard and the author in [2,3] for projective surfaces that admit Zariski dense entire curves, or equivalently, for those in which entire curves must be algebraically degenerate. A notion of "refined Kodaira" dimension was given in [20,21] whose vanishing yields the same characterization in the case of quasiprojective surfaces and is shown there to characterize all surfaces (modulo the Lang conjecture) that admit identically vanishing Kobayashi metric, including all K3 surfaces. Campana, at the same time, has coined the apt terminology of being "special" to conjecturally characterize in [8] the same class of varieties, though he was apparently more motivated by the arithmetic Lang conjectures among other sources. He has also defined a "refined Kodaira" dimension there that defines his original notion of "special," though he later changed it in [8] to conform more to the notion of "refined Kodaira" dimension mentioned first above. The two definitions of "refined Kodaira" dimension were shown to be equivalent in [22]. We now quickly introduce this notion of "special" as given in [21, 22] in order to explain some of our main results in [27].

Let X be a complex projective manifold and $\mathrm{Div}'(X)$ the set of codimension-one subvarieties of X. An orbifold structure on X is a $\mathbb{Q}$-divisor of the form

$$A = \sum_i (1 - 1/m_i) D_i$$

where $1 \leq m_i \in \mathbb{Q} \cup \{\infty\}$ and $D_i \in \mathrm{Div}'(X)$ for all i. We denote X with its orbifold structure by $X \setminus A$ and we set $K_{X\setminus A} := K_X(A)$ to be the orbifold canonical $\mathbb{Q}$-bundle. We set

$$m(A \cap D_i) = m(D_i \cap A) = m_i$$

and call it the multiplicity of the orbifold $X \setminus A$ (or simply, the orbifold multiplicity) at D_i. Note that when A is a (reduced) normal crossing divisor, the orbifold $X \setminus A$ is nothing but a log-manifold (X, A) whose birational geometry is dictated precisely by the proper birational geometry of the complement of A in X.

We now define the "base" Kodaira dimension of a rational map from an orbifold following [22]. Let $f\colon X \dashrightarrow Y$ be a rational map of complex projective manifolds and let w be a rational section of K_Y. If f is dominant, then f^*w defines a rational section of Ω_X^m with $m = \dim Y$, and hence determines in the standard way a saturated rank-one subsheaf L of Ω_X^m, which is easily seen to be unique in the birational equivalence class of f (it is even unchanged after composing with a dominant map from Y to a variety of the same dimension as Y). We recall that a saturated subsheaf of a locally free sheaf $\mathcal{S}$ is one that is not contained in any larger subsheaf of the same rank, and that it is reflexive. It follows that such a subsheaf, if it is rank-one, is locally free (see, for example, [30]). Hence, we can even define L without the dominant condition on f by setting $m = \dim f(X)$ and replacing Y by a desingularization of the algebraic image of f in general. Now f gives rise to an orbifold rational map in the category of orbifolds if an orbifold structure A is imposed on X. We denote this orbifold map by f^∂ and the orbifold $X \setminus A$ by X^∂ if A is implicit.

Definition 5.1. Let $f\colon X \dashrightarrow Y$ be a rational map giving rise to an invertible sheaf L on X as defined above. Let A be an orbifold structure on X giving rise to an orbifold rational map that we denote by

$$f^\partial = f|_{X\setminus A}\colon X \setminus A \dashrightarrow Y.$$

Define the vertical part of A with respect to f by

$$A \cap f = \sum \left\{ \left(1 - \frac{1}{m(D \cap A)}\right) D \;\Big|\; f(D) \neq f(X), D \in \mathrm{Div}'(X) \right\}.$$

We set $L_f^\partial = L(A \cap f)$, which is a $\mathbb{Q}$-invertible sheaf, and we define the Kodaira dimension of the orbifold rational map f^∂ by

$$\kappa(f, A) = \kappa(f^\partial) := \kappa(L_f^\partial).$$

Recall that a dominant rational map is called almost holomorphic if its general fibres are well defined. It is called a rational fibration if it becomes a fibration, i.e., connected, after resolving its indeterminancies.

Definition 5.2. Notation being as above, we call the orbifold rational map f^∂ (*base-wise*) *of general type* (or simply *base-general(-typical)*) if

$$\kappa(f^\partial) = \dim(f) > 0,$$

where $\dim(f)$ is the dimension of the algebraic image of f. We call the orbifold $X \setminus A$ *special* if it admits no base-general orbifold rational map and *general-typical* or *of general type* if the identity map restricted to the orbifold is base general. If f is a rational fibration, we say that $f^\partial = f|_{X\setminus A}$ is *base-special* if $X \setminus A$ has no orbifold rational map that is base-general and that factors through f; We will also consider the obvious generalization of this notion to dominant rational maps via Stein factorization. If f is an almost holomorphic fibration, we say that f^∂ is *special* (respectively *general-typical*) if its general fibres endowed with the orbifold structure given by the restriction of A are special orbifolds (repectively orbifolds of general type).

In the case where A is reduced and normal crossing, it is easily seen that these notions are, in the obvious manner, proper birational invariants of the open subset $X \setminus A$ and of the restriction of f to it.

Now let $f\colon X \to Y$ be a fibration with X and Y projective and smooth and let A be an orbifold structure on X. Then the induced orbifold fibration $f^\partial = f|_{X\setminus A}$ imposes an orbifold structure on Y as follows. Given $D \in \mathrm{Div}'(Y)$, we may write $f^*D = \sum_i m_i D_i$ with $m_i \in \mathbb{N}$ and $D_i \in \mathrm{Div}'(X)$. Then we define the (minimum) multiplicity of f^∂ over D by

$$m(D, f^\partial) = \min\{m_i\, m(D_i \cap A) \mid f(D_i) = D\}.$$

Definition 5.3. With the notation as given above, the $\mathbb{Q}$-divisor

$$D(f^\partial) = D(f|_{X\setminus A}) = D(f, A) := \sum \left\{ \left(1 - \frac{1}{m(D, f^\partial)}\right) D \;\middle|\; D \in \mathrm{Div}'(Y) \right\}$$

on Y defines the *orbifold base* $Y \setminus D(f^\partial)$ of $f^\partial = f|_{X\setminus A}$.

We now apply our setup to the quasi-Albanese map. Let X be a complex quasiprojective manifold, T a semi-Abelian variety and $u\colon X \to T$ an algebraic morphism. Let $\overline{T}$ be a smooth equivariant compactification of T, i.e., $\overline{T}$ is smooth and admits an algebraic action by T—an example being the compactification of T in the exact sequence (2.1) via the compactification $\mathbb{C}^k \subset (\mathbb{P}^1)^k$. Then one can observe, see [34], that $\overline{T} \setminus T$ is a normal crossing divisor and that $\overline{\Omega}_T$ is a trivial bundle over $\overline{T}$ (via simultaneous equivariant resolution of singularities for example). By the resolution of singularity theorem (see [5, 13]), there is a compactification $\overline{X}$ of X with normal crossing boundary divisor A such that u extends to a morphism $\bar{u}\colon \overline{X} \to \overline{T}$.

Definition 5.4. We call $\bar{u}$ as above a *natural compactification* of u. We will set $\bar{u}^\partial = \bar{u}|_{\overline{X}\setminus A}$ and, in the case where u is a fibre space (i.e., $\bar{u}$ is a fibration), we set $D(u) := D(\bar{u}^\partial)|_X$.

We note that $D(u)$ is a $\mathbb{Q}$-divisor on X that is independent of the natural compactification $\bar{u}$ of u chosen, since two such compactifications are always dominated by a third such compactification. By the same token, the notions of being special, being general-typical, being base-special and being base-general(-typical) are defined for u (independent of the natural compactifications).

Observe that if X is of log-general type and u is generically finite, then u is base general. Hence, Theorem 1.4 is reduced to the following proposition.

Proposition 5.5. *Let X be a smooth complex quasiprojective variety admitting a Zariski dense entire holomorphic curve. Then its (quasi-)Albanese map $\alpha\colon X \to T_0 = \mathrm{Alb}(X)$ is a fibre space and α is base special.*

The proof of this proposition uses two key ingredients, the first algebraic and the second holomorphic. We summarize these ingredients below.

The first ingredient is the following proposition.

Proposition 5.6. *Let X be a complex quasiprojective manifold with quasi-Albanese map $f\colon X \to T_0 = \mathrm{Alb}(X)$. If f is not base special, then there is a proper semi-abelian subvariety T' of T_0 such that if $g\colon T_0 \to T = T_0/T'$ is the quotient map, then $h = g \circ f$ is base general and $\dim T > 0$. Furthermore, if $D = D(h)_{\mathrm{red}}$, then $\mathcal{O}(\overline{D})$ on $\overline{T}$ is big for any equivariant compactification $\overline{T}$ of T.*

This proposition follows quite easily from the structure theorem of Kawamata [16] and the following basic lemma.

Lemma 5.7. *Let $f\colon X \to Y$ and $g\colon Y \to T$ be fibrations with orbifold structures A on X and $D(f, A)$ on Y and t a general point on T. Assume A is a normal crossing divisor. If f^{∂} and $Y_t^{\partial} = Y_t \setminus D(f|_{X_t^{\partial}})$ are special, then so is $X_t^{\partial} := X_t \setminus A_t$. If $T \setminus D(g \circ f, A)$ is special, then f^{∂} is base-special if f_t^{∂} is.*

This lemma, in turn, follows readily from the next proposition ([22, Proposition 6.5]), which is the key, if not the most important, proposition for understanding varieties in terms of special varieties.[1]

Proposition 5.8. *Let X^{∂} be a (smooth) orbifold, $f^{\partial}\colon X^{\partial} \to T$ a special orbifold fibration and $h^{\partial}\colon X^{\partial} \dashrightarrow Z$ a base-general orbifold rational map. Then $h^{\partial} = k \circ f^{\partial}$ for some rational map $k\colon T \dashrightarrow Z$ and, for any such, $k^{\partial} := k|_{T \setminus D(f^{\partial})}$ is base-general. In particular, if $T \setminus D(f^{\partial})$ is special, then f^{∂} is base special and hence X^{∂} is special.*

For the details of all of these, most of which are elementary, we refer to [27].

The second ingredient consists of the Second Main Theorem in Nevanlinna theory with truncation to level one for semi-Abelian varieties due to Noguchi–Winkelmann–Yamanoi in [35, 37], as well as a few other results due to the same authors. We will detail this theory below before proving Proposition 5.5.

We first recall some relevant definitions and facts from Nevanlinna theory. We will follow Section 2 of [37]. Let T be a complex manifold, ω a real smooth $(1,1)$-form and $\gamma\colon \mathbb{C} \to T$ a holomorphic map. Then the order function of γ with respect to ω is defined by

$$T_{\gamma}(r;\omega) = \int_1^r \frac{\mathrm{d}t}{t} \int_{|z|<t} f^*\omega \quad (r > 1). \tag{5.1}$$

If T is Kähler and ω, ω' are d-closed real $(1,1)$-forms in the same cohomology class $[\omega]$, then

$$T_{\gamma}(r;\omega) = T_{\gamma}(r;\omega') + O(1).$$

[1] This proposition is also found in [9] in a slightly weaker form that is not directly applicable for us.

Hence we may set, up to $O(1)$-terms,

$$T_\gamma(r;[\omega]) = T_\gamma(r;\omega). \tag{5.2}$$

Let $L \to T$ be a Hermitian line bundle. As its Chern class is a real $(1,1)$-class, we may set

$$T_\gamma(r;L) = T_\gamma\bigl(r;c_1(L)\bigr).$$

We will denote by $\mathcal{O}_T(D)$ the line bundle determined by a divisor D on T via a standard abuse of notation and set $T_\gamma(r;D) = T_\gamma(r;\mathcal{O}_T(D))$. In terms of a standard Weil function for a subscheme W of T (see [37, Section 2]), we can define $T_\gamma(r;W)$ similarly, which we will also denote by $T_\gamma(r;\mathcal{I}_W)$, where $\mathcal{I}_W$ is the ideal subsheaf of $\mathcal{O}_T$ defining W. This is because, ideal sheaves pull back to ideal sheaves on $\mathbb{C}$ which then define effective divisors on $\mathbb{C}$. Henceforth, we will identify ideal sheaves with their effective divisors on $\mathbb{C}$.

Let $E = \sum_{z\in\mathbb{C}}(\operatorname{ord}_z E)z$ be an effective divisor on $\mathbb{C}$, $S \subset \mathbb{C}$ and $l \in \mathbb{N} \cup \{\infty\}$. Then the above sum is a sum over a discrete subset of $\mathbb{C}$. Hence the sum is finite when restricted to the disk $\mathbb{D}_t$ of radius $t > 0$ and so

$$n(t;E) = \deg_{\mathbb{D}_t} E := \sum_{z\in\mathbb{D}_t} \operatorname{ord} E$$

makes sense. We define the restriction of E to S truncated to order l by

$$E_{S,l} = \sum_{z\in S} \min(\operatorname{ord}_z E, l)z$$

and set $n_l(t;E,S) = n(t;E_{S,l})$. Then the counting functions of E with (respectively without) truncation to order l are given by

$$N_l(r;E,S) = \int_1^r \frac{n_l(t;E,S)}{t}\,\mathrm{d}t, \qquad N_l(r;E) = N_l(r;E,\mathbb{C})$$

(respectively $N(r;E) = N_\infty(r;E)$).

A well-known consequence of these definitions via the classical Jensen formula is the First Main Theorem:

$$N(r;\gamma^*\mathcal{I}_W) \le T_\gamma(r;W) + O(1), \tag{5.3}$$

where W is a subscheme of T and $\gamma^*\mathcal{I}_W$, as an ideal subsheaf of $\mathcal{O}_\mathbb{C}$, is identified with a divisor on $\mathbb{C}$. By the linearity of n_l and hence of N_l with respect to the third variable, if T is a disjoint union of U and V, then

$$N_l(r;\gamma^*\mathcal{I}_W) = N_l\bigl(r;\gamma^*\mathcal{I}_W,\gamma^{-1}(U)\bigr) + N_l\bigl(r;\gamma^*\mathcal{I},\gamma^{-1}(V)\bigr).$$

Also if $Z \subset E$ is an inclusion of reduced algebraic subsets of T and $Z = \operatorname{supp}\mathcal{I}$, then

$$N(r;\gamma^*\mathcal{I}_Z) \ge N(r;(\gamma^*\mathcal{I}_Z)_{\mathrm{red}}) = N_1(r;\gamma^*\mathcal{I}_Z) = N_1(r;\gamma^*\mathcal{I}_E,\gamma^{-1}(Z)). \tag{5.4}$$

By definition, we also have for D a strictly effective divisor on T that $T_\gamma(r;D) \ge 0$ if the image of γ is not in D_{red}. This fact along with Kodaira's lemma (Lemma 2.5) and linearity of T_γ with respect to to the second variable yields easily the following (which is Lemma 2.3 of [37]):

Lemma 5.9. *Suppose that T is a complex projective manifold with a big divisor A. Then*

$$T_\gamma(r) = O\bigl(T_\gamma(r;A)\bigr).$$

(Here by convention $T_\gamma(r) := T_\gamma(r;H)$ for an ample divisor H on T.)

We will need the following two theorems of Noguchi – Winkelmann – Yamanoi.

Theorem 5.10 ([36]). *Let X be a normal complex quasiprojective variety admitting a Zariski dense entire holomorphic curve. Let $f\colon X \to T$ be a finite morphism to a semi-Abelian variety. Then f is an étale morphism onto a semi-Abelian subvariety.*

Theorem 5.11 ([37]). *Let T be a semi-abelian variety and $\gamma\colon \mathbb{C} \to T$ a holomorphic map with Zariski dense image. Let E be a divisor on T and $\overline{E}$ be its Zariski closure in an equivariant compactification of T. Let $\mathcal{I}$ be an ideal subsheaf of $\mathcal{O}_T$ such that $\mathcal{O}_T/\mathcal{I}$ is supported on a codimension-two subvariety of T. Then we have:*

$$N(r;\gamma^*\mathcal{I}) \le \epsilon T_\gamma(r)\|_\epsilon \quad \text{for all } \epsilon > 0, \tag{5.5}$$

$$T_\gamma(r;\overline{E}) \le N_1(r;\gamma^*E) + \epsilon T_\gamma(r;\overline{E})\|_\epsilon \quad \text{for all } \epsilon > 0, \tag{5.6}$$

where "$\|_\epsilon$" stands for the inequality to hold for every $r > 1$ outside a Borel set of finite Lebesgue measure that depends on ϵ.

Corollary 5.12. *Let X be a smooth complex quasiprojective variety admitting a Zariski dense entire holomorphic curve. Then the (quasi-)Albanese map f of X is a fibre space.*

Proof of corollary. This is an easy argument applying Theorem 5.10 to the Stein factorization of a natural compactification of f. □

We are finally ready to prove Proposition 5.5.

Proof of Proposition 5.5. The first part of the proposition is just Corollary 5.12. We may thus assume that α is a fibre space but not base special. The proposition is proved once we reach a contradiction with the existence of an entire map $\gamma_0\colon \mathbb{C} \to X$ with Zariski dense image in T_0. By Proposition 5.6, there is a quotient morphism $g : T_0 \to T$ of semi-Abelian varieties such that the fibre space $h = g \circ f\colon X \to T$ induces an orbifold base of general type with $\dim T > 0$, that is, given any equivariant compactification $\overline{T}$ of T, $\mathcal{O}(\overline{D})$ is big, where $D = D(h)_{\mathrm{red}}$. Let $\gamma = h \circ \gamma_0$. As γ has Zariski dense image in T, Theorem 5.11 gives

$$T_\gamma(r;\overline{D}) \le N_1(r;\gamma^*D) + \epsilon T_\gamma(r;\overline{D})\|_\epsilon \quad \text{for all } \epsilon > 0. \tag{5.7}$$

From the definition of $D(h)$, we see that outside the exceptional locus E of h, the effective divisor $D_0 = h^*(D)$ is nowhere-reduced along $h^{-1}(D)$. Since $Z = h(E)$ is of co-dimension two or higher, we have from Theorem 5.11 that

$$N(r;\gamma^*Z) \le \epsilon T_\gamma(r)\|_\epsilon \text{for all } \epsilon > 0. \tag{5.8}$$

As $\gamma^*(D) = \gamma_0^*(D_0)$ is then an effective divisor that is nowhere-reduced outside $\gamma^{-1}(Z)$, we have by (5.4) that

$$N(r;\gamma^*(D)) \ge 2N_1\big(r;\gamma^*D,\gamma^{-1}(T\setminus Z)\big) \ge 2N_1\big(r;\gamma^*(D)\big) - 2N(r;\gamma^*\mathcal{I}_Z). \tag{5.9}$$

Also, since $\overline{D}$ is big, $T_\gamma(r) = O(T_\gamma(r;\overline{D})$ by Lemma 5.9. That is, there exists a constant $C > 0$ (depending on the ample divisor used to define $T_\gamma(r)$) such that

$$T_\gamma(r) \le CT_\gamma(r;\overline{D}).$$

Couple all these with the First Main Theorem, equation (5.3), gives

$$\begin{aligned}T_\gamma(r;\overline{D}) + O(1) \geq N(r;\gamma^* D) &\geq 2N_1(r;\gamma^* D) - 2N(r;\gamma^* Z)\\ &\geq (1-\epsilon)2T_\gamma(r;\overline{D}) - 2\epsilon C T_\gamma(r;\overline{D})\|_\epsilon\\ &\geq 2(1-\epsilon-\epsilon C)T_\gamma(r;\overline{D})\|_\epsilon,\end{aligned}$$

valid for all $\epsilon > 0$. This gives a contradiction for $\epsilon > 0$ sufficiently small. □

References

1. D. Abramovich, *Subvarieties of semiabelian varieties*, Compositio Math. **90** (1994), no. 1, 37–52.
2. C. Birkenhake and H. Lange, *Complex abelian varieties*, 2nd ed., Grundlehren Math. Wiss., vol. 302, Springer, Berlin, 2004.
3. G. T. Buzzard and S. S.-Y. Lu, *Algebraic surfaces holomorphically dominable by* $\mathbf{C}^2$, Invent. Math. **139** (2000), no. 3, 617–659.
4. ———, *Double sections, dominating maps, and the Jacobian fibration*, Amer. J. Math. **122** (2000), no. 5, 1061–1084. MR1781932 (2001m:32053)
5. E. Bierstone and P. D. Milman, *Canonical desingularization in characteristic zero by blowing up the maximum strata of a local invariant*, Invent. Math. **128** (1997), no. 2, 207–302, DOI 10.1007/s002220050141. MR1440306 (98e:14010)
6. W. Barth, C. Peters, and A. Van de Ven, *Compact complex surfaces*, Ergeb. Math. Grenzgeb. (3), vol. 4, Springer, Berlin, 1984.
7. F. A. Bogomolov, *Holomorphic tensors and vector bundles on projective manifolds*, Izv. Akad. Nauk SSSR Ser. Mat. **42** (1978), no. 6, 1227–1287 (Russian); English transl., Math. USSR-Izv. **13** (1979), no. 3, 499–555.
8. F. Campana, *Orbifolds, special varieties and classification theory*, Ann. Inst. Fourier (Grenoble) **54** (2004), no. 3, 499–630, available at `arXiv:math/0110051`.
9. ———, *Orbifoldes géométriques spéciales et classification biméromorphe des variétés kählériennes compactes*, available at `arXiv:0705.0737`.
10. O. Debarre, *Tores et variétés abéliennes complexes*, Cours Spec., vol. 6, Soc. Math. France, Paris, 1999.
11. G. Dethloff and S. S.-Y. Lu, *Logarithmic surfaces and hyperbolicity*, Ann. Inst. Fourier (Grenoble) **57** (2007), no. 5, 1575–1610.
12. M. Green and P. Griffiths, *Two applications of algebraic geometry to entire holomorphic mappings*, The Chern Symposium 1979 (Berkeley, CA, 1979) (W. Y. Hsiang, S. Kobayashi, I. M. Singer, A. Weinstein, J. Wolf, and H. H. Wu, eds.), Springer, New York, 1980, pp. 41–74.
13. H. Hironaka, *Resolution of singularities of an algebraic variety over a field of characteristic zero*. I, Ann. of Math. (2) **79** (1964), 109–203; II, 205–326.
14. S. Iitaka, *Algebraic geometry*, Grad. Texts in Math., vol. 76, Springer, New York, 1982.
15. Y. Kawamata, *On Bloch's conjecture*, Invent. Math. **57** (1980), no. 1, 97–100.
16. ———, *Characterization of abelian varieties*, Compositio Math. **43** (1981), no. 2, 253–276.
17. J. Kollár and T. Matsusaka, *Riemann–Roch type inequalities*, Amer. J. Math. **105** (1983), no. 1, 229–252.
18. S. Lang, *Hyperbolic and Diophantine analysis*, Bull. Amer. Math. Soc. (N.S.) **14** (1986), no. 2, 159–205.
19. A. Langer, *Logarithmic orbifold Euler numbers of surfaces with applications*, Proc. London Math. Soc. (3) **86** (2003), no. 2, 358–396.
20. S. S.-Y. Lu, *Multiply marked Riemann surfaces and the Kobayashi pseudometric on algebraic manifolds* (2000), preprint.
21. ———, *The Kobayashi pseudometric on algebraic manifold and a canonical fibration*, available at `arXiv:math/0206170`.
22. ———, *A refined Kodaira dimension and its canonical fibration*, available at `arXiv:math/0211029`.
23. ———, *On surfaces of general type with maximal Albanese dimension*, J. Reine Angew. Math. **641** (2010), 163–175.
24. ———, *Non-hyperbolic curves in quasiprojective surfaces*, preprint.

25. S. S.-Y. Lu and Y. Miyaoka, *Bounding curves in algebraic surfaces by genus and Chern numbers*, Math. Res. Lett. **2** (1995), no. 6, 663–676.
26. ———, *Bounding codimension-one subvarieties and a general inequality between Chern numbers*, Amer. J. Math. **119** (1997), no. 3, 487–502.
27. S. S.-Y. Lu and J. Winkelmann, *Quasiprojective varieties admitting Zariski dense entire holomorphic curves*, Forum Math, to appear.
28. Y. Miyaoka, *On the Chern numbers of surfaces of general type*, Invent. Math. **42** (1977), 225–237.
29. ———, *The orbibundle Miyaoka – Yau – Sakai inequality and an effective Bogomolov – McQuillan theorem*, Publ. Res. Inst. Math. Sci. **44** (2008), no. 2, 403–417.
30. Y. Miyaoka and T. Peternell, *Geometry of higher-dimensional algebraic varieties*, DMV Sem., vol. 26, Birkhäuser, Basel, 1997.
31. J. Noguchi, *Lemma on logarithmic derivatives and holomorphic curves in algebraic varieties*, Nagoya Math. J. **83** (1981), 213–233.
32. ———, *Holomorphic curves in algebraic varieties.*, Hiroshima Math. J. **7** (1977), no. 3, 833–853; Supplement, **10** (1980), no. 1, 229–231.
33. J. Noguchi and T. Ochiai, *Geometric function theory in several complex variables*, Iwanami Shoten, Tokyo, 1984 (Japanese); English transl. in Transl. Math. Monogr., vol. 80, Amer. Math. Soc., Providence, RI, 1990.
34. J. Noguchi and J. Winkelmann, *Bounds for curves in abelian varieties*, J. Reine Angew. Math. **572** (2004), 27–47.
35. J. Noguchi, J. Winkelmann, and K. Yamanoi, *The second main theorem for holomorphic curves into semi-abelian varieties*, Acta Math. **188** (2002), no. 1, 129–161.
36. ———, *Degeneracy of holomorphic curves into algebraic varieties*, J. Math. Pures Appl. (9) **88** (2007), no. 3, 293–306.
37. ———, *The second main theorem for holomorphic curves into semi-abelian varieties.* II, Forum Math. **20** (2008), no. 3, 469–503.
38. F. Sakai, *Semistable curves on algebraic surfaces and logarithmic pluricanonical maps*, Math. Ann. **254** (1980), no. 2, 89–120.
39. K. Ueno, *Classification theory of algebraic varieties and compact complex spaces*, Lecture Notes in Math., vol. 439, Springer, Berlin, 1975.
40. O. Zariski, *The theorem of Riemann – Roch for high multiples of an effective divisor on an algebraic surface*, Ann. of Math. (2) **76** (1962), 560–615.

Département de mathématiques, Université du Québec à Montréal, C.P. 8888, succ. Centre-ville, Montréal, QC H3C 3P8, Canada

E-mail address: lu.steven@uqam.ca

Centre de Recherches Mathématiques
CRM Proceedings and Lecture Notes
Volume **54**, 2011

Locally Nilpotent Derivations of Affine Domains

L. Makar-Limanov

To Peter Russell who asks the right questions

Abstract. Any locally nilpotent derivation of an affine domain is equivalent to a restriction of a Jacobian type derivation of a polynomial ring.

1. Introduction

Let us recall that a locally nilpotent derivation is a generalization of a partial derivative of a polynomial ring. It is a derivation of a ring which sends to zero a given element of the ring after sufficiently many applications

Locally nilpotent derivations (let us call them lnd) are useful if rather elusive objects. Though we do not have them at all on "majority" of rings, when we have them, they are rather hard to find and it is even harder to find all of them or to give any qualitative statements. We do not know much even for polynomial rings. It is, of course, easy enough to understand the situation for a polynomial ring with one variable. It is not easy to describe all lnds for polynomial rings in two variables; in fact it requires considerable effort (see [14, 16]). For three variables, the kernel of a nonzero lnd is a polynomial ring in two variables ([15]) and there is a description of all weighted homogeneous derivations, i.e., derivations which send forms, homogeneous relative to a given set of variables supplied with given positive weights to homogeneous forms ([2, 6, 8]). On polynomial rings with more than four variables, there are examples of lnds with nonfinitely generated kernels ([4, 7, 9, 17]). Though it is not known whether such lnds exist in the case of four variables, it is know that there is no uniform bound on the number of generators of kernels in this case ([5]).

2010 *Mathematics Subject Classification.* 13N15, 14R20.

The author is supported by NSA and NSF grants and by the Fulbright United States – Israel Educational foundation. It is also a pleasure to acknowledge the hospitality of several institutions the author enjoyed while working on this project: the Faculty of Mathematics and Computer Science of Nicolaus Copernicus University in Torun, Poland, the Department of Mathematics of University of Hong Kong, China, the Max Plank Institute of Mathematics in Bonn, Germany, and the Weizmann Institute of Science where he is presently a Fulbright Scholar.

This is the final form of the paper.

There are several classes of surfaces for which we know everything as far as lnds are concerned ([3, 12–14, 19]) and a class of threefolds for which we know a lot ([10, 11]), and that is all.

The purpose of this paper is to provide a standard form for an lnd on the affine domains. This form is somewhat analogous to a matrix representation of a linear operator. Hopefully, this representation of lnd will be helpful in the future research.

2. Definitions and claim

Let $\mathbb{C}$ be the field of complex numbers and let A be a domain over $\mathbb{C}$. A $\mathbb{C}$-linear mapping $\partial\colon A \to A$ is called a derivation if it distributes a product of elements of A according to the Leibnitz law: $\partial(ab) = \partial(a)b + a\partial(b)$.

Constants $A^\partial = \{a \in A \mid \partial(a) = 0\}$ of ∂ form a subalgebra of A.

A derivation is locally nilpotent if for any $a \in A$ there exists a natural $n = n(a)$ for which $\partial^n(a) = 0$. Let us denote the set of all lnds of A by $\mathrm{LND}(A)$.

Let $\mathbb{C}_n = \mathbb{C}[x_1, \dots, x_n]$ be the polynomial ring with n generators. A class of derivations on $\mathbb{C}_n$ can be obtained as follows. Define $\partial(g) = \mathrm{J}(f_1, \dots, f_{n-1}, g)$ where $\mathrm{J}(f_1, \dots, f_{n-1}, g)$ is the Jacobian of the elements $f_1, \dots, f_{n-1}, g \in \mathbb{C}_n$, i.e., the determinant of the corresponding Jacobi matrix. These are so called Jacobian derivations. A Jacobian derivation is also a derivation in every parameter f_i, which helps to make computations

Now we are ready to formulate the main result.

Theorem 2.1. *Let I be a prime ideal of $\mathbb{C}_n$, R be the factor ring $\mathbb{C}_n/I$, and $\pi\colon \mathbb{C}_n \to R$ be the corresponding projection. Let $\partial \in LND(R)$. Then there exist elements $f_1, \dots, f_{n-1} \in \mathbb{C}_n$ and nonzero elements $a, b \in R^\partial$ for which $a\partial\big(\pi(f)\big) = b\pi\big(\mathrm{J}(f_1, \dots, f_{n-1}, f)\big)$ for any $f \in C_n$.*

The proof consists of eight lemmas. To make it less boring for the reader, here is an outline of what is going to happen.

In the first five lemmas we will see that it is possible to find k elements $i_1, \dots, i_k \in I$ so that $\mathrm{J}(i_1, \dots, i_k, f_{k+1}, \dots, f_n) \in I$ if and only if the elements $f_{k+1}, \dots, f_n$ are algebraically dependent over I (this is probably obvious to a geometer). The derivation ∂ appears only in the last three lemmas.

3. The first five lemmas

Let I be a prime ideal of $\mathbb{C}_n$. Choose a nonzero element $p \in I$ of the minimal possible total degree. We are not concerned, of course, with the case $p \in \mathbb{C}$. Up to renumbering of generators $x_1, \dots, x_n$ we may assume that p contains x_1. If we consider p as a polynomial in x_1 with coefficients in $\mathbb{C}[x_2, \dots, x_n]$, then the coefficients of p with positive degrees of x_1 do not belong to I since their degrees are too small. By looking at all the elements of I as polynomials of x_1 we can find an element $p_1 \in I$ satisfying the following two conditions:

The coefficient q_1 of the highest degree of x_1 of p_1 is not in I.

The degree d of p_1 relative to x_1 is the smallest degree possible, provided that the first condition is satisfied.

The degree d is positive because otherwise $p_1 \in I$ is the coefficient of x_1^0, and the first condition is not satisfied.

Let $I_1 = I \cap \mathbb{C}[x_2, \ldots, x_n]$. Now find $p_2 \in I_1$ with the same property relative to x_2 (again up to a possible renumbering) and so on until we reach k for which $I_k = 0$. In the process, we will obtain polynomials $p_1, p_2, \ldots, p_k$.

Lemma 3.1. *Any elements $f_k, \ldots, f_n \in \mathbb{C}_n$ are algebraically dependent over I.*

PROOF. We may assume that $f_k, \ldots, f_n$ are algebraically independent since otherwise the statement is clear. For a natural number N consider the linear subspace V_N of $\mathbb{C}_n$ of all polynomials of $f_k, \ldots, f_n$ with the total degree relative to $f_k, \ldots, f_n$ not exceeding N. The linear dimension of V_N over $\mathbb{C}$ is at least $(N/n)^{n-k+1}$ since the monomials $\prod_{l=k}^{n} f_l^{i(l)}$ belong to V_N if all $i(l) \leq N/n$ and are linearly independent because $f_k, \ldots, f_n$ are algebraically independent.

Denote by deg the total degree of elements of $\mathbb{C}_n$ relative to $x_1, \ldots, x_n$ and by $\deg_j$ the degree relative to x_j. If $D = \max\big(\deg(f_k), \ldots, \deg(f_n)\big)$ then $\deg(f) \leq DN$ for any $f \in V_N$. Also let $e_i = \deg(p_i)$.

Take $q = ax_1^m + \cdots$ with $\deg(q) \leq DN$. Recall that $p_1 = q_1 x_1^d + \cdots$. (Here $\cdots$ are terms of smaller degrees in x_1.) Let $r_{1,1} = q_1 q - a x_1^{m-d} p_1$ if $m \geq d$. Then $\deg_1(r_{1,1}) < \deg_1(q)$. If $m < d$ take $r_{1,1} = q_1 q$. Since $\deg(a x_1^{m-d}) \leq \deg(q)$ and $\deg(q_1) < \deg(p_1)$ in both cases $\deg(r_{1,1}) \leq DN + e_1$. Since $DN \geq \deg_1(q)$, after at most DN steps like this we will obtain r_1 with $\deg_1(r_1) < \deg_1(p_1)$ and $\deg(r_1) \leq DN(1+e_1)$. Now apply the same procedure to r_1 and p_2. After at most $DN(1+e_1)$ steps we obtain r_2 with $\deg_2(r_2) < \deg_2(p_2)$ and $\deg(r_2) \leq DN(1+e_1)(1+e_2)$. Also $\deg_1(r_2) \leq \deg_1(r_1)$ since $\deg_1(q_2) = \deg_1(p_2) = 0$. Proceeding with r_2, p_3, etc., we will get r_k with $\deg(r_k) \leq DN(1+e_1)\cdots(1+e_k)$ and $\deg_i(r_k) < \deg_i(p_i)$.

Take $\mathbf{m} = q_1 q_2^{(1+e_1)} \ldots q_k^{(1+e_1)\cdots(1+e_{k-1})}$. We observed that $\mathbf{m}^{DN} q = \sum \alpha_i p_i + r_k$ where $\deg_j(r_k) < \deg_j(p_j)$ for $j = 1, \ldots, k$ and $\deg(r_k) \leq DN(1+e_1)\cdots(1+e_k)$ if $\deg(q) \leq DN$. Denote r_k by $r_k(q)$. It is clear that when N is fixed, $q \to r_k(q)$ is a linear mapping of V_N onto a subspace R_N of $\mathbb{C}_n$. Let us estimate the linear dimension of R_N. By construction, $\deg_i(r_k) < \deg_i(p_i)$ and this bound does not depend on N. Also $\deg\big(r_k(q)\big) \leq DN(1+e_1)\cdots(1+e_k)$. For a monomial $x_1^{i_1} \ldots x_n^{i_n}$ of $r_k(q)$ powers $i_j < \deg_j(p_j)$ for $j = 1, \ldots, k$ and $i_j \leq DN(1+e_1)\cdots(1+e_k)$ for $j = k+1, \ldots, n$. The number of monomials with these restrictions does nor exceed $\big(\prod_{j=1}^{k} \deg_j(p_j)\big)[D(1+e_1)\cdots(1+e_k)]^{n-k} N^{n-k}$ and gives an upper bound for the dimension of R_N.

Hence, $\dim(V_N) > \dim(R_N)$ for a sufficiently large N and a linear mapping from V_N to R_N should have a kernel. This implies that there is a nonzero $f \in V_N$ with $r_k(f) = 0$. So $\mathbf{m}^{DN} f = \sum \alpha_i p_i \in I$ and since I is prime and all $q_i \notin I$, we conclude that f, which is a polynomial in $f_k, \ldots, f_n$, is in I. □

Remark 3.2. If $f \in I$ then $r_k(f) = 0$ for any N. Indeed, in this case $r = r_k(f) \in I$. Since $\deg_1(r) < \deg_1(p_1)$, from the definition of p_1 all coefficients of r as a polynomial in x_1 must be in I_1. Looking at these coefficients as polynomials in x_2 and using $\deg_2(r) < \deg_2(p_2)$ and the definition of p_2, we can see that they, in their turn, have all the coefficients in I_2, etc.. Thus $r = 0$ since $I_k = 0$.

Lemma 3.3. $\mathrm{J}(p_1, p_2, \ldots, p_k, x_{k+1}, \ldots, x_n) \notin I$.

PROOF. Indeed,

$$\mathrm{J}(p_1, p_2, \ldots, p_k, x_{k+1}, \ldots, x_n) = \frac{\partial p_1}{\partial x_1} \frac{\partial p_2}{\partial x_2} \cdots \frac{\partial p_k}{\partial x_k}$$

since only p_1 depends on x_1, only p_2 depends on x_2, etc.. But $\partial p_i/\partial x_i \notin I$ for all i by definition of p_i, so $(\partial p_1/\partial x_1)(\partial p_2/\partial x_2)\cdots(\partial p_k/\partial x_k) \notin I$ since I is prime. □

Lemma 3.4. *$\mathrm{J}(f_1,\dots,f_{k+1},f_{k+2},\dots,f_n) \in I$ if $f_1,\dots,f_{k+1} \in I$.*

PROOF. By Remark 3.2 we can find a monomial $\mathbf{M} \notin I$ such that $\mathbf{M}f_i = \sum \alpha_{i,j}p_j$ for $i \le k+1$. Therefore,

$$\begin{aligned}\mathrm{J}(\mathbf{M}f_1,\dots,\mathbf{M}f_{k+1},f_{k+2},\dots,f_n)&\\ &= \mathbf{M}^{k+1}\,\mathrm{J}(f_1,\dots,f_{k+1},f_{k+2},\dots,f_n)+\Delta\\ &= \sum \mathrm{J}(\alpha_{1,j_1}p_{j_1},\dots,\alpha_{k+1,j_{k+1}}p_{j_{k+1}},f_{k+2},\dots,f_n).\end{aligned}$$

Each summand $\mathrm{J}(\alpha_{1,j_1}p_{j_1},\dots,\alpha_{k+1,j_{k+1}}p_{j_{k+1}},f_{k+2},\dots,f_n) \in I$ since there is only k different p_i's. Also $\Delta \in I$ because it is a linear combination of Jacobians with coefficients containing f_i's where $i \le k+1$. Thus,

$$\begin{aligned}\mathbf{M}^{k+1}\,\mathrm{J}(f_1,\dots,f_{k+1},f_{k+2},\dots,f_n)&\\ &= \sum \mathrm{J}(\alpha_{1,j_1}p_{j_1},\dots,\alpha_{k+1,j_{k+1}}p_{j_{k+1}},f_{k+2},\dots,f_n)-\Delta \in I\end{aligned}$$

and $\mathrm{J}(f_1,\dots,f_{k+1},f_{k+2},\dots,f_n) \in I$ since I is prime. □

Lemma 3.5. *$\mathrm{J}(f_1,\dots,f_k,f_{k+1},\dots,f_n) \in I$ if $f_1,\dots,f_k \in I$ and $f_{k+1},\dots,f_n$ are dependent over I.*

PROOF. Let P be a nonzero polynomial of $f_{k+1},\dots,f_n$ with value in I of the minimal total degree possible. We may assume (up to renumbering) that P depends on f_n. Hence, $\mathrm{J}(f_1,\dots,f_k,f_{k+1},\dots,f_{n-1},P) \in I$ by Lemma 3.4 and $\mathrm{J}(f_1,\dots,f_k,f_{k+1},\dots,f_n) \in I$ since I is prime, $\partial P/\partial f_n \notin I$, and

$$\mathrm{J}(f_1,\dots,f_k,f_{k+1},\dots,f_{n-1},P) = \frac{\partial P}{\partial f_n}\,\mathrm{J}(f_1,\dots,f_k,f_{k+1},\dots,f_n). \qquad \square$$

Lemma 3.6. *$\mathrm{J}(p_1,\dots,p_k,f_{k+1},\dots,f_n) \notin I$ if $f_{k+1},\dots,f_n$ are independent over I.*

PROOF. Assume that $\mathrm{J}(p_1,\dots,p_k,x_{k+1},\dots,x_{k+i},f_{k+i+1},\dots,f_n) \notin I$ and that i is the minimal possible. From Lemma 3.3 we know that $i \le n-k$. The elements $x_{k+1},\dots,x_{k+i},f_{k+i+1},\dots,f_n$ are algebraically independent over I by Lemma 3.5 and $x_{k+1},\dots,x_{k+i},f_{k+i},f_{k+i+1},\dots,f_n$ are algebraically dependent over I by Lemma 3.1. Let us take a nonzero polynomial P of $x_{k+1},\dots,x_{k+i},f_{k+i},f_{k+i+1},\dots,f_n$ with value in I of the minimal total degree possible. By the assumption of the Lemma, $f_{k+i},f_{k+i+1},\dots,f_n$ are algebraically independent over I, thus P depends on f_{k+i} and (up to renumbering) on x_{k+i}. By Lemma 3.4

$$\mathrm{J}(p_1,\dots,p_k,x_{k+1},\dots,x_{k+i-1},P,f_{k+1+i},\dots,f_n) \in I.$$

Therefore

$$\begin{aligned}\frac{\partial P}{\partial x_{k+i}}\,\mathrm{J}(p_1,\dots,p_k,x_{k+1},\dots,x_{k+i},f_{k+i+1},\dots,f_n)&\\ +\frac{\partial P}{\partial f_{k+i}}\,\mathrm{J}(p_1,\dots,p_k,x_{k+1},\dots,x_{k+i-1},f_{k+i},\dots,f_n) &\in I.\end{aligned}$$

Since $\mathrm{J}(p_1,\dots,p_k,x_{k+1},\dots,x_{k+i},f_{k+i+1},\dots,f_n) \notin I$, $\partial P/\partial x_{k+i} \notin I$ and $\partial P/\partial f_{k+i} \notin I$, we can see that $\mathrm{J}(p_1,\dots,p_k,x_{k+1},\dots,x_{k+i-1},f_{k+i},\dots,f_n) \notin I$. To avoid a contradiction we must assume that $\mathrm{J}(p_1, \dots, p_k, x_{k+1}, \dots, x_{k+i}, f_{k+i+1}, \dots, f_n)$ does not contain x_j's at all, which proves the lemma. □

4. Additional definitions and facts about lnd

Let A be a domain of characteristic zero admitting a nonzero locally nilpotent derivation ∂. Then there exists an element $t \in A$ for which $\partial(t) \in A^\partial \setminus 0$. For $a \in A$, we can find a "denominator" $b \in A^\partial \setminus 0$ so that $ba \in A^\partial[t]$.

Derivation ∂ can be extended on $F = \mathrm{Frac}(A)$ by $\partial(ab^{-1}) = \partial(a)b^{-1} - a\partial(b)b^{-2}$. Denote this extension also by ∂. Element $s = t/\partial(t)$ is a *slice*, i.e., $\partial(s) = 1$, s is transcendental over $F^\partial = \mathrm{Frac}(A^\partial)$, $A \subset F^\partial[s]$, and $F = F^\partial(s)$. Since the transcendence degree of F over F^∂ is one, we have $\mathrm{trdeg}(A^\partial) = \mathrm{trdeg}(A) - 1$.

A nonzero lnd ∂ determines a degree function on A by $\deg_\partial(a) = \max(n \mid \partial^n(a) \neq 0)$. Two lnds ∂_1, ∂_2 of A are *equivalent* if they determine the same degree function on A. If $A^{\partial_1} = A^{\partial_2}$ then ∂_1 and ∂_2 are equivalent.

All these facts can be found in [18].

5. Remaining lemmas

Recall that $R = \mathbb{C}_n/I$, π is the projection of $\mathbb{C}_n$ on R, and $\partial \in \mathrm{LND}(R)$. Thus, all of the above is applicable to our setting. In particular the first five lemmas imply that the transcendence degree of R is $n-k$. Accordingly, the transcendence degree of R^∂ is $n-k-1$.

We will be looking at the elements of R also as polynomials from $F^\partial[s]$, where F is the field of fractions of R, $s \in F$, $\partial(s) = 1$, and denote by deg the degree relative to s (which is the degree corresponding to ∂).

Let $i_1, \dots, i_k \in I$ and $f_{k+1}, \dots, f_{n-1} \in \mathbb{C}_n$. Consider a derivation $D(f) = \mathrm{J}(i_1, \dots, i_k, f_{k+1}, \dots, f_{n-1}, f)$. It follows from Lemma 3.4 that $\pi\big(D(f)\big) = \pi\big(D(f + i)\big)$ if $i \in I$. Hence we can define a derivation ϵ on R by $\epsilon\big(\pi(f)\big) = \pi\big(D(f)\big)$.

Take some algebraically independent $r_{k+1}, \dots, r_{n-1} \in R^\partial$. It is possible because the transcendence degree of R^∂ is $n-k-1$. Let $q_{k+1}, \dots, q_{n-1} \in \mathbb{C}_n$ be any elements for which $\pi(q_i) = r_i$. Extend on F a derivation $\epsilon(\mathbf{i}, \mathbf{r})\big(\pi(f)\big) = \pi\big(\mathrm{J}(i_1, \dots, i_k, q_{k+1}, \dots, q_{n-1}, f)\big)$. From Lemma 3.4 we know that $\epsilon(\mathbf{i}, \mathbf{r})(g)$ does not depend on the particular choice of $q_{k+1}, \dots, q_{n-1}$. From Lemma 3.6 we know that for some choices of $\mathbf{i}$ these derivations are nonzero.

Lemma 5.1. *Nonzero derivations $\epsilon(\mathbf{i}, \mathbf{r})$ and $\epsilon(\mathbf{i}, \mathbf{s})$ are linearly dependent over F^∂.*

PROOF. The claim is obvious if $\mathbf{r}$ and $\mathbf{s}$ are the same. Let us assume that $\epsilon(\mathbf{i}, \mathbf{r})$ and $\epsilon(\mathbf{i}, \mathbf{s})$ do not satisfy the claim and that the number of common elements of $\mathbf{r}$ and $\mathbf{s}$ is the maximal possible under this condition. Up to renumbering, we may assume that $s_{n-1} \notin \{r_{k+1}, \dots, r_{n-1}\}$. The elements $r_{k+1}, \dots, r_{n-1}, s_{n-1}$ are algebraically dependent. Let us take a nonzero polynomial dependence P of $r_{k+1}, \dots, r_{n-1}, s_{n-1}$ of the minimal total degree possible. P depends on s_{n-1} and on at least one $r_j \notin \{s_{k+1}, \dots, s_{n-1}\}$ since the elements $\{s_{k+1}, \dots, s_{n-1}\}$ are algebraically independent. We may assume up to renumbering that it is r_{n-1}. Recall that $\pi(q_i) = r_i$ and take a $q \in \mathbb{C}_n$ for which $\pi(q) = s_{n-1}$. Hence $P(q_{k+1}, \dots, q_{n-1}, q) \in I$ and $I \ni$

$\mathrm{J}(i_1, \dots, i_k, q_{k+1}, \dots, q_{n-2}, P, f)$ by Lemma 3.4. Therefore

$$\pi\left(\frac{\partial P}{\partial q_{n-1}} \mathrm{J}(i_1, \dots, i_k, q_{k+1}, \dots, q_{n-1}, f) + \frac{\partial P}{\partial q} \mathrm{J}(i_1, \dots, i_k, q_{k+1}, \dots, q_{n-2}, q, f)\right)$$
$$= \pi\left(\frac{\partial P}{\partial q_{n-1}}\right)\epsilon(\mathbf{i}, \mathbf{r})\big(\pi(f)\big) + \pi\left(\frac{\partial P}{\partial q}\right)\epsilon(\mathbf{i}, r_{k+1}, \dots, r_{n-2}, s_{n-1})\big(\pi(f)\big) = 0.$$

Both $\pi(\partial P/\partial q_{n-1})$ and $\pi(\partial P/\partial q)$ are polynomials of ∂-constants $r_{k+1}, \dots, r_{n-1}$, s_{n-1}. So $\pi(\partial P/\partial q_{n-1}), \pi(\partial P/\partial q) \in R^\partial \setminus 0$ by the definition of P and $\epsilon(\mathbf{i}, \mathbf{r})$ and $\epsilon(\mathbf{i}, r_{k+1}, \dots, r_{n-2}, s_{n-1})$ are nonzero derivations linearly dependent over F^∂. Since sets $\{s_{k+1}, \dots, s_{n-1}\}$ and $\{r_{k+1}, \dots, r_{n-2}, s_{n-1}\}$ have more common elements than $\{s_{k+1}, \dots, s_{n-1}\}$ and $\{r_{k+1}, \dots, r_{n-1}\}$, the derivations $\epsilon(\mathbf{i}, r_{k+1}, \dots, r_{n-2}, s_{n-1})$ and $\epsilon(\mathbf{i}, \mathbf{s})$ are also linearly dependent over F^∂ which brings us to a contradiction. □

Remark 5.2. Since $\mathbf{r}$ is a transcendence basis of R^∂, the kernel of $\epsilon(\mathbf{i}, \mathbf{r})$ is F^∂ by Lemma 3.5.

Lemma 5.1 shows that the derivations $\epsilon(\mathbf{i}, \mathbf{r})$ essentially do not depend on $\mathbf{r}$. So let us fix a basis $\mathbf{r}$ and omit it from the notation for ϵ: $\epsilon(\mathbf{i}, \mathbf{r}) = \epsilon(\mathbf{i})$. Take $s \in F$ satisfying $\partial(s) = 1$. Then $\epsilon(\mathbf{i})(s) = \epsilon(\mathbf{i})\big(t/\partial(t)\big) = \big(1/\partial(t)\big)\epsilon(\mathbf{i})(t) \in F^\partial[s]$. If $\epsilon(\mathbf{i})(s) = 0$ then $\epsilon(\mathbf{i})$ is the zero derivation since the kernel of this derivation contains $F^\partial[s]$. From Lemma 3.6 we know that $\epsilon(\mathbf{i})$ is a nonzero derivation for some choices of $\mathbf{i}$. Hence we can choose $\mathbf{i}$ so that $\epsilon(\mathbf{i}) \neq 0$ and $\deg_s\big(\epsilon(\mathbf{i})(s)\big)$ is the minimal possible. Let us take such an $\mathbf{i}$ and put $g = \epsilon(\mathbf{i})(s)$.

Lemma 5.3. *$\epsilon(\mathbf{j})(s)$ is divisible by g as an element of $F^\partial[s]$ for any $j_1, \dots, j_k \in I$.*

PROOF. Denote $\epsilon(\mathbf{h})(s)$ by $\Delta(h_1, \dots, h_k)$. Since

$$\Delta(h_1, \dots, h_k) = \pi(\mathrm{J}(h_1, \dots, h_k, r_{k+1}, \dots, r_{n-1}, s)$$

it is skew-symmetric. Besides,

$$\begin{aligned}\Delta(fh_1, h_2, \dots, h_k) &\\ &= \pi\big(\mathrm{J}(fh_1, h_2, \dots, h_k, r_{k+1}, \dots, r_{n-1}, s)\big)\\ &= \pi\big(h_1 \mathrm{J}(f, \dots, h_k, r_{k+1}, \dots, r_{n-1}, s) + f\,\mathrm{J}(h_1, \dots, h_k, r_{k+1}, \dots, r_{n-1}, s)\big)\\ &= \pi(f)\pi\big(\mathrm{J}(h_1, \dots, h_k, r_{k+1}, \dots, r_{n-1}, s)\big)\\ &= \pi(f)\Delta(h_1, \dots, h_k).\end{aligned}$$

This can be written, with some slight abuse of notations, as $\Delta(f'h_1, h_2, \dots, h_k) = f'\Delta(h_1, \dots, h_k)$ where $f' = \pi(f)$ because any co-image of f' in the left side gives the same right side. Of course, $\Delta(h_1, h_2, \dots, f'h_j, \dots, h_k) = f'\Delta(h_1, \dots, h_k)$ for any j since Δ is skew-symmetric.

For any $i \in I$, $\delta(i) = \Delta(i_1, \dots, i_{k-1}, i)$ is divisible by g. To verify this consider g and $\delta(i)$ as elements of $F^\partial[s]$ and choose $u, v \in F^\partial[s]$ so that $w = ug + v\delta(i)$ is the greatest common divisor of g and $\delta(i)$. If g does not divide $\delta(i)$ then $\deg(w) < \deg(g)$. We may assume that $u, v \in R$ since they can be multiplied by any element of R^∂. Then $\Delta(i_1, \dots, i_{k-1}, ui_k + vi) = w$ which contradicts the definition of g.

If $\epsilon(\mathbf{k})(s)$ is not always divisible by g, choose a pair $\mathbf{i}$, $\mathbf{j}$ so that $g = \epsilon(\mathbf{i})(s)$, $p = \epsilon(\mathbf{j})(s)$ is not divisible by g, and the number d of the common elements of

$\mathbf{i}$ and $\mathbf{j}$ is the maximal possible. Without loss of generality we may assume that $j_k \notin \{i_1, \dots, i_k\}$ and that $i_k \notin \{j_1, \dots, j_k\}$.

As we already know, $\Delta(i_1, \dots, i_{k-1}, j_k) = rg$ for some $r \in F^\partial[s]$ and we may assume that $r \in R$, replacing j_k by fj_k where $f \in R^\partial$, if necessary. For $j = j_k - ri_k$ we have $\Delta(i_1, \dots, i_{k-1}, j) = 0$. Now, $\Delta(j_1, \dots, j_{k-1}, j + i_k) = \Delta(j_1, \dots, j_k) + (1 - r)\Delta(j_1, \dots, j_{k-1}, i_k)$. The element $\Delta(j_1, \dots, j_{k-1}, i_k)$ is divisible by g since the number of common elements of $\{i_1, \dots, i_k\}$ and $\{j_1, \dots, j_{k-1}, i_k\}$ is $d + 1$. On the other hand, $\Delta(i_1, \dots, i_{k-1}, j + i_k) = g$ since $\Delta(i_1, \dots, i_{k-1}, j) = 0$. The sets $\{j_1, \dots, j_{k-1}, j+i_k\}$ and $\{i_1, \dots, i_{k-1}, j+i_k\}$ also have $d+1$ common elements, thus g divides $\Delta(j_1, \dots, j_{k-1}, j+i_k)$. Therefore $\Delta(j_1, \dots, j_k) = \Delta(j_1, \dots, j_{k-1}, j + i_k) - (1 - r)\Delta(j_1, \dots, j_{k-1}, i_k)$ is divisible by g contrary to our assumption. Hence the lemma is proved. □

Lemma 5.4. $g \in F^\partial \setminus 0$.

PROOF. Let us show that $1 = \pi\big(\mathrm{J}(x_1, \dots, x_n)\big)$ is divisible by g in $F^\partial[s]$. Write $\pi(x_a) = \sum f_{a,b}s^b$ where $f_{a,b} \in F^\partial$. As we know, it is possible to find a $v \in R^\partial \setminus 0$ so that $v\pi(x_a) = \sum g_{a,b}t^b$ where $g_{a,b} \in R^\partial$ and $t \in R$. We can lift these expressions into $\mathbb{C}_n$: $Vx_a = \sum G_{a,b}T^b + i_a$ where V, $G_{a,b}$, and T are corresponding co-images and $i_a \in I$. Let $A = \mathbb{C}_n[V^{-1}]$ be a subring of the field of fractions of $\mathbb{C}_n$ with denominators which are powers of V. Projection π can be defined on A and it will take A to $R[v^{-1}]$, which is a subring of $F^\partial[s]$.

So $\mathrm{J}(x_1, \dots, x_n) = \sum \mathrm{J}(y_1, \dots, y_n)$ where y_a is either $(G_{a,b}/V)T^b$ or i_a/V. Since $\mathrm{J}(\dots, HT^b, \dots) = T^b\,\mathrm{J}(\dots, H, \dots) + bT^{b-1}H\,\mathrm{J}(\dots, T, \dots)$ and

$$\mathrm{J}(\dots, G/V, \dots) = V^{-1}\,\mathrm{J}(\dots, G, \dots) - GV^{-2}\,\mathrm{J}(\dots, V, \dots),$$

we can write $\mathrm{J}(x_1, \dots, x_n) = \sum u_\mu\,\mathrm{J}(z_1, \dots, z_n)$ where z_j is either $G_{a,b}$, or T, or i_a and $u_\mu \in \mathbb{C}[T, G_{1,0}, \dots, G_{n,N}, i_1, \dots, i_n, V^{-1}]$. Since $\pi(u_\mu) \in F^\partial[s]$, it suffices to show that g divides $\pi\big(\mathrm{J}(z_1, \dots, z_n)\big)$.

Assume that $\pi(\mathrm{J}(z_1, \dots, z_n))$ is not divisible by g. A Jacobian is zero if T appears in it in two positions, thus there is at most one T among $z_1, \dots, z_n$. Also, if more than k of z_i's are in I, then $\pi(\mathrm{J}(z_1, \dots, z_n)) = 0$ by Lemma 3.4. If exactly k positions are filled by elements from I, then g divides $\pi(\mathrm{J}(z_1, \dots, z_n))$ by Lemma 5.3. Rewrite $z_1, \dots, z_n$ so that the elements from I are in the beginning and T is in the last position, if at all. Then $\mathrm{J}(z_1, \dots, z_n) = \mathrm{J}(i_1, \dots, i_m, h_{m+1}, \dots, h_{n-1}, U)$, where $i_j \in I$, $\pi(h_j) \in R^\partial$, $\pi(U) \in R^\partial \cup t$, and $m < k$. Assume also that m is the maximal possible. Since $\pi(h_j) \in R^\partial$ and since $m < k$, $h_{m+1}, \dots, h_{n-1}$ are algebraically dependent over I because the transcendence degree of F^∂ is $n - k - 1$. Let P be a nonzero polynomial of $h_{m+1}, \dots, h_{n-1}$ with value in I and of the minimal total degree possible. We may assume (up to renumbering) that P depends on h_{m+1}. Since m was the maximal possible,

$$\pi\big(\mathrm{J}(i_1, \dots, i_m, P, h_{m+2}, \dots, h_{n-1}, U)\big) = \pi\left(\frac{\partial P}{\partial h_{m+1}}\,\mathrm{J}(i_1, \dots, i_m, h_{m+1}, \dots, h_{n-1}, U)\right)$$

is divisible by g.

Therefore, contrary to our assumption, $\pi\big(\mathrm{J}(i_1, \dots, i_m, h_{m+1}, \dots, h_{n-1}, U)\big)$ is divisible by g because $\pi(\partial P/\partial h_{m+1}) \in R^\partial \setminus 0$. Hence g is a unit in $F^\partial[s]$, i.e., $g \in F^\partial$. □

Now, $\epsilon(\mathbf{i}) = \epsilon(\mathbf{i})(s)\partial$ for any $\epsilon(\mathbf{i})$ since $\epsilon(\mathbf{i})(s) = \epsilon(\mathbf{i})(s)\partial(s)$ and the kernels of these derivations are F^{∂}. Hence if $\epsilon(\mathbf{i})(s) = g$ then $\epsilon = \epsilon(\mathbf{i})$ is equivalent to ∂, which proves the theorem.

Remark 5.5. Some forms of this theorem have appeared before. In [14] it was proved for any lnd of $\mathbb{C}_n$, and in [1] a more precise result was found for $n = 3$. Also, in [10] a somewhat confusing form of it was proved for hypersurfaces. A similar exposition appeared as a preprint in the Max Plank Institute of Mathematics (MPIM2004-92).

It should be mentioned that for hypersurfaces and other complete intersections the choice of elements from the corresponding ideal is essentially unique.

The theorem claims that an lnd can be presented by a Jacobian only up to a coefficient from F^{∂}. The following example shows that we cannot get rid of the coefficient.

Example 5.6. Let ∂ be an lnd on $\mathbb{C}_4$ given by $\partial(z) = y$, $\partial(y) = x$, $\partial(x) = a$, and $\partial(a) = 0$. Since ∂ is a homogeneous operator, any invariant of ∂ is the sum of homogeneous invariants. It is clear that there is just one invariant of degree 1. So $J(f_1, f_2, f_3, x) \neq a$ if $f_i \in \ker(\partial)$. Indeed, f_i may be assumed homogeneous and then only one of them can have degree 1. Therefore $\deg\big(J(x, f_1, f_2, f_3)\big) > 1$.

Acknowledgments. Many colleagues are familiar with preliminary versions of the proof and helped improve the exposition. I would like to single out Tatiana Bandman, Daniel Daigle, and Shulim Kaliman for very helpful suggestions. I also want to thank the listeners of the course in the Weizmann Institute for their remarks.

References

1. D. Daigle, *On some properties of locally nilpotent derivations*, J. Pure Appl. Algebra **114** (1997), no. 3, 221–230.
2. ———, *On kernels of homogeneous locally nilpotent derivations of $k[X,Y,Z]$*, Osaka J. Math. **37** (2000), no. 3, 689–699.
3. ———, *Locally nilpotent derivations and Danielewski surfaces*, Osaka J. Math. **41** (2004), no. 1, 37–80.
4. D. Daigle and G. Freudenburg, *A counterexample to Hilbert's fourteenth problem in dimension 5*, J. Algebra **221** (1999), no. 2, 528–535.
5. ———, *A note on triangular derivations of $k[X_1, X_2, X_3, X_4]$*, Proc. Amer. Math. Soc. **129** (2001), no. 3, 657–662.
6. D. Daigle and P. Russell, *On weighted projective planes and their affine rulings*, Osaka J. Math. **38** (2001), no. 1, 101–150.
7. H. G. J. Derksen, *The kernel of a derivation*, J. Pure Appl. Algebra **84** (1993), no. 1, 13–16.
8. G. Freudenburg, *Actions of $\mathbf{G}_a$ on $\mathbf{A}^3$ defined by homogeneous derivations*, J. Pure Appl. Algebra **126** (1998), no. 1-3, 169–181.
9. ———, *A counterexample to Hilbert's fourteenth problem in dimension six*, Transform. Groups **5** (2000), no. 1, 61–71.
10. S. Kaliman and L. Makar-Limanov, *On the Russell – Koras contractible threefolds*, J. Algebraic Geom. **6** (1997), no. 2, 247–268.
11. L. Makar-Limanov, *On the hypersurface $x + x^2y + z^2 + t^3 = 0$ in $\mathbf{C}^4$ or a $\mathbf{C}^3$-like threefold which is not $\mathbf{C}^3$*, Israel J. Math. **96** (1996), part B, 419–429.
12. ———, *On the group of automorphisms of a surface $x^ny = P(z)$*, Israel J. Math. **121** (2001), 113–123.
13. ———, *Locally nilpotent derivations on the surface $xy = p(z)$*, Proceedings of the Third International Algebra Conference (Tainan, 2002) (Y. Fong, L.-S. Shiao, and E. Zelmanov, eds.), Kluwer Acad. Publ., Dordrecht, 2003, pp. 215–219.

14. ———, *Locally nilpotent derivations, a new ring invariant and applications*, preprint.
15. M. Miyanishi, *Noncomplete algebraic surfaces*, Lecture Notes in Math., vol. 857, Springer, Berlin, 1981.
16. R. Rentschler, *Opérations du groupe additif sur le plan affine*, C. R. Acad. Sci. Paris Sér. A-B **267** (1968), A384–A387.
17. P. Roberts, *An infinitely generated symbolic blow-up in a power series ring and a new counterexample to Hilbert's fourteenth problem*, J. Algebra **132** (1990), no. 2, 461–473.
18. A. van den Essen, *Polynomial automorphisms and the Jacobian conjecture*, Progr. Math., vol. 190, Birkhäuser, Basel, 2000.
19. J. Wilkens, *On the cancellation problem for surfaces*, C. R. Acad. Sci. Paris Sér. I Math. **326** (1998), no. 9, 1111–1116.

DEPARTMENT OF MATHEMATICS, WAYNE STATE UNIVERSITY, DETROIT, MI 48202, USA

Current address: Department of Mathematics, Faculty of Mathematics and Computer Science The Weizmann Institute of Science POB 26, Rehovot 76100, ISRAEL

E-mail address: `lml@math.wayne.edu`

Centre de Recherches Mathématiques
CRM Proceedings and Lecture Notes
Volume **54**, 2011

Equivariant Derivations and Additive Group Actions

Kayo Masuda

ABSTRACT. Let G be a reductive algebraic group defined over an algebraically closed field of characteristic 0. Let X be an affine variety with an algebraic G-action. We are interested in the G_{a}-actions on X commuting with the G-action. We observe G-equivariant (locally nilpotent) derivations in various settings from the viewpoint of the invariant theory.

1. Introduction

Let k be an algebraically closed field of characteristic 0, which is the ground field. The action of the additive group G_{a} on an affine algebraic variety gives us important information on the underlying variety. Let G be a reductive algebraic group and let $X = \operatorname{Spec} B$ be an affine G-variety. We observe G_{a}-actions on X commuting with the G-action. It is well-known that G_{a}-actions on X correspond bijectively to locally nilpotent derivations on B. We say that a derivation δ on B is *G-equivariant* if $\delta \cdot g = g \cdot \delta$ for every $g \in G$. The G_{a}-action on X induced by a locally nilpotent derivation δ commutes with the G-action if and only if δ is G-equivariant. Let $\pi\colon X \to X /\!\!/ G = \operatorname{Spec} B^G$ be the algebraic quotient where B^G is the ring of G-invariants. Every G_{a}-action on X commuting with the G-action induces a G_{a}-action on $X /\!\!/ G$. However, a G_{a}-action on $X /\!\!/ G$ does not necessarily lift up to a (G-equivariant or not) G_{a}-action on X. In this article, we investigate G-equivariant (locally nilpotent) derivations in various settings from the viewpoint of the invariant theory. After reviewing G-actions and locally nilpotent derivations in Section 2, we observe G-equivariant (locally nilpotent) derivations on G-representation spaces in Sections 3 and 4. In Section 4, we focus on the case where G is finite. We see in examples that a (locally nilpotent) derivation on B^G does not necessarily lift up to a (locally nilpotent) derivation on B. In Section 5, we show that the G-equivariant G_{a}-action characterizes a smooth acyclic complex affine G-variety X when the dimension of $X /\!\!/ G$ is one or two (Theorems 5.1 and 5.2). In the last section, we consider the case where G is abelian and observe semi-invariant locally nilpotent derivations associated to G-equivariant embeddings.

2010 *Mathematics Subject Classification.* Primary 14R10; Secondary 14R20, 13A50.
Key words and phrases. Equivariant derivation, locally nilpotent derivation, equivariant embedding.
Supported by Grant-in-Aid for Scientific Research (C) 18540045, JSPS.
This is the final form of the paper.

Let V and $\widetilde{V}$ be G-representation spaces of dimension m and n, respectively. A G-equivariant embedding $V \hookrightarrow \widetilde{V}$ of G-varieties is abbreviated to a G-embedding. We say that a G-embedding $\varphi\colon V \to \widetilde{V}$ is *G-rectifiable* if there exists a system of semi-invariant coordinate functions $f_1, \ldots, f_n$ on $\widetilde{V}$ such that the image $\varphi(V)$ is defined by the G-stable ideal $(f_{m+1}, \ldots, f_n)$. Forgetting the G-action, a G-rectifiable embedding $V \to \widetilde{V}$ is a rectifiable embedding $\mathbb{A}^m \to \mathbb{A}^n$. It is known in [6] that the embeddings $\mathbb{A}^m \hookrightarrow \mathbb{A}^n$ bijectively correspond to the sequences $\Delta = (\delta_1, \ldots, \delta_m)$ of mutually commuting, locally nilpotent derivations δ_i's of some type on a polynomial ring with a linear action of the algebraic torus G_m. This result extends G-equivariantly (Theorem 6.2, cf. [7]). For a given G-embedding $\varphi\colon V \to \widetilde{V}$, there exsits a sequence $\Delta_\varphi = (\delta_1, \ldots, \delta_m)$ associated to φ where δ_i's are mutually commuting, semi-invariant locally nilpotent derivations of some type on a polynomial ring B with a linear $(G \times G_m)$-action. Let $A = \bigcap_{i=1}^m B^{\delta_i}$ where B^{δ_i} denotes the kernel of δ_i. Then the G-embedding φ is G-rectifiable if and only if $Y = \operatorname{Spec} A$ is isomorphic to a $(G \times G_m)$-representation space (Theorem 6.3, cf. [7]). If there exists a G-embedding which is rectifiable, but not G-rectifiable, it gives rise to a counterexample to the Linearization Problem and the Equivariant Cancellation Problem (See Section 6).

2. Preliminaries

Let $X = \operatorname{Spec} B$ be an affine variety with an algebraic G-action. Throughout this article, G is a reductive algebraic group. Then the affine domain B has a G-action and decomposes to $B = \bigoplus_{\omega \in \Omega} B_\omega$ where Ω is the set of isomorphism classes of irreducible G-modules and B_ω is the isotypic component of type ω. Each B_ω is a finitely generated module over B^G. Let δ be a G-equivariant k-derivation on B. Then it follows that $\delta(B_\omega) \subset B_\omega$. In particular, δ restricts to a k-derivation $\delta|_{B^G}$ on B^G. If δ is locally nilpotent, then $\delta|_{B^G}$ is locally nilpotent as well. The kernel B^δ of δ has a G-action. It holds that $(B^\delta)^G = (B^G)^\delta$. The restriction $\delta|_{B_\omega}$ is a B^G-module derivation on B_ω, i.e., $\delta(b_1 + b_2) = \delta(b_1) + \delta(b_2)$ for $b_1, b_2 \in B_\omega$ and $\delta(ab) = \delta(a)b + a\delta(b)$ for $a \in B^G$ and $b \in B_\omega$. We denote by $\operatorname{Der}(B)$ the B-module of derivations on B and by $\operatorname{Der}^G(B)$ the B^G-module of G-equivariant derivations on B.

Suppose that δ is locally nilpotent. The G_{a}-action on X induced by δ is fixed-point free if and only if the ideal of B generated by the image $\delta(B)$ contains 1. Suppose further that δ has a slice $s \in B$, i.e., $\delta(s) = 1$. Then the G_{a}-action induced by δ is fixed-point free. Note that we may assume that $s \in B^G$ since δ is G-equivariant. Then s is a slice of $\delta|_{B^G}$ as well.

We recall some properties of locally nilpotent derivations.

Lemma 2.1. *Let δ be a locally nilpotent derivation on B.*

(1) (Freudenburg [3]) *Suppose that there exist $b_1, \ldots, b_n \in B$ such that $\delta(b_i) \in b_{\sigma(i)}B$ for each i where σ is a permutation on n letters. Then in each orbit of σ, there is an i such that $\delta(b_i) = 0$. In particular, if $\delta(b) \in bB$, then $\delta(b) = 0$.*

(2) (cf. [6]) *Suppose that δ has a slice $s \in B$. Then s is transcendental over B^δ, and $B = B^\delta[s]$. The kernel B^δ is given by the image of the Dixmier map $\phi_s\colon B \to B$, which is the algebra homomorphism defined by*

$$\phi_s(b) = \sum_{i \geq 0} \frac{(-1)^i}{i!} \delta^i(b) s^i \quad \textit{for } b \in B.$$

Then the kernel of ϕ_s is the ideal $(s) \subset B$. Suppose further that δ is G-equivariant. Then δ has a G-invariant slice $s_0 \in B^G$. It follows that $B^G = (B^G)^\delta[s_0]$ and the Dixmier map ϕ_{s_0} is G-equivariant.

Let $\pi\colon X \to X /\!\!/ G$ be the quotient morphism which is defined by the inclusion $B^G \hookrightarrow B$. Then π induces an embedding of the fixed point locus X^G into $X /\!\!/ G$. The G-action on X is called *fix-pointed* if $\pi|_{X^G}$ is an isomorphism, i.e., every closed orbit of X is a fixed point. It is known by Bass and Haboush [2] that if X is smooth and the G-action on X is fix-pointed then $X \cong X^G \times V$ for a G-representation space V. Hence if $X^G \cong X /\!\!/ G$ is the affine space, then X is isomorphic to the affine space, say, $\mathbb{A}^n$ and the G-action on X is linearizable, i.e., isomorphic to a linear action in the automorphisms of $\mathbb{A}^n$.

By the slice theorem (Luna [5]), there is a finite stratification of $X /\!\!/ G = \bigcup_i Y_i$ into locally closed subvarieties Y_i such that the isotropy groups of closed orbits in $\pi^{-1}(Y_i)$ are all conjugate to a fixed reductive subgroup H_i. Furthermore, if $\overline{Y_i} \supset Y_j$ then H_i is conjugate to a subgroup of H_j. The unique open dense stratum of $X /\!\!/ G$ is called the principal stratum and the corresponding isotropy group a principal isotropy group.

Let us consider the case where G is abelian. Then it follows that $B = \bigoplus_{\chi \in \Omega} B_\chi$ where Ω is the set of characters of G and $B_\chi = \{b \in B \mid g \cdot b = \chi(g)b \text{ for all} g \in G\}$. An element of B_χ is called semi-invariant of weight χ. A derivation δ on $B = \bigoplus_{\chi \in \Omega} B_\chi$ is called semi-invariant of weight ω if $\delta(B_\chi) \subset B_{\chi+\omega}$ for every χ. Suppose that a locally nilpotent derivation δ on B which is semi-invariant of weight $-\omega$ has a slice $s \in B$. Then we may assume that s is semi-invariant of weight ω.

When G is a finite group, then B is a finite G-module over B^G. We consider the case where B is a finite module over a subring A of B. Let K (resp. L) be the quotient field of A (resp. B). Then L is a simple extension of K and L is written as $L = K(\theta)$ for some $\theta \in L$. Let δ be a derivation on A. Then δ extends uniquely to a derivation δ_K on K. Let $F(X)$ be the minimal polynomial of θ over K. Then it is well-known that a derivation δ_K lifts uniquely to a derivation Δ on L such that $\Delta(\theta) = -F^{\delta_K}(\theta)/F'(\theta)$, where $F^{\delta_K}(X)$ is the polynomial with all the coefficients of $F(X)$ replaced by their δ_K-images. The derivation Δ on L does not necessarily restrict to a derivation on B, i.e., $\Delta(B) \subset B$.

As for the derivation Δ, the following is known.

Theorem 2.2. *Let B be a normal affine domain and let A be a subring of B such that B is a finite A-module. Let δ be a derivation on A and let Δ be the derivation on the quotient field L of B such that $\Delta|_A = \delta$. Then the following assertions hold.*

(1) (Vasconcelos [10]) *If $\Delta(B) \subset B$ is satisfied, then Δ is locally nilpotent provided so is δ.*

(2) ([8]) *Suppose that δ is locally nilpotent. Suppose further that there exists a nonzero ideal $\mathfrak{a}$ of A satisfying the conditions :*

(i) *The ideal $\mathfrak{a}$ has height at least two.*

(ii) *The quotient morphism $\pi\colon \operatorname{Spec} B \to \operatorname{Spec} A$ is étale outside $V(\mathfrak{a})$.*

Then $\Delta(B) \subset B$ and Δ is locally nilpotent.

(3) (Scheja – Storch [9]) *Let $\mathfrak{R}$ be the radical of the annihilator* $\operatorname{Ann}(\Omega_{B/A})$ *and let $\mathfrak{b} = A \cap \mathfrak{R}$. Then $\Delta(B) \subset B$ if and only if $\delta(\mathfrak{p}) \subset \mathfrak{p}$ for every height 1 prime*

divisor $\mathfrak{p}$ *of* $\mathfrak{b}$*. In particular, if* $\Omega_{B/A} = (0)$*, i.e.,* B *is unramified over* A*, then* Δ *satisfies* $\Delta(B) \subset B$*, i.e.,* δ *lifts to a derivation* Δ *on* B*.*

If a derivation Δ on B restricts to a derivation δ on A, we call Δ a lifting of δ. When $A = B^G$ with G finite, the lifting Δ is G-equivariant (cf. Lemma 4.1).

A G-equivariant derivation Δ on B which restricts to a derivation δ on B^G is called a lifting of δ.

3. Equivariant derivations on G-modules

In this section, we observe G-equivariant derivations on G-modules.

Let W be a G-representation space of dimension n with the representation $\rho\colon G \to GL(W)$ and let $x_1, \ldots, x_n$ be the coordinate functions of W. The coordinate ring B of W is $B = k[x_1, \ldots, x_n]$ with the G-action given by $g \cdot (x_1, \ldots, x_n) = (x_1, \ldots, x_n)\rho^*(g)$ for $g \in G$ where $\rho^*(g) = {}^t\rho(g^{-1})$. The G-module $W^* = kx_1 + \cdots + kx_n$ is the dual module of W. The coordinate ring B has a decomposition $B = \bigoplus_{d \geq 0} S^d(W^*)$ where $S^d(W^*)$ is the dth symmetric algebra of W^*. Each $S^d(W^*)$ is a G-module and decomposes to isotypic components.

Let δ be a derivation on B of a form

$$\delta = f_1 \partial_{x_1} + \cdots + f_n \partial_{x_n}$$

where $f_i \in B$. We decompose f_i to $f_i = \sum_{j=0}^r f_{ij}$ for $1 \leq i \leq n$ where $f_{ij} \in S^j(W^*)$ is a homogeneous polynomial of degree j. Then it follows that $\delta = \sum_{j=0}^r \delta_j$ where $\delta_j = f_{1j}\partial_{x_1} + \cdots + f_{nj}\partial_{x_n}$ is a homogeneous derivation on B. Let $V_j = kf_{1j} + \cdots + kf_{nj}$ be the submodule of $S^j(W^*)$.

Lemma 3.1. (1) *The derivation* δ *is* G*-equivariant if and only if*

$$g \cdot (f_{1i}, \ldots, f_{ni}) = (f_{1i}, \ldots, f_{ni})\rho^*(g)$$

for all $g \in G$ *and* $0 \leq i \leq r$*.*

(2) *Suppose that* δ *is* G*-equivariant. Then the* G*-module* V_i *is isomorphic to a* G*-submodule of* W^**. Hence if* W *is irreducible, then* $V_i = \{0\}$ *or* $V_i \cong W^*$*.*

Proof. (1) The derivation δ is G-equivariant if and only if $(g \cdot \delta)(x_i) = (\delta \cdot g)(x_i)$ for every i and all $g \in G$, which is equivalent to that $g \cdot (f_1, \ldots, f_n) = (f_1, \ldots, f_n)\rho^*(g)$ for all $g \in G$. Since the G-action preserves the degree, the assertion follows.

(2) The G-module V_i is a G-submodule of $S^i(W^*)$. Let $q_i\colon W^* \to V_i$ be the surjection defined by $q_i(x_j) = f_{ji}$ for $1 \leq j \leq n$. Then by (1), q_i is G-equivariant. Hence the assertion follows. □

Suppose that W is irreducible. Let $\omega \in \Omega$ be the isomorphism class of W^*. By Lemma 3.1, it follows that for $\delta \in \mathrm{Der}^G(B)$, $k\delta(x_1) + \cdots + k\delta(x_n) = \sum_i V_i$ is a G-submodule of B_ω.

In the following in this section, we assume that δ is G-equivariant.

We consider the case where $\delta = \delta_0$. Then $\delta = c_1\partial_{x_1} + \cdots + c_n\partial_{x_n}$ for $c_i \in k$. Since δ is G-equivariant, it follows that ${}^t(c_1, \ldots, c_n)$ is the eigenvector with value 1 of $\rho(g)$ for all $g \in G$. Hence ${}^t(c_1, \ldots, c_n) \in W^G$ and we have the following.

Lemma 3.2. *If* $\delta = c_1\partial_{x_1} + \cdots + c_n\partial_{x_n}$ *is* G*-equivariant where* $c_i \in k$ *for* $1 \leq i \leq n$*, then* ${}^t(c_1, \ldots, c_n) \in W^G$*. As a consequence, if* $W^G = \{0\}$*, then* δ *is trivial. In particular, if* W *is an irreducible* G*-module, then* $\delta = 0$*.*

By Lemma 3.2, we have the following.

Proposition 3.3. *Suppose that* $W^G = \{0\}$. *Then there exists no* G*-equivariant* G_a*-action on* W *which is fixed-point free.*

PROOF. Assume that there were a G-equivariant, fixed-point free, G_a-action on W. Let $\delta = f_1\partial_{x_1} + \cdots + f_n\partial_{x_n}$ be the G-equivariant locally nilpotent derivation on B corresponding to the G_a-action. Since the G_a-action is fixed-point free, it follows that $h_1f_1 + \cdots + h_nf_n = 1$ for $h_i \in B$. Decompose $\delta = \sum_{j=0}^{r} \delta_j$ where δ_j is a homogeneous derivation. Then by Lemma 3.2, δ_0 is trivial, i.e., each degree of f_i is positive. However, this contradicts to that $h_1f_1 + \cdots + h_nf_n = 1$. Hence the assertion follows. □

Next, consider the case where $\delta = \delta_1$, i.e., $\delta = f_1\partial_{x_1} + \cdots + f_n\partial_{x_n}$ where each $f_i \in B$ is linear in $x_1, \ldots, x_n$. Then by Lemma 3.1, it follows that the G-module $V_1 = kf_1 + \cdots + kf_n$ is isomorphic to a G-submodule of W^*.

Lemma 3.4. *Suppose that* $\delta = f_1\partial_{x_1} + \cdots + f_n\partial_{x_n}$ *is a* G*-equivariant derivation on* B *where each* $f_i \in B$ *is linear in* $x_1, \ldots, x_n$. *If* W *is irreducible, then* δ *is of a form*

$$\delta = c(x_1\partial_{x_1} + \cdots + x_n\partial_{x_n})$$

for $c \in k$.

PROOF. Suppose that δ is nontrivial and write $f_i = \sum_{j=1}^{n} x_j c_{ji}$ for $c_{ji} \in k$. By Lemma 3.1, it follows that $V_1 = kf_1 + \cdots + kf_n = W^*$. The $n \times n$ matrix $C = (c_{ji})$ gives a G-equivariant endomorphism of W^*. Since W^* is irreducible, C is a scalar matrix. Hence the assertion follows. □

We observe G-equivariant derivations on B when the dimension of $W /\!\!/ G$ is small.

First, we consider the case $\dim W /\!\!/ G = 0$, i.e., $B^G = k$. Then every isotypic component B_σ is a finite G-module over k and isomorphic to a direct sum of some copies of V where V is an irreducible G-module such that $\sigma = [V]$. Hence if W is irreducible, then $\mathrm{Der}^G(B)$ is a finite free module over k.

Lemma 3.5. *Let* W *be an irreducible* G*-module with* $\dim W /\!\!/ G = 0$ *and let* δ *be a* G*-equivariant derivation on* B. *Suppose that the multiplicity of* W^* *in the* G*-module* B_ω *for* $\omega = [W^*]$ *is one. Then* δ *is of a form*

$$\delta = c(x_1\partial_{x_1} + \cdots + x_n\partial_{x_n})$$

for $c \in k$. *If* δ *is locally nilpotent, then* δ *is trivial.*

PROOF. Since the multiplicity of W^* in B_ω is one, it follows that $B_\omega = W^* = kx_1 + \cdots + kx_n$. Hence the assertion follows from Lemma 3.1. If δ is locally nilpotent, then δ is trivial by Lemma 2.1(1). □

Example 3.1. Let $G = \mathrm{SL}_n$ for $n \geq 2$ and let $W = R_1$ be the standard SL_n-representation space. Let $B = k[x_1, \ldots, x_n]$ be the coordinate ring of R_1. Then $B^G = k$ and B decomposes to $B = \bigoplus_{d \geq 0} S^d(R_1^*) = \bigoplus_{d \geq 0} R_d^*$ where $R_d^* = S^d(R_1^*)$ is an irreducible G-module. For $\omega = [R_1^*]$, $B_\omega = R_1^*$. Hence $\delta \in \mathrm{Der}^G(B)$ is of a form $\delta = c(x_1\partial_{x_1} + \cdots + x_n\partial_{x_n})$ for $c \in k$ and there is no nontrivial G-equivariant G_a-action on R_1.

Next, consider the case where $W /\!\!/ G$ is of one dimension. Then it follows that $W /\!\!/ G \cong \mathbb{A}^1$, i.e., $B^G = k[t]$ for some homogeneous polynomial $t \in B^G$. We have two cases. The first case is the case of the fix-pointed action, i.e., $W^G = k^1$ and $W = k^1 \oplus W'$ where W' is a G-submodule with the trivial quotient. In this case, we have $B = B'[t]$ where $B' = k[y_1, \ldots, y_{n-1}]$ is the coordinate ring of W' and the coordinate ring of W^G is $k[t]$. Then the $k[t]$-module $\mathrm{Der}^G(B)$ is a sum of $\mathrm{Der}(k[t])$ and $\mathrm{Der}^G(B') \otimes_k k[t]$. Hence every derivation on B^G lifts up to a G-equivariant derivation on B. The second case is that $W^G = \{0\}$. Then t is a homogeneous polynomial of degree $d \geq 2$. Since $B^G = k[t]$ is a principal ideal domain, it follows that $B_\omega = k[t] \otimes_k S$ where S is a free G-submodule of B. Let δ be a derivation on B^G. Then $\delta = f(t)(\mathrm{d}/\mathrm{d}t)$ for $f(t) \in k[t]$. If δ is locally nilpotent, then it follows that $\delta = c(\mathrm{d}/\mathrm{d}t)$ for $c \in k$.

Theorem 3.6. *Let* $\dim W /\!\!/ G = 1$ *and* $W^G = \{0\}$. *Then there exist no* G_{a}*-actions on* W *which restrict to nontrivial* G_{a}*-actions on* $W /\!\!/ G$.

Proof. As observed above, $B^G = k[t]$ with $d = \deg t > 1$ and a nontrivial locally nilpotent derivation on B^G is of a form $c(\mathrm{d}/\mathrm{d}t)$ for $c \in k^*$. Let Δ be a locally nilpotent derivation on B such that $\Delta|_{B^G} = c(\mathrm{d}/\mathrm{d}t)$. Write $\Delta = f_1\partial_{x_1} + \cdots + f_n\partial_{x_n}$ for $f_i \in B$. Then $c = \Delta(t) = f_1(\partial_{x_1}t) + \cdots + f_n(\partial_{x_n}t)$. The degree of $\partial_{x_i}t$ is $d-1 > 0$ unless $\partial_{x_i}t = 0$, while the one of c is zero. This is a contradiction. Hence the assertion follows. □

Remark. There is no derivation on B which restricts to a locally nilpotent derivation on $B^G = k[t]$.

Example 3.2. Let G be a complex orthogonal group $\mathrm{O}(2) = \mathbb{C}^* \rtimes \mathbb{Z}/2\mathbb{Z}$. Let $B = \mathbb{C}[x, y]$ be the coordinate ring of the representation space W_n $(n \geq 1)$, on which G acts by

$$\lambda \cdot (x, y) = (\lambda^n x, \lambda^{-n} y), \quad \tau \cdot (x, y) = (y, x)$$

where $\lambda \in \mathbb{C}^*$ and τ is the generator of $\mathbb{Z}/2\mathbb{Z}$. Then $B^G = \mathbb{C}[t]$ where $t = xy$ and W_n has one-dimensional quotient. For $\omega = [W_n^*]$, $B_\omega = \mathbb{C}[t] \otimes_{\mathbb{C}} W_n^*$. Hence every G-equivariant derivation Δ on B is of a form

$$\Delta = f(t)(x\partial_x + y\partial_y)$$

for $f(t) \in B^G$. As a consequence, there is no nontrivial G-equivariant locally nilpotent derivation on B. Further, by the above remark, there is no derivation on B which restricts to a locally nilpotent derivation $\mathrm{d}/\mathrm{d}t$ on B^G.

4. G-equivariant derivations with G finite

In this section, we observe G-equivariant derivations on G-modules with G finite.

Let G be a finite group and let W be a G-representation space. As remarked in Section 2, the coordinate ring B of W is a finite G-module over B^G. Let δ be a derivation on B^G. Then δ uniquely extends to a derivation δ_K on the quotient field K of B^G. Further, the derivation δ_K on K uniquely lifts up to a derivation Δ on the quotient field L of B. The following is verified.

Lemma 4.1 ([8]). (1) *Let* Δ *be a derivation on* L. *Then* $\Delta(K) \subset K$ *if and only if* $g \cdot \Delta = \Delta \cdot g$ *for every* $g \in G$.

(2) *Let Δ be a derivation on B. Then $\Delta(B^G) \subset B^G$ if and only if $g \cdot \Delta = \Delta \cdot g$ for every $g \in G$.*

By Lemma 4.1, if a derivation δ on B^G lifts up to a derivation Δ on B, then Δ is G-equivariant. By Theorem 2.2, the following holds.

Theorem 4.2. *Suppose that the following are satisfied.*

(1) *The principal isotropy group of W is trivial.*

(2) *The complement of the principal stratum has codimension ≥ 2 in W/G.*

Then every G_a-action on W/G lifts up to a G-equivariant G_a-action on W.

Example 4.1. Let $G = \mathbb{Z}/n\mathbb{Z}$ be a cyclic group of order n which acts on $B = \mathbb{C}[x, y]$ by

$$\sigma \cdot (x, y) = (\zeta x, \zeta^d y),$$

where σ is a generator of G, ζ is a primitive n-th root of unity and d is an integer such that $0 < d < n$ and $(d, n) = 1$. Then the isotropy group of every closed point of $W = \operatorname{Spec} B$ except the origin is trivial. The G-invariant subring B^G is generated by monomials $x^i y^j$ such that $i + dj \equiv 0 \pmod{n}$. The stratification of W/G is $W/G = (W/G - \{\bar{0}\}) \cup \{\bar{0}\}$ where $\bar{0}$ is the image of the origin of W by the quotient morphism. Since the principal isotropy group is trivial, it follows from Theorem 4.2 that every G_a-action on W/G lifts up to a G-equivariant G_a-action on W. The following is known.

Theorem 4.3 ([8]). *In the notation above, the following hold.*

(1) *Let $\Delta \in \operatorname{Der}^G(B)$. Then*

$$\Delta = (a_1 x + a_2 y^{d'})\partial_x + (b_1 x^d + b_2 y)\partial_y$$

where $a_1, a_2, b_1, b_2 \in B^G$ and d' is an integer such that $0 < d' < n$ and $dd' \equiv 1 \pmod{n}$.

(2) *Let δ' be an arbitrarily chosen, locally nilpotent derivation on B^G and let Δ' be the locally nilpotent derivation on B which lifts δ'. Then, after a suitable change of coordinates x, y of B, Δ' is given by $\Delta' = f(x^n)x^d\partial_y$ with $f(x^n) \in \mathbb{C}[x^n]$.*

Example 4.2. Let G be a dihedral group $D_d = \mathbb{Z}/d\mathbb{Z} \rtimes \mathbb{Z}/2\mathbb{Z}$ for an odd prime integer d. Let $B = \mathbb{C}[x, y]$ be the coordinate ring of the representation space W_1, on which G acts by

$$\sigma \cdot (x, y) = (\zeta x, \zeta^{-1} y), \qquad \tau \cdot (x, y) = (y, x)$$

where σ is a generator of $\mathbb{Z}/d\mathbb{Z}$, τ the generator of $\mathbb{Z}/2\mathbb{Z}$, and ζ is a d-th primitive root of unity. Then $B^G = \mathbb{C}[s, t]$ where $s = x^d + y^d$ and $t = xy$. The stratification of $W_1/G \cong \mathbb{A}^2$ is

$$\mathbb{A}^2 = (\mathbb{A}^2 - V) \cup (V - \{\bar{0}\}) \cup \{\bar{0}\}$$

where $V = \{(s, t) \mid s^2 - 4t^d = 0\}$. The isotropy groups corresponding to the strata $\mathbb{A}^2 - V$, $V - \{\bar{0}\}$, $\{\bar{0}\}$ are the trivial group, $\mathbb{Z}/2\mathbb{Z}$ (the second factor), G, respectively. We have the following result.

Theorem 4.4 ([8]). (1) *Let $\Delta \in \operatorname{Der}^G(B)$. Then*

$$\Delta = f_1(x\partial_x + y\partial_y) + \sum_{i=1}^{l} f_{2i}(y^{id-1}\partial_x + x^{id-1}\partial_y)$$

where $f_1, f_{2i} \in B^G$ and $l \geq 1$.

(2) *There is no nontrivial G-equivariant locally nilpotent derivation on B.*

By Lemma 4.1 and Theorem 4.4, any G_{a}-action on $W_1/G \cong \mathbb{A}^2$ does not lift up to a G_{a}-action on W_1.

Example 4.3. Let G be the symmetric group S_n on n letters, which acts on the polynomial ring $B = \mathbb{C}[x_1, \ldots, x_n]$ in the standard way such that $\sigma(x_i) = x_{\sigma(i)}$ for $\sigma \in S_n$. Then $B^G = \mathbb{C}[s_1, \ldots, s_n] = \mathbb{C}[t_1, \ldots, t_n]$, where s_i and t_i are the i-th elementary symmetric polynomials

$$s_i = \sum_{j_1 < \cdots < j_i} x_{j_1} x_{j_2} \cdots x_{j_i}, \quad t_i = \sum_{j=1}^{n} x_j^i.$$

Let $W = \operatorname{Spec} B$. Then $W^G = \mathbb{C}$ and $W/G \cong \mathbb{A}^n$. The principal isotropy group is trivial and the complement of the principal stratum has codimension 1.

Theorem 4.5 ([8])**.** *The B^G-module $\operatorname{Der}^G(B)$ is freely generated by*

$$\Delta_i = x_1^i \partial_{x_1} + \cdots + x_n^i \partial_{x_n}$$

for $0 \leq i \leq n-1$. If $\Delta \in \operatorname{Der}^G(B)$ is locally nilpotent, then $\Delta = f\Delta_0$ where f is an element of B^G such that $\Delta_0(f) = 0$.

Let $\delta_0 = \Delta_0|_{B^G}$. Then the locally nilpotent derivation δ_0 on $B^G = \mathbb{C}[t_1, \ldots, t_n]$ has a slice $s = t_1/n$. By Theorem 4.5, a G_{a}-action on $W/G \cong \mathbb{A}^n$ is liftable onto W if and only if it is given by the locally nilpotent derivation $f\delta_0$ for $f \in (B^G)^{\delta_0}$.

5. Equivariant derivations on affine G-domains

In this section, we observe G-equivariant G_{a}-actions on smooth complex affine G-varieties when the dimension of the algebraic quotient is small.

Let $X = \operatorname{Spec} B$ be a smooth complex affine G-variety. Let Δ be a nontrivial G-equivariant locally nilpotent derivation on B. Then Δ restricts to a locally nilpotent derivation $\delta = \Delta|_{B^G}$ on B^G. Note that $(B^\Delta)^G = (B^G)^\delta$.

Suppose that $\dim X /\!\!/ G = 0$, i.e., $B^G = \mathbb{C}$. Then it follows from [2] that X is isomorphic to a G-representation space W. Hence this case reduces to the case where X is a G-representation space.

Suppose that $\dim X /\!\!/ G = 1$. If X is acyclic, then it follows that $B^G = \mathbb{C}[t]$ for some $t \in B^G$ (cf. [4]). The locally nilpotent derivation δ on B^G is of a form $\delta = c(\mathrm{d}/\mathrm{d}t)$ for $c \in \mathbb{C}$. The G-equivariant G_{a}-action on X given by Δ induces a nontrivial G_{a}-action on $X /\!\!/ G \cong \mathbb{A}^1$ if and only if Δ has a slice.

Theorem 5.1. *Suppose that $X = \operatorname{Spec} B$ is a smooth, acyclic complex affine G-variety with one-dimensional quotient. Then X is isomorphic to a G-representation space $W \oplus \mathbb{C}$ if and only if there exist a G-equivariant G_{a}-action on X which induces a nontrivial G_{a}-action on $X /\!\!/ G$. Here, W is a G-representation space such that $W /\!\!/ G$ is a point and $\mathbb{C}$ is a one-dimensional trivial G-representation space.*

PROOF. "Only if" part is obvious. We show that "if" part. As observed above, there is a G-equivariant locally nilpotent derivation Δ on B having a slice $t \in B^G$ and $B = B^\Delta[t]$. Let $Y = \operatorname{Spec} B^\Delta$. Then Y is a smooth, acyclic G-variety since $X \cong Y \times \mathbb{A}^1$. Since $(B^\Delta)^G = (B^G)^\delta = \mathbb{C}$, it follows from [2] that Y is isomorphic to a G-representation space W such that $W /\!\!/ G$ is a point. Hence $X \cong Y \times \mathbb{A}^1 \cong W \oplus \mathbb{C}$ and the assertion follows. □

Suppose that $\dim X /\!\!/ G = 2$. If Δ has a slice, then $Y = \operatorname{Spec} B^{\Delta}$ has one-dimensional quotient since $\dim(B^{\Delta})^G = \dim(B^G)^{\delta} = 1$.

Theorem 5.2. *Suppose that $X = \operatorname{Spec} B$ is a smooth, acyclic complex affine G-variety with two-dimensional quotient. Then X is isomorphic to a G-representation space $W \oplus \mathbb{C}^2$ if and only if there exists a locally nilpotent derivation on B^G having a slice and every G_{a}-action on $X /\!\!/ G$ lifts up to a G-equivariant G_{a}-action on X. Here, W is a G-representation space such that $W /\!\!/ G$ is a point.*

PROOF. We show that "if" part. Let δ be a locally nilpotent derivation on B^G with a slice $s \in B^G$. Then δ lifts up to a G-equivariant locally nilpotent derivation Δ on B. Since Δ has a slice $s \in B^G$, it follows that $B = B^{\Delta}[s]$. Hence $Y = \operatorname{Spec} B^{\Delta}$ is a smooth, acyclic G-variety with one-dimensional quotient. Thus $(B^{\Delta})^G = \mathbb{C}[t]$ for $t \in B^{\Delta}$, and $B^G = \mathbb{C}[t,s]$. The locally nilpotent derivation $\mathrm{d}/\mathrm{d}t$ on B^G lifts up to a G-equivariant locally nilpotent derivation Δ' on B. Since $\Delta'(s) = ds/dt = 0$, Δ' induces a G-equivariant G_{a}-action on the G-subvariety Y of X. It follows from Theorem 5.1 that $Y \cong W \oplus \mathbb{C}$ for a G-representation space W such that $\dim W /\!\!/ G = 0$. Hence $X \cong Y \times \mathbb{A}^1 \cong W \oplus \mathbb{C}^2$. □

6. Semi-invariant locally nilpotent derivations associated to equivariant embeddings

In this section, we observe semi-invariant locally nilpotent derivations associated with G-equivariant embeddings where G is abelian. Throughout this section, G is an abelian reductive algebraic group.

First, we make some observations on G-equivariant embeddings. Let V and $\widetilde{V}$ be G-representation spaces, respectively. Since G is abelian, V (resp. $\widetilde{V}$) is a direct sum of one-dimensional G-representation spaces. The coordinate ring S (resp. R) of V (resp. $\widetilde{V}$) is a polynomial ring with semi-invariant coordinates. Let $\varphi\colon V \hookrightarrow \widetilde{V}$ be a G-equivariant embedding of G-varieties, which we abbreviate to a G-embedding. We may assume that φ maps the origin of V to the origin of $\widetilde{V}$. Since φ induces a G-equivariant injective homomorphism of the tangent space at the origin of V into the tangent space at the origin of $\widetilde{V}$, it follows that $\widetilde{V} = V \oplus V'$ for some G-representation space V'. Let $S = k[v_1, \dots, v_m]$ and $R = k[X_1, \dots, X_n]$ where v_i $(1 \le i \le m)$ is semi-invariant of weight α_i and X_j $(1 \le j \le n)$ of weight β_j. Then we may and assume $\beta_i = \alpha_i$ for $1 \le i \le m$. For two G-embeddings φ and φ' of V into $\widetilde{V}$, φ' is called G-equivalent to φ if there is a G-equivariant automorphism γ of $\widetilde{V}$ such that $\varphi' = \gamma \circ \varphi$. A G-embedding $\varphi\colon V \to \widetilde{V}$ is called *G-rectifiable* if there exists a system of semi-invariant coordinate functions $f_1, \dots, f_n$ on $\widetilde{V}$ such that the image $\varphi(V)$ is defined by the G-stable ideal $(f_{m+1}, \dots, f_n)$. If $\varphi\colon V \to \widetilde{V}$ is G-equivalent to the standard G-embedding $V \hookrightarrow V \oplus V' = \widetilde{V}$, then φ is G-rectifiable. Forgetting the G-action, a G-rectifiable embedding $V \to \widetilde{V}$ is a rectifiable embedding $\mathbb{A}^m \to \mathbb{A}^n$.

Now, we construct semi-invariant locally nilpotent derivations associated with $\varphi\colon V \to \widetilde{V}$. Let B be a polynomial ring with a linear G-action

$$B = k[u, v_1, \dots, v_m, x_1, \dots, x_n]$$

where the weight of u, v_i $(1 \le i \le m)$, x_j $(1 \le j \le n)$ is 0, α_i, β_j, respectively. As a G-variety, $X = \operatorname{Spec} B$ is isomorphic to $L \oplus V \oplus \widetilde{V}$ where L is the trivial

G-representation space of one dimension. We give B a $\mathbb{Z}$-grading by

$$\deg u = -1, \quad \deg v_i = 0, \quad \deg x_j = d$$

for $1 \le i \le m$ and $1 \le j \le n$ where d is a fixed positive integer. There is a bijective correspondence between algebraic actions of an algebraic torus G_m on $\operatorname{Spec} B$ and $\mathbb{Z}$-graded k-algebra structures on B. Hence there is a G_m-action on X corresponding to this grading, which we call the T-action where $T = G_m$. The T-action on X commutes with the G-action. Note that

$$B^T = k[v_1, \dots, v_m, u^d x_1, \dots, u^d x_n]$$

and the algebraic quotient $X /\!\!/ T = \operatorname{Spec} B^T$ is G-equivariantly isomorphic to $V \oplus \widetilde{V}$. It follows that $B = \bigoplus_{(\chi,i)\in\Omega\oplus\mathbb{Z}} B_{\chi,i}$ where $B_{\chi,i} = \{b \in B \mid (g,t)\cdot b = t^i\chi(g)b$ for all $(g,t) \in G \times T\}$. The weight of u, v_i $(1 \le i \le m)$, x_j $(1 \le j \le n)$ is $(0,-1)$, $(\alpha_i, 0)$, (β_j, d), respectively. As a $(G \times T)$-representation space,

$$X = L \oplus V_0 \oplus V_{-d} \oplus V'_{-d}$$

where

$$L = \operatorname{Spec} k[u], \quad V_0 = \operatorname{Spec} S, \quad V_{-d} = \operatorname{Spec} k[x_1, \dots, x_m],$$
$$V'_{-d} = \operatorname{Spec} k[x_{m+1}, \dots, x_n].$$

Note that $V_0 = V = V_{-d}$ as a G-representation space. Let $\Phi\colon R \to S$ be the G-equivariant surjection associated with the G-embedding $\varphi\colon V \to \widetilde{V}$. Let f_j be the image of X_j by $\Phi\colon R \to S$ for $1 \le j \le n$. We define a derivation δ_i on B by

$$\delta_i = u^d \partial_{v_i} + (\partial_{v_i} f_1)\partial_{x_1} + \cdots + (\partial_{v_i} f_n)\partial_{x_n}$$

for $1 \le i \le m$. Then δ_i is locally nilpotent and semi-invariant of weight $(-\alpha_i, -d)$, i.e., $\delta_i(B_{\chi,j}) \subset B_{\chi-\alpha_i, j-d}$ for every $(\chi, j) \in \Omega \oplus \mathbb{Z}$ and $\delta_i\delta_l = \delta_l\delta_i$ for any i and l. We consider the sequence $\Delta_\varphi = (\delta_1, \dots, \delta_m)$ of mutually commuting semi-invariant locally nilpotent derivations δ_i's defined above. The sequence $\Delta_\varphi = (\delta_1, \dots, \delta_m)$ has a semi-invariant slice system $(s_1, \dots, s_m)$, i.e., each δ_i has a semi-invariant slice $s_i \in B$ such that $\delta_i(s_l) = 0$ for $i \ne l$. In fact, since Φ is a surjection, there exists a polynomial $F_i \in k[X_1, \dots, X_n]$ such that

$$F_i(f_1, \dots, f_n) = v_i$$

for $1 \le i \le m$. Let $s_i \in B$ be an element satisfying

$$F_i(f_1 - u^d x_1, \dots, f_n - u^d x_n) = v_i - u^d s_i.$$

Since $\delta_l(f_j - u^d x_j) = 0$ for $1 \le l \le m$ and $1 \le j \le n$, it follows that $\delta_l(v_i - u^d s_i) = 0$, i.e., $\delta_i(s_i) = 1$ and $\delta_l(s_i) = 0$ for $i \ne l$. We can take s_i to be semi-invariant of weight (α_i, d). Hence $(s_1, \dots, s_m)$ is a semi-invariant slice system of Δ_φ. We call Δ_φ the sequence associated to the G-embedding φ. In the following, a sequence $\Delta = (\delta_1, \dots, \delta_m)$ implies a sequence of mutually commuting locally nilpotent derivations δ_i's on B such that each δ_i is of the form

$$(*) \qquad \delta_i = u^d \partial_{v_i} + f_{i1}\partial_{x_1} + \cdots + f_{in}\partial_{x_n}$$

with $f_{ij} \in B^T$ and semi-invariant of weight $(-\alpha_i, -d)$ for $1 \le i \le m$.

Conversely, for a given sequence $\Delta = (\delta_1, \dots, \delta_m)$ having a slice system, there exists a G-embedding $\varphi_\Delta\colon V \to \widetilde{V}$ associated to Δ ([7]).

Proposition 6.1 ([7], **cf.** [6]). *Let $\varphi\colon V \to \widetilde{V}$ be a G-embedding and let $\Delta_\varphi = (\delta_1, \dots, \delta_m)$ be the sequence associated to φ. Then the G-embedding φ_{Δ_φ} associated to Δ_φ is G-equivalent to φ.*

For two sequences $\Delta_1 = (\delta_1^{(1)}, \dots, \delta_m^{(1)})$ and $\Delta_2 = (\delta_1^{(2)}, \dots, \delta_m^{(2)})$, we say that Δ_1 and Δ_2 are G-equivalent if there is a $(G \times T)$-equivariant $k[v_1, \dots, v_m]$-automorphism ψ of B which satisfies $\delta_i^{(2)} \circ \psi = \psi \circ \delta_i^{(1)}$ for every i. The following is known.

Theorem 6.2 ([7], **cf.** [6]). *There is a bijective correspondence between the G-equivalence classes of G-embeddings $V \to \widetilde{V}$ and the G-equivalence classes of sequences $\Delta = (\delta_1, \dots, \delta_m)$ having slice systems.*

Let $\Delta = (\delta_1, \dots, \delta_m)$ be a sequence having a semi-invariant slice system $(s_1, \dots, s_m)$. Let

$$A := \bigcap_{i=1}^{m} B^{\delta_i}.$$

Then A inherits the $(G \times T)$-action on B. Let $Y = \operatorname{Spec} A$. Since Δ has a semi-invariant slice system $(s_1, \dots, s_m)$, it follows from Lemma 2.1 that $B = A[s_1, \cdots, s_m]$, i.e.,

$$(1) \qquad Y \times V_{-d} \cong X = L \oplus V_0 \oplus V_{-d} \oplus V'_{-d}.$$

Note that Y is smooth. If Y is isomorphic to a $(G \times T)$-representation space, then it follows from (1) that Y is necessarily isomorphic to the $(G \times T)$-representation space

$$W := L \oplus V_0 \oplus V'_{-d}.$$

Here, we recall that a locally nilpotent derivation having a slice induces an algebraic G_m-action. For $1 \le i \le m$, δ_i and its slice s_i induce an algebraic G_m-action on X, which we call the T_i-action where $T_i = G_m$ for $1 \le i \le m$. The T_i-action on X corresponds to the $\mathbb{Z}$-grading on $B = A[s_1, \dots, s_m]$ such that $\deg s_i = 1$ and $\deg a = 0$ for $a \in B^{\delta_i} - \{0\}$. The subalgebra B^{T_i} is equal to $B^{\delta_i} \cong B/(s_i)$. Since the T_i-actions commute with each other, the m-dimensional torus $T' = T_1 \times \cdots \times T_m$ acts on X and $B^{T'} = A$. Since the $(G \times T)$-action commutes with every T_i-action, $G \times T \times T'$ acts on X. Note that $Y = X /\!\!/ T'$ since $A = B^{T'}$. The algebraic quotient $X /\!\!/ T' = \operatorname{Spec} B^{T'}$ is $(G \times T)$-equivariantly isomorphic to the fixed-point locus $X^{T'} = \operatorname{Spec} B/(s_1, \dots, s_m)$.

The following is verified.

Theorem 6.3 ([7]). *Let $\Delta = (\delta_1, \dots, \delta_m)$ be a sequence having a slice system. Then the following are equivalent.*

(1) *Y is isomorphic to the $(G \times T)$-representation space W.*
(2) *A is a polynomial ring with semi-invariant coordinates with respect to the $(G \times T)$-action.*
(3) *The G-embedding $\varphi_\Delta\colon V \to \widetilde{V}$ associated to Δ is G-rectifiable.*
(4) *The $(G \times T \times T')$-action on X is linearizable.*

The case that G is trivial is shown in [6] (cf. [1]). If the associated G-embedding φ_Δ is G-rectifiable, then it follows from Theorem 6.3 that $Y \cong W$, namely, the equivariant cancellation holds.

Suppose that there is a G-embedding $\varphi\colon V \to \widetilde{V}$ which is rectifiable, but not G-rectifiable. Let $\Delta_\varphi = (\delta_1, \ldots, \delta_m)$ be the sequence associated to φ and let $Y = \operatorname{Spec} A$ where $A = \bigcap_{i=1}^m B^{\delta_i}$. Then Y is isomorphic to a T-representation space, hence the affine space of dimension $n+1$, but not isomorphic to a $(G \times T)$-representation space. Namely, the $(G \times T)$-action on $Y \cong \mathbb{A}^{n+1}$ is nonlinearizable.

Note that W is isomorphic to $L \oplus \widetilde{V}$ as a G-representation space.

Proposition 6.4 ([7]). *Let Δ be a sequence having a slice system. Suppose that the G-action on X is fix-pointed and $V^G = \{0\}$. Then $Y \cong L \oplus \widetilde{V}$ as a G-representation space.*

Proof. Since the G-action on Y is fix-pointed as well, it follows that $Y /\!\!/ G \cong Y^G$. Since $V^G = \{0\}$, it follows from (1) that $Y^G \cong L \oplus (V'_{-d})^G$, hence $Y /\!\!/ G$ is an affine space. Therefore Y is isomorphic to a G-representation space, hence $Y \cong L \oplus \widetilde{V}$. □

Remark. Under the assumption in Proposition 6.4, the G-action on $Y_i = \operatorname{Spec} A_i$ is fix-pointed where $A_i = B^{\delta_i}$. Hence it follows that every Y_i is isomorphic to a G-representation space.

Suppose that G is an r-dimensional torus $(k^*)^r$. Then B is $\mathbb{Z}^r$-graded. If every component of $\deg v_i$ is positive and every component of $\deg x_j$ is nonnegative, then the G-action on X is fix-pointed and $V^G = \{0\}$. By Proposition 6.4, it follows that $Y \cong L \oplus \widetilde{V}$.

References

1. T. Asanuma, *Non-linearizable algebraic k^*-actions on affine spaces*, Invent. Math. **138** (1999), no. 2, 281–306.
2. H. Bass and W. Haboush, *Linearizing certain reductive group actions*, Trans. Amer. Math. Soc. **292** (1985), no. 2, 463–482.
3. G. Freudenburg, *Algebraic theory of locally nilpotent derivations*, Encyclopaedia Math. Sci., vol. 136, Springer, Berlin, 2006.
4. H. Kraft and G. W. Schwarz, *Reductive group actions with one-dimensional quotient*, Inst. Hautes Études Sci. Publ. Math. **76** (1992), 1–97.
5. D. Luna, *Slices étales*, Sur les groupes algébriques, Bull. Soc. Math. France, Mém. 33, Soc. Math. France, Paris, 1973, pp. 81–105.
6. K. Masuda, *Homogeneous locally nilpotent derivations having slices and embeddings of affine spaces*, J. Algebra **321** (2009), no. 6, 1719–1733.
7. ———, *Equivariant embeddings and semi-invariant locally nilpotent derivations*, preprint.
8. K. Masuda and M. Miyanishi, *Lifting of the additive group scheme actions*, Tohoku Math. J. (2) **61** (2009), no. 2, 267–286.
9. G. Scheja and U. Storch, *Fortsetzung von Derivationen*, J. Algebra **54** (1978), no. 2, 353–365.
10. W. V. Vasconcelos, *Derivations of commutative noetherian rings*, Math. Z. **112** (1969), 229–233.

Department of Mathematical Sciences, Faculty of Science and Technology, Kwansei Gakuin University, 2-1 Gakuen, Sanda 669-1337, Japan

E-mail address: kayo@kwansei.ac.jp

Centre de Recherches Mathématiques
CRM Proceedings and Lecture Notes
Volume **54**, 2011

Frobenius Sandwiches of Affine Algebraic Surfaces

M. Miyanishi

Abstract. Given an affine algebraic surface X, we consider a normal algebraic surface sandwiched between X and $X^{(p)}$, which is a quotient X/δ by a regular vector field δ on X. We investigate the case where X is the affine plane $\mathbb{A}^2$ and the quotient X/δ is endowed with an $\mathbb{A}^1$-fibration or some mild singularities. We also construct a family of surfaces of general type which are sandwiched by the Frobenius morphism of an abelian surface.

1. Introduction

Throughout the present article, k stands for the ground field which is an algebraically closed field of positive characteristic p. Let R be an affine k-domain, which signifies a finitely generated integral k-domain. Let $\varphi\colon k \to k$ be the pth-power endomorphism $\lambda \to \lambda^p$. Then there is an injective k-algebra homomorphism $\varphi_R\colon R \otimes_k (k, \varphi) \to R$ defined by $\varphi_R(a \otimes \lambda) = \lambda a^p$ for $a \in R$ and $\lambda \in k$. The image of φ_R is denoted by $R^{(p)}$. In short, $R^{(p)}$ is the k-subalgebra of R generated by $\{a^p \mid a \in R\}$, and called the Frobenius image of R. In terms of schemes, let X be an algebraic k-scheme. Then there exists the Frobenius morphism $F_X\colon X \to X \times_{\operatorname{Spec} k} (\operatorname{Spec} k, {}^a\varphi)$. We denote $X \times_{\operatorname{Spec} k} (\operatorname{Spec} k, {}^a\varphi)$ by $X^{(p)}$.

Given a normal affine k-domain R, we generalize the Jacobson theory of purely inseparable extensions of exponent one and establish a Galois correspondence between the set $\mathcal{S}$ of normal affine k-domains B satisfying $R^{(p)} \subseteq B \subseteq R$ and the set $\mathcal{L}$ of saturated p-Lie subalgebras of $\operatorname{Der}_k(R)$ (see Lemma 2.3). If R is a two-dimensional polynomial ring $k[x,y]$, a nonzero, saturated p-Lie subalgebra of $\operatorname{Der}_k(R)$, not identical to $\operatorname{Der}_k(R)$, is generated by a p-closed, regular vector field $\delta = f\partial/\partial x + g\partial/\partial y$ as an R-module, where $f, g \in R$ and $\gcd(f,g) = 1$. The corresponding normal subalgebra B of R is given as $\operatorname{Ker}\delta$, whence $X := \operatorname{Spec} B$ is the quotient of $\mathbb{A}^2$ by δ, $\mathbb{A}^2/\delta$ by notation, which is called a *plane sandwich*. We are interested in the singularities of $\mathbb{A}^2/\delta$ and in determining exactly when $\mathbb{A}^2/\delta$ is endowed with an $\mathbb{A}^1$-fibration induced by the pencil of translates of a given variable for $\mathbb{A}^2$. If δ is of multiplicative type, i.e., $\delta^p = \delta$, then $\mathbb{A}^2/\delta$ has cyclic quotient

2010 *Mathematics Subject Classification.* Primary 14R05; Secondary 14J29.

Key words and phrases. polynomial ring, purely inseparable covering, normal subalgebra, p-closed derivation.

Supported by Grant-in-Aid for Scientific Research (C), JSPS.

This is the final form of the paper.

singularity, though $\mathbb{A}^2/\delta$ may have nonrational singularity when δ is of additive type, i.e., $\delta^p = 0$. Given a p-closed regular vector field $\delta = f\partial/\partial x + g\partial/\partial y$ such that $\gcd(f, g) = 1$, the plane sandwich $\mathbb{A}^2/\delta$ is smooth if and only if the curves $f = 0$ and $g = 0$ have no intersection points (cf. [18] and Lemma 2.6). There are nonrational examples of plane sandwiches (see Theorem 3.4).

As for the existence of $\mathbb{A}^1$-fibrations, we treat the case where $\mathbb{A}^2/\delta$ is smooth and is endowed with an $\mathbb{A}^1$-fibration which lifts to an $\mathbb{A}^1$-fibration on the upper $\mathbb{A}^2$ (see Theorems 3.7 and 3.8). In this case, if the plane sandwich is a hypersurface, we can also write down explicitly the form of δ.

In Section 4, we discuss a problem of Abhyankar – Moh asking whether, given a polynomial $f \in k[x, y]$, the curve $f = \lambda$ is isomorphic to $\mathbb{A}^1$ for every $\lambda \in k$ provided the curve $f = 0$ is isomorphic to $\mathbb{A}^1$. We consider some necessary (and partly sufficient) conditions for the problem to be answered positively (see Theorem 4.2), and a generalization of an example of Nagata (see Theorem 4.4) which is interesting also for its connection with the topic of forms of $\mathbb{A}^1$.

In general, given a normal algebraic surface V, we call a normal algebraic surface X a *Frobenius sandwich* of V if X splits the Frobenius morphism $F_V : V \to X \to V^{(p)}$. Such an object will be a target to be explored in the future.

In Section 5, we give a family of projective smooth surfaces of general type which are sandwiched by an abelian surface when rational curves are contracted to normal singular points, though the result is a bit apart from the above stream of arguments (see Theorem 5.1).

The present article is in part a survey summarizing the known results and in part an announcement of new research results.

The author is indebted to the referee for pointing out numerous places where the first version of the article could be improved and suggesting appropriate improvements. Theorem 4.4 is one of those improvements, and is due to the referee. The author would like to express his deep appreciation to the referee.

2. Galois correspondence

Some results are partially treated in [3, 15]. We begin with recalling the infinitesimal group schemes α_p and μ_p. These are defined as the kernels of the Frobenius endomorphisms of G_a and G_m, respectively, and have the same scheme structure $\operatorname{Spec} k[t]/(t^p)$, where we identify t with its residue class modulo (t^p). The group scheme structure of α_p (resp. μ_p) is given by the comultiplication Δ, the augmentation ε and the coinverse η on the algebra $k[t]/(t^p)$ corresponding respectively to the multiplication, the unit and the inverse of the group scheme, which are defined by

$$\Delta(t) = t \otimes 1 + 1 \otimes t, \quad \varepsilon(t) = 0 \text{ and } \eta(t) = -t$$

$$\left(\text{resp. } \Delta(t) = t \otimes 1 + 1 \otimes t + t \otimes t, \quad \varepsilon(t) = 0 \text{ and } \eta(t) = \sum_{i=1}^{p-1} (-1)^i t^i\right).$$

Let $X = \operatorname{Spec} R$ be an affine k-scheme. An α_p-action (resp. μ_p-action) $\sigma : \alpha_p \times X \to X$ (resp. $\sigma : \mu_p \times X \to X$) is defined in the standard way and described by the associated k-algebra homomorphisms $\sigma^* : R \to R[t]$ with $t^p = 0$ satisfying the conditions:

$$(2.1) \qquad (\sigma^* \otimes 1_{k[t]}) \cdot \sigma^* = (1_R \otimes \Delta) \cdot \sigma^* \quad \text{and} \quad (1_R \otimes \varepsilon) \cdot \sigma^* = 1_R,$$

where $1_{k[t]}$ and 1_R denote the identity morphisms of $k[t]$ and R. Write

$$\sigma^*(x) = \sum_{i=0}^{p-1} \delta_i(x)t^i \quad \text{for } x \in R, \tag{2.2}$$

where δ_i is a k-linear endomorphism of the k-module R. Then the first relation in (2.1) implies:

In the case of α_p,

$$\sum_{i,j=0}^{p-1} \delta_j\delta_i(x) \otimes t^j \otimes t^i = \sum_{\substack{i,j=0\\ i+j<p}}^{p-1} \binom{i+j}{j} \delta_{i+j}(x) \otimes t^j \otimes t^i,$$

whence $\delta_j\delta_i = \binom{i+j}{j}\delta_{i+j}$ and hence $\delta_j\delta_i = 0$ whenever $i + j \geq p$.

In the case of μ_p,

$$\sum_{i,j=0}^{p-1} \delta_j\delta_i(x) \otimes t^j \otimes t^i = \sum_{l=0}^{p-1} \delta_l(x) \otimes (t \otimes 1 + 1 \otimes t + t \otimes t)^l,$$

whence, by comparison of the $t^j \otimes t$ terms in both sides, we obtain

$$\delta_j\delta_1 = (j+1)\delta_{j+1} + j\delta_j \quad (0 \leq j < p-1), \qquad \delta_{p-1}\delta_1 = (p-1)\delta_{p-1}. \tag{2.3}$$

Furthermore, the second relation in (2.1) yields $\delta_0 = 1_R$ in both cases. Since σ^* is a k-algebra homomorphism, δ_1 is a k-derivation of R. Set $\delta := \delta_1$. In the case of α_p, we have $\delta_i = \delta^i/i!$, $\delta^p = 0$ and $\sigma^*(x) = \sum_{i=0}^{p-1}(1/i!)\delta^i(x) \otimes t^i$. In the case of μ_p, the relation (2.3) gives rise to

$$\delta_{j+1} = \frac{1}{(j+1)!}\prod_{i=0}^{j}(\delta - i) \quad \text{for } 0 \leq j < p-1$$

and

$$\prod_{i=0}^{p-1}(\delta - i) = -\frac{1}{(p-1)!}\prod_{i=0}^{p-1}(\delta - i) = -\delta_{p-1}\big(\delta - (p-1)\big) = 0.$$

Hence $\delta^p = \delta$, and

$$\sigma^*(x) = x + \sum_{j=0}^{p-2} \frac{1}{(j+1)!}\left(\prod_{i=0}^{j}(\delta - i)\right)(x) \otimes t^{j+1}. \tag{2.4}$$

A k-derivation δ of R satisfying $\delta^p = 0$ (resp. $\delta^p = \delta$) is said to be of *additive type* (resp. of *multiplicative type*). Conversely, given a k-derivation δ of R such that $\delta^p = 0$ (resp. $\delta^p = \delta$), the k-algebra homomorphism $\sigma^* \colon R \to R \otimes k[t]$ defined by $\sigma^*(x) = \sum_{i=0}^{p-1}(1/i!)\delta^i(x) \otimes t^i$ (resp. (2.4)) gives rise to an α_p-action (resp. μ_p-action) σ on X. A k-derivation δ of an affine k-domain R is synonymous with a regular vector field on the affine k-scheme $X = \operatorname{Spec} R$.

Summarizing the above arguments, we obtain the following result.

Lemma 2.1. *Let R be an affine k-algebra and let $X = \operatorname{Spec} R$. Then an α_p-action (resp. μ_p-action) σ on X corresponds bijectively to a k-derivation δ of additive (resp. multiplicative) type.*

Let now R be an affine k-domain and let $K = Q(R)$ be the quotient field of R. Let $R^{(p)}$ be the k-subalgebra of R generated by $\{a^p \mid a \in R\}$, which is identified with $R \otimes_k (k, \varphi)$, and let $K^{(p)} = Q(R^{(p)})$. Let $\Omega_{R/k}$ be the R-module of k-differentials of R, which is a finitely generated R-module, and let $\mathrm{Der}_k(R) = \mathrm{Hom}_R(\Omega_{R/k}, R)$. Then $\mathrm{Der}_k(K) = \mathrm{Hom}_K(\Omega_{K/k}, K) \cong \mathrm{Der}_k(R) \otimes_R K$. For $\delta, \delta' \in \mathrm{Der}_k(R)$, define the bracket product $[\delta, \delta'] = \delta\delta' - \delta'\delta$ and the pth power iteration of δ, respectively. It is easy to see that $[\delta, \delta']$ and δ^p belong to $\mathrm{Der}_k(R)$. We express these facts by saying that $\mathrm{Der}_k(R)$ is a p-Lie algebra of derivations. Similarly, the K-module $\mathrm{Der}_k(K)$ is a p-Lie algebra over K. Since k is perfect, $\mathrm{Der}_k(K, K) = \mathrm{Der}(K, K)$ $(k \subseteq K^p)$. In this case our p-Lie algebra is exactly what Jacobson [7] called a *restricted K-Lie algebra*.

An R-submodule L of $\mathrm{Der}_k(R)$ is a p-Lie subalgebra if it is closed under the bracket product and the pth power iteration. If L is an R-module generated by a single element δ, then L being a p-Lie subalgebra is equivalent to the condition that $\delta^p = f\delta$ for some element $f \in R$. We then call δ to be *p-closed*. A p-Lie subalgebra L of $\mathrm{Der}_k(R)$ is *saturated* if $f\delta \in L$ for nonzero $f \in R$ and $\delta \in \mathrm{Der}_k(R)$ implies $\delta \in L$. The following result is a well-known Galois theory due to Jacobson [7].

Lemma 2.2. *With the above notations, there is a bijective correspondence between the set $\mathcal{M}$ of intermediate fields M with $K^{(p)} \subseteq M \subseteq K$ and the set $\mathcal{L}$ of p-Lie subalgebras L of* $\mathrm{Der}_k(K)$. *The correspondence is given by*

$$M \mapsto L(M) = \{\delta \in \mathrm{Der}_k(K) \mid \delta|_M = 0\}$$

and

$$L \mapsto M(L) = \{\xi \in K \mid \delta(\xi) = 0 \textit{ for all } \delta \in L\}.$$

We can generalize this result to the following extent. This result is implicitly written in [15, 18]. In the statement and proof of Lemma 2.3, we abuse notation as follows: Let C be a k-domain. For each $\delta \in \mathrm{Der}_k(C)$, we denote also by δ its unique extension to an element of $\mathrm{Der}_k\big(Q(C)\big)$. The context will make clear whether δ is being regarded as a derivation of C or $Q(C)$.

Lemma 2.3. *Suppose that R is normal. Then there is a bijective correspondence between the set $\mathcal{B}$ of normal k-subalgebras B with $R^{(p)} \subseteq B \subseteq R$ and the set $\mathcal{L}_R$ of saturated p-Lie subalgebras L of* $\mathrm{Der}_k(R)$. *The correspondence is given by*

$$B \mapsto L(B) := L\big(Q(B)\big) \cap \mathrm{Der}_k(R)$$

and

$$L \mapsto R \cap M(L_K), \ L_K = L \otimes_R K.$$

PROOF. Given a normal k-subalgebra B of R with $R^{(p)} \subseteq B$, $L(B)$ is a p-Lie subalgebra of $\mathrm{Der}_k(R)$. We shall show that it is saturated. In fact, if $f\delta \in L(B)$ with nonzero $f \in R$ and $\delta \in \mathrm{Der}_k(R)$, then $\delta \in L\big(Q(B)\big) \cap \mathrm{Der}_k(R)$, whence $\delta \in L(B)$. Set $B' = \{a \in R \mid \delta(a) = 0 \text{ for all } \delta \in L(B)\}$. Suppose $a \in B'$. Since $L\big(Q(B)\big) = L(B) \otimes_R K$, we have $a \in M\left(L\big(Q(B)\big)\right) = Q(B)$ by the Jacobson theory. Since a is integral over B and B is normal, we have $a \in B$. Hence $B' \subseteq B$. The converse $B \subseteq B'$ is clear. So, $B = B'$.

Conversely, let L be a saturated p-Lie subalgebra of $\mathrm{Der}_k(R)$. Then $Q\big(R \cap M(L_K)\big) = M(L_K)$. In fact, $Q\big(R \cap M(L_K)\big) \subseteq M(L_K)$ is clear. If $\xi = ba^{-1} \in M(L_K)$ with $a, b \in R$, then $\xi = (a^{p-1}b)a^{-p}$ with $a^p, a^{p-1}b \in R \cap M(L_K)$.

Hence $R \cap M(L_K)$ is a normal subalgebra of R containing $R^{(p)}$. Furthermore, we have

$$\begin{aligned} L(R \cap M(L_K)) &= L\left(Q(R \cap M(L_K))\right) \cap \mathrm{Der}_k(R) \\ &= L(M(L_K)) \cap \mathrm{Der}_k(R) = L_K \cap \mathrm{Der}_k(R) = L. \end{aligned} \qquad \square$$

Let $n = \dim R$. Suppose that R is regular. Let $\delta \in \mathrm{Der}_k(R)$ be nonzero. Then, for any maximal ideal $\mathfrak{m}$ of R, we can write $\mathrm{Der}_k(R)_{\mathfrak{m}} = \bigoplus_{i=1}^n R_{\mathfrak{m}}(\partial/\partial x_i)$, where $\{x_1, \ldots, x_n\}$ is a local system of parameters of $R_{\mathfrak{m}}$. As an element of $\mathrm{Der}_k(R)_{\mathfrak{m}}$, write $\delta = \sum_{i=1}^n f_i(\partial/\partial x_i)$ with $f_i \in R_{\mathfrak{m}}$. If the ideal $I(\delta; R_{\mathfrak{m}}) := (f_1, \ldots, f_n)$ has height ≥ 2, we say that δ is *reduced* at $\mathfrak{m}$. In general, there exists an element $h \in R_{\mathfrak{m}}$ such that $h^{-1}\delta$ is reduced at $\mathfrak{m}$. Moving $\mathfrak{m}$ over all maximal ideals of R, we find an effective divisor (δ) on $X = \mathrm{Spec}\, R$ such that (δ) is generated by h near the point $\mathfrak{m}$. In accordance with [18], we call (δ) the *divisorial part* of δ. If $(\delta) = 0$, we simply say that δ is reduced.

Lemma 2.4. *Suppose that R is regular and factorial and that $\mathrm{Der}_k(R)$ is a free R-module of rank n. Let L be a saturated p-Lie subalgebra of $\mathrm{Der}_k(R)$ of rank one, i.e., $\dim_K L \otimes_R K = 1$. Then there exists an element δ of L such that $L = R\delta$, δ is reduced and δ is p-closed. Conversely, if δ is a reduced, p-closed k-derivation of R, then $R\delta$ is a saturated p-Lie subalgebra of $\mathrm{Der}_k(R)$.*

PROOF. Since $\mathrm{Der}_k(R)$ is R-free, we may write $\mathrm{Der}_k(R) = \bigoplus_{i=1}^n R\delta_i$. Since $\dim_K L \otimes_R K = 1$, $L \otimes_R K = K\delta$ with $\delta \in L$. If we write $\delta = \sum_{i=1}^n f_i \delta_i$ with $f_i \in R$, we may assume that $\gcd(f_1, \ldots, f_n) = 1$. We shall show that $L = R\delta$. In fact, let $\delta' \in L$. Then there exist elements $a, b \in R$ such that $a\delta' = b\delta$ and $\gcd(a, b) = 1$. This implies that $a \mid bf_i$ for $1 \leq i \leq n$. Since $\gcd(a, b) = 1$, we have $a \mid f_i$ for every i and hence $a \mid \gcd(f_1, \ldots, f_n)$. So, a is invertible and $\delta' \in R\delta$. Hence $L = R\delta$. Since L is a p-Lie subalgebra, $\delta^p = f\delta$ with $f \in R$, i.e., δ is p-closed. By the choice of δ, it follows that δ is reduced. The converse is proved by a similar argument. $\square$

We now look into conections between local properties of a p-closed k-derivation δ and the singularity of $\mathrm{Ker}\, \delta$.

Lemma 2.5. *Let $(\mathcal{O}, \mathfrak{m})$ be a geometric local ring over k and let $K = Q(\mathcal{O})$. Let δ be a p-closed k-derivation of K, i.e., $\delta^p = \lambda\delta$ with $\lambda \in K$. Let $K_0 = \mathrm{Ker}\, \delta$, let $\mathcal{O}_0 = K_0 \cap \mathcal{O}$ and let $\mathfrak{m}_0 = K_0 \cap \mathfrak{m}$. Then the following assertions hold.*

1. *$(\mathcal{O}_0, \mathfrak{m}_0)$ is a geometric local ring over k and $K_0 = Q(\mathcal{O}_0)$.*
2. *$\mathcal{O}$ is a finitely generated $\mathcal{O}_0$-module. If $\mathcal{O}$ is normal, so is $\mathcal{O}_0$.*
3. *Let $\widehat{\mathcal{O}}$ be the $\mathfrak{m}$-adic completion of $\mathcal{O}$. Then δ extends naturally to a p-closed k-derivation $\widehat{\delta}$ of $\widehat{\mathcal{O}}$ provided $\delta(\mathcal{O}) \subseteq \mathcal{O}$. Furthermore, $(\widehat{\mathcal{O}})_0 := \mathrm{Ker}\, \widehat{\delta}$ is the $\mathfrak{m}_0$-adic completion of $\mathcal{O}_0$.*

PROOF. Multiplying δ by an element a^p with $a \in \mathcal{O}$, we may assume that $\delta(\mathcal{O}) \subseteq \mathcal{O}$. Let $\mathcal{O}^{(p)}$ be the subalgebra of $\mathcal{O}$ generated by $\{b^p \mid b \in \mathcal{O}\}$ over k. Then $\mathcal{O}^{(p)} \subseteq \mathcal{O}_0$, whence $\mathcal{O}$ is integral over $\mathcal{O}_0$. Write $\mathcal{O} = k[a_1, \ldots, a_n]_{\mathfrak{p}}$ for some prime ideal $\mathfrak{p}$. Then it is clear that $\mathcal{O} = \mathcal{O}^{(p)}[a_1, \ldots, a_n]$. Hence $\mathcal{O}$ is a finitely generated $\mathcal{O}_0$-module.

Let $a \in \mathcal{O}_0 - \mathfrak{m}_0$. Then $a^{-1} \in \mathcal{O}$ and $\delta(a^{-1}) = 0$. Hence $a^{-1} \in \mathcal{O}_0$. This implies that $(\mathcal{O}_0, \mathfrak{m}_0)$ is a local ring. Since $\mathcal{O}_0$ is an $\mathcal{O}^{(p)}$-submodule of a finitely generated $\mathcal{O}^{(p)}$-module $\mathcal{O}$, it follows that $(\mathcal{O}_0, \mathfrak{m}_0)$ is a geometric local ring. If $\mathcal{O}$ is normal, $\mathcal{O}_0$ is clearly normal.

Note finally that $\delta(\mathfrak{m}^r) \subseteq \mathfrak{m}^{r-1}$ for every $r > 0$. Hence δ induces a k-derivation $\widehat{\delta}$ on the completion $\widehat{\mathcal{O}}$ such that $\widehat{\delta}\mid_{\mathcal{O}} = \delta$. On the other hand, $\widehat{\mathcal{O}} = \mathcal{O} \otimes_{\mathcal{O}_0} \widehat{\mathcal{O}}_0$. Since we have an exact sequence of $\mathcal{O}_0$-modules

$$0 \to \mathcal{O}_0 \to \mathcal{O} \xrightarrow{\delta} \mathcal{O},$$

we obtain an exact sequence of $\widehat{\mathcal{O}}_0$-modules

$$0 \to \widehat{\mathcal{O}}_0 \to \widehat{\mathcal{O}} \xrightarrow{\widehat{\delta}} \widehat{\mathcal{O}}.$$

This implies that $(\widehat{\mathcal{O}})_0 = \widehat{\mathcal{O}}_0$. □

Hereafter, whenever we consider a geometric local ring $(\mathcal{O}, \mathfrak{m})$ over k, we tacitly assume that $\mathcal{O}/\mathfrak{m} = k$. Let R be a regular affine k-domain, let δ be a reduced p-closed k-derivation on R and let $B = \operatorname{Ker}\delta$. Let $(\mathcal{O}, \mathfrak{m})$ be a geometric local ring of R, i.e., $\mathcal{O}$ is the localization of R with respect to a maximal ideal. With δ extended to $K = Q(R)$, it is clear that $\delta(\mathcal{O}) \subseteq \mathcal{O}$. Let $\mathcal{O}_0 = \operatorname{Ker}\delta\mid_{\mathcal{O}}$. We recall the following result [18, Theorem 1 (p. 1208) and Corollary to Theorem 4 (p. 1215)].

Lemma 2.6. *With the above notations and assumptions, suppose further that* $\dim R = 2$. *Then* $\mathcal{O}_0$ *is a regular local ring if and only if the ideal* $I(\delta; \mathcal{O})$ *is the unit ideal.*

In fact, Theorem 1 in [18] implies the "if" part, and Corollary [*ibid.*] implies the "only if" part.

If δ is reduced and $I(\delta; \mathcal{O})$ is not the unit ideal, $I(\delta; \mathcal{O})$ has height ≥ 2. If $\dim \mathcal{O} = 2$, then $\mathcal{O}_0$ has isolated singularity by Lemma 2.6. The following result asserts that if δ is of multiplicative type then the singularity of $\mathcal{O}_0$ is very mild.

Theorem 2.7. *Let* $\mathcal{O}, \delta$ *and* $\mathcal{O}_0$ *be as above, with* $\dim \mathcal{O} = 2$. *Assume that* δ *is of multiplicative type and* $\operatorname{ht} I(\delta; \mathcal{O}) = 2$. *Then* $\mathcal{O}_0$ *has analytically cyclic quotient singularity.*

Proof. By [18, Theorem 2], the induced k-derivation $\widehat{\delta}$ on $\widehat{\mathcal{O}}$ is written as $\widehat{\delta} = \alpha x(\partial/\partial x) + \beta y(\partial/\partial y)$ with respect to a local system of parameters $\{x, y\}$ of $\mathcal{O}$, where $\alpha, \beta \in \mathbb{F}_p$. Since $\operatorname{ht} I(\delta; \mathcal{O}) \geq 2$, it follows that $\alpha\beta \neq 0$. Then there exists $d \in \mathbb{F}_p^*$ such that $\beta = \alpha d$. Hence $\widehat{\mathcal{O}}_0$ is the kernel of a k-derivation $x(\partial/\partial x) + dy(\partial/\partial y)$. Then $\widehat{\mathcal{O}}_0 = k[\![x^p, y^p, x^i y^j \mid i + dj \equiv 0 \pmod{p}, 0 < i < p]\!]$, which is the completion of a local ring $k[x, y]_{(x,y)} \cap k(x^p, y^p, y/x^d)$ having cyclic quotient singularity of type (p, d) (see [16]). □

Meanwhile, if δ is not of multiplicative type, say, of additive type for example, the singularity of $\mathcal{O}_0$ tends to become more complicated (see Example 3.10).

3. Plane sandwiches

Throughout this section, R denotes a polynomial ring $k[x, y]$ in two variables. Certain basic results were obtained in [6, 16]. Our purpose in this section is to supplement these results. We begin with an easy remark in the case $p = 2$.

Lemma 3.1. *Let* δ *be a reduced,* p*-closed* k*-derivation of* $k[x, y]$. *Assume that* $p = 2$. *Then the following results hold.*

(1) δ *is of additive type if and only if* $\delta = f(\partial/\partial x) + g(\partial/\partial y)$, *where* $f, g \in k[x^p, y^p]$.

(2) *δ is of multiplicative type if and only if $\delta = (x+f)(\partial/\partial x)+(y+g)(\partial/\partial y)$, where $f, g \in k[x^p, y^p]$.*

If $p \neq 2$, the "only if" part of Lemma 3.1 no longer holds (see Theorem 3.4 and Example 3.5). It is an interesting problem to determine k-derivations of $k[x, y]$ of additive type or of multiplicative type for $p \geq 3$.

We call a normal subalgebra B (or $\operatorname{Spec} B$ by abuse of terminology) of $k[x, y]$ a *plane sandwich* if $B \supset k[x^p, y^p]$ and B has rank p as a free $k[x^p, y^p]$-module. This terminology resembles that in [6], where it is used for towers of polynomial rings factoring the Frobenius of $k[x, y]$. A plane sandwich B is called a *hypersurface* if there exists $f \in k[x, y]\setminus k[x^p, y^p]$ such that $B = k[x^p, y^p, f]$. For any plane sandwich B, the quotient field $Q(B)$ is an intermediate field $k(x^p, y^p) \subsetneqq Q(B) \subsetneqq k(x, y)$. Hence it is a simple extension $Q(B) = k(x^p, y^p)(f)$, where $f \in B \setminus k[x^p, y^p]$. Hence B is birational to a hypersurface $k[x^p, y^p, f]$, and $B = k[x^p, y^p, f]$ if and only if $k[x^p, y^p, f]$ is normal.

Lemma 3.2. *Let $B = k[x^p, y^p, f]$ be a subalgebra of $k[x, y]$. Then the following assertions hold.*

(1) *The hypersurface B is a plane sandwich if and only if f_x and f_y have no common divisors.*
(2) *Suppose that B is a plane sandwich. Then a corresponding reduced, p-closed k-derivation of $k[x, y]$ is $\delta = f_y(\partial/\partial x) - f_x(\partial/\partial y)$.*

Proof. (1) It suffices to show that B is normal if and only if f_x and f_y have no common divisors. Note that B is the coordinate ring of a hypersurface defined by an equation $Z^p = f^{(p)}(X, Y)$ with $X = x^p, Y = y^p, Z = f(x, y)$ and $f^{(p)}(x^p, y^p) = f(x, y)^p$. This follows readily from the Jacobian criterion of smoothness and the Serre criterion of normality for a hypersurface.

(2) Let $\delta' = f_y(\partial/\partial x) - f_x(\partial/\partial y)$. Since f_x and f_y have no common divisors, δ' is reduced. Let $B' = \operatorname{Ker}\delta'$. Then $f \in B'$, whence $B \subseteq B'$. Since $\delta' \neq 0$, the subfield of $k(x, y)$ annihilated by δ' (extended naturally to $k(x, y)$) is equal to $Q(B)$. Hence $B' = Q(B) \cap k[x, y] = B$ because B is normal, and $\delta' \in \operatorname{Der}_k(R) \cap L\big(Q(B)\big) = L(B)$. Since $L(B) = R\delta$ and δ' is reduced, $\delta' = c\delta$ with $c \in R^* = k^*$. (See Lemmas 2.3 and 2.4). □

Remarks 3.3. (1) A plane sandwich is a hypersurface if $p = 2$, but this is not necessarily the case if $p \geq 3$ (cf. [16, Lemma 1.5 and the example following it]). For example, if $B = \operatorname{Ker}\delta$ with respect to a k-derivation of multiplicative type $\delta = x(\partial/\partial x) + dy(\partial/\partial y)$ with $d \in \mathbb{F}_p^*$ and $p \geq 3$, B is a hypersurface if and only if $d = -1$ (cf. [16, Lemma 3.2]).

(2) For $f \in k[x, y]$, let $\delta = f_y(\partial/\partial x) - f_x(\partial/\partial y)$. Then δ is p-closed if f_x and f_y have no common divisors by Lemma 3.2. In fact, without assuming that f_x and f_y have no common divisors, $\delta^p = h\delta$ holds for some h in the ring of δ-constants (cf. [11, Theorem 1.4(1)]).

(3) Suppose $p = 2$. Let $f \in k[x, y]$ and write $f = f_0+f_1x+f_2y+f_3xy$ with $f_i \in k[x^2, y^2]$. Let g be the greatest common divisor of $f_x = f_1 + f_3y$ and $f_y = f_2 + f_3x$. Is g an element of $k[x^2, y^2]$?[1] If this is the case, we have another proof of the fact that a plane sandwich is a hypersurface. In fact, let $f_i = gh_i$ with $h_i \in k[x^2, y^2]$

[1]The referee communicated to us that g is, in fact, an element of $k[x^2, y^2]$.

and let $h = h_1 x + h_2 y + h_3 xy$. Then $\delta = g\delta'$, where $\delta' = h_y(\partial/\partial x) - h_x(\partial/\partial y)$ is a reduced, p-closed derivation, which entails that $B = k[x^2, y^2, h]$.

In [16, Theorems 4.1 and 5.1], an example of a plane sandwich is given, which is a regular, factorial, irrational k-algebra. If $p > 2$, it is a hypersurface $w^p = (u^p v^p + 1)u + v^{p+1}$ and if $p = 2$, it is of the form $w^2 = (\alpha(u^2)v^2 + 1)u + v^3$, where $\alpha(u^2) \in k[u^2]\backslash k[u^4]$. We shall give another example of an irrational plane sandwich, which is based on a result in [12].

Theorem 3.4. *Let* $p = 3$ *and let* $B = k[x, y, t]/\big(y^2 = x^3 + \varphi(t)\big)$, *where* $\varphi(t) \in k[t]$. *Let* $d = \deg \varphi(t)$ *and let* $m = [d/6]$. *If* $d \geq 7$, *we suppose that the following two conditions are satisfied.*

(i) *For every root* α *of* $\varphi'(t)$, $v_\alpha\big(\varphi(t) - \varphi(\alpha)\big) \leq 5$, *where* v_α *is a valuation of* $k(t)$ *with* $v_\alpha(t - \alpha) = 1$.

(ii) *If* $\varphi(t) - \varphi(\alpha) = a(t-\alpha)^3 +$*(terms of degree* > 3 *in* $t - \alpha$*) for some root* α *of* $\varphi'(t)$ *and* $a \in k^*$, *then* $v_\alpha(\varphi(t) - \varphi(\alpha) - a(t-\alpha)^3) \leq 5$.

Then the following assertions hold.

(1) B *is normal, and a point* $(x, y, t) = (\alpha, \beta, \gamma)$ *is a singular point of* Spec B *if and only if* $\beta = 0$, $\varphi'(\gamma) = 0$ *and* $\alpha^3 + \varphi(\gamma) = 0$.

(2) Spec B *is* k*-rational iff* $m = 0$, *birational to a* $K3$*-surface iff* $m = 1$, *and birational to a surface with Kodaira dimension one iff* $m > 1$.

(3) *Write* $t = \tau^3$ *and* $\varphi(t) = \psi(\tau)^3$. *Set* $u = y\big(x + \psi(\tau)\big)^{-1}$. *With these notations,* $y = u^3$, $x = u^2 - \psi(\tau)$ *and* $B = k[\tau^3, u^3, u^2 - \psi(\tau)]$. *Hence* B *is a plane sandwich, and the corresponding* k*-derivation is* $\delta = u(\partial/\partial\tau) - \psi'(\tau)(\partial/\partial u)$, *where* $\delta^3 = -\psi''(\tau)\delta$.

(4) Spec B *is an* α_p*-quotient* $\mathbb{A}^2/\alpha_p$ *iff* $\psi''(\tau) = 0$, *and a* μ_p*-quotient* $\mathbb{A}^2/\mu_p$ *iff* $\psi''(\tau) = -1$.

PROOF. Straightforward. For the assertion (2), refer to [12]. □

Example 3.5. (1) Let $\varphi(t) = t^{10} - t^4 + 1$. Then the conditions (i) and (ii) are satisfied, and Spec B is birational to a $K3$-surface. Since $\psi''(\tau) = 0$ in this case, Spec B is an α_p-quotient $\mathbb{A}^2/\alpha_p$.

(2) Let $\varphi(t) = t^{10} - t^4 + t^2 + 1$. Then the conditions (i) and (ii) are satisfied, and Spec B is birational to a $K3$-surface. Since $\psi''(\tau) = -1$, Spec B is a μ_p-quotient $\mathbb{A}^2/\mu_p$.

We consider an α_p-quotient of $\mathbb{A}^2$ which is k-rational. It seems difficult to give a characterization of derivations δ of $k[x, y]$ of additive type, if $p > 2$. A derivation $\delta = f(x^p, y^p)(\partial/\partial x) - g(x^p, y^p)(\partial/\partial y)$ is clearly a derivation of additive type, where $f(x^p, y^p), g(x^p, y^p) \in k[x^p, y^p]$. The quotient $\mathbb{A}^2/\alpha_p$ is given as Spec B, where $B = \operatorname{Ker} \delta$. But the quotient has very bad singularities. We shall give below two results which ensure that the quotient is k-rational.

Proposition 3.6. *Let* δ *be the above* k*-derivation of additive type. Suppose that* $f(x^p, y^p)$ *and* $g(x^p, y^p)$ *are homogeneous polynomials of the same degree without common divisors. Then* $B = \operatorname{Ker} \delta$ *is* k*-rational.*

PROOF. Let $F = g(x^p, y^p)x + f(x^p, y^p)y$. Then $\delta(F) = 0$, $F_x = g(x^p, y^p)$ and $F_y = f(x^p, y^p)$. Hence $B = k[x^p, y^p, F]$ by Lemma 3.2, which is the coordinate ring of a hypersurface

$$Z^p = g(X, Y)^p X + f(X, Y)^p Y,$$

where $X = x^p, Y = y^p$ and $Z = F$. Let $d = \deg f(X,Y) = \deg g(X,Y)$. Set $Z = X^d Z_1$ and $Y = XY_1$. Then we obtain

$$X = Z_1^p / \{g(1,Y_1)^p + f(1,Y_1)^p Y_1\},$$

which implies that $Q(B) = k(Y_1, Z_1)$. □

An $\mathbb{A}^1$-fibration on a normal affine surface V is a surjective morphism $\rho : V \to C$ onto a smooth algebraic curve such that $\rho^{-1}(U)$ is isomorphic to the direct product $U \times \mathbb{A}^1$ for a nonempty open set U of C. Hence some closed fiber F of ρ may be a singular fiber. Namely, F may be reducible or nonreduced. Then F_{red} is a disjoint union of the affine lines, some of which may carry single cyclic quotient singularity of V. We refer the reader to [14, Chapter 3] for these results. Although the proof in [14] is given over the complex field $\mathbb{C}$, it is not difficult to extend it to the case of our ground field k of positive characteristic.

Given a nonzero polynomial $f(x) \in k[x]$, we denote by $\sqrt{f(x)}$ the monic polynomial $\prod_{i=1}^n (x - \alpha_i)$ when $f(x) = c\prod_{i=1}^n (x-\alpha_i)^{m_i}$ has n distinct linear factors. Furthermore $f^{(p)}(x)$ is the polynomial defined by $f^{(p)}(x^p) = f(x)^p$. If $f(x)$ is a constant c, we set $\sqrt{f(x)} = 1$.

Theorem 3.7. *Let $a(x) \in k[x]$ be a polynomial such that $a(x)$ has distinct roots $\alpha_1, \ldots, \alpha_n$ and $\sqrt{a(x)}$ is a factor of $\sqrt{a'(x)}$. Let $b(x) \in k[x]$ be a polynomial such that $b'(x)$ is prime to $a(x)$. Let $\delta = a(x)(\partial/\partial x) - \big(a'(x)y + b'(x)\big)(\partial/\partial y)$. Let $B = \ker \delta$. Then the following assertions hold.*

(1) *$B = k[x^p, y^p, a(x)y + b(x)]$, which is the coordinate ring of a smooth hypersurface V defined by $Z^p = a^{(p)}(X)Y + b^{(p)}(X)$.*

(2) *The inclusion $k[X] \hookrightarrow k[X,Y,Z]$ defines an $\mathbb{A}^1$-fibration $\rho\colon V \to \mathbb{A}^1$ such that all fibers of ρ are isomorphic to $\mathbb{A}^1$ when reduced and that the fibers over the points $X = \alpha_1^p, \ldots, X = \alpha_n^p$ are multiple fibers with multiplicity p.*

(3) *Let $\varphi\colon \mathbb{A}^2 \to V$ and $\psi\colon V \to \mathbb{A}^2$ be the morphisms defined by the inclusions $B \hookrightarrow k[x,y]$ and $k[x^p, y^p] \hookrightarrow B$, respectively. Let $\rho_0\colon \mathbb{A}^2 \to \mathbb{A}^1$ be the $\mathbb{A}^1$-fibration on the lower $\mathbb{A}^2$ defined by the inclusion $k[x^p] \hookrightarrow k[x^p, y^p]$. Then the fibration ρ is the lifting of ρ_0 by ψ.*

PROOF. (1) Let $Z = a(x)y + b(x)$. Then $\delta(Z) = 0$. Let $X = x^p$ and $Y = y^p$. Then we have a relation $Z^p = a^{(p)}(X)Y + b^{(p)}(X)$. In order to show that $B = k[X,Y,Z]$, it suffices to show that the hypersurface V defined by the equation $Z^p = a^{(p)}(X)Y + b^{(p)}(X)$ is smooth. This follows from the conditions on $a(x)$ and $b(x)$.

(2) Let $\beta \in k - \{\alpha_1^p, \ldots, \alpha_n^p\}$. The fiber of ρ over the point $X = \beta$ is then defined by $Z^p = a^{(p)}(\beta)Y + b^{(p)}(\beta)$. Since $a^{(p)}(\beta) \neq 0$, it is isomorphic to the affine line with coordinate Z. If $X = \alpha_i^p$ $(1 \le i \le n)$, the fiber of ρ over this point is defined by $\big(Z - b(\alpha_i)\big)^p = 0$. Hence it is the affine line with coordinate Y and multiplicity p.

(3) The assertion is clear. □

The converse of Theorem 3.7 holds. Namely, we have the following result.

Theorem 3.8. *Let V be a smooth plane sandwich which we assume to be a hypersurface. Assume further that V satisfies the following conditions.*

(i) *V has an $\mathbb{A}^1$-fibration $\rho\colon V \to \mathbb{A}^1 = \operatorname{Spec} k[X]$, where $X = x^p$.*

(ii) *Let* $\varphi_0 \colon \mathbb{A}^1 \to \mathbb{A}^1$ *be the Frobenius endomorphism defined by* $k[X] \hookrightarrow k[x]$. *The normalization of* $V \times_{\mathbb{A}^1} (\mathbb{A}^1, \varphi_0)$ *is isomorphic to the affine plane* $\mathbb{A}^2 = \operatorname{Spec} k[x,y]$ *with the specified* x *as one of the coordinates.*

Then the coordinate ring B *of* V *is equal to* $k[x^p, y^p, Z]$ *with* $Z = a(x)y + b(x)$, *where* $a(x)$ *and* $b(x)$ *are elements of* $k[x]$ *satisfying the assumptions in Theorem* 3.7. *Hence* $\delta = a(x)(\partial/\partial x) - \big(a'(x)y + b'(x)\big)(\partial/\partial y)$ *is the corresponding reduced,* p*-closed* k*-derivation.*

PROOF. By the condition (ii), every fiber of ρ is the image of the affine line, whence it is irreducible. As remarked above, any singular fiber of ρ is a disjoint union of the affine lines when the fiber is given the reduced structure. Hence a fiber of ρ is a smooth fiber if it is irreducible and reduced. If ρ has no singular fibers, then V is the affine plane $\operatorname{Spec} B$ with $B = k[x^p, Z]$, where $Z = cy + b(x)$ with $c \in k^*$. So, the conclusion holds with c as $a(x)$. Henceforth, we assume that ρ has $n > 0$ multiple fibers. Let $F_1, \ldots, F_n$ be all multiple fibers of ρ which lie over the points $X = \alpha_1^p, \ldots, X = \alpha_n^p$, respectively. Since ρ restricted to $V - \bigcup_{i=1}^n F_i$ is a trivial $\mathbb{A}^1$-bundle over $\mathbb{A}^1 - \{\alpha_1^p, \ldots, \alpha_n^p\}$, we have

$$B\Big[\prod_{i=1}^n (X - \alpha_i^p)^{-1}\Big] = k\Big[x^p, \prod_{i=1}^n (x^p - \alpha_i^p)^{-1}, w\Big],$$

where $w \in B$. By the condition (ii) again, it holds that

$$B\Big[x, \prod_{i=1}^n (X - \alpha_i^p)^{-1}\Big] = k\Big[x, \prod_{i=1}^n (x - \alpha_i)^{-1}, w\Big] = k\Big[x, \prod_{i=1}^n (x - \alpha_i)^{-1}, y\Big].$$

Hence we may write $w = Cy + D$ with $C, D \in k[x, \prod_{i=1}^n (x - \alpha_i)^{-1}]$ and C an invertible element. Meanwhile, B is written as $B = k[x^p, y^p, Z]$, and w is written as a polynomial in Z with coefficients in $k[x^p, y^p]$. Hence we may assume that Z is a linear polynomial $Z = a(x)y + b(x)$ with $a(x), b(x) \in k[x]$. If there are pth power terms contained in Z, we can safely discard them. Hence the defining equation of V is $Z^p = a^{(p)}(X)Y + b^{(p)}(X)$ with $X = x^p$ and $Y = y^p$. Let β be a root of $a(x)$. Then $a^{(p)}(\beta^p) = 0$ and $\big(Z - b(\beta)\big)^p = 0$. Hence the fiber of ρ over the point $X = \beta^p$ is a multiple fiber. This implies $\beta \in \{\alpha_1, \ldots, \alpha_n\}$. If $a(\alpha_i) \neq 0$ for some i, the fiber would be a reduced fiber, a contradiction. So, it follows that $\alpha_1, \ldots, \alpha_n$ exhaust the roots of $a(x)$. Hence we may replace w by Z.

Since $B = k[x^p, y^p, Z]$ is a plane sandwich, the corresponding p-closed derivation is given as in the satement (cf. Lemma 3.2). Since V is smooth, we can show that $a(x)$ and $b(x)$ satisfy the conditions as stated. □

Example 3.9. (1) Let $f(x) = \prod_{i=1}^n (x - \alpha_i)$ with distinct elements $\alpha_1, \ldots, \alpha_n$ of k, let $a(x) = f(x^p)$ and let $b(x) = x$. Then $a'(x) = 0$ and $b'(x) = 1$. Hence $\delta = f(x^p)(\partial/\partial x) - (\partial/\partial y)$ and $B = k[X, Y, Z]$ with a relation $Z^p = f(X)^p Y + X$ satisfy the conditions of Theorem 3.7. The derivation δ is of additive type. Hence the hypersurface $V := \operatorname{Spec}(B)$ is an α_p-quotient $\mathbb{A}^2/\alpha_p$.

(2) If $p = 2$ in Theorem 3.8, δ is of multiplicative type if and only if $a'(x) = 1$. Hence $a(x) = x + a_0(x^p)$. Then $\sqrt{a(x)}$ is not a factor of $a'(x)$. Hence V is not obtained as a μ_p-quotient $\mathbb{A}^2/\mu_p$. This seems to be the case for all prime p.

The following example shows that the α_p-quotient $\mathbb{A}^2/\alpha_p$ may have nonrational singularity.

Example 3.10. Suppose $p = 2$. For an integer $n > 0$, let $\delta = x^{2n}(\partial/\partial x) + y^{2n}(\partial/\partial y)$. Then $\delta^2 = 0$ and $\operatorname{Ker}\delta = k[x^2, y^2, x^{2n}y + y^{2n}x]$. Hence $\operatorname{Ker}\delta$ is the coordinate ring of a hypersurface V defined by $Z^2 = X^{2n}Y + Y^{2n}X$, where $X = x^2$, $Y = y^2$ and $Z = x^{2n}y + y^{2n}x$. The unique singular point of V is $(X, Y, Z) = (0, 0, 0)$, and the exceptional curves by its minimal resolution consist of nonsingular rational curves $C_0, C_1, \ldots, C_{2n+1}$ such that $(C_0^2) = -(n+1)$, $(C_i^2) = -2$ and $(C_0 \cdot C_i) = 1$ for $1 \leq i \leq 2n+1$, and $(C_i \cdot C_j) = 0$ otherwise. The fundamental cycle is $Z_0 = 2C_0 + (C_1 + \cdots + C_{2n+1})$ with $p_a(Z_0) = n - 1$. Hence the singularity is not rational if $n \geq 2$. If $n = 1$, it is a rational double point of type D_4 (see Artin [1]).

4. Problem of Abhyankar – Moh

The following problem is attributed to Abhyankar – Moh. The author was communicated by T. T. Moh in 1985 in the following form.

Problem of Abhyankar – Moh. Let $f(x, y)$ be a polynomial in x and y such that the plane curve $f = 0$ is isomorphic to $\mathbb{A}^1_k$. Is the curve $f = \lambda$ also isomorphic to $\mathbb{A}^1_k$ for every $\lambda \in k$?

There is an article by Moh himself [17] which investigates the behavior of the curve $f = 0$ at infinity. We shall approach this problem from a different point of view. Statements (1) $\sim$ (4) below, under the assumption that $f = 0$ defines a plane curve with one place at infinity, are due to Ganong [5] (see also [13]).

Lemma 4.1. *Let f be as in the above problem. Let $\Lambda = \{C_\lambda \mid \lambda \in k\}$ be a pencil of curves on $\mathbb{A}^2_k = \operatorname{Spec} k[x, y]$ with C_λ defined by $f = \lambda$, and let $\overline{\Lambda} = \{\overline{C}_\lambda \mid \lambda \in k \cup (\infty)\}$ be the closure of Λ on $\mathbb{P}^2_k$ with $\mathbb{A}^2_k$ embedded into $\mathbb{P}^2_k$ in the natural way, where $\overline{C}_\infty = dl_\infty$ with $d = \deg f$ and l_∞ the line at infinity. Let $\sigma\colon V \to \mathbb{P}^2_k$ be the minimal sequence of blowings-up which eliminate the base points of $\overline{\Lambda}$, and let $\rho\colon V \to \mathbb{P}^1_k$ be the morphism induced by the proper transform $\sigma'\overline{\Lambda}$. Further, let S be the last (-1) curve of σ. Then the following assertions hold.*

(1) *All the exceptional curves of σ but S are the components of the fiber $\rho^*(P_\infty)$ containing the proper transform of $\overline{C}_\infty$.*

(2) *The morphism ρ restricted on S is a purely inseparable morphism of degree $q = p^n$ for some $n \geq 0$.*

(3) *The multiplicity sequence of the generic fiber F_η of ρ at the point at infinity which is the one for $F_\eta \otimes_{k(\eta)} \overline{k(\eta)}$ at the unique point at infinity with an algebraic closure $\overline{k(\eta)}$ of $k(\eta)$ coincides with the multiplicity sequence of a general fiber F_λ of ρ at the unique point at infinity.*

(4) *For every $\lambda \in k$, the curve C_λ is irreducible and reduced, although C_λ may not have one place at infinity.*

(5) *Let $\widetilde{V}$ be the normalization of $V \times_{\mathbb{P}^1} (S, \rho_S)$ and let $\theta\colon \widetilde{V} \to V$ be the normalization morphism. Then the morphism $\widetilde{\rho}\colon \widetilde{V} \to S$ induced by ρ has smooth general fibers provided $\widetilde{\rho}$ has $\Sigma = \theta^*(S)_{\mathrm{red}}$ as a cross-section.*

Proof. We prove the first part of the assertion (4). For the second part, the readers are referred to [4]. Let T be the second-last exceptional curve of σ. Then $(T \cdot S) = 1$. Hence the fiber at infinity $\rho^*(P_\infty)$ of ρ is of the form $q\Gamma$, where $\operatorname{Supp}(\Gamma) = \sigma^{-1}(l_\infty)_{\mathrm{red}} - S$ by the assertion (1). For $P \in \mathbb{P}^1_k$, let ν_P be the number of irreducible components of $\rho^*(P)$. Then we have

$$\operatorname{rank}\operatorname{Pic}(V) = 2 + \sum_{P\in\mathbb{P}^1}(\nu_P - 1),$$

while $\operatorname{rank}\operatorname{Pic}(V) = 1 + \nu_{P_\infty}$ by the construction of V and the assertion (1). This implies that $\nu_P = 1$ for every $P \in \mathbb{P}^1_k$ other than P_∞. Hence C_λ is irreducible if $\lambda \neq \infty$. Suppose that C_λ is nonreduced for some $\lambda \neq \infty$. Then $f - \lambda = g^m$ with $g \in k[x,y]$ and $m \geq 2$. Since $(\rho^*(P_\lambda)\cdot S) = q = m(G\cdot S)$, where G is the closure of the curve $g = 0$ on V. Hence m is a power of p, say $m = p^r$ with $r \geq 1$. Then $f = g^{p^r} + \lambda = \left(g + \lambda^{p^{-r}}\right)^{p^r}$, which is a contradiction because f is an irreducible polynomial.

(5) Since $\widetilde{V}$ is normal, the singular locus of $\widetilde{V}$ is a finite set if it is not empty. Let $\widetilde{F}$ be a general fiber of $\widetilde{\rho}\colon \widetilde{V} \to S$. Since $\widetilde{\rho}$ has a cross-section Σ by the assumption, $\widetilde{F}$ meets Σ transversally in one point which is then a smooth point. Since general members of the pencil Λ in the statement are smooth, the affine part $\widetilde{F} - \widetilde{F}\cap\Sigma$ is also smooth. Hence $\widetilde{F}$ is smooth. □

The problem of Abhyankar – Moh remains open as this article is written. The impediment to giving a positive answer lies in a possibility that an affine plane curve $g = 0$ of positive geometric genus and with one place at infinity may degenerate to an affine plane curve $f = g - \lambda = 0$ which happens to be isomorphic to the affine line. A result significantly different from the following, but also involving the statements (1) and (2), is proven in [4], and appears in the article by Ganong in the present volume — the reader may profitably compare the following with that result.

Theorem 4.2. *Let f be the same as in the statement of the problem of Abhyankar – Moh. Consider the following four conditions.*

(1) *For every $\lambda \in k$, the curve C_λ defined by $f = \lambda$ is isomorphic to $\mathbb{A}^1_k$.*
(2) *For almost all $\lambda \in k$, the curve C_λ defined by $f = \lambda$ is isomorphic to $\mathbb{A}^1_k$.*
(3) *The generic fiber $C_\eta := \operatorname{Spec} k[x,y]\otimes_{k[f]} k(f)$ is a form of $\mathbb{A}^1_{k(f)}$.*
(4) *With the notations in Lemma 4.1, $\widetilde{\rho}\colon \widetilde{V} \to S$ is a $\mathbb{P}^1$-fibration and Σ is a cross-section of $\widetilde{\rho}$, where the fiber $\widetilde{\rho}^*(\widetilde{P}_\infty)$ is the unique degenerate fiber. Hence $\widetilde{V} - \Sigma\cup\widetilde{\rho}^*(\widetilde{P}_\infty)$ is isomorphic to $\mathbb{A}^2_k$.*

Then the conditions (2) *and* (3) *are equivalent, and the condition* (1) *or* (4) *implies the condition* (2).

PROOF. (2) $\Leftrightarrow$ (3) follows from the assertion (3) of Lemma 4.1.

(1) $\Rightarrow$ (2) is obvious.

(4) $\Rightarrow$ (2). Note that the curve C_λ is smooth for almost all $\lambda \in k$ because the curve C_0 defined by $f = 0$ is smooth. It is clear that the curve C_λ is dominated by $F_{\widetilde{P}} - \Sigma\cap F_{\widetilde{P}}$ which is isomorphic to $\mathbb{A}^1_k$, where the point $P = \rho_S(\widetilde{P})$ corresponds to $\lambda \in k$. So, C_λ is a rational curve with one place at infinity. Hence it is isomorphic to $\mathbb{A}^1$ if C_λ is smooth. □

The following example, due to M. Nagata was perhaps the first example of a pencil of lines $\{f = \lambda\}$ in which f is not a variable, i.e., such that there is no g in $k[x,y]$ with $k[x,y] = k[f,g]$.

Example 4.3. Let $f(x,y) \in k[x,y]$ be given by a parametrization

$$x = t + t^{pn} \quad \text{and} \quad y = t^{p^2} \qquad \text{with } (p,n) = 1 \text{ and } n > 1.$$

Then the following assertions hold.

(1) For every $\lambda \in k$, $f = \lambda$ is isomorphic to $\mathbb{A}^1_k$.

(2) Let $\Lambda = \{C_\lambda \mid \lambda \in k\}$ be the pencil of curves C_λ on $\mathbb{A}^2_k$ defined by $f = \lambda$. Let $\varphi : \mathbb{A}^2_k \to \mathbb{A}^2_k$ be the purely inseparable morphism defined by $\varphi^* : k[x^p, y] \hookrightarrow k[x,y]$. Then $\Lambda = \varphi^* L$, where L is a pencil of lines on $\mathbb{A}^2_k$.

(3) If $p = 2$ and $n = 3$, the generic fiber $\operatorname{Spec} k[x,y] \otimes_{k[f]} k(f)$ is a form of $\mathbb{A}^1_{k(f)}$ with arithmetic genus 1, and the fiber $\rho^*(P_\infty)$ of the morphism $\rho \colon V \to \mathbb{P}^1_k$ is a degenerate fiber of type II^* according to the classification of Kodaira.

Here we give some remarks to elucidate the assertion (2). The polynomial $f(x,y)$ is given as $f = (x^p - y^n)^p - y$. Let $\xi = x^p - y^n$ and $\eta = (x^p - y^n)^p - y$. Then $y = \xi^p - \eta$ and $x^p = \xi + (\xi^p - \eta)^n$. Hence $k[x^p, y] = k[\xi, \eta]$. Let L be the pencil $\{\eta = \lambda \mid \lambda \in k\}$. Then $\Lambda = \varphi^* L$ because $\eta = f(x,y)$. The fact that f is not a variable follows from van der Kulk theorem because $\deg_t x = pn$ and $\deg_t y = p^2$ and either one does not divide the other.

The following theorem is a generalization of the above example of Nagata whose proof applies also to the proof of the nontriviality of the generic fiber $\operatorname{Spec} k(f)[x,y]$ of Λ as an $\mathbb{A}^1$-form in the above example. For the concepts related to, and results about forms of $\mathbb{A}^1$, we refer the reader to [2, 8, 9]. For example, see [9] for the definition of *height* of such a form.

Theorem 4.4. *Let C be an affine plane curve defined by a parametrization*

$$x = t + \theta(t^{p^r}), \quad y = t^{p^s},$$

where s, r are positive integers with $s > r$ and $\theta(t) \in k[t]$ with $n = \deg \theta(t)$ not divisible by p and greater than 1. Then the following assertions hold.

(1) *The curve C is defined by $u = 0$, where*

$$u := y + \theta^{(p^s)}(y^{p^r}) - x^{p^s}.$$

(2) *The curve C_α defined by $u = \alpha$ is isomorphic to $\mathbb{A}^1$ for every $\alpha \in k$.*

(3) *The generic fiber $C_\eta = \operatorname{Spec} k(u)[x,y]$ is a nontrivial form of $\mathbb{A}^1$ with height less than or equal to $s - r$.*

Proof. Put $j = s - r$. Let $v = x^{p^j}$ and $w = y^{p^j}$. Then we have

$$u = y + (\theta^{(p^{s-r})}(y) - v)^{p^r} = y + \beta^{p^r},$$

where $\beta = \theta^{(p^{s-r})}(y) - v$. Then it is easy to verify that u and β are polynomials in y and v, and that y is a polynomial in u and β. So, $v = \theta^{(p^{s-r})}(y) - \beta$ is also a polynomial in u and β. This implies that $k[u,\beta] = k[y,v]$, which we denote by R'. Then $k[v,w][u] = k[u,v,w] \subseteq R'$, and $Q(k[u,v,w])$ and $Q(R')$ are extension fields of $k(x^{p^j}, y^{p^j})$ of degree p^j. Hence the inclusion $k[u,v,w] \subset R'$ is birational. On the other hand, since $R'^{p^j} \subseteq k[u,v,w]$, R' is integral over $k[u,v,w]$. But $k[u,v,w]$ is the coordinate ring of the hypersurface

$$u^{p^j} = w + \theta^{(p^{s+j})}(w^{p^r}) - v^{p^s},$$

which is smooth. So, $k[u,v,w] = R'$ defines the affine plane, and u is a variable in R'. Hence $k(u)[v,w]$ defines an affine line over $k(u)$. This implies that $k(u^{p^{-j}})[x,y]$

defines an affine line over $k(u^{p^{-j}})$. It follows that C_η is a form of $\mathbb{A}^1$ of height less than or equal to j.

The curve C_η is not isomorphic to $\mathbb{A}^1$, for otherwise one of the t-degrees of x and y divides the other by van der Kulk theorem. The assertion (2) is easy to verify. □

5. Abelian sandwiches

In this section, we shall generalize an example in [10, Example 6.2] and construct a family of singular projective surfaces X such that X is sandwiched between an abelian surface and its Frobenius image and that the minimal resolution $\widetilde{X}$ of X is a minimal algebraic surface of general type. The characteristic of the ground field k is assumed to be 2 throughout this section.

Let E_1 be the elliptic curve defined by $u^2+u=x^3$ and let E_2 be its copy. We write the defining equation of E_2 by $v^2+v=y^3$. Let $A=E_1\times E_2$. Set

$$\delta=(x+x^{2n})\frac{\partial}{\partial x}+(y+y^{2n})\frac{\partial}{\partial y}$$

with $n>0$. Then $\delta(x)=x+x^{2n}$, $\delta(y)=y+y^{2n}$, $\delta(u)=x^2(x+x^{2n})$ and $\delta(v)=y^2(y+y^{2n})$. It is straightforward to see that $\delta^2=\delta$. We prove the following result.

Theorem 5.1. *Let $X=A/\delta$ and let $\widetilde{X}$ be the minimal resolution of singularity. Then $\widetilde{X}$ is a minimal surface of general type such that $(K_{\widetilde{X}}{}^2)=8n^2+8n-2$ and $q_{\widetilde{X}}=2$.*

The proof will be given in the rest of the present section. First of all, we consider the singular points of X and their minimal resolutions. Let P be a closed point of A and let $(\mathcal{O}_P,\mathfrak{m}_P)$ be the local ring at P. Denote by $\mathcal{O}_P^\delta$ the subring $\operatorname{Ker}\delta\cap\mathcal{O}_P$. By virtue of Lemma 2.6, $\mathcal{O}_P^\delta$ is smooth unless

(a) P is a point at finite distance and P is a singular point of δ, i.e., $x+x^{2n}=0$ and $y+y^{2n}=0$, or
(b) P is a point at infinity, i.e., $u^{-1}\in\mathfrak{m}_P$ or $v^{-1}\in\mathfrak{m}_P$.

There are $16n^2$ points of type (a). Namely, for each pair (α,β) with $\alpha+\alpha^{2n}=0$ and $\beta+\beta^{2n}=0$, we have four points $P_{\alpha\beta}^{(\gamma\delta)}=\{(\alpha,\gamma),(\beta,\delta)\}$, where $\gamma^2+\gamma=\alpha^3$ and $\delta^2+\delta=\beta^3$. At one of these points $P=P_j$ $(1\le j\le 16n^2)$, we may assume, after linear changes of variables $(x,u)\mapsto(x+\alpha,u+\gamma)$ and $(\beta,\delta)\mapsto(y+\beta,v+\delta)$, that $\widehat{\mathcal{O}}_P=k[\![x,y]\!]$ and $\delta=(x+x^{2n})(\partial/\partial x)+(y+y^{2n})(\partial/\partial y)$. Then $(\widehat{\mathcal{O}}_{P_i})^\delta=k[\![x^2,y^2,(x+x^{2n})(y+y^{2n})]\!]$. Hence the analytic type of singularity at the point $\pi(P)$ is the one for a hypersurface $Z^2=(X+X^{2n})(Y+Y^{2n})$ at the point $(X,Y,Z)=(0,0,0)$, where $\pi\colon A\to X$ is the quotient morphism and where we put $X=x^2,Y=y^2$ and $Z=(x+x^{2n})(y+y^{2n})$. Then the singularity of X at $\pi(P)$ is a rational double point of type A_1.

Consider the case P is of type (b). Since the homogenization of $u^2+u=x^3$ is $u^2z+uz^2=x^3$, E_1 has a unique one-place point $Q:=(x/u=0,1/u=0)$ at infinity. Set $s=x/u$ and $\xi=1/u$. Then $\xi+\xi^2=s^3$. Hence $\partial/\partial x=\left(\xi+s^3/\xi\right)(\partial/\partial s)=\partial/\partial s$ because $\xi+s^3/\xi=1$ and

$$(x+x^{2n})\frac{\partial}{\partial x}=\frac{s(\xi^{2n-1}+s^{2n-1})}{\xi^{2n}}\frac{\partial}{\partial s}.$$

We consider the completion $\widehat{\mathcal{O}}_P$ which contains $\widehat{\mathcal{O}}_Q$. In $\widehat{\mathcal{O}}_Q$, we have an expansion

$$\xi = s^3 + s^6 + s^{12} + s^{24} + s^{48} + \cdots = s^3(1 + s^3 + s^9 + \cdots).$$

Hence we have

$$(x + x^{2n})\frac{\partial}{\partial x} = \left\{\frac{1 + s^{4n-2}(1 + s^3 + \cdots)^{2n-1}}{s^{4n}(1 + s^3 + \cdots)^{2n}}\right\}\frac{\partial}{\partial s}.$$

If $P = (Q, R)$ with $R \neq Q$, then

$$\delta = \left\{\frac{1 + s^{4n-2}(1 + s^3 + \cdots)^{2n-1}}{s^{4n}(1 + s^3 + \cdots)^{2n}}\right\}\frac{\partial}{\partial s} + \left\{\frac{s^{4n}(1 + s^3 + \cdots)^{2n}}{1 + s^{4n-2}(1 + s^3 + \cdots)^{2n-1}} \cdot (y + y^{2n})\right\}\frac{\partial}{\partial y}$$

and $\mathcal{O}_P^\delta$ is regular by Lemma 2.6.

Suppose $P = P_\infty := (Q, Q)$. Then we have

$$\delta = \left\{\frac{1 + s^{4n-2}(1 + s^3 + \cdots)^{2n-1}}{s^{4n}(1 + s^3 + \cdots)^{2n}}\right\}\frac{\partial}{\partial s} + \left\{\frac{1 + t^{4n-2}(1 + t^3 + \cdots)^{2n-1}}{t^{4n}(1 + t^3 + \cdots)^{2n}}\right\}\frac{\partial}{\partial t}, \tag{5.1}$$

where $Q = (t = 0, \eta = 0)$ with $t = y/v$, $\eta = 1/v$ and $\eta + \eta^2 = t^3$.

Lemma 5.2. *$\mathcal{O}_{P_\infty}^\delta$ is a normal local ring whose singularity has the same analytic type as the one for the singular point of a hypersurface $Z^2 = S^{4n+1} + T^{4n+1}$. The exceptional divisor of the minimal resolution of the singularity has the same configuration as in Example* 3.10 *with n there replaced by $2n$.*

PROOF. Note that $\delta\big((x + x^n)(y + y^n)\big) = 0$ and

$$\begin{aligned}(x + x^n)(y + y^n) &= \left(\frac{s}{\xi} + \left(\frac{s}{\xi}\right)^{2n}\right)\left(\frac{t}{\eta} + \left(\frac{t}{\eta}\right)^{2n}\right) \\ &= \frac{\{1 + s^{4n-2}(1 + s^3 + \cdots)^{2n-1}\}\{1 + t^{4n-2}(1 + t^3 + \cdots)^{2n-1}\}}{(st)^{4n}\big((1 + s^3 + \cdots)(1 + t^3 + \cdots)\big)^{2n}}.\end{aligned}$$

Let

$$Z' = \{1 + s^{4n-2}(1 + s^3 + \cdots)^{2n-1}\}\{1 + t^{4n-2}(1 + t^3 + \cdots)^{2n-1}\}$$

and $Z = Z' + 1 + s^{4n-2} + t^{4n-2}$. Then $\delta(Z) = 0$ and

$$Z = s^{4n+1} + t^{4n+1} + \text{terms of higher degree in } s, t.$$

Then one can readily show that $(\widehat{\mathcal{O}}_{P_\infty})^\delta = k[\![s^2, t^2, Z]\!]$ and this complete local ring is isomorphic to the complete local ring of the singular point of a hypersurface

$$Z^2 + S^{4n+1} + T^{4n+1} + \text{terms of higher degree in } S, T, \tag{5.2}$$

where $S = s^2$ and $T = t^2$.

Let $\sigma\colon \tilde{A} \to A$ be the composite of blowings-up with centers at P_i ($1 \leq i \leq 16n^2$) and the point P_∞ as well as $(4n+1)$ infinitely near points Q_j ($1 \leq j \leq 4n+1$) given by the equation $(t/s)^{4n+1} + 1 = 0$. Let $G_i = \sigma^{-1}(P_i)$ and $\sigma^{-1}(P_\infty) = F_0 + \sum_{j=1}^{4n+1} F_j$. Let $A^{(p)}$ (resp. $\tilde{A}^{(p)}$) be the Frobenius image of A (resp. $\tilde{A}$). Then we have the morphism $\sigma^{(p)}\colon \tilde{A}^{(p)} \to A^{(p)}$ as the Frobenius image and $\sigma^{(p)}$ is the composite of blowings-up of the corresponding points $P_i^{(p)}$ $(1 \le i \le 16n^2)$, $P_\infty^{(p)}$ and its $(4n+1)$ infinitely near points. Thus we obtain the exceptional curves ${\sigma^{(p)}}^{-1}(P_i^{(p)}) = G_i^{(p)}$ and ${\sigma^{(p)}}^{-1}(P_\infty^{(p)}) = F_0^{(p)} + \sum_{j=1}^{4n+1} F_j^{(p)}$.

Let $\mu\colon \widetilde{X} \to X$ be the minimal resolution of singularities. Then we have the following commutative diagram

$$\begin{array}{ccccc} \tilde{A} & \xrightarrow{\ \widetilde{\varphi}\ } & \widetilde{X} & \xrightarrow{\ \widetilde{\psi}\ } & \tilde{A}^{(p)} \\ \Big\downarrow{\scriptstyle \sigma} & & \Big\downarrow{\scriptstyle \mu} & & \Big\downarrow{\scriptstyle \sigma^{(p)}} \\ A & \xrightarrow{\ \varphi\ } & X & \xrightarrow{\ \psi\ } & A^{(p)} \end{array}$$

where $\widetilde{\psi} \cdot \widetilde{\varphi}$ (resp. $\psi \cdot \varphi$) is the Frobenius morphism $F_{\tilde{A}}$ (resp. F_A) of $\tilde{A}$ (resp. A). Let $\overline{G}_i = \widetilde{\varphi}(G_i)$ $(1 \le i \le 16n^2)$ and $\overline{F}_j = \widetilde{\varphi}(F_j)$ $(0 \le j \le 4n+1)$. Then it is readily verified that $\widetilde{\psi}^*(G_i^{(p)}) = \overline{G}_i$, $\widetilde{\varphi}^*(\overline{G}_i) = 2G_i$ for $1 \le i \le 16n^2$, $\widetilde{\psi}^*(F_0^{(p)}) = 2\overline{F}_0$, $\widetilde{\varphi}^*(\overline{F}_0) = F_0$ and $\widetilde{\psi}^*(F_j^{(p)}) = \overline{F}_j$, $\widetilde{\varphi}^*(\overline{F}_j) = 2F_j$ for $1 \le j \le 4n+1$. This entails that $(\overline{G}_i^{\,2}) = -2$ $(1 \le i \le 16n^2)$, $(\overline{F}_0^{\,2}) = -(2n+1)$ and $(\overline{F}_j^{\,2}) = -2$ $(1 \le j \le 4n+1)$. We obtain thus the stated configuration of the exceptional locus of the minimal resolution at the singular point $\pi(P_\infty)$. □

In the computation below, we use the following information: Let C_i be the proper transforms of two transversal fibers through P_∞ of the natural projections $A \to E_{3-i}$ for $i = 1, 2$. Let $\overline{C}_i = \widetilde{\varphi}(C_i)$ and $C_i^{(p)} = \widetilde{\psi}(\overline{C}_i)$. Then $\widetilde{\psi}^*(C^{(p)}) = \overline{C}_i$, $\widetilde{\varphi}^*(\overline{C}_i) = 2C_i$, $(C_i^{\,2}) = ({C_i^{(p)}}^2) = -1$ and $(\overline{C}_i^{\,2}) = -2$.

Let $\tilde{\delta}$ be the rational k-vector field δ extended to $\tilde{A}$. We shall look into the divisor $(\tilde{\delta})$ which is the divisorial part of $\tilde{\delta}$.

(1) At $P = P_i$ $(1 \le i \le 16n^2)$, we may assume that $x = y = 0$. Let $\tau = y/x$. Then

$$\tilde{\delta} = x\left\{(1 + x^{2n})\frac{\partial}{\partial x} + x^{2n-2}(\tau + \tau^{2n})\frac{\partial}{\partial \tau}\right\}.$$

In fact, $\tilde{\delta} = \alpha(\partial/\partial x) + \beta(\partial/\partial \tau)$, and α and β are determined by $\alpha = \tilde{\delta}(x)$ and $\beta = \tilde{\delta}(\tau)$. Hence $\alpha = x + x^{2n}$ and $\beta = x^{2n-1}(\tau + \tau^{2n})$. Thus the divisor $(\tilde{\delta})$ on $\tilde{A}$ has G_i with coefficient 1.

(2) At $P = P_\infty$, δ is expressed in terms of s, t in (5.1). Let $\tau = t/s$. Then $\tilde{\delta}$ is expressed as $\tilde{\delta} = A(\partial/\partial s) + B(\partial/\partial \tau)$, where

$$A = \frac{1 + s^{4n-2}(1 + s^3 + \cdots)^{2n-1}}{s^{4n}(1 + s^3 + \cdots)^{2n}}$$

and

$$B = \frac{\tau\{1 + s^{4n-2}(1 + s^3 + \cdots)^{2n-1}\}}{s^{4n+1}\tau^{4n}(1 + s^3 + \cdots)^{2n}(1 + (\tau s)^3 + \cdots)^{2n}} + \frac{s^2\{1 + (\tau s)^{4n-2}(1 + (\tau s)^3 + \cdots)^{2n-1}\}}{s^{4n+1}\tau^{4n}(1 + s^3 + \cdots)^{2n}(1 + (\tau s)^3 + \cdots)^{2n}}.$$

After these computations, we obtain

$$(\tilde{\delta})_{\tilde{A}} = \sum_{i=1}^{16n^2} G_i - (4n+1)F_0 - 4n\sum_{j=1}^{4n+1} F_j - 4n(C_1 + C_2).$$

On the other hand, we have the following relation on the canonical divisors by [18, Corollary 1, p. 1213]

$$K_{\tilde{A}} = \widetilde{\varphi}^*(K_{\widetilde{X}}) + (\tilde{\delta})_{\tilde{A}}.$$

Since

$$K_{\tilde{A}} = \sum_{i=1}^{16n^2} G_i + F_0 + 2\sum_{j=1}^{4n+1} F_j,$$

we have

$$\widetilde{\varphi}^* K_{\widetilde{X}} = (4n+2)F_0 + (4n+2)\sum_{j=1}^{4n+1} F_j + 4n(C_1 + C_2).$$

Hence, by taking the above ramification data into account, we obtain

$$2K_{\widetilde{X}} = (8n+4)\overline{F}_0 + (4n+2)\sum_{j=1}^{4n+1} \overline{F}_j + 4n(\overline{C}_1 + \overline{C}_2).$$

This gives rise to the numerical equivalence

$$K_{\widetilde{X}} \equiv (4n+2)\overline{F}_0 + (2n+1)\sum_{j=1}^{4n+1} \overline{F}_j + 2n(\overline{C}_1 + \overline{C}_2).$$

In particular, $(K_{\widetilde{X}}{}^2) = 8n^2+8n-2$. It is straightforward to see that $\widetilde{X}$ is a minimal surface with $q_{\widetilde{X}} = 2$. This completes the proof of Theorem 5.1.

References

1. M. Artin, *On isolated rational singularities of surfaces*, Amer. J. Math. **88** (1966), 129–136.
2. T. Asanuma, *Purely inseparable k-forms of affine algebraic curves*, Affine Algebraic Geometry (Seville, 2003) (J. Gutierrez, V. Shpilrain, and J.-T. Yu, eds.), Contemp. Math., vol. 369, Amer. Math. Soc., Providence, RI, 2005, pp. 31–46.
3. M. Demazure and P. Gabriel, *Groupes algébriques*. Tome I: *Géométrie algébrique, généralités, groupes commutatifs*, Masson, Paris, 1970.
4. R. Ganong, *On plane curves with one place at infinity*, Ph.D. Thesis, McGill University, 1978.
5. ———, *On plane curves with one place at infinity*, J. Reine Angew. Math. **307/308** (1979), 173–193.
6. R. Ganong, *Plane Frobenius sandwiches*, Proc. Amer. Math. Soc. **84** (1982), no. 4, 474–478.
7. N. Jacobson, *Lectures in abstract algebra*. Vol. III: *Theory of fields and Galois theory*, D. Van Nostrand Co., Inc., Princeton, NJ – Toronto, ON – London – New York, 1964.
8. T. Kambayashi, M. Miyanishi, and M. Takeuchi, *Unipotent algebraic groups*, Lecture Notes in Math., vol. 414, Springer, Berlin, 1974.
9. T. Kambayashi and M. Miyanishi, *On forms of the affine line over a field*, Lectures in Math., vol. 10, Kinokuniya, Tokyo, 1977.
10. T. Katsura and Y. Takeda, *Quotients of abelian and hyperelliptic surfaces by rational vector fields*, J. Algebra **124** (1989), no. 2, 472–492.
11. J. Lang, *The divisor class group of the surface* $Z^{p^n} = G(X, Y)$ *over fields of characteristic* $p > 0$, J. Algebra **84** (1983), no. 2, 391–410.
12. M. Miyanishi, *Unirational quasi-elliptic surfaces in characteristic* 3, Osaka J. Math. **13** (1976), no. 3, 513–522.
13. ———, *Curves on rational and unirational surfaces*, Tata Inst. Fund. Res. Lectures on Math. and Phys., vol. 60, Tata Inst. Fund. Res., Bombay, 1978.
14. ———, *Open algebraic surfaces*, CRM Monogr. Ser., vol. 12, Amer. Math. Soc., Providence, RI, 2001.
15. M. Miyanishi and T. Nomura, *Finite group scheme actions on the affine plane*, J. Pure Appl. Algebra **71** (1991), no. 2-3, 249–264.
16. M. Miyanishi and P. Russell, *Purely inseparable coverings of exponent one of the affine plane*, J. Pure Appl. Algebra **28** (1983), no. 3, 279–317.

17. T. T. Moh, *On the classification problem of embedded lines in characteristic p*, Algebraic Geometry and Commutative Algebra, Vol. I (H. Hijikata, H. Hironaka, M. Maruyama, H. Matsumura, M. Miyanishi, T. Oda, and K. Ueno, eds.), Kinokuniya, Tokyo, 1988, pp. 267–279.
18. A. N. Rudakov and I. R. Shafarevich, *Inseparable morphisms of algebraic surfaces*, Izv. Akad. Nauk SSSR Ser. Mat. **40** (1976), no. 6, 1269–1307 (Russian); English transl., Math. USSR-Izv. **40** (1976), no. 6, 1205–1237.
19. P. Russell, *Factoring the Frobenius morphism of an algebraic surface*, Algebraic Geometry, Bucharest 1982 (Bucharest, 1982) (L. LBădescu and D. Popescu, eds.), Lecture Notes in Math., vol. 1056, Springer, Berlin, 1984, pp. 366–380.

Research Center for Mathematical Sciences, Kwansei Gakuin University, 2-1 Gakuen, Sanda, Hyogo 669–1337, Japan
E-mail address: miyanisi@kwansei.ac.jp

Centre de Recherches Mathématiques
CRM Proceedings and Lecture Notes
Volume **54**, 2011

Automorphism Groups of Koras – Russell Threefolds of the First Kind

L. Moser-Jauslin

Abstract. Koras – Russell threefolds are certain smooth contractible complex hypersurfaces in $\mathbb{A}^4$ which are not algebraically isomorphic to $\mathbb{A}^3$. One of the important examples is the cubic Russell threefold, defined by the equation $x^2y + z^2 + t^3 + x = 0$. In [1], the automorphism group of the Russell cubic threefold was studied. It was shown, in particular, that all automorphisms of this hypersurface extend to automorphisms of the ambient space. This has several interesting consequences, including the fact that one can find another hypersurface which is isomorphic to the Russell cubic, but such that the two hypersurfaces are inequivalent. In the present article, we will discuss how some of these results can be generalized to the class of Koras – Russell threefolds of the first kind.

1. Introduction

In this article, we will study the automorphism groups of certain Koras – Russell threefolds. These varieties were first introduced by Koras and Russell when they were proving their remarkable result that all algebraic actions of $\mathbb{C}^*$ on affine three-space are linearizable [8] (see also [4]). In order to prove this, they studied hyperbolic $\mathbb{C}^*$-actions on more general smooth contractible threefolds. This led them to introduce a set of threefolds which are smooth affine and contractible, however not isomorphic to $\mathbb{A}^3$. These varieties are known as Koras – Russell threefolds. One of the families of these varieties is given by hypersurfaces in $\mathbb{A}^4_{\mathbb{C}} = \mathbb{A}^4$ of the following form. Let $d \geq 2$, and let $2 \leq k < l$ with k and l relatively prime. We denote by $X_{d,k,l}$ the zero set of $P_{d,k,l} = x^dy + z^k + t^l + x$. That is, we have that $X_{d,k,l} = V(P_{d,k,l})$, the zero set of $P_{d,k,l}$. We will call these Koras – Russell threefolds of the first kind. The case where $d = k = 2$ and $l = 3$ is known as the Russell cubic. In order to prove the linearizability of $\mathbb{C}^*$-actions on $\mathbb{A}^3$, it was necessary to show that none of the Koras – Russell threefolds are isomorphic to $\mathbb{A}^3$. For some of these varieties, this could be done with geometric invariants. However, for the Russell cubic, this problem was quite difficult. It was finally shown to be distinct

2010 *Mathematics Subject Classification.* 14R10.
Key words and phrases. polynomial automorphism groups; Koras – Russell threefolds; equivalence of hypersurfaces.
The author is a member of the ANR program COMPLEXE (ANR 08-JCJC- 0130-01).
This is the final form of the paper.

from $\mathbb{A}^3$ using the Makar-Limanov invariant [9]. This tool has since become a very important and useful tool to study affine algebraic varieties.

It was shown by Makar-Limanov and Kaliman [5] that all of the Koras – Russell threefolds have non-trivial Makar-Limanov invariant, and are therefore not isomorphic to $\mathbb{A}^3$. The varieties of the first family have Makar-Limanov invariant isomorphic to a polynomial ring in one variable. These are the varieties we look at in this article.

Many authors, in studying exotic structures, are particularly interested in varieties with trivial Makar-Limanov invariant. However, here, and in several previous works, we take just the opposite approach. The fact that the Makar-Limanov invariant is not trivial gives strong restrictions on the automorphism group, which allows us to prove several interesting properties of these varieties.

Consider the variety $X_{2,2,3}$. As stated before, it is known as the Russell cubic threefold. In [1], in collaboration with A. Dubouloz and P.-M. Poloni, we studied the automorphism group of the Russell cubic. (In that article, it was called the Koras – Russell cubic threefold.) We showed in particular, that when the Russell cubic is embedded into $\mathbb{A}^4$ as a hypersurface defined by $P = x^2y + z^2 + t^3 + x$, then all automorphisms extend to automorphisms of $\mathbb{A}^4$. This leads to several interesting consequences. First of all, we gave an example of a polynomial Q of four variables whose zero set is isomorphic to the Russell cubic, but such that not all automorphisms extend to automorphisms of $\mathbb{A}^4$. In other words, we find inequivalent embeddings of the Russell threefolds in $\mathbb{A}^4$. Also, we showed that the Russell threefold has a particular point (the origin of $\mathbb{A}^4$) which is fixed by all automorphisms.

Our goal here is to show that most of these results hold for all Koras – Russell threefolds of the first kind. Some of the arguments from [1] carry over with few changes, however, in order to study the automorphism group in the generalized setting, we need another result concerning the construction of certain automorphisms (see Lemma 3.1).

In this article, we will review some of the main ideas of the previous results, and we will show how they can be generalized for this class of Koras – Russell threefolds.

Acknowledgements. I would like to thank David Wright for helpful conversations concerning Lemma 3.1.

2. The Makar-Limanov invariant and the Derksen invariant

In this section, we will give the definition of the Makar-Limanov invariant for an affine variety, and describe the result which is the starting point of our study.

Let X be an affine irreducible complex variety with coordinate ring $\mathbb{C}[X]$. A locally nilpotent derivation ∂ on $\mathbb{C}[X]$ is a $\mathbb{C}$-derivation of the algebra $\mathbb{C}[X]$ such that for any $f \in \mathbb{C}[X]$, there exists $n \in \mathbb{N}$ with $\partial^n(f) = 0$. We write $\partial \in \mathrm{LND}(\mathbb{C}[X])$. It is well-known that there is a bijective correspondence between LND's on $\mathbb{C}[X]$ and algebraic actions of the additive complex group $\mathbb{G}_a$ on $\mathbb{C}[X]$. The action corresponding to ∂ is given by the map $\exp(t\partial)$ from $\mathbb{C}[X]$ to $\mathbb{C}[X] \otimes \mathbb{C}[t]$. This, in turn, corresponds to an algebraic $\mathbb{G}_a$-action on X. An element $P \in \mathbb{C}[X]$ is in the kernel of an LND, if and only if the hypersurfaces of X defined $P =$ a constant are all stable by the corresponding action.

Definition 2.1. The *Makar-Limanov invariant* $\mathrm{ML}(\mathbb{C}[X])$ is the subring of $\mathbb{C}[X]$ defined as the intersection of the kernels of all locally nilpotent derivations on $\mathbb{C}[X]$. One says that the Makar-Limanov invariant is *trivial* if it is equal to $\mathbb{C}$. The *Derksen invariant* is the subring of $\mathbb{C}[X]$ generated by all the kernels of LND's on $\mathbb{C}[X]$. One says that the Derksen invariant is *trivial* if it is equal to the ring $\mathbb{C}[X]$.

It is clear, for example, that $\mathbb{C}[x, y, z]$ has a trivial Makar-Limanov invariant and trivial Derksen invariant, because the intersection of the kernels of the three locally nilpotent derivations $\partial_x = \partial/\partial x$, $\partial_y = \partial/\partial y$ and $\partial_z = \partial/\partial z$ is already $\mathbb{C}$, and the ring generated by these three kernels is $\mathbb{C}[x, y, z]$.

We define now the Koras – Russell threefolds of the first type. Given $d \geq 2$, and $2 \leq k < l$ with k and l relatively prime, we denote by $X_{d,k,l}$ the zero set of $P_{d,k,l} = x^d y + z^k + t^l + x$. That is, we have that $X_{d,k,l} = V(P_{d,k,l})$.

For any such variety, one can consider $\mathbb{C}[X_{d,k,l}]$ as a subring of $\mathbb{C}[x, z, t, x^{-1}]$, where $y = -(z^k + t^l + x)/x^d$.

It was shown in [5] that the Makar-Limanov invariant is given by $\mathbb{C}[x]$. In fact, one can show also that the Derksen invariant is given by $\mathbb{C}[x, z, t]$ (see [6, proof of Theorem 9.1 and Example 9.1]). Any automorphism of $X_{d,k,l}$ must preserve the two invariants. This allows us to describe the automorphism groups in terms of certain subgroups of the group of automorphisms of $\mathbb{C}[x, z, t]$. This will be described in detail in Section 4.

3. The special automorphism group of $\mathbb{C}[x]/(x^d)[z, t]$

Before studying the automorphism group of the Koras – Russell threefolds, we look at some properties of a related group of automorphisms.

Let $R = \mathbb{C}[x]$. Consider the Poisson bracket on $R[z, t]$ given by $\{f, g\} = f_z g_t - g_z f_t$. (Here, $f_z = \partial f/\partial z$ and $f_t = \partial f/\partial t$ for any $f \in R[z, t]$.) The ring $R[z, t]$ endowed with the Poisson bracket is a Lie algebra, and, in particular, the Poisson bracket satisfies the Jacobian identity.

For any $H \in R[z, t]$, we can define a derivation D_H on $R[z, t]$ by defining $D_H(f) = \{H, f\} = H_z f_t - H_t f_z$. Then the map $\widehat{\varphi_{wH}} = \exp(wD_H)$ is a well-defined automorphism of $R[z, t][\![w]\!]$ which fixes all elements of R and fixes w. Its inverse is $\widehat{\varphi_{-wH}}$. Note that wD_H is a derivation, but not, in general, locally nilpotent. Therefore $\widehat{\varphi_{wH}}$ does not give an automorphism of the polynomial ring $\mathbb{C}[x, z, t, w]$. However, over the formal power series in w, we do get an automorphism. The determinant of the Jacobian is therefore an invertible function in $R[z, t][\![w]\!]$. We can even say more.

Lemma 3.1. *For any* $H \in \mathbb{C}[x, z, t]$*. Then the determinant of the Jacobian of* $\widehat{\varphi_{wH}}$ *is* 1.

Remark 3.2. The automorphism $\widehat{\varphi_H}$ is a formal version of a Hamiltonian flow for the Hamiltonian H, and the lemma simply expresses the fact that Hamiltonian flows preserve volume.

This result has been proven in a more general setting in [11, Corollary 2.4.9; 12, Corollary 2.4]. We will give the main argument here.

PROOF. To simplify the notation, we denote $\varphi = \widehat{\varphi_{wH}}$, and $D = D_H$. We have that the determinant of the Jacobian is given by $\det(\mathrm{Jac}) = \{\varphi(z), \varphi(t)\}$. Now we calculate this value:

$$\begin{aligned}\det(\mathrm{Jac}) &= \sum_{l=0}^{\infty}\frac{w^l}{l!}D^l(z)_z\sum_{m=0}^{\infty}\frac{w^m}{m!}D^m(t)_t - \sum_{l=0}^{\infty}\frac{w^l}{l!}D^l(z)_t\sum_{m=0}^{\infty}\frac{w^m}{m!}D^m(t)_z \\ &= \sum_{n=0}^{\infty}\frac{w^n}{n!}\sum_{k=0}^{n}\binom{n}{k}((D^kz)_z(D^{n-k}t)_t - (D^kz)_t(D^{n-k}t)_z) \\ &= \sum_{n=0}^{\infty}\frac{w^n}{n!}\sum_{k=0}^{n}\binom{n}{k}\{D^kz, D^{n-k}t\}\end{aligned}$$

Now let $\Sigma_n = \sum_{k=0}^{n}\binom{n}{k}\{D^kz, D^{n-k}t\}$. By using the Jacobian identity and the definition of the derivation D, one finds that $D(\Sigma_n) = \Sigma_{n+1}$. Also, $\Sigma_0 = 1$. We therefore find by induction that $\Sigma_n = 0$ if $n \geq 1$, and therefore $\det(\mathrm{Jac}) = 1$. □

Throughout this article, we fix an integer $d \geq 2$. Denote by $\overline{R}$ the quotient ring $R/(x^d)$. We call an automorphism of $\overline{R}[z,t]$ which fixes $\overline{R}$ *special* if the determinant of its Jacobian is a non-zero constant. We call $\mathrm{SAut}_{\overline{R}}\,\overline{R}[z,t]$ the group of special automorphisms. We will use Lemma 3.1 to determine a set of generators of the subgroup of special automorphisms of $\overline{R}[z,t]$ which are congruent to the identity modulo (x).

For any $H \in \mathbb{C}[x,z,t]$, $\widehat{\varphi_{wH}}$ induces an automorphism $\overline{\varphi}^d_{xH} = \overline{\varphi}_{xH}$ of $\overline{R}[z,t] \cong \mathbb{C}[x,z,t][\![w]\!]/(w-x, w^d)$. In other words, $\overline{\varphi}_{xH}$ is the exponential of the locally nilpotent derivation xD_H of $\overline{R}[z,t]$. By the lemma, we have that $\overline{\varphi}_{xH}$ is special with Jacobian determinant equal to 1.

For $j = 1,\dots,d$, we define:

$$\begin{aligned}\mathcal{A}_j &= \{\varphi \in \mathrm{Aut}_R\,R[z,t] \mid \varphi \equiv \mathrm{id} \bmod (x^j)\} \\ \bar{\mathcal{A}}^d_j = \bar{\mathcal{A}}_j &= \{\overline{\varphi} \in \mathrm{SAut}_{\overline{R}}\,\overline{R}[z,t] \mid \overline{\varphi} \equiv \mathrm{id} \bmod (x^j)\}.\end{aligned}$$

Notation 3.3. Throughout this article, the integer d remains fixed. When we consider automorphisms of $\mathcal{A}_j$ modulo (x^d), we simplify the notation of $\bar{\mathcal{A}}^d_j$ to $\bar{\mathcal{A}}_j$. There will be one proof (for corollary 6.1) where will consider $\bar{\mathcal{A}}^d_j$ and $\bar{\mathcal{A}}^{d+1}_j$, and we will need to distinguish between the two different cut-off points of the series in x. Thus, for that case, we will specify the two different notations.

It is clear that $\bar{\mathcal{A}}_d = \{\mathrm{id}\}$, and that $\bar{\mathcal{A}}_{j+1}$ is a normal subgroup of $\bar{\mathcal{A}}_j$.

It was shown in [15] that the natural map $\pi = \pi_d\colon \mathrm{Aut}_R[z,t] \to \mathrm{SAut}_{\overline{R}}\,\overline{R}[z,t]$ given by restriction modulo (x^d) is surjective. In other words, any special automorphism modulo (x^d) lifts to an automorphism of the polynomial ring $R[z,t]$. From this result, one sees that the group $\mathcal{A}_1/\mathcal{A}_d \cong \bar{\mathcal{A}}_1$ and that for $j = 1,\dots,d-1$, we have that $\mathcal{A}_j/\mathcal{A}_{j+1} \cong \bar{\mathcal{A}}_j/\bar{\mathcal{A}}_{j+1}$. The automorphisms of $\mathcal{A}_1$ have Jacobian determinant equal to 1.

Proposition 3.4. *For $j = 1,\dots,d-1$, we have that $\bar{\mathcal{A}}_j/\bar{\mathcal{A}}_{j+1} \cong \mathcal{A}_j/\mathcal{A}_{j+1}$ is congruent to the additive group of the ideal $(z,t) \in \mathbb{C}[z,t]$.*

Proof. For $j = 1,\dots,d-1$, we will show that $\bar{\mathcal{A}}_j/\bar{\mathcal{A}}_{j+1}$ is congruent to the additive group given by the ideal $(z,t) \subset \mathbb{C}[z,t]$. To do this, we construct a surjective group homomorphism $\Phi_j\colon \bar{\mathcal{A}}_j \to (z,t) \subset \mathbb{C}[z,t]$ whose kernel is $\bar{\mathcal{A}}_{j+1}$. If $\overline{\varphi} \in \bar{\mathcal{A}}_j$ then, by using the condition that the jacobian determinant of $\overline{\varphi}$ is one, it is easy to see that there exists $h \in (z,t) \subset \mathbb{C}[z,t]$ such that $\overline{\varphi}(z) \equiv z + h_t x^j \bmod (x^{j+1})$ and $\overline{\varphi}(t) \equiv t - h_t x^j \bmod (x^{j+1})$. We define the map Φ_j from $\bar{\mathcal{A}}_j/\bar{\mathcal{A}}_{j+1}$ to (z,t) by $\Phi_j(\overline{\varphi}) = h$. We now find a preimage for any $h \in (z,t)$ to show that Φ_j is surjective. We consider $\overline{\varphi}_{x^j h} = \exp(x^j D_h)$ the corresponding automorphism of $\overline{R}[z,t]$. We have that $\Phi_j(\overline{\varphi}_{x^j h}) = h$, and by Lemma 3.1, the Jacobian determinant is 1, and therefore, we have found a preimage of h in $\bar{\mathcal{A}}_j$. □

In particular, we have shown the following. For any $h \in (z,t)$ and any $j = 1, \dots, d-1$, choose any automorphism $\varphi_{x^j h}$ which restricts to $\overline{\varphi}_{x^j h}$ modulo (x^d). (The choice is not unique.)

Corollary 3.5. *Let $\mathcal{A}_1$ and $\bar{\mathcal{A}}_1$ be defined as above. Then:*

- *$\bar{\mathcal{A}}_1$ is generated by the set $\{\overline{\varphi}_{x^j h} \mid j \in \{1, \dots, d-1\}; h \in (z,t)\}$; and*
- *$\mathcal{A}_1$ is generated by $\mathcal{A}_d$ and the set $\{\varphi_{x^j h} \mid j \in \{1, \dots, d-1\}; h \in (z,t)\}$.*

Instead of $\varphi_{x^j h}$ one could also choose any other automorphism which is equivalent to it modulo (x^{j+1}).

4. The automorphism group of the Koras – Russell threefolds

Let $X = X_{d,k,l}$ be one of the Koras – Russell threefolds defined above. We will study the automorphism group $\mathcal{A}$ of $\mathbb{C}[X]$.

As was done in [1], using the Makar-Limanov invariant and the Derksen invariant, we see that any automorphism $\tilde{\varphi}$ of $\mathbb{C}[X]$ restricts to an automorphism φ of $\mathbb{C}[x,z,t]$ which stabilizes the ideal (x) and the ideal $I = (x^d, z^k + t^l + x)$. Also, any automorphism of this form of $\mathbb{C}[x,z,t]$ lifts to a unique automorphism of $\mathbb{C}[X]$. Thus the automorphism group of $\mathbb{C}[X]$ is isomorphic to the subgroup of the automorphism group of $\mathbb{C}[x,z,t]$ of those automorphims that stabilize (x) and I. Using the $\mathbb{C}^*$-action on X, we can show that $\mathcal{A}$ is isomorphic to the semi-direct product $\mathcal{A}_1(d; z^k + t^l + x) \rtimes \mathbb{C}^*$, where $\mathcal{A}_1(d; z^k + t^l + x)$ is the subgroup of R-automorphisms of $R[z,t]$ which are congruent to the identity modulo (x), and which stabilize I. Thus, to study the automorphism group, we must determine $\mathcal{A}_1(d; z^k + t^l + x)$.

$$\mathcal{A}_1(d; z^k + t^l + x) = \{\varphi \in \operatorname{Aut}_R R[z,t] \mid \varphi(I) = I \text{ and } \varphi \equiv \mathrm{id} \bmod (x)\}.$$

Terminology 4.1. If $\tilde{\varphi}$ is an automorphism of $\mathbb{C}[X]$ which restricts to an automorphism φ of $\mathbb{C}[x,z,t]$, we say that φ *lifts to the automorphism* $\tilde{\varphi}$. From what we have seen, an automorphism lifts if and only if (x) and I are preserved. Also, the lifting is unique. Throughout this article, given an automorphism φ of $\mathbb{C}[x,z,t]$ which preserves the ideals (x) and I, we will denote by $\tilde{\varphi}$ the unique lift to an automorphism of $\mathbb{C}[X]$. More precisely, if $\varphi(z^k + t^l + x) = a(z^k + t^l + x) + bx^d$ with $a, b \in \mathbb{C}[x,z,t]$, then we have that $\tilde{\varphi}(y) = ay - b$. The inverse $\tilde{\varphi}^{-1}$ is given by the lift $\widetilde{\varphi^{-1}}$ of the inverse of φ.

First, we consider a more general setting. Throughout this section, we fix $r \in \mathbb{C}[x,z,t]$ such that the zero set of $r_0 = r(0,z,t)$ is a connected curve in $\mathbb{A}^2$, and r_0 has no multiple factors. For example, we can choose $r = z^k + t^l + x$ where k and l are relatively prime. We study certain subgroups of the automorphism group of affine three space. Let $I_{d,r}$ be the ideal generated by r and x^d. Now we define the following groups.

$$\mathcal{A}_j(d;r) = \{\varphi \in \mathcal{A}_j \mid \varphi(I_{d,r}) = I_{d,r}\}$$

$$\bar{\mathcal{A}}_j(r) = \{\overline{\varphi} \in \bar{\mathcal{A}}_j \mid \overline{\varphi}(r) \in (r) \subset \overline{R}[z,t]\}$$

(In the definition of the second group, we consider r as an element of $\overline{R}[z,t]$.)

Note that $\mathcal{A}_d(d;r) = \mathcal{A}_d$. Also, note that $\pi^{-1}(\bar{\mathcal{A}}_j(r)) = \mathcal{A}_j(d;r)$. From this, it is easy to see that $\mathcal{A}_1(d;r)/\mathcal{A}_d \cong \bar{\mathcal{A}}_1(r)$, and that $\mathcal{A}_j(d;r)/\mathcal{A}_{j+1}(d;r) \cong \bar{\mathcal{A}}_j(r)/\bar{\mathcal{A}}_{j+1}(r)$ for $j = 1, \ldots, d-1$.

Proposition 4.2. *For $j = 1, \ldots, d-1$, the group*

$$\mathcal{A}_j(d;r)/\mathcal{A}_{j+1}(d;r) \cong \overline{\mathcal{A}}_j(r)/\bar{\mathcal{A}}_{j+1}(r)$$

is isomorphic to the additive group $\mathbb{C}[z,t]$.

PROOF. We use Lemma 3.1 together with the idea of the proof of the first case, for the Russell cubic given in [1], to construct a surjective homomorphism Ψ_j from $\bar{\mathcal{A}}_j(r)$ to $\mathbb{C}[z,t]$ whose kernel is $\bar{\mathcal{A}}_{j+1}(r)$.

Let $\overline{\varphi}$ be an automorphism in $\bar{\mathcal{A}}_j(r)$. By calculating the Jacobian of the automorphism as before, we have that there exists $h_0 \in \mathbb{C}[z,t]$ such that

$$\overline{\varphi}(z) \equiv z + x^j(h_0)_t \bmod (x^{j+1})$$

$$\overline{\varphi}(t) \equiv t - x^j(h_0)_z \bmod (x^{j+1}).$$

Now $\overline{\varphi}(r) \equiv r + x^j\{r_0, h_0\} \bmod (x^{j+1})$, and since the ideal (r) of $R[z,t]$ is stable, and $j-1 < d$, we have that $\{r_0, h_0\}$ is in the ideal (r_0) of $\mathbb{C}[z,t]$ generated by r_0. Now, using the fact that r_0 has no multiple factors and that the zero set of r_0 in $\mathbb{A}^2$ is connected, this implies that, up to a constant, $h_0 \in (r_0)$. This comes from the fact that, by the hypotheses on r_0, the restriction to the zero set of r_0 of the function defined by h_0 is constant. We may therefore assume that h_0 is in the ideal generated by r_0. (see Proposition 3.6 of [1]). If $h_0 = \alpha r_0$, we define $\Psi_j(\overline{\varphi}) = \alpha$. This gives a homomorphism from $\bar{\mathcal{A}}_j(r)$ to the additive group $\mathbb{C}[z,t]$, whose kernel is exactly $\bar{\mathcal{A}}_{j+1}(r)$. Now we are left to show that Ψ_j is surjective. For any $\alpha \in \mathbb{C}[z,t]$, we find a preimage in $\bar{\mathcal{A}}_j(r)$. Let $h = \alpha r$, and let $\overline{\varphi}$ be equal to $\overline{\varphi}_{x^j h}$. We have that $\overline{\varphi} \in \bar{\mathcal{A}}_j$, and since r is a factor of h, we have that the ideal $\overline{\varphi}(r) \in (r) \subset \overline{R}[z,t]$. Also, since $r \equiv r_0 \bmod (x)$, we have that $\Psi_j(\overline{\varphi}) = \alpha$. □

As before, for any $h \in R[z,t]$ and any $j \in \{1, \ldots, d-1\}$, we choose an arbitrary preimage $\varphi_{x^j h} \in \pi^{-1}(\overline{\varphi}_{x^j h}) \subset \mathcal{A}_1$. Note that, since $\pi^{-1}(\bar{\mathcal{A}}_j(r)) = \mathcal{A}_j(d;r)$, we have that $\varphi_{x^j h} \in \mathcal{A}_j(d;r)$.

Remark 4.3. Note the difference between the proof of this proposition and the proof of Proposition 3.4. In the previous section, we could choose generators of $\mathcal{A}_j/\mathcal{A}_{j+1}$ of the form $\varphi_{x^j h}$ where h is independant of x. Now, we need to find elements of $\mathcal{A}_j$ which preserve the ideal (r). In order to do this, we choose h in the ideal (r). In particular, h depends on x. If $\alpha \in \mathbb{C}[z,t]$, let $h = h(x,z,t) = \alpha r$ and $h_0 = h_0(z,t) = \alpha r_0 - h(0,0,0)$. Then $h_0 \in (z,t) \subset \mathbb{C}[z,t]$, and $\varphi_{x^j h}$ is congruent to $\varphi_{x^j h_0}$ modulo $\mathcal{A}_{j+1}$.

As an immediate consequence of the proposition, we have the following set of generators of $\mathcal{A}_1(d;r)$:

Corollary 4.4. *Suppose that the zero set of r_0 is connected, and that r_0 has no multiple factors. Then*

- *the group $\bar{\mathcal{A}}_1(r)$ is generated by the set of automorphisms of the form $\overline{\varphi}_{x\gamma r}$, for $\gamma \in \overline{R}[z,t]$;*

- *the group* $\mathcal{A}_1(d;r)$ *is generated by* $\mathcal{A}_d$ *and the set of automorphisms of the form* $\varphi_{x\gamma r}$, *for* $\gamma \in R[z,t]$.

5. Extension of automorphisms to $\mathbb{A}^4$

Terminology 5.1. Suppose $\tilde{\varphi}$ is an automorphism of $\mathbb{C}[X]$. We say that Φ *defines an extension of* $\tilde{\varphi}$ if Φ is an automorphism of $\mathbb{C}[x,y,z,t]$ which preserves the ideal generated by $P = P_{d,k,l}$, and $\tilde{\varphi}$ is the induced automorphism of $\mathbb{C}[X]$. In geometric terms, this simply means that the automorphism of X corresponding to $\tilde{\varphi}$ extends to an automorphism of $\mathbb{A}^4$. This extension is not necessarily unique.

The hypersurfaces $X_{d,k,l}$ are embedded in affine four-space. In this section, we will show that all automorphisms of $X_{d,k,l}$ extend to automorphisms of the ambient four-space. This section uses a technique, developed in [13], used also in [10], and was adapted for the case of the Russell cubic in [1]. The idea is simply to lift families of automorphisms. More precisely, we use the fact that we can construct automorphisms of all the fibers of the polynomial $P_{d,k,l}$ in a continuous way.

Theorem 5.2. *All automorphisms of* $X_{d,k,l}$ *extend to automorphisms of* $\mathbb{A}^4$.

Proof. We set $r = z^k + t^l + x$. First of all, the action of $\mathbb{C}^*$ extends to $\mathbb{A}^4$, and thus we only must check the automorphisms lifted from elements in $\mathcal{A}_1(d;r)$.

Now, if $\varphi \in \mathcal{A}_d$, then we can easily see that the automorphism $\tilde{\varphi}$ of $\mathbb{C}[X]$ extends to an automorphism of $\mathbb{C}[x,y,z,t]$. Indeed, if $\varphi(r) = r + bx^d$ with $b \in \mathbb{C}[x,z,t]$, then $\tilde{\varphi}(y) = y - b$. We now give an automorphism Φ of $\mathbb{C}[x,y,z,t]$ which defines an extension of $\tilde{\varphi}$. We choose $\Phi|_{\mathbb{C}[x,z,t]} = \varphi$, and $\Phi(y) = y - b$. This is an automorphism, because the image of y differs from y by a polynomial in the other variables.

Now, we use that $\mathcal{A}_1(d;r)$ is generated by $\mathcal{A}_d$ and automorphisms of the form $\varphi_{x\gamma r}$ where γ is in $\mathbb{C}[x,z,t]$. We must show that we can pick an extension for any automorphism of the form $\tilde{\varphi} = \tilde{\varphi}_{x\gamma r}$ where $\varphi = \varphi_{x\gamma r}$. This is not at all obvious, because if we try to use a similar argument to the one above, we find that $\varphi(r) = ar + bx^d$, where $b \in \mathbb{C}[x,z,t]$ and $a \in \mathbb{C}[x,z,t]$. Then if we try to define $\Phi|_{\mathbb{C}[x,z,t]} = \varphi$, and $\Phi(y) = ay - b$, we have an endormorphism of $\mathbb{C}[x,y,z,t]$ which is not an automorphism, since a is not invertible in $\mathbb{C}[x,z,t]$. We therefore must adapt this endomorphism by adding appropriate elements of the ideal $(P) \in \mathbb{C}[x,y,z,t]$ to construct an automorphism. This is done exactly as in the proof of Lemma 4.2 of [1]. We describe the process here.

Let $r_\lambda = r - \lambda$ for any $\lambda \in \mathbb{C}$. We will construct a family of automorphisms φ_λ such that the following properties hold.

(1) For each $\lambda \in \mathbb{C}$, $\varphi_\lambda \in \mathcal{A}_1(d; r - \lambda)$;
(2) For $\lambda = 0$, φ_0 is congruent to φ modulo $\mathcal{A}_d$; and
(3) The map $\mathbb{C} \to \mathcal{A}_1$ given by $\lambda \mapsto \varphi_\lambda$ is polynomial in λ.

Let $H = \gamma(r - \lambda) \in \mathbb{C}[\lambda, x, z, t]$, and let D_H be the $\overline{R}[\lambda]$-derivation on $\overline{R}[\lambda][z,t]$ given by $D_H(g) = H_z g_t - H_t g_z$. We have that xD_H is locally nilpotent as a derivation on $\overline{R}[\lambda][z,t]$. Let $\overline{F} = \exp(xD_H)$. It is a special automorphism in $\mathrm{Aut}_{\overline{R}[\lambda]} \overline{R}[\lambda][z,t]$, by lemma 3.1, and, therefore, by the result of [15], there exists and automorphism $F \in \mathrm{Aut}_{R[\lambda]} R[\lambda, z, t]$ which restricts to $\overline{F}$ modulo (x^d). For any $\lambda \in \mathbb{C}$, let $\varphi_\lambda(g) = F(g) \in R[\lambda, z, t]$ for any $g \in R[z,t]$. By construction, these automorphisms satisfy the three properties above.

Now we construct Φ. First, we show that F extends to an automorphism $\widetilde{F}$ of the ring $R[\lambda][y,z,t]/(P-\lambda)$, which contains $R[\lambda][z,t]$ as a subring. Note that, by construction, $\overline{F}(r-\lambda)$ belongs to the ideal $(r-\lambda)$, and therefore, $F(r-\lambda)$ is in the ideal $(r-\lambda, x^d) \subset R[\lambda][z,t]$. Suppose that $F(r-\lambda) = (r-\lambda)\alpha + x^d\beta$ with α and β elements of $R[\lambda][z,t]$. Then we extend the automorphism F to an endomorphism $\widetilde{F}$ of $R[\lambda][z,t,y]/(P-\lambda)$ by posing $\widetilde{F}|_{R[\lambda][z,t]} = F$ and $\widetilde{F}(y) = y\alpha - \beta$. (One can check easily that this makes sense, since $\widetilde{F}(P-\lambda) = (P-\lambda)\alpha = 0 \in R[\lambda](z,t,y]/(P-\lambda)$.) Note that $\widetilde{F}$ is the unique lifting of the automorphism F of $R[\lambda][z,t]$ to an endomorphism of $R[\lambda][y,z,t]/(P-\lambda)$. Now we remark that $\widetilde{F}$ is in fact an automorphism. To see this, note that if G is the inverse of F, then $\widetilde{G}$ is the inverse of $\widetilde{F}$.

Now the projection $R[\lambda,y,z,t] \to R[y,z,t]$ which sends λ to P induces an isomorphism between $R[\lambda,y,z,t]/(P-\lambda)$ and $R[y,z,t] = \mathbb{C}[x,y,z,t]$. Thus $\widetilde{F}$ induces an automorphism Φ of $\mathbb{C}[x,y,z,t]$. By the construction above, $\Phi(P) = (P)$, and for any $\lambda \in \mathbb{C}$, we have that Φ induces the automorphism $\tilde{\varphi}_\lambda$ on $\mathbb{C}[x,y,z,t]/(P-\lambda)$.

(Another way to see the construction is that we have simply replaced λ by P in the automorphism $\widetilde{F}$. Thus, in the geometric setting, $\tilde{\varphi}_\lambda$ induces an automorphism ϕ_λ of $V(P-\lambda) \subset \mathbb{A}^4$ for any λ, and the map of $\mathbb{A}^4$ induced by Φ restricts to ϕ_λ on $V(P-\lambda)$ for any λ.) □

6. The fixed point and some comments on inequivalent hypersurfaces

In [1], we showed that all automorphims of the Russell cubic threefold have a common fixed point, namely, the origin of $\mathbb{A}^4$. We will now show that this result also holds for all Koras–Russell threefolds of the first kind. The proof is not the same as in [1].

Corollary 6.1. *Any automorphism of $X = X_{d,k,l}$ fixes the origin.*

Proof. The $\mathbb{C}^*$-action on X fixes the origin, so we now consider automorphisms lifted from $\mathcal{A}_1(d;r)$, where $r = z^k + t^l + x$. Note that the ideal (x,z,t) is stable by any automorphism φ of $\mathcal{A}_1(d;r)$, since the origin is the unique singular point of the zero set of the ideal (x,r) in $\mathbb{C}[x,z,t]$. We know that φ lifts to an automorphism $\tilde{\varphi}$ of $\mathbb{C}[X]$ determined as follows. If $\varphi(r) = ar + bx^d$ with $a,b \in \mathbb{C}[x,z,t]$, then $\tilde{\varphi}(y) = ay - b$. We thus are left to check that $b \in (x,z,t)$, that is, has no constant term. First consider any automorphism φ of $\mathcal{A}_d$. In this case, φ is congruent to the identity modulo (x^d), and thus we can choose $a = 1$. We have that $\varphi(r) = r + bx^d$, where $b \in \mathbb{C}[x,z,t]$, and $\varphi(z) \equiv z + h_t x^d \bmod (x^{d+1})$ and $\varphi(t) \equiv t - h_z x^d \bmod (x^{d+1})$. Then we find that $b(0,z,t) = \{r_0, h\} \in ((r_0)_z, (r_0)_t)$. Since $r_0 = z^k + t^l$ and $2 \leq k, l$, we have that $b \in (x,z,t)$.

Now we know that the group of automorphisms is generated by $\mathcal{A}_d$ and ones of the form $\varphi_{x\gamma r}$ where $\gamma \in \mathbb{C}[x,z,t]$. Thus we are left to consider the automorphisms of the form $\varphi_{x\gamma r}$. Remember that $\varphi_{x\gamma r}$ was chosen as an arbitrary pre-image of $\overline{\varphi}_{x\gamma r} \in \bar{\mathcal{A}}_1$ by the homomorphism $\pi = \pi_d$. Thus, we may assume that $\varphi_{x\gamma r}$ is in $\mathcal{A}_1(d+1;r)$. In other words, we can choose $\varphi_{x\gamma r}$ as an element in the pre-image of $\overline{\varphi}^{d+1}_{x\gamma r}$ by the homomorphism $\pi_{d+1}\colon \mathcal{A}_1 \to \bar{\mathcal{A}}_1^{d+1}$. This gives that $\varphi_{x\gamma r}(r)$ is in the ideal (r, x^{d+1}), and therefore, we may choose $b \in (x)$. □

Definition 6.2. Let X and Z be two affine varieties. An embedding $\varphi\colon X \to Z$ is an injective morphism whose image is closed, and which induces an isomorphism between X and $\varphi(X)$. Two embeddings $\varphi_j \colon X \to Z$ $(j = 1, 2)$ are said to be *equivalent* if there exists an automorphism Φ of Z such that $\Phi \circ \varphi_1 = \varphi_2$.

Definition 6.3. Two subvarieties X_1 and X_2 of an affine variety Z are said to be *equivalent* if there exists an automorphism Φ of Z such that $\Phi(X_1) = X_2$. If X_1 and X_2 are hypersurfaces of $Z = \mathbb{A}^n$, then we say they are *equivalent hypersurfaces.*

These two notions are related, but they are not identical. To be more precise, we consider what happens in the case of two isomorphic subvarieties X_1 and X_2 of Z, where $\phi \colon X_1 \to X_2$ is an isomorphism. Let $i_j \colon X_j \to Z$ be the inclusion maps. Then one can ask of the embeddings i_1 and $i_2 \circ \phi$ of X_1 into Z are equivalent as embeddings. The answer is affirmative if and only if the isomorphism ϕ extends to an automorphism of Z. One can also ask if X_1 and X_2 are equivalent as subvarieties. Here, the answer is affirmative if and only if there exists an isomorphism between X_1 and X_2 which extends to an automorphism of the ambient space. In particular, if two isomorphic subvarieties are not equivalent, then there exist inequivalent embeddings of the given subvariety. Finally, if all automorphisms of X_1 extend to automorphisms of Z, then the two notions become the same: X_1 is equivalent as a subvariety of Z to X_2 if and only if i_1 and $i_2 \circ \phi$ are equivalent embeddings.

Many examples of inequivalent hypersurfaces exist in the literature. See, for example, [2, 3, 14]. However, it is a long-standing open problem to know if there is a hypersurface of $\mathbb{A}^n$ isomorphic to $\mathbb{A}^{n-1}$ but not equivalent to a hyperplane. For this reason, it is interesting to find inequivalent hypersurfaces isomorphic to varieties that are, in some sense, close to affine space. In [7], for example, Kaliman and Zaidenberg gave examples of acyclic surfaces which admit nonequivalent embeddings in three-space. The obstruction to equivalence is of a topological nature. One can also ask what happens for Koras–Russell threefolds. In [1], we give an example of two inequivalent hypersufaces isomorphic to the Russell cubic. The methods are purely algebraic, because, in particular, they become equivalent as analytic embeddings. This leads to the natural question if all Koras–Russell threefolds have inequivalent representations as hypersurfaces. The method used for the Russell cubic only works in special cases.

One way to show that two hypersurfaces $V(P)$ and $V(Q)$ are inequivalent is to show that the fibers of P and Q are different. In other words, if $V(P)$ and $V(Q)$ are equivalent as hypersurfaces, then for all $c \in \mathbb{C}^*$, there exists $c' \in \mathbb{C}^*$ such that $V(P - c) \cong V(Q - c')$. So if one can find non-isomorphic fibers, the hypersurfaces are inequivalent. This is, however, not a necessary condition. In [10] and in [13], there are many examples of Danielewski hypersurfaces of $\mathbb{A}^3$ for which all fibers are isomorphic, but the hypersurfaces are inequivalent. The methods used there for surfaces can be partially adapted to the case of Koras–Russell threefolds. This was how the Russell cubic was treated. A similar method can be used to find inequivalent embeddings for all Koras–Russell threefolds of the first kind. More precisely, the equations $x^d y + z^k + t^l + x$ and $x^d y + (1 + x^{d-1})(z^k + t^l + x)$ define isomorphic varieties which are inequivalent as hypersurfaces. For a more general study, work is in progress.

References

1. A. Dubouloz, L. Moser-Jauslin, and P.-M. Poloni, *Inequivalent embeddings of the Koras–Russell cubic 3-fold*, Michigan Math. J. **59** (2010), no. 3, 679–694.

2. G. Freudenburg and L. Moser-Jauslin, *Embeddings of Danielewski surfaces*, Math. Z. **245** (2003), no. 4, 823–834.
3. Sh. Kaliman, *Extensions of isomorphisms between affine algebraic subvarieties of k^n to automorphisms of k^n*, Proc. Amer. Math. Soc. **113** (1991), no. 2, 325–334.
4. Sh. Kaliman, M. Koras, L. Makar-Limanov, and P. Russell, *$\mathbf{C}^*$-actions on $\mathbf{C}^3$ are linearizable*, Electron. Res. Announc. Amer. Math. Soc. **3** (1997), 63–71 (electronic).
5. Sh. Kaliman and L. Makar-Limanov, *On the Russell – Koras contractible threefolds*, J. Algebraic Geom. **6** (1997), no. 2, 247–268.
6. Sh. Kaliman and L. Makar-Limanov, *AK-invariant of affine domains*, Affine Algebraic Geometry (T. Hibi, ed.), Osaka Univ. Press, Osaka, 2007, pp. 231–255.
7. Sh. Kaliman and M. Zaidenberg, *Affine modifications and affine hypersurfaces with a very transitive automorphism group*, Transform. Groups **4** (1999), no. 1, 53–95.
8. M. Koras and P. Russell, *$\mathbf{C}^*$-actions on $\mathbf{C}^3$: the smooth locus of the quotient is not of hyperbolic type*, J. Algebraic Geom. **8** (1999), no. 4, 603–694.
9. L. Makar-Limanov, *On the hypersurface $x + x^2y + z^2 + t^3 = 0$ in $\mathbf{C}^4$ or a $\mathbf{C}^3$-like threefold which is not $\mathbf{C}^3$*, Israel J. Math. **96** (1996), part B, 419–429.
10. L. Moser-Jauslin and P.-M. Poloni, *Embeddings of a family of Danielewski hypersurfaces and certain $\mathbf{C}^+$-actions on $\mathbf{C}^3$*, Ann. Inst. Fourier (Grenoble) **56** (2006), no. 5, 1567–1581.
11. A. Nowicki, *Polynomial derivations and their rings of constants*, Uniwersytet Mikołaja Kopernika, Toruń, 1994.
12. ———, *The divergence of polynomial derivations*, available at `http://www-users.mat.umk.pl/~anow/ps-dvi/diverg.pdf`.
13. P.-M. Poloni, *Sur les plongements des hypersurfaces de Danielewski*, Ph.D. Thesis, Univ. de Bourgogne, 2008.
14. V. Shpilrain and J.-T. Yu, *Embeddings of hypersurfaces in affine spaces*, J. Algebra **239** (2001), no. 1, 161–173.
15. A. van den Essen, S. Maubach, and S. Vénéreau, *The special automorphism group of $R[t]/(t^m)[x_1, \ldots, x_n]$ and coordinates of a subring of $R[t][x_1, \ldots, x_n]$*, J. Pure Appl. Algebra **210** (2007), no. 1, 141–146.

Institut de Mathématiques de Bourgogne UMR 5584 - CNRS, Université de Bourgogne, 9 av. Alain Savary - B.P. 47870, 21078 Dijon Cedex, France

E-mail address: `lucy.moser-jauslin@u-bourgogne.fr`

Centre de Recherches Mathématiques
CRM Proceedings and Lecture Notes
Volume **54**, 2011

Recent Progress in the Geometry of $\mathbb{Q}$-Acyclic Surfaces

Karol Palka

Abstract. We give a survey of results on the geometry of complex algebraic $\mathbb{Q}$-acyclic surfaces including some recent results.

1. Introduction

In this article we want to take the reader to the rich and yet not fully explored world of plane-like complex algebraic surfaces. We hope our survey will give a general reader a taste of methods used and will serve as an update on recent results for experts. All varieties considered are complex algebraic.

The story begins with a surprising discovery of Ramanujam [36] of a contractible affine surface nonisomorphic to $\mathbb{C}^2$, one of the many nontrivial *smooth homotopy planes*, i.e., smooth contractible surfaces. Ramanujam discovered an important characterization of $\mathbb{C}^2$ by showing that it is the unique smooth homotopy plane which is simply connected at infinity. Since then affine algebraic geometers began the study of smooth (and more generally normal) varieties with the same Betti numbers as $\mathbb{C}^2$, so-called *$\mathbb{Q}$-homology planes.* One of the motivations was the search for more practical characterizations of the complex plane (the computation of the fundamental group at infinity of an affine variety is usually very difficult). In the face of the topological simplicity the intriguing algebraic side of $\mathbb{Q}$-acyclic surfaces is more clearly visible. Because of their homological similarity to the plane, smooth $\mathbb{Q}$-homology planes, and especially smooth homotopy planes, play today an important role as a source of examples or counterexamples when studying working hypotheses as well as more challenging conjectures. They accompany us when studying exotic structures on $\mathbb{C}^n$'s (see [46]), the Cancellation Conjecture[1] (which was the motivation for Ramanujam and has been proved in dimension two by Fujita and Miyanishi [8]), the Jacobian Conjecture[2] (see [24, Section 5.2]) and others.

2010 *Mathematics Subject Classification.* Primary 14R05; Secondary 14J17, 14J26.
Key words and phrases. Acyclic surface, homology plane, $\mathbb{Q}$-homology plane.
The author was supported by Polish Grant NCN 201 608640.
This is the final form of the paper.

[1]The m-dimensional Cancellation Conjecture. If $X \times \mathbb{C}^n \cong \mathbb{C}^{n+m}$ then $X \cong \mathbb{C}^m$.

[2]The Jacobian Conjecture. A polynomial map $f\colon \mathbb{C}^n \to \mathbb{C}^n$ with nowhere vanishing Jacobian determinant $|J(f)|$ has a polynomial inverse.

Today, after almost forty years, mainly due to the tools of the theory of open surfaces, the scheme of the classification, and in most cases the classification itself, are settled. Recently the author had the pleasure of adding his part to the story.

Although we try to avoid notions which are not well-known to any well-versed algebraic geometer, we need to refer the reader for the basics of the theory of open surfaces to [23]. Since the Section 3.4 loc. cit. is a review of smooth $\mathbb{Q}$-homology planes, we concentrate mainly on the singular case, stating the updated results so to include the smooth case if possible (see also [24]).

Acknowledgements. The author would like to thank Professors P. Russell and S. Lu for the invitation to Montréal. He also thanks Prof. P. Russell for stimulating discussions.

2. Preliminaries

A *normal pair* $(\overline{X}, D)$ consists of a complete normal surface $\overline{X}$ and a divisor D contained in the smooth locus of $\overline{X}$ which has smooth components and simple normal crossings (an snc-divisor). A normal pair $(\overline{X}, D)$ is said to be smooth if $\overline{X}$ is smooth. If X is a normal surface then an embedding $\iota\colon X \to \overline{X}$, where $(\overline{X}, \overline{X} \setminus X)$ is a normal pair, is called a *normal completion* of X. A normal completion is a *smooth completion* if $\overline{X}$ is smooth. Two normal completions $\iota_j\colon X \to \overline{X}_j$, $j = 1, 2$ are isomorphic if there exists a morphism $f\colon \overline{X}_1 \to \overline{X}_2$ for which $f \circ \iota_1 = \iota_2$. A surface with isolated singularities is called *logarithmic* if each of its singular points is of analytical type $\mathbb{C}^2/G$ for some finite subgroup $G < \mathrm{GL}(2, \mathbb{C})$. By an *$n$-curve* we mean a smooth rational curve with self-intersection n.

An *affine ruling* (a $\mathbb{P}^1$-ruling, a $\mathbb{C}^{(n*)}$*-ruling*) is a morphism from a surface onto a smooth curve with a general fiber isomorphic to $\mathbb{C}^1$ (respectively to $\mathbb{P}^1$, $\mathbb{C}^1$ with n points deleted). All these morphisms are called *rational rulings*.

Definition 2.1. Let $p\colon X \to B$ be a rational ruling of a normal surface X. A triple $(\overline{X}, D, \bar{p})$ is called *a completion of* p if and only if $(\overline{X}, D)$ is a normal completion of X and $\bar{p}\colon \overline{X} \to \overline{B}$ is a $\mathbb{P}^1$-ruling onto a smooth curve $\overline{B} \supseteq B$ extending p. Such a completion is *minimal* (we say also that $(\overline{X}, D)$ is $\bar{p}$-minimal) if it does not dominate any other completion of p.

If an snc-divisor T (or rather its dual graph) is a chain and $T = T_1 + T_2 + \cdots + T_n$ is its decomposition into irreducible components so that $T_i \cdot T_{i+1} = 1$ for $i = 1, \ldots, n-1$, we then write $T = [-T_1^2, \ldots, -T_n^2]$. As long as T is not considered as a twig attached to some other divisor containing T there is no preferred choice of the tip (T_1 or T_n) of T, so in this case $T = [-T_n^2, \ldots, -T_1^2]$ as well. If T is a twig of some fixed bigger divisor then by convention we always choose the tip of T which is a tip of this bigger divisor as T_1.

The Iitaka dimension of a divisor F on a smooth complete surface $\overline{X}$ (hence projective by the theorem of Zariski) is defined as

$$\kappa(F) = \sup_{n>0} \dim \mathrm{Im}(\Phi_{|nF|}) \in \{-\infty, 0, 1, 2\},$$

where $\Phi_{|nF|}\colon \overline{X} \dashrightarrow \mathbb{P}^N$ is the mapping given by the linear system $|nF|$. Then the (logarithmic) Kodaira dimension of a smooth open surface X can be defined by taking some smooth completion $(\overline{X}, D)$ of X and by putting

$$\bar{\kappa}(X) = \kappa(K_{\overline{X}} + D),$$

where $K_{\overline{X}}$ is a canonical divisor on $\overline{X}$. This is well-known to be independent of the smooth completion (see [16] for the properties of $\kappa(F)$). The Kodaira dimension of a singular surface is defined to be the Kodaira dimension of any resolution.

If R is an snc-divisor on a complete surface and $Q(R)$ is the intersection matrix of R then we define the *discriminant* of R by $d(R) = d\big(-Q(R)\big)$.

3. Balanced and standard completions

One of the basic steps to take when dealing with an open surface is to construct a completion and a boundary. Of course, these are not unique, but we want to bring to the attention of the reader some normalizing conditions, which make them more unique and more useful in practice. We came to this problem when trying to distinguish (or to find an isomorphism) between some $\mathbb{Q}$-acyclic surfaces. This type of analysis was done at least partially by many authors, with the most complete treatment in terms of weighted dual graphs in [2] and [5]. Here we present the necessary results using partially our own terminology. For simplicity we restrict ourselves to divisors whose dual graphs contain no loops (forests).

Definition 3.1. A rational chain $D = [a_1, \dots, a_n]$ is *balanced* if $a_1, \dots, a_n \in \{0, 2, 3, \dots\}$ or if $D = [1]$. A reduced snc-forest D is *balanced* if all rational chains contained in D which do not contain branching components of D are balanced. A normal pair (X, D) is *balanced* if D is balanced.

The word *balanced* stands here for the property that on one hand we do not allow nonbranching (-1)-curves, but on the other hand we do not allow nonbranching b-curves with positive b. The following operation is responsible for nonuniqueness of balanced completions of a given surface.

Definition 3.2. Let (X, D) be a normal pair. Let L be a 0-curve which is a nonbranching component of D. Make a blowup of a point $c \in L$, such that $c \in L \cap (D - L)$ if $L \cdot (D - L) = 2$ and contract the proper transform of L. The resulting pair (X', D'), where D' is the reduced direct image of the total transform of D is called an *elementary transform of* (X, D). The point $c \in L$ is the *center* of the transformation. A composition of elementary transformations of D and its subsequent elementary transforms is called *a flow inside D.*

For example, taking $T = [0, 0, a_1, \dots, a_n]$ one can obtain by a flow exactly the chains $[0, b, a_1, \dots, a_n]$, $[a_1, \dots, a_{k-1}, a_k - b, 0, b, a_{k+1}, \dots, a_n]$ and $[a_1, \dots, a_n, b, 0]$, where $1 \leq k \leq n$ and $b \in \mathbb{Z}$. In particular, it is easy to see when two rational chains differ by a flow. The following result is the main property of balanced completions, it follows from [5, Corollary 3.36].

Proposition 3.3. *Any normal surface which admits a normal completion with a forest as a boundary has a balanced completion. Two such completions differ by a flow inside the boundary. In particular, all balanced boundaries of a given surface are isomorphic as curves.*

Of course, two balanced boundaries of a given surface are in general nonisomorphic as weighted curves (weights are here the self-intersections of their components). One introduces normalizing conditions to deal with this. After [5] we state the following definition.

Definition 3.4. Let D be an snc-divisor. The connected components of the divisor which remains after subtracting all nonrational and all branching components of D are called the *segments of* D. An snc-divisor is *standard* if and only if any of its connected components is either $[1]$ or has all segments of type $[0]$, $[0,0]$, $[0,0,0]$, $[0,0,a_1,\dots,a_n]$ or $[a_1,\dots,a_n]$ with $a_1,\dots,a_n \geq 2$.

It follows from the example above and from the Hodge index theorem that each balanced chain can be carried to a standard form by a flow.

Example 3.5. The affine plane $\mathbb{C}^2$ has a balanced completion $(\mathbb{F}_2, T)$, where $\mathbb{F}_2$ is the second Hirzebruch surface and $T = [0,2]$. It follows that each balanced boundary of $\mathbb{C}^2$ is of type $[0,a]$ for some $a \geq 2$, it is standard if and only if $a = 0$. Note that the isomorphism type of the boundary can change when we admit nonbalanced completions, for example the boundary of $\mathbb{C}^2$ embedded in $\mathbb{P}^2$ is $[-1]$.

More generally, it follows from the above considerations about a flow inside a chain that if a surface has a completion with some rational chain as a boundary then there are at most two weighted graphs which can be dual graphs of a standard boundary of this surface. If there are two then one of them is of type $T = [0,0,a_1,\dots,a_n]$ with some $a_1,\dots,a_n \geq 2$ and the other is *the reversion* of the first one, $T^{\mathrm{rev}} = [0,0,a_n,\dots,a_1]$. Given one weighted dual graph of some standard boundary of some surface each other can be easily described, which gives a method of distinguishing between many open surfaces.

In general two standard boundaries of a given surface can differ by more than the reversion of segments. It is useful to introduce the following, more restrictive normalizing conditions.

Definition 3.6. A balanced snc-forest D is *strongly balanced* if and only if it is standard and either D contains no segments of type $[0]$, $[0,0,0]$ or for at least one of such segments there is a component $B \subseteq D$ intersecting it, such that $B^2 = 0$. A normal pair (X, D) for which D is a forest is *strongly balanced* if D is strongly balanced.

4. Basic properties

Let R be a ring. An *R-homology plane* is a normal surface X with $H^*(X,R) \cong R$. (This is a bit nonstandard, as usually R-homology planes are defined as smooth by definition). We say that X is a *homotopy plane* if $\pi_i(X) \cong 0$ for $i > 0$. A $\mathbb{Z}$-homology plane with trivial π_1 is a homotopy plane by the theorem of Hurewicz. By a theorem of Whitehead homotopy planes are contractible.

Example 4.1. Let $G < \mathrm{GL}(2,\mathbb{C})$ be a finite subgroup without pseudoreflections. Then $\mathbb{C}^2/G$ is a singular logarithmic homotopy plane for which the smooth locus has negative Kodaira dimension. Indeed, the linear contraction of $\mathbb{C}^2$ to $0 \in \mathbb{C}^2$ descends to a contraction of the quotient and the quotient morphism $\mathbb{C}^2 - \{0\} \to \mathbb{C}^2 - \{0\}/G$ is étale, so $\bar{\kappa}(\mathbb{C}^2 - \{0\}/G) = \bar{\kappa}(\mathbb{C}^2 - \{0\}) = -\infty$.

We now fix the notation for the rest of the paper. Let $\epsilon\colon S \to S'$ be a minimal snc-resolution of a singular $\mathbb{Q}$-homology plane S'. Denote the exceptional divisor of ϵ by $\widehat{E}$. Let $(\overline{S}, D)$ be a smooth completion of S and let S_0 be the smooth locus of S'. Since S' is normal, its singular locus consists of a finite number of points.

For topological spaces $A \subseteq X$ we write $H_i(X, A)$ for $H_i(X, A, \mathbb{Q})$ and $b_i(X, A)$ for $\dim H_i(X, A)$.

Logarithmic $\mathbb{Q}$-homology planes are known to be affine by an argument of Fujita (cf. [9, 2.4]). They are also known to be rational due to Gurjar – Pradeep – Shastri [14, 15, 35]. On the other hand, the following example shows that nonlogarithmic $\mathbb{Q}$-homology planes can be nonrational.

Example 4.2. Let $C \subseteq \mathbb{P}^n$ be a projectively normal embedding of a smooth projective curve. Then the affine cone over C is normal and contractible. It has a standard cylinder resolution, for which the exceptional divisor is isomorphic to C. In case C is not rational, the cone is nonrational and nonlogarithmic.

In general we have the following result (cf. [30, Corollary 3.2, Proposition 3.4]).

Theorem 4.3. *Every $\mathbb{Q}$-homology plane is affine and birationally ruled.*

As for the affiness we note that the above mentioned argument of Fujita works as long as one assumes the inclusion $\iota\colon D \cup \widehat{E} \to \overline{S}$ induces an isomorphism $H_2(\iota)\colon H_2(D \cup \widehat{E}) \to H_2(\overline{S})$. One can prove that this condition is always satisfied. If $\widehat{E}$ is a rational forest this can be seen as follows. First note that $H_i(\overline{S}, D \cup \widehat{E}) \cong H^{4-i}(S_0)$ by the Lefschetz duality. One can prove that $b_2(S_0) = b_1(S_0) = b_1(\widehat{E})$, so if $\widehat{E}$ is a rational forest then $b_3(\overline{S}, D \cup \widehat{E}) = b_2(\overline{S}, D \cup \widehat{E}) = 0$ and $H_2(\iota)$ is an isomorphism. For $b_1(\widehat{E}) \neq 0$ the argument is more complicated. For the proof of ruledness we refer to loc. cit.

Corollary 4.4. *If S' is a singular $\mathbb{Q}$-homology plane then its boundary is connected and the homology groups $H_i(S', \mathbb{Z})$ vanish for $i \geq 2$.*

Proof. Since S' is affine, by [17] its boundary is connected and S' is homotopy equivalent to a CW-complex of real dimension at most two, hence $H_2(S', \mathbb{Z})$ is torsionfree and $H_3(S', \mathbb{Z}) = H_4(S', \mathbb{Z}) = 0$. Since $b_2(S') = 0$, we get $H_2(S', \mathbb{Z}) = 0$. □

The rationality of smooth $\mathbb{Q}$-homology planes has strong consequences for vector bundles.

Theorem 4.5. *Let S' be a smooth $\mathbb{Q}$-homology plane. Then any vector bundle over S' is a sum of a line bundle and a trivial vector bundle. Moreover,* $\operatorname{Pic} S' \cong H_1(X, \mathbb{Z})$, *so if S' is a $\mathbb{Z}$-homology plane then all vector bundles over S' are trivial.*

Proof. The first part of the theorem is a result of Murthy [29] valid for smooth affine surfaces which are birationally ruled. Since S' is rational, we have $\operatorname{Pic} S' \cong H^2(S', \mathbb{Z})$. The groups $H_1(S', \mathbb{Z})$ and $H_2(S', \mathbb{Z})$ are finite, so by the universal coefficient theorem $H^2(S', \mathbb{Z}) \cong H_1(S', \mathbb{Z})$. □

We note that the question whether a vector bundle over any smooth contractible threefold is necessarily trivial is open.

5. Case $\bar{\kappa}(S_0) = -\infty$ and moduli

Let us start with recalling some known results in case S_0 has negative Kodaira dimension. This is a strong assumption, which in particular forces S' to be logarithmic (and hence rational). This can be seen as follows.

Suppose S_0 is affine-ruled. The affine ruling extends to a $\mathbb{P}^1$-ruling of some smooth completion $(\overline{S}, D+\widehat{E})$ of S_0, where $\widehat{E}$ is the exceptional divisor of some (not necessarily minimal) snc-resolution of singularities of S'. The unique section of this extension contained in the boundary is in fact contained in D. Indeed, if it is contained in $\widehat{E}$, then D is vertical for this ruling (i.e., all its components intersect trivially with fibers), so since the homology classes of components of $D+\widehat{E}$ generate $H_2(\overline{S})$, the intersection form on $\overline{S}$ is semi-negative definite, contradicting the Hodge index theorem. Thus the affine ruling of S_0 extends to an affine ruling of S', so S' has at most cyclic singularities by [22, Theorem 1]. There is no bound on the number of singularities, $H_1(S', \mathbb{Z})$ can be any finite abelian group. The extension has a unique fiber contained in the boundary (the fiber can be assumed to be smooth) and each singular fiber contains a unique component not contained in $D \cup \widehat{E}$ (see [25, Sections 1, 2] for more details). A strongly balanced boundary of S' is unique if it is branched and it is unique up to reversion if it is a chain. There can be infinitely many strongly balanced completions (see [30, Example 4.6]). Note that from Proposition 1.2 loc. cit. if follows that $\mathbb{C}^2$ is the only smooth $\mathbb{Z}$-homology plane of negative Kodaira dimension (see [8] for the first proof of this) and this characterization implies the positive solution of the two-dimensional Cancellation Conjecture.

Now if S_0 is not affine-ruled (this can happen only if S' is singular) then it follows from an important structure theorem by Miyanishi – Tsunoda [27] that it contains an open subset U with a very special $\mathbb{C}^*$-ruling called a *Platonic fibration.* In fact one shows that $S_0 = U$ (cf. [19, Theorem 3.1]), which implies that $S' \cong \mathbb{C}^2/G$ for a finite small noncyclic subgroup $G < \mathrm{GL}(2, \mathbb{C})$. The strongly balanced completion of S' is unique.

It is well known that there are only finitely many *fake projective planes*, i.e., smooth projective surfaces with Betti numbers of $\mathbb{P}^2$ (cf. [1, Section V.1]). This is clearly not the case for $\mathbb{Q}$-homology planes (the name *fake affine planes* is not used). Moreover, there are arbitrarily high-dimensional nontrivial families of $\mathbb{Q}$-homology planes with the same weighted boundary. The following example of a family of singular $\mathbb{Q}$-homology planes is a modification of a similar example in [6, Example 4.16]. We sketch the arguments.

Example 5.1. Let $p\colon \mathbb{F}_1 \to B \cong \mathbb{P}^1$ be the $\mathbb{P}^1$-ruling of the first Hirzebruch surface and let N be a positive integer. Choose $N+3$ distinct points

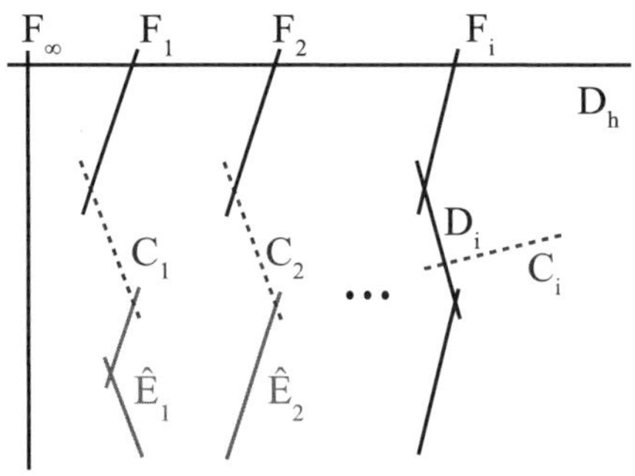

FIGURE 1. Singular fibers in Example 5.1

$x_\infty, x_1, x_2, \ldots, x_{N+2}$ on the negative section. Blow up successively over each x_i so to produce singular fibers with reductions $F_\infty = [0]$, $F_1 = [3,1,2,2]$, $F_2 = [2,1,2]$, $F_i' = [2,1,2]$ for $i = 2, \ldots, N+2$ lying respectively over $x_\infty, x_1, \ldots, x_{N+2}$. Denote the resulting surface by V and the proper transform of the negative section by D_h. Note there are chains $\widehat{E}_1 = [2,2]$ and $\widehat{E}_2 = [2]$ contained in F_1 and F_2 respectively. We can assume that $\widehat{E}_1, \widehat{E}_2$ do not intersect D_h. For each $i \in \{3, \ldots, N\}$ choose a point y_i on the (-1)-curve D_i of F_i' and blow up once. For $i \in \{1, \ldots, N\}$ denote the unique (-1)-curve of F_i by C_i. Let $\overline{S}_y$, where $y = (y_3, \ldots, y_{N+2})$, be the resulting surface and put $D = F_\infty + D_h + (F_1 - C_1 - \widehat{E}_1) + (F_2 - C_2 - \widehat{E}_2) + \sum_{i=3}^N (F_i - C_i)$. One checks that the surface S_y' obtained by the contraction of $\widehat{E}_1$ and $\widehat{E}_2$ on $\overline{S} - D$ is a singular $\mathbb{Q}$-homology plane. Clearly, the above family S_y' is N-dimensional. Now if $S_y' \cong S_z'$ then the isomorphism lifts to $\overline{S} - D$ and then, since F_∞ is the only 0-curve in D, extends to $\overline{S}_y - F_\infty \cong \overline{S}_z - F_\infty$ by Proposition 3.3, which in turn descends to an automorphism of $U = \mathbb{F}_1 - F_\infty - D_h \cong \mathbb{C}^2$ fixing fibers. However, if x, y are respectively the horizontal and vertical coordinate on U, each automorphism of U fixing fibers can be written as $(x,y) \to \big(x, \lambda y + P(x)\big)$ for some $P[x] \in \mathbb{C}[x]$ and its lifting to V acts by λ^2 on D_i in some coordinates on D_i (the multiplicity of D_i in the fiber is 2). Thus if we consider an $(N-1)$-dimensional subfamily with fixed $y_3 \in D_3$ then $\lambda^2 = 1$, so the mentioned action on each D_i is trivial, hence $y = z$ and different members of this subfamily are nonisomorphic.

6. Logarithmic Bogomolov – Miyaoka – Yau

When studying $\mathbb{Q}$-homology planes we often use the logarithmic version of the Bogomolov – Miyaoka – Yau inequality proved by Kobayashi (cf. [18]). Usually this inequality is stated for surfaces of general type in terms of the so-called *strongly minimal model.* This is not necessary and in fact the inequality works for surfaces of nonnegative Kodaira dimension. For example, the following lemma has been proved in [31, Corollary 2.5] as an easy corollary from an inequality of Bogomolov – Miyaoka – Yau type proved by Langer (cf. [20]). The latter generalizes in particular both the inequality of Kobayashi and an inequality of Miyaoka [28]. For the notion of the Zariski decomposition and of the bark $\operatorname{Bk} D$ of an effective snc-divisor D see [23, Section 2.3]. Put $D^\# = D - \operatorname{Bk} D$.

Lemma 6.1. *Let (X, D) be a smooth pair with $\kappa(K_X + D) \geq 0$. Then:*

(i)

$$3\chi(X - D) + \tfrac{1}{4}\big((K_X + D)^-\big)^2 \geq (K_X + D)^2.$$

(ii) *For each connected component of D, which is a connected component of $\operatorname{Bk} D$ (hence contractible to a quotient singularity) denote by G_P the local fundamental group of the respective singular point P. Then*

$$\chi(X - D) + \sum_P \frac{1}{|G_P|} \geq \frac{1}{3}(K_X + D^\#)^2.$$

We now illustrate the usefulness of the second inequality in our context.

Corollary 6.2. *Let S_0 be the smooth locus of a singular $\mathbb{Q}$-homology plane S' and let $\widehat{E}_1, \ldots, \widehat{E}_q$ be the connected components of the exceptional divisor of the snc-minimal resolution.*

(i) *If $\bar{\kappa}(S_0) = 2$ then S' is logarithmic and $q = 1$.*

(ii) *If* $\bar{\kappa}(S_0) = 0$ *or* 1 *then either* $q = 1$, *or* $q = 2$ *and* $\widehat{E}_1 = \widehat{E}_2 = [2]$.

PROOF. Let (S_m, D_m) be the almost minimal model of $(\overline{S}, D + \widehat{E})$. For the notion of almost minimality see [23, 2.3.11]. Since S' is affine, $S_m - D_m$ is isomorphic to an open subset of S_0 satisfying $\chi(S_m - D_m) \leq \chi(S_0) = 1 - q$. Let Q be the set of singular points which have been created by contracting connected components of D as in Lemma 6.1(ii). Since $\#Q \leq q$, the above inequality gives

$$\frac{1}{3}\left((K_{S_m} + D_m)^+\right)^2 \leq \chi(S_m - D_m) + \sum_{P \in Q} \frac{1}{|G_P|} \leq 1 - q + \frac{\#Q}{2} \leq 1 - \frac{q}{2}.$$

Now if $\bar{\kappa}(S_0) = 2$ then $\left((K_{S_m} + D_m)^+\right)^2 > 0$ and we get $q = 1$ and $0 < \sum_{P \in Q} 1/|G_P|$, so there is a unique singular point on S' and it is of quotient type. If $q > 1$ and $\bar{\kappa}(S_0) = 0, 1$ then $\left((K_{S_m} + D_m)^+\right)^2 = 0$ and we get $q = 2$ and $1 = 1/|G_{P_1}| + 1/|G_{P_2}|$, so $|G_{P_1}| = |G_{P_2}| = 2$. □

7. Exceptional $\mathbb{Q}$-homology planes

The structure theorem for $\mathbb{Q}$-homology planes with smooth locus of nongeneral type is based on general structure theorems for open surfaces.

Definition 7.1. A $\mathbb{Q}$-homology plane for which the smooth locus is neither of general type, nor $\mathbb{C}^1$- or $\mathbb{C}^*$-ruled is *exceptional.*

Now the mentioned structure theorems lead to the fact that for exceptional S' one has $\bar{\kappa}(S_0) = 0$. It was proved by Fujita (cf. [9, 8.64]) that each exceptional smooth $\mathbb{Q}$-homology plane is up to isomorphism one of three surfaces called $Y\{3,3,3\}$, $Y\{2,4,4\}$ and $Y\{2,3,6\}$ (the Fujita's surfaces of type $H[k,-k]$ with $k \geq 1$ are $\mathbb{C}^*$-ruled). The snc-minimal boundary of $Y\{a,b,c\}$ is a rational fork (a tree with one branching component), such that its three maximal twigs consist of (-2)-curves and have discriminants equal to a, b, c respectively. A singular exceptional $\mathbb{Q}$-homology plane having a boundary with this property will be denoted by $SY\{a,b,c\}$. Together with the description of exceptional singular $\mathbb{Q}$-homology planes in [31] we have the following theorem (to be precise one still needs to prove that smooth $Y\{a,b,c\}$'s are not $\mathbb{C}^*$-ruled, but this can be done as in loc. cit.).

Theorem 7.2. *If* S' *is a* $\mathbb{Q}$*-homology plane with smooth locus* S_0 *of nongeneral type then* S_0 *is affine-ruled or* $\mathbb{C}^*$*-ruled or* S' *is up to isomorphism one of five exceptional* $\mathbb{Q}$*-homology planes having smooth locus of Kodaira dimension zero: smooth* $Y\{3,3,3\}$, $Y\{2,4,4\}$, $Y\{2,3,6\}$ *and singular* $SY\{3,3,3\}$, $SY\{2,4,4\}$. *For the last two surfaces* $\bar{\kappa}(S') = 0$ *and the singular locus consists of a unique point of Dynkin type* A_2 *and* A_1, *respectively.*

Since the exceptional $\mathbb{Q}$-homology planes admit boundaries with no nonbranching 0-curves, their snc-minimal (which are also balanced) completions are unique. None of the exceptional surfaces is a $\mathbb{Z}$-homology plane. Writing this review the author noticed that a posteriori there is another (different than the one discovered in loc. cit.) nice description of exceptional singular $\mathbb{Q}$-homology planes. Namely, $Y\{2,4,4\}$ and $Y\{3,3,3\}$ have automorphism groups $\mathbb{Z}_2$ and $\mathbb{Z}_3$ respectively (the automorphism group of $Y\{2,3,6\}$ is trivial) and the actions have unique fixed points. The quotients are two nonisomorphic $\mathbb{Q}$-homology planes with smooth loci of Kodaira dimension zero. Suppose, say, the quotient $S' = Y\{3,3,3\}/\operatorname{Aut} Y\{3,3,3\}$ is

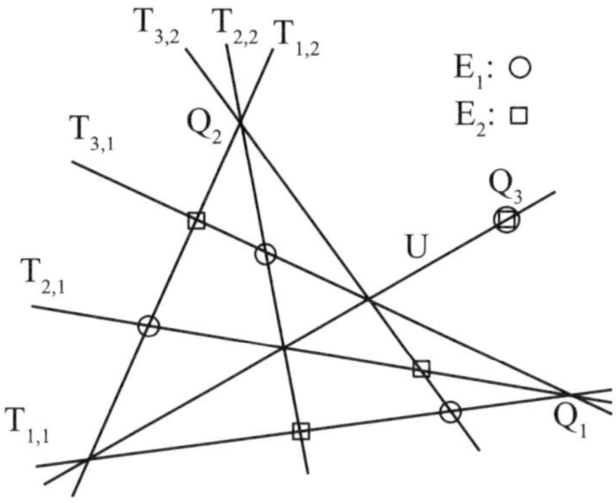

FIGURE 2. Singular Y{3,3,3}, dual Hesse configuration.

not exceptional. Then its smooth locus S_0 is $\mathbb{C}^*$-ruled, so the Stein factorization of the pull-back of this $\mathbb{C}^*$-ruling gives a $\mathbb{C}^*$-ruling of the complement of the fixed point of $Y\{3,3,3\}$. Since $Y\{3,3,3\}$ is exceptional, the closures of the fibers meet in the fixed point, hence the closures of the fibers of the $\mathbb{C}^*$-ruling of S_0 meet in the singular point of S'. Thus the last $\mathbb{C}^*$-ruling does not extend to a ruling of S'. Since the singularity is cyclic, it follows from [30, Theorem 5.4] that $\bar{\kappa}(S_0) = -\infty$, a contradiction. (One can also get a contradiction with the fact that $Y\{3,3,3\}$ does not contain infinitely many contractible curves, cf. Theorem 10.1).

We recall here the construction of $SY\{3,3,3\}$, mainly because of its beautiful connection with classical geometry. A *projective configuration of type* (a_c, b_d) is an arrangement of b lines in a projective space and a points on these lines, such that each point belongs to c lines and each line contains d points. Clearly, $ac = bd$ for such a configuration.

Example 7.3. Up to a projective automorphism there exists a unique projective configuration $\mathfrak{H}$ of type $(12_3, 9_4)$ (the uniqueness is easy to show using information on the automorphism group of the configuration which we give below). The dual configuration $\mathfrak{H}^*$ is the famous Hesse configuration $(9_4, 12_3)$ of flexes of an elliptic curve. It is known that $\mathrm{Aut}(\mathfrak{H}) \cong \mathrm{Aut}(\mathfrak{H}^*)$ has order 216 and is isomorphic to the group of special affine transformations of $\mathbb{F}_3^2$, i.e., $\mathbb{F}_3^2 \rtimes / \mathrm{SL}(2, \mathbb{F}_3)$ (cf. [3, Section 4]). Choose three points Q_1, Q_2, Q_3 of $\mathfrak{H}$ none two of which lie on a common line of $\mathfrak{H}$ and choose some line U incident to Q_3. By taking the dual choice in $\mathfrak{H}$ we check using linear algebra that $\mathrm{Aut}(\mathfrak{H})$ acts transitively on the set of such choices and the stabilizer $\Gamma < \mathrm{Aut}(\mathfrak{H})$ has order three. For $i = 1, 2$ let $T_{1,i}$, $T_{2,i}$, $T_{3,i}$ be the lines incident to Q_i and let E_1, E_2, U be the lines incident to Q_3. Blow up once in each of eight points not incident to U. Let $\overline{S}$ be the resulting complete surface. We denote divisors and their proper transforms by the same letters. Let B be the exceptional curve over Q_2, put $D = B + T_{1,1} + T_{1,2} + T_{2,1} + T_{2,2} + T_{3,1} + T_{3,2}$ and $\widehat{E} = E_1 + E_2$. Clearly, all components of $D - B + \widehat{E}$ are (-2)-curves, D and $\widehat{E}$ are disjoint and D is a fork. Put $S' = (\overline{S} - D)/\widehat{E}$. One can show that $\mathrm{Aut}(S') \cong \Gamma$.

To see that S' is $\mathbb{Q}$-acyclic note that $b_1(\overline{S}) = 0$ and $b_2(\overline{S}) = b_2(D \cup \widehat{E}) = 9$. Now since $d(D + \widehat{E}) \neq 0$, the natural morphism $H_2(D \cup \widehat{E}) \to H_2(\overline{S})$ is an isomorphism. Put $S = \overline{S} \setminus D$. The homology exact sequence of the pair $(\overline{S}, D)$ and the Lefschetz duality give $b_1(S)=b_3(S)=b_4(S)=0$ and $b_2(S)=\#\widehat{E}$. We know from the above that $H_2(\widehat{E})\to H_2(S)$ is a monomorphism, so the homology exact sequence of the pair $(S,\widehat{E})$ gives that S' is $\mathbb{Q}$-acyclic.

We check easily that in both cases $K_{\overline{S}}+D^{\#}=K_{\overline{S}}+D^{\#}+\widehat{E}^{\#}$ intersects trivially with all components of $D+\widehat{E}$, hence $\bar{\kappa}(S_0)=\bar{\kappa}(S')=0$. See [31, Section 5] for an explicit realization of $\mathfrak{H}$ and for a proof that the constructed $\mathbb{Q}$-homology plane does not admit a $\mathbb{C}^*$-ruling.

8. Nonlogarithmic $\mathbb{Q}$-homology planes

Let us start with a generalization of Example 4.2.

Example 8.1. Let U be an affine cone over a projectively normal curve and let G be a finite group acting on it so that the vertex of the cone is a fixed point, the action is free on its complement and respects the set of lines of the cone. Then the quotient $S'=U/G$ is a normal contractible surface.

What is surprising is that all nonlogarithmic $\mathbb{Q}$-homology planes arise in the above way. Namely, we have the following theorem (see [30, Corollary 5.8]).

Theorem 8.2. *Every singular $\mathbb{Q}$-homology plane containing a nonquotient singularity is a quotient of an affine cone over a smooth projective curve by an action of a finite group which is free off the vertex of the cone and respects the set of lines through the vertex. In particular, it is contractible, has negative Kodaira dimension, has a unique singular point and its smooth locus is $\mathbb{C}^*$-ruled. The snc-minimal completion of such a surface is unique.*

Proof (sketch). By Lemma 6.2 and Theorem 7.2 the smooth locus of a nonlogarithmic $\mathbb{Q}$-homology plane S' is $\mathbb{C}^1$- or $\mathbb{C}^*$-ruled. By the results of Section 5, $\bar{\kappa}(S_0)\geq 0$, so in fact S_0 is $\mathbb{C}^*$-ruled. One can show (cf. Theorem 3.6 loc. cit.) that if this ruling extends to a $\mathbb{C}^*$-ruling of S' then S' is necessarily logarithmic, so we can further assume that this is not the case. This means that there is a $\mathbb{P}^1$-ruling $p:\overline{S}\to\mathbb{P}^1$ of some completion $(\overline{S},D+\widehat{E})$ of S_0 as before, such that for a general fiber f we have $f\cdot D=f\cdot\widehat{E}=1$. We can assume that the completion is p-minimal. Then each singular fiber of p is a so-called *columnar fiber*, which means that it is a rational snc-chain of discriminant zero, its components have negative self-intersections, it contains a unique (-1)-curve C_i which is also a unique S_0-component of the fiber (i.e., a component not contained in $D\cup\widehat{E}$) and it is intersected by the horizontal components of D and $\widehat{E}$ in tips. It follows that D and $\widehat{E}$ contain unique branching components D_h and $\widehat{E}_h$. In particular, the boundary contains no nonbranching 0-curves, so S' has a unique snc-minimal (hence also balanced) normal completion. Write the singular fibers as $D_i+C_i+E_i$, where $D_i\subseteq D$ and $E_i\subseteq\widehat{E}$ (see Figure 3). Contracting singular fibers we get a $\mathbb{P}^1$-bundle over a complete curve, so all nonlogarithmic singular $\mathbb{Q}$-homology planes can be reconstructed starting from such a bundle by producing columnar singular fibers, taking out D and contracting $\widehat{E}$. The mentioned bundle admits a usual $\mathbb{C}^*$-action fixing pointwise the images of D_h and $\widehat{E}_h$. This action induces a $\mathbb{C}^*$-action on S' with the singular point as the unique fixed point. Thus by [34, Theorem 1.1] S' is a quotient of an affine cone over a smooth projective curve by an action of a finite group. In fact without knowing this global description the contractibility

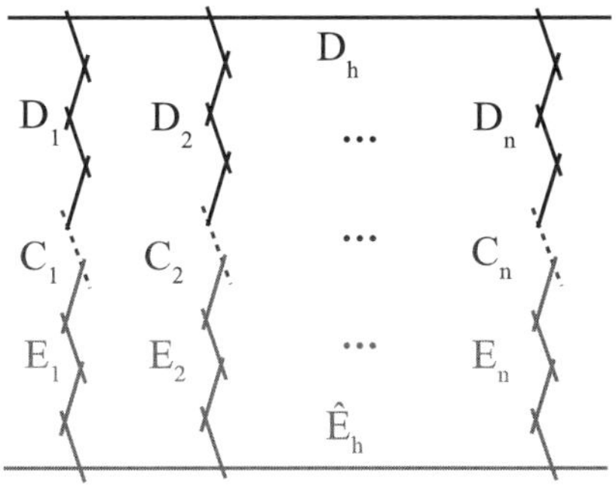

FIGURE 3. Construction of nonlogarithmic $\mathbb{Q}$-homology planes.

follows also from [9, 5.9, 4.19] and the Whitehead theorem, as one can show that $\pi_1(\widehat{E}_h) \to \pi_1(S)$ is an isomorphism and $\pi_1(S) \to \pi_1(S')$ is an epimorphism. Since $S = \overline{S} - D$ is affine-ruled, we get $\bar{\kappa}(S') = -\infty$. □

We see that one can obtain any nonlogarithmic $\mathbb{Q}$-homology plane by creating its snc-minimal completion $\overline{S}$ of the resolution as in the proof above by starting with a nontrivial $\mathbb{P}^1$-bundle over some smooth complete curve B and by blowing up some number n of columnar fibers. Then exactly one of two forks, say $\widehat{E}$, separated by vertical (-1)-curves is negative definite, hence by [10] it contracts analytically to a singularity. What is quite surprising here is that this contraction is always algebraic (cf. the contraction criterion [30, Corollary 2.6]). Suppose $\widehat{E}_h \cong B \cong \mathbb{P}^1$. Although $\widehat{E}$ is a rational tree, the singularity does not have to be a rational singularity. Indeed, using Artin's criterion one can show that if $\widehat{E}_h^2 + n \leq 0$ then S' has a rational and if $\widehat{E}_h^2 + n \geq 2$ a nonrational singularity (cf. [30, Corollary 5.8]). The case $\widehat{E}_h^2 + n = 1$ is more subtle (cf. [34, Corollary 5.8]).

9. $\mathbb{C}^*$-ruled $\mathbb{Q}$-homology planes

By the results described in previous sections the classification of $\mathbb{Q}$-homology planes of nongeneral type reduces now to the classification of logarithmic (and hence rational) $\mathbb{Q}$-homology planes which are $\mathbb{C}^*$-ruled, or in other words, for which the smooth locus is $\mathbb{C}^*$-ruled and the ruling extends to a $\mathbb{C}^*$-ruling of S'. Note that the existence of a $\mathbb{C}^*$-ruling implies that $\bar{\kappa}(S_0) \neq 2$ by the 'easy addition theorem' (cf. [23, 2.1.5]). This case was analyzed in [25], where one can find a description of singular fibers and a computation of $\bar{\kappa}(S_0)$ and $H_1(S', \mathbb{Z})$ in terms of these fibers. We note here that in [30, Theorem 6.8] we have redone some incorrect computations of $\bar{\kappa}(S_0)$ from loc. cit. (identified then with the Kodaira dimension of S') and we have computed $\bar{\kappa}(S')$ too. We discuss two issues here.

First, it is practically useful to know when a $\mathbb{C}^{(n*)}$-ruled surface is a $\mathbb{Q}$-homology plane. To formulate a criterion we need to recall the definition of some numbers characterizing rational rulings. Having a fixed $\mathbb{P}^1$-ruling of a smooth complete surface X and a reduced divisor T we define

$$\Sigma_{X-T} = \sum_{F \not\subseteq T} (\sigma(F) - 1),$$

where $\sigma(F)$ is the number of $(X-T)$-components (i.e., irreducible components not contained in T) of a fiber F (cf. [9, 4.16]). T_h is the horizontal part of T, which consists of components of T intersecting nontrivially with a general fiber. If $T_h = 0$ then T is *vertical.* The numbers h and ν are defined respectively as $\#T_h$ and as the number of fibers contained in T.

Lemma 9.1. *Let $(\overline{S}, T)$ be a smooth pair and let $p\colon \overline{S} \to \mathbb{P}^1$ be a $\mathbb{P}^1$-ruling. Assume the following conditions are satisfied:*

(i) *there exists a unique connected component D of T which is not vertical,*
(ii) *D is a rational tree,*
(iii) $\Sigma_{\overline{S}-T} = h + \nu - 2$,
(iv) $d(D) \neq 0$.

Then the surface S' defined as the image of $\overline{S} - D$ after contraction of connected components of $T - D$ to points is a rational $\mathbb{Q}$-homology plane and p induces a rational ruling of S'. Conversely, if $p'\colon S' \to B$ is a rational ruling of a rational $\mathbb{Q}$-homology plane S' (not necessarily singular) then any completion $(\overline{S}, T, p)$ of the restriction of p' to the smooth locus of S' has the above properties.

The conditions (iii) – (iv) are equivalent to the fact that $H_2(D \cup \widehat{E}) \to H_2(\overline{S})$ is an isomorphism, similar criteria were used by many authors. What is important, in case of a $\mathbb{C}^*$-ruling the most problematic condition (iv) can be replaced by an easier and more geometric condition (cf. [30, Lemma 6.1]).

Second, there is a question of uniqueness of a $\mathbb{C}^*$-ruling of S'. Let us assume S' is singular. In case $\bar{\kappa}(S_0) = 1$ it is easy to prove that there is a unique $\mathbb{C}^*$-ruling of S' and it is induced by the $\mathbb{C}^*$-ruling of S_0 given by the linear system of some multiple of the logarithmic canonical divisor of S_0. In case $\bar{\kappa}(S_0) = 0$ generically there are two $\mathbb{C}^*$-rulings of S', but there may be zero (exceptional $\mathbb{Q}$-homology planes), one or three as well (cf. Theorem 6.12 loc. cit.). We will see that this information is important for example when computing the number of contractible curves on S'.

Assume that S' admits a $\mathbb{C}^*$- ruling but no affine-ruling. Then it has a unique balanced completion, unless it admits an untwisted $\mathbb{C}^*$-ruling with base $\mathbb{C}^1$ (untwisted means that for some completion of the surface the ruling extends to a $\mathbb{P}^1$-ruling for which the horizontal part of the boundary consists of two irreducible components). In the latter case S' has infinitely many balanced completions, but there are exactly two which are strongly balanced (cf. Theorem 6.11 loc. cit.).

10. Contractible curves

It is known that the logarithmic Bogomolov – Miyaoka – Yau inequality imposes restrictions on the number of topologically contractible curves on S'. According to the author's knowledge this number is known except the cases when S' is nonlogarithmic or when S' is singular and $\bar{\kappa}(S_0) = 0$ (see [12, 13, 44, 45]). In the nonlogarithmic case it is easily seen to be infinity by Theorem 8.2 and, as we sketch below, in the last case it can be deduced from the knowledge on the number (and types) of $\mathbb{C}^*$-rulings of S_0 [30, Theorem 6.12]. The final result is as follows.

Theorem 10.1. *Let l be the number of topologically contractible curves on a $\mathbb{Q}$-homology plane S'. Let S_0 be the smooth locus of S'. Then:*

(i) *if $\bar{\kappa}(S_0) = 2$ then $l = 0$,*

(ii) *if* S' *is exceptional (hence* $\bar{\kappa}(S') = \bar{\kappa}(S_0) = 0$*) then* $l = 0$,
(iii) *if* S' *is nonlogarithmic (hence* $\bar{\kappa}(S') = -\infty$, $\bar{\kappa}(S_0) = 0, 1$*) then* $l = \infty$,
(iv) *if* $\bar{\kappa}(S_0) = -\infty$ *then* $l = \infty$,
(v) *in other cases (*S' *is* $\mathbb{C}^*$*-ruled and* $\bar{\kappa}(S_0) = 0, 1$*)* $l = 1$ *or* 2.

PROOF (sketch). If $\bar{\kappa}(S_0) = -\infty$ then S' is affine-ruled or isomorphic to $\mathbb{C}^2/G$ for some $G < \mathrm{GL}(2, \mathbb{C})$, so $l = \infty$. If S' is nonlogarithmic then by Theorem 8.2 $l = \infty$. We can therefore assume that $\bar{\kappa}(S_0) \geq 0$ and S' is logarithmic. Suppose S' contains a contractible curve L. It follows from the logarithmic Bogomolov–Miyaoka–Yau inequality that $\bar{\kappa}(S_0 - L) \leq 1$ (cf. [12]). We show that $S_0 - L$ is $\mathbb{C}^*$-ruled. This is the case if $\bar{\kappa}(S_0 - L) = 1$, as the ruling is given by some multiple of the logarithmic canonical divisor or $S_0 - L$. We may therefore assume $\bar{\kappa}(S_0 - L) = 0$. Since S' is rational, $\mathrm{Pic}(S_0) = \mathrm{Coker}(\mathrm{Pic}(D + \widehat{E}) \to \mathrm{Pic}\, \overline{S})$ is torsion, so the class of L in $\mathrm{Pic}(S_0)$ is torsion. Then there is a morphism $f\colon S_0 - L \to \mathbb{C}^*$. Taking its Stein factorization one gets a $\mathbb{C}^*$-ruling of $S_0 - L$, so we are done. Since $\bar{\kappa}(S_0) \geq 0$, any $\mathbb{C}^*$-ruling of $S_0 - L$ extends to a $\mathbb{C}^*$-ruling of S_0. Since S' is logarithmic, each $\mathbb{C}^*$-ruling of S_0 extends in turn to a $\mathbb{C}^*$-ruling of S'. Therefore, any contractible curve on S' is vertical for some $\mathbb{C}^*$-ruling of S'. In particular, if $l > 0$ then S', and hence S_0, is necessarily $\mathbb{C}^*$-ruled, hence S' cannot be exceptional. The analysis of fibers (Suzuki's formula) leads to the corollary that there exist one or two contractible vertical curves for a given $\mathbb{C}^*$-ruling. Now if $\bar{\kappa}(S_0) = 1$ then this ruling is unique (given by a multiple of the logarithmic canonical divisor of S_0), hence $l = 1, 2$. Consider now the case $\bar{\kappa}(S_0) = 0$. Here the problem is more difficult, as there may be more $\mathbb{C}^*$-rulings of S'. In case S' is smooth it was shown in [13] that $l = 1$. In case S' is singular we have computed the number and types of possible $\mathbb{C}^*$-rulings of S_0 in [30, Theorem 6.12](one can do the same if S' is smooth in a similar way). Since this number is finite we see that l is finite and nonzero. Looking more closely at the computations one deduces that $l \leq 2$ (cf. Corollary 6.13. loc. cit). □

Example 10.2. Let $\pi\colon S \to \mathbb{C}^2$ be the restriction of the projection $(x, y, z) \to (x, y)$ to the surface $S = \{(x, y, z) \in \mathbb{C}^3 : z^n = f(x, y)\}$, where $n \geq 2$. Then π is a branched cover with the curve $C = \{(x, y) \in \mathbb{C}^2 : f(x, y) = 0\}$ as the branch locus. Suppose S is a smooth $\mathbb{Q}$-homology plane. Then C is smooth and we have $1 - \chi(C) = 1 - \chi(\pi^{-1}(C)) = 1 - \chi(S) + \chi(S - \pi^{-1}(C)) = n\chi(\mathbb{C}^2 - C) = n(1 - \chi(C))$, hence $\chi(C) = 1$. If U is any smooth affine curve then it is noncomplete, so its Euler characteristic is smaller than the Euler characteristic of the smooth completion $\overline{U}$, hence $\chi(U) \leq 1 - 2g(\overline{U})$, where g is the genus. In case U is irreducible it follows that $\chi(U) \geq 0$ only if $U \cong \mathbb{C}^1$ or $U \cong \mathbb{C}^*$. In particular, some component of $C \cong \pi^{-1}(C)$ is an affine line, so by Theorem 10.1 S is not of general type. In fact either $\bar{\kappa}(S) = -\infty$ or S is $\mathbb{C}^*$-ruled. Smooth $\mathbb{Q}$-homology planes of this kind have been classified in [21].

11. Smooth $\mathbb{Q}$-homology planes of general type

Let S' be a smooth $\mathbb{Q}$-homology plane of general type, i.e., $\bar{\kappa}(S') = 2$. We know already that S' is rational. By Theorem 10.1 S' contains no contractible curves, which implies that its snc-minimal completion $(\overline{S}, D)$ is almost minimal. We can assume that this completion is balanced. Since S' is neither $\mathbb{C}^1$- nor $\mathbb{C}^*$-ruled, D contains no nonbranching 0-curves, so any flow inside D is trivial. It follows that the balanced completion of S' is unique. As for now there is no classification, but there are some partial results.

The first example of such a surface which was shown to be nonisomorphic to $\mathbb{C}^2$ is the famous example of Ramanujam ([36]). We know now infinitely many examples of this kind. One method of construction of $\mathbb{Q}$-homology planes of general type is to use $\mathbb{C}^{(n*)}$-rulings with $n \geq 2$. For $n = 2$ this was done in [26]. Another method was used by tom Dieck and Petrie.

Definition 11.1. Let $(\overline{S}, D)$ be a completion of S' for which there exists a birational morphism $f : (\overline{S}, D) \to (\mathbb{P}^2, f_*D)$. Then f_*D is called a *plane divisor* of S'. If f_*D is a sum of lines then it is a *linear plane divisor* and we say that S' comes from the *line arrangement* f_*D.

In [39] it was noticed that an inequality of Hirzebruch bounds the number of types of possible linear plane divisors for smooth $\mathbb{Q}$-homology planes, a list of these divisors has been given. In [42] a general algorithm for recovering smooth $\mathbb{Q}$-homology planes (in fact countable series of them) starting from a given rational divisor on a minimal rational complete surface is described. The conjecture that all smooth $\mathbb{Q}$-homology planes of general type have linear plane divisors is not true by an example of tom Dieck (cf. [40]). Tom Dieck's smooth $\mathbb{Z}$-homology plane has a nontrivial automorphism group (it is necessarily finite, as S' is of general type), so is also a counterexample to the earlier conjecture of Petrie [33].

Example 11.2. Let $C \subseteq \mathbb{P}^2$ be an irreducible curve. Recall that a singular point $p \in C$ is a *cusp* if C is locally irreducible at p. Since $C \hookrightarrow \mathbb{P}^2$ induces a monomorphism on $H_2(\cdot, \mathbb{Q})$, by the Lefschetz duality and by the long exact sequence of the pair $(\mathbb{P}^2, C)$ the Betti numbers of $S_C = \mathbb{P}^2 \setminus C$ are $b_2(S_C) = b_1(C)$ and $b_i(S_C) = 0$ for $i = 1$ and $i > 2$. Assume that C is rational and cuspidal, i.e., it has only cusps as singularities. Then S_C is a smooth $\mathbb{Q}$-homology plane (these conditions are in fact equivalent). By [43] if C has two cusps then $\bar{\kappa}(S_C) \geq 0$ and if it has more than two cusps then S_C is of general type. The literature on plane cuspidal curves is rich, see for example [4, 7] and references there.

We now list some conjectural properties of smooth $\mathbb{Q}$-homology planes.

Conjecture. Let S' be a smooth $\mathbb{Q}$-homology plane of general type and let $(\overline{S}, D)$ be its minimal smooth completion.

(A) S' has a plane divisor consisting of lines and conics.
(B) S' admits a $\mathbb{C}^{(3*)}$-ruling.
(C) $(K_{\overline{S}} + D)^2 = -2$, or equivalently $K_{\overline{S}} \cdot (K_{\overline{S}} + D) = 0$.
(D) S' is rigid and has unobstructed deformations.
(E) The set of all possible Eisenbud – Neumann diagrams for smooth $\mathbb{Q}$-homology planes is finite.

In [41] conjectures (A) – (C) have been stated and verified for all known smooth $\mathbb{Z}$-homology planes. Sugie [38] analyzed $\mathbb{C}^{(n*)}$-rulings on smooth $\mathbb{Q}$-homology planes and classified possible singular fibers for $n = 2$. Flenner and Zaidenberg [6, Proposition 6.12] have shown that part (D), which implies (C), holds for smooth $\mathbb{Q}$-homology planes of general type having linear plane divisors. See [47] for (D), (E) and related conjectures.

Recently, the following result has been proved [11].

Theorem 11.3. *Let S' be a smooth homotopy plane of general type. Then* $\operatorname{Aut} S'$ *is cyclic, its action on S' has a unique fixed point and is free off this point.*

It is not known whether $\operatorname{Aut} S'$ has to be cyclic for a singular $\mathbb{Q}$-homology plane with smooth locus of general type.

12. Smooth locus of general type

Let S' be a singular $\mathbb{Q}$-homology plane with smooth locus S_0 of general type, i.e., $\bar{\kappa}(S_0) = 2$. By Lemma 6.2 S' has a unique singular point and this point is of quotient type. This means that $\widehat{E}$ is either a chain if this point is a cyclic singularity or a negative definite rational fork if not. Since S' is logarithmic, it is rational. By Theorem 10.1 it contains no topologically contractible curves, which implies that the snc-minimal completion $(\overline{S}, D + \widehat{E})$ of S_0 is almost minimal. Arguing as above we see that in fact this snc-minimal completion is unique and does not contain nonbranching 0-curves. Again, there is no classification, but there are some partial results.

Note that due to the existence of the transfer homomorphism for branched coverings (cf. [37]) the quotient of a smooth $\mathbb{Q}$-homology plane of general type by its automorphism group is a $\mathbb{Q}$-homology plane (with smooth locus of general type).

A priori there is no restriction on the Kodaira dimension of S'. However, refining the methods of Koras – Russell [19] M. Koras and the author have obtained the following theorem (this is also a part of the thesis of the author written under the supervision of M. Koras, cf. [32]).

Theorem 12.1. *Singular $\mathbb{Q}$-homology planes with smooth locus of general type have nonnegative Kodaira dimension.*

The cases $\bar{\kappa}(S') = 0, 1$ and especially the case $\bar{\kappa}(S') = 2$ have not been analyzed systematically as for now. Also the number of known examples is much smaller than in the case of smooth $\mathbb{Q}$-homology planes. For example, similarly as in the smooth case, one could ask if S_0 has a completion $(\overline{S}, D+\widehat{E})$ admitting a birational morphism $f\colon \overline{S} \to \mathbb{P}^2$ with $f_*(D + \widehat{E})$ a sum of lines and conics. Even the case when $f_*(D + \widehat{E})$ is a sum of lines has not been studied.

In the known examples of singular $\mathbb{Z}$-homology planes of general type the local fundamental group of the singular point (which has order equal to $d(\widehat{E})$) is cyclic of order not bigger than six. As for now the following result in this direction has been proved ([11]):

Theorem 12.2. *A singular $\mathbb{Z}$-homology plane with smooth locus of general type has a cyclic quotient singularity.*

References

1. W. P. Barth, K. Hulek, C. A. M. Peters, and A. Van de Ven, *Compact complex surfaces*, 2nd ed., Ergeb. Math. Grenzgeb. (3), vol. 4, Springer, Berlin, 2004.
2. D. Daigle, *Classification of weighted graphs up to blowing-up and blowing-down*, available at `arXiv:math/0305029`.
3. I. V. Dolgachev, *Abstract configurations in algebraic geometry*, The Fano Conference (Turin, 2002) (A. Collino, A. Conte, and M. Marchisio, eds.), Univ. Torino, Turin, 2004, pp. 423–462.
4. J. Fernández de Bobadilla, I. Luengo, A. Melle-Hernández, and A. Némethi, *On rational cuspidal plane curves, open surfaces and local singularities*, Singularity Theory (Marseille, 2005) (D. Chéniot, N. Dutertre, C. Murolo, D. Trotman, and A. Pichon, eds.), World Sci. Publ., Hackensack, NJ, 2007, pp. 411–442.

5. H. Flenner, S. Kaliman, and M. Zaidenberg, *Birational transformations of weighted graphs*, Affine Algebraic Geometry (T. Hibi, ed.), Osaka Univ. Press, Osaka, 2007, pp. 107–147.
6. H. Flenner and M. Zaidenberg, *$\mathbb{Q}$-acyclic surfaces and their deformations*, Classification of Algebraic Varieties (L'Aquila, 1992) (C. Ciliberto, E. L. Livorni, and A. J. Sommese, eds.), Contemp. Math., vol. 162, Amer. Math. Soc., Providence, RI, 1994, pp. 143–208.
7. ———, *On a class of rational cuspidal plane curves*, Manuscripta Math. **89** (1996), no. 4, 439–459.
8. T. Fujita, *On Zariski problem*, Proc. Japan Acad. Ser. A Math. Sci. **55** (1979), no. 3, 106–110.
9. ———, *On the topology of noncomplete algebraic surfaces*, J. Fac. Sci. Univ. Tokyo Sect. IA Math. **29** (1982), no. 3, 503–566.
10. H. Grauert, *Über Modifikationen und exzeptionelle analytische Mengen*, Math. Ann. **146** (1962), 331–368.
11. R. V. Gurjar, M. Koras, M. Miyanishi, and P. Russell, *A homology plane of general type can have at most a cyclic quotient singularity*, available at `arXiv:1012.4120`.
12. R. V. Gurjar and M. Miyanishi, *Affine lines on logarithmic $\mathbf{Q}$-homology planes*, Math. Ann. **294** (1992), no. 3, 463–482.
13. R. V. Gurjar and A. J. Parameswaran, *Affine lines on $\mathbf{Q}$-homology planes*, J. Math. Kyoto Univ. **35** (1995), no. 1, 63–77.
14. R. V. Gurjar and C. R. Pradeep, *$\mathbf{Q}$-homology planes are rational.* III, Osaka J. Math. **36** (1999), no. 2, 259–335.
15. R. V. Gurjar, C. R. Pradeep, and A. R. Shastri, *On rationality of logarithmic $\mathbf{Q}$-homology planes.* II, Osaka J. Math. **34** (1997), no. 3, 725–743.
16. S. Iitaka, *Algebraic geometry*, Grad. Texts in Math., vol. 76, Springer, New York, 1982.
17. K. K. Karčjauskas, *A generalized Lefschetz theorem*, Funkcional. Anal. i Priložen. **11** (1977), no. 4, 80–81 (Russian); English transl., Functional Anal. Appl. **11** (1977), no. 4, 312–313.
18. R. Kobayashi, *Uniformization of complex surfaces*, Kähler Metric and Moduli Spaces (T. Ochiai, ed.), Adv. Stud. Pure Math., vol. 18, Academic Press, Boston, MA, 1990, pp. 313–394.
19. M. Koras and P. Russell, *Contractible affine surfaces with quotient singularities*, Transform. Groups **12** (2007), no. 2, 293–340.
20. A. Langer, *Logarithmic orbifold Euler numbers of surfaces with applications*, Proc. London Math. Soc. (3) **86** (2003), no. 2, 358–396.
21. A. Maharana, *Q-homology planes as cyclic covers of $\mathbb{A}^2$*, J. Math. Soc. Japan **61** (2009), no. 2, 393–425.
22. M. Miyanishi, *Singularities of normal affine surfaces containing cylinderlike open sets*, J. Algebra **68** (1981), no. 2, 268–275.
23. ———, *Open algebraic surfaces*, CRM Monogr. Ser., vol. 12, Amer. Math. Soc., Providence, RI, 2001.
24. ———, *Recent developments in affine algebraic geometry: from the personal viewpoints of the author*, Affine Algebraic Geometry (T. Hibi, ed.), Osaka Univ. Press, Osaka, 2007, pp. 307–378.
25. M. Miyanishi and T. Sugie, *Homology planes with quotient singularities*, J. Math. Kyoto Univ. **31** (1991), no. 3, 755–788.
26. ———, *$\mathbf{Q}$-homology planes with $\mathbf{C}^{**}$-fibrations*, Osaka J. Math. **28** (1991), no. 1, 1–26.
27. M. Miyanishi and S. Tsunoda, *Noncomplete algebraic surfaces with logarithmic Kodaira dimension $-\infty$ and with nonconnected boundaries at infinity*, Japan. J. Math. (N.S.) **10** (1984), no. 2, 195–242.
28. Y. Miyaoka, *The maximal number of quotient singularities on surfaces with given numerical invariants*, Math. Ann. **268** (1984), no. 2, 159–171.
29. M. P. Murthy, *Vector bundles over affine surfaces birationally equivalent to a ruled surface*, Ann. of Math. (2) **89** (1969), 242–253.
30. K. Palka, *On the classification of singular $\mathbf{Q}$-acyclic surfaces*, available at `arXiv:0806.3110v3`.
31. ———, *Exceptional singular $\mathbf{Q}$-homology planes*, Ann. Inst. Fourier (Grenoble), available at `arXiv:0909.0772`.
32. K. Palka and M. Koras, *Singular $\mathbf{Q}$-homology planes of negative Kodaira dimension have smooth locus of non-general type*, available at `arXiv:1001.2256`.
33. T. Petrie, *Algebraic automorphisms of smooth affine surfaces*, Invent. Math. **95** (1989), no. 2, 355–378.

34. H. Pinkham, *Normal surface singularities with C^* action*, Math. Ann. **227** (1977), no. 2, 183–193.
35. C. R. Pradeep and A. R. Shastri, *On rationality of logarithmic* **Q***-homology planes.* I, Osaka J. Math. **34** (1997), no. 2, 429–456.
36. C. P. Ramanujam, *A topological characterisation of the affine plane as an algebraic variety*, Ann. of Math. (2) **94** (1971), 69–88.
37. L. Smith, *Transfer and ramified coverings*, Math. Proc. Cambridge Philos. Soc. **93** (1983), no. 3, 485–493.
38. T. Sugie, *Singular fibers of homology planes with pencils of rational curves*, Mem. Fac. Educ. Shiga Univ. III Nat. Sci. **49** (1999), 29–40.
39. T. tom Dieck, *Linear plane divisors of homology planes*, J. Fac. Sci. Univ. Tokyo Sect. IA Math. **37** (1990), no. 1, 33–69.
40. ———, *Symmetric homology planes*, Math. Ann. **286** (1990), no. 1-3, 143–152.
41. ———, *Optimal rational curves and homotopy planes*, Bol. Soc. Mat. Mexicana (2) **37** (1992), no. 1-2, 115–138.
42. T. tom Dieck and T. Petrie, *Homology planes and algebraic curves*, Osaka J. Math. **30** (1993), no. 4, 855–886.
43. I. Wakabayashi, *On the logarithmic Kodaira dimension of the complement of a curve in P^2*, Proc. Japan Acad. Ser. A Math. Sci. **54** (1978), no. 6, 157–162.
44. M. G. Zaidenberg, *Isotrivial families of curves on affine surfaces, and the characterization of the affine plane*, Izv. Akad. Nauk SSSR Ser. Mat. **51** (1987), no. 3, 534–567 (Russian); English transl., Math. USSR-Izv. **30** (1988), no. 3, 503–532.
45. ———, *Additions and corrections to the paper "Isotrivial families of curves on affine surfaces, and the characterization of the affine plane"*, Izv. Akad. Nauk SSSR Ser. Mat. **55** (1991), no. 2, 444–446 (Russian); English transl., Math. USSR-Izv. **38** (1992), no. 2, 435–437.
46. ———, *Exotic algebraic structures on affine spaces*, Algebra i Analiz **11** (1999), no. 5, 3–73 (Russian); English transl., St. Petersburg Math. J. **11** (2000), no. 5, 703–760.
47. ———, *Selected problems*, available at `arXiv:math/0501457`.

Department of Mathematics and Statistics, McGill University, 805 Sherbrooke St. West, Montréal, QC H3A 2K6, Canada and Institute of Mathematics, Polish Academy of Sciences, ul. Śniadeckich 8, 00-956 Warsaw, Poland

E-mail address: `palka@impan.pl`

Centre de Recherches Mathématiques
CRM Proceedings and Lecture Notes
Volume **54**, 2011

On the Makar-Limanov, Derksen Invariants, and Finite Automorphism Groups of Algebraic Varieties

Vladimir L. Popov

To Peter Russell on the occasion of his 70th birthday

ABSTRACT. A simple method of constructing a big stock of algebraic varieties with trivial Makar-Limanov invariant is described, the Derksen invariant of some varieties is computed, the generalizations of the Makar-Limanov and Derksen invariants are introduced and discussed, and some results on the Jordan property of automorphism groups of algebraic varieties are obtained.

Introduction

The subject matter of this note are automorphism groups of algebraic varieties.

In Section 1 I discuss the Makar-Limanov and Derksen invariants. As is known, they have been first introduced as the means for distinguishing the Koras – Russell threefolds from affine spaces. Since then studying varieties with certain properties of these invariants (for instance, with trivial Makar-Limanov invariant) became an independent line of research, see, e.g., [7, 10, 12] and references therein. At the conference *Affine Algebraic Geometry*, June 1–5, 2009, Montréal, I was surprised to find that a simple general method of constructing a big stock of such varieties remained unnoticed by the experts. In Section 1 I expand my comment on this point made after one of the talks and give the related proofs and some illustrating examples. Then I consider the Derksen invariant and show that in many cases in presence of an algebraic group action it coincides with the coordinate algebra. At the end of this section I introduce and discuss the natural generalizations of the Makar-Limanov and Derksen invariants. In Section 2 some results on the Jordan property of automorphism groups of algebraic varieties are obtained.

Conventions and notation. Below variety means algebraic variety. All varieties are taken over an algebraically closed field k of characteristic zero. I use the standard conventions of [3] and [39] and the following notation.

2010 *Mathematics Subject Classification.* 14L30.

Supported by grants РФФИ 11-01-00185-a, НШ–4713.2010.1, and the program *Contemporary Problems of Theoretical Mathematics* of the Russian Academy of Sciences, Branch of Mathematics.

This is the final form of the paper.

A^* is the group of units of the commutative ring A with identity.
$\mathbf{M}_{n\times m}$ is the affine space of all $n \times m$-matrices with entries in k.
$\mathbf{A}^1_*$ is the punctured affine line $\mathbf{A}^1 \setminus \{0\}$.
$\mathbf{Z}_{>0}$ is the set of positive integers.
$|M|$ is the number of elements of the set M.
$\operatorname{Rad} G$ is the radical of the linear algebraic group G.
$\operatorname{Rad}_u G$ is the unipotent radical of the linear algebraic group G.
(G, G) is the commutator subgroup of the group G.
$k[X]$ is the k-algebra of regular function on the variety X.
$k(X)$ is the field of rational function on the irreducible variety X.
$\mathrm{T}_{x,X}$ is the tangent space to the variety X at the point $x \in X$.
$\mathrm{Aut}(X)$ is the automorphism group of the variety X.
$\mathrm{Bir}(X)$ is the group of birational automorphisms of the irreducible variety X.

Given the varieties X and Y (not necessarily affine), $k[X]$ and $k[Y]$ are naturally identified with the k-subalgebras of $k[X\times Y]$. Recall that then $k[X\times Y]$ is generated by $k[X]$ and $k[Y]$ and, moreover, $k[X \times Y] = k[X] \otimes_k k[Y]$, see [40]. If A and B are the k-subalgebras of resp. $k[X]$ and $k[Y]$, then the subalgebra of $k[X\times Y]$ generated by A and B is $A \otimes_k B$.

Below action of an algebraic group on an algebraic variety means algebraic action. Homomorphism of algebraic groups means algebraic homomorphism.

Let X be a variety endowed with an action of an algebraic group G. Then the natural homomorphism $\varphi\colon G \to \mathrm{Aut}(X)$ defined by this action is called *algebraic* and $\varphi(G)$ is called the *algebraic subgroup* of $\mathrm{Aut}(X)$. If φ is injective, $\varphi(G)$ is identified with G by means of φ.

Acknowledgements. I am grateful to I. Dolgachev, Yu. Prokhorov, and Yu. Zarhin for discussions on automorphism groups of surfaces.

1. The Makar-Limanov and Derksen invariants

1.1. The Makar-Limanov invariant. Recall that the *Makar-Limanov invariant* of a variety X is the following k-subalgebra of $k[X]$:

$$\mathrm{ML}(X) := \bigcap_H k[X]^H \tag{1.1}$$

where H in (1.1) runs over the images of all homomorphisms $\mathbf{G}_a \to \mathrm{Aut}(X)$.

Below is described a simple method of constructing varieties whose Makar-Limanov invariant is trivial (i.e., equal to k). The starting point is

Lemma 1.1. *For every connected linear algebraic group G, the following are equivalent*:

(i) *G has no nontrivial characters*;
(ii) *G is generated by one-dimensional unipotent subgroups*;
(iii) *G is generated by unipotent elements*;
(iv) $\operatorname{Rad} G = \operatorname{Rad}_u G$.

PROOF. Let G_0 be the subgroup of G generated by all one-dimensional unipotent subgroups of G; it is normal and, by [39, 2.2.7], closed. Since $\operatorname{char} k = 0$, for every nonidentity unipotent element $u \in G$, the closure of $\{u^n \mid n \in \mathbb{Z}\}$ is a one-dimensional unipotent subgroup of G (cf., e.g., [25, Chapter 3, Section 2, no. 2,

Theorem 1]). Hence G_0 coincides with the subgroup generated by all unipotent element of G. This yields (ii) $\Leftrightarrow$ (iii).

Since homomorphisms of algebraic groups preserve Jordan decompositions, G_0 is contained in the kernel of every character of G and every element of the G/G_0 is semisimple. The latter yields that G/G_0 is a torus (cf., e.g., [3, I.4.6]). Hence G has no nontrivial characters if and only if $G = G_0$. This proves (i) $\Leftrightarrow$ (ii).

Since $\operatorname{char} k = 0$, there is a reductive subgroup L in G such that G is the semidirect product of $\operatorname{Rad}_u G$ and L (cf., e.g. [25, Chapter 6, Section 4]). Let Z and Z^0 be resp. the center of L and the identity component of Z. Put $H := (L, L) \operatorname{Rad}_u G$. Then Z^0 is a torus, $F := (L, L) \cap Z^0$ is finite, $L = Z^0(L, L)$, and H is connected and normal. Being connected semisimple, (L, L) has no nontrivial characters. Hence H is generated by unipotent elements. This yields $H \subseteq G_0$. As G/H is isomorphic to Z^0/F and the latter is a torus, all elements of G/H are semisimple. Hence $H = G_0$. Thus, (i) holds if and only if Z^0 is the identity. Since $\operatorname{Rad} G = Z^0 \operatorname{Rad}_u G$, this proves (i) $\Leftrightarrow$ (iv). □

Corollary 1.2. $\mathrm{ML}(X) = \bigcap_{H \subseteq \mathrm{Aut}(X)} k[X]^H$, *where H runs over all connected linear algebraic subgroups of* $\mathrm{Aut}(X)$ *that have no nontrivial characters.*

Theorem 1.3. *Let X be a variety and let G be a connected linear algebraic subgroup of* $\mathrm{Aut}(X)$ *that has no nontrivial characters. Then*

$$\mathrm{ML}(X) \subseteq k[X]^G. \tag{1.2}$$

PROOF. From Lemma 1.1 we infer that $k[X]^G = \bigcap_H k[X]^H$ where H runs over all one-parameter unipotent subgroups of G. This and (1.1) imply (1.2). □

Corollary 1.4. *Maintain the notation of Theorem 1.3. If G has no nontrivial characters and $k[X]^G = k$, then $\mathrm{ML}(X) = k$.*

Since there are no nonconstant invariant functions on orbit closures, this yields the following.

Corollary 1.5. *Maintain the notation of Theorem 1.3. If G has no nontrivial characters and X is the closure of a G-orbit, then $\mathrm{ML}(X) = k$.*

Corollary 1.6. *Let G be a connected algebraic group that has no nontrivial characters. Let H be a reductive subgroup of G. Then G/H is an irreducible affine variety with trivial Makar-Limanov invariant.*

PROOF. As G acts on G/H transitively, Corollary 1.5 yields $\mathrm{ML}(G/H) = k$. By [3, Theorem 6.8] and [31, Theorem 4.9] reductivity of H implies that G/H is affine. □

The following generalizes Corollary 1.5.

Theorem 1.7. *Let X be a variety endowed with an action of a connected linear algebraic group G. Let d be the dimension of the center of $G/\operatorname{Rad}_u G$. If X contains a dense G-orbit, then*

$$\operatorname{tr}\deg_k \mathrm{ML}(X) \leqslant d. \tag{1.3}$$

PROOF. Let X be the closure of the G-orbit of a point $x \in X$. The morphism $G \to X$, $g \mapsto g \cdot x$, is G-equivariant with respect to the action of G on itself by left translations. Since its image is dense in X, the corresponding comorphism is a G-equivariant embedding of the k-algebras

$$k[X] \hookrightarrow k[G]. \tag{1.4}$$

Let L, Z, Z^0, F, and H be as in the proof of Lemma 1.1. From (1.4) and Theorem 1.3 we then infer that

$$\mathrm{ML}(X) \subseteq k[X]^H \hookrightarrow k[G]^H. \tag{1.5}$$

Since G/H is isomorphic to Z^0/F and $\dim Z^0/F = \dim Z^0 = \dim Z = d$, we have $\dim G/H = d$. As $k[G]^H$ is isomorphic to $k[G/H]$, this and (1.5) imply the claim. □

Corollary 1.8. *Let X be the closure in $\mathbf{P}^n$ of an orbit of a connected algebraic subgroup G in $\mathrm{Aut}(\mathbf{P}^n)$. Let $\widehat{X} \subseteq k^{n+1}$ be the affine cone over X. Then $\operatorname{tr deg}_k \mathrm{ML}(\widehat{X}) \leqslant d+1$, where d is the dimension of the center of $G/\operatorname{Rad}_u G$.*

PROOF. Let $\widehat{G}$ be the pullback of G with respect to the natural projection $\mathbf{GL}_{n+1} \to \mathrm{Aut}(\mathbf{P}^n)$. Then $\widehat{X}$ is the closure of a $\widehat{G}$-orbit in k^{n+1} and the dimension of the center of $\widehat{G}/\operatorname{Rad}_u \widehat{G}$ is $d+1$; whence the claim by Theorem 1.7. □

Lemma 1.9. *For any varieties X_1 and X_2,*

$$\mathrm{ML}(X_1 \times X_2) \subseteq \mathrm{ML}(X_1) \otimes_k \mathrm{ML}(X_2). \tag{1.6}$$

PROOF. Take an element $f \in \mathrm{ML}(X_1 \times X_2)$. Since $k[X_1 \times X_2]$ is generated by $k[X_1]$ and $k[X_2]$, there is a decomposition

$$f = \sum_{i=1}^{n} s_i t_i, \quad s_1, \dots, s_n \in k[X_1],\ t_1, \dots, t_n \in k[X_2]. \tag{1.7}$$

We may (and shall) assume that $t_1, \dots, t_n$ in (1.7) are linearly independent over k. As $k[X_1 \times X_2] = k[X_1] \otimes_k k[X_2]$, then they are also linearly independent over $k[X_1]$.

Consider an action α of $\mathbf{G}_a$ on X_1. Then $k[X_1]$ is stable and $k[X_2]$ is pointwise fixed with respect to the diagonal action of $\mathbf{G}_a$ on $X_1 \times X_2$ determined by α and trivial action on X_2. For every element $g \in \mathbf{G}_a$ and this diagonal action, (1.1) and (1.7) imply that

$$\sum_{i=1}^{n} s_i t_i = f = g \cdot f = \sum_{i=1}^{n} (g \cdot s_i) t_i. \tag{1.8}$$

Since $t_1, \dots, t_n$ are linearly independent over $k[X_1]$, we infer from (1.8) that every s_i is invariant with respect to α. As α is arbitrary, (1.1) implies that $s_1, \dots, s_n \in \mathrm{ML}(X_1)$. Hence f is decomposed as

$$f = \sum_{i=1}^{m} s'_i t'_i, \quad s'_1, \dots, s'_m \in \mathrm{ML}(X_1),\ t'_1, \dots, t'_m \in k[X_2], \tag{1.9}$$

where $s'_1, \dots, s'_m$ are linearly independent over k. The same argument as above then yields $t'_1, \dots, t'_m \in \mathrm{ML}(X_2)$. Now (1.6) follows from (1.9). □

Corollary 1.10. *For any varieties X_1 and X_2, the following are equivalent:*

(i) $\mathrm{ML}(X_1)$ *and* $\mathrm{ML}(X_2)$ *lie in* $\mathrm{ML}(X_1 \times X_2)$;

(ii) $\mathrm{ML}(X_1 \times X_2) = \mathrm{ML}(X_1) \otimes_k \mathrm{ML}(X_2)$.

Corollary 1.11. *If* $\mathrm{ML}(X_1) = k$ *and* $\mathrm{ML}(X_2) = k$, *then* $\mathrm{ML}(X_1 \times X_2) = k$.

Corollary 1.12. *Let* X_1 *and* X_2 *be the varieties such that* $\mathrm{ML}(X_1)$ *and* $\mathrm{ML}(X_2)$ *are generated by units. Then* $\mathrm{ML}(X_1 \times X_2) = \mathrm{ML}(X_1) \otimes_k \mathrm{ML}(X_2)$.

PROOF. This follows from Corollary 1.10 since $k[X_1]^*$ and $k[X_2]^*$ lie in $k[X_1 \times X_2]^*$ and $k[X_1 \times X_2]^* \subset \mathrm{ML}(X_1 \times X_2)$, cf. [13, 1.4]. □

Definition 1.13. A variety is called *toral* if it is isomorphic to a closed subvariety of a linear algebraic torus.

Note that closed subvarieties and products of toral varieties are toral.

Lemma 1.14. *Let* X *be an affine variety.*

(a) *The following are equivalent:*
   (a_1) X *is toral;*
   (a_2) $k[X]$ *is generated by* $k[X]^*$.

(b) *For every finite subgroup* G *of* $\mathrm{Aut}(X)$, *there is a covering of* X *by* G*-stable open toral sets.*

(c) *If* X *is toral, then*
   (c_1) *for every unipotent linear algebraic group* H, *every algebraic homomorphism* $\varphi\colon H \to \mathrm{Aut}(X)$ *is trivial;*
   (c_2) $\mathrm{ML}(X) = k[X]$.

PROOF. (a) Every character of a linear algebraic torus T is an element of $k[T]^*$ and $k[T]^*$ is the k-linear span of the set of all characters [3, Section 8.2]; this and Definition 1.13 imply (a_1) $\Rightarrow$ (a_2).

Conversely, if (a_2) holds, let $k[X] = k[f_1, \ldots, f_n]$ for some $f_i \in k[X]^*$. Then $\iota\colon X \to \mathbf{A}^n$, $x \mapsto \big(f_1(x), \ldots, f_n(x)\big)$, is a closed embedding since X is affine. The standard coordinate functions on $\mathbf{A}^n$ do not vanish on $\iota(X)$ since f_i does not vanish on X. Hence $\iota(X) \subset (\mathbf{G}_m)^n$. This proves ($a_2$) $\Rightarrow$ (a_1) and completes the proof of (a).

(b) Let x be a point of X. We have to show that x is contained in a G-stable open toral subset of X. Let $k[X] = k[h_1, \ldots, h_s]$. Replacing h_i by $h_i + \alpha_i$ for an appropriate $\alpha_i \in k$, we may (and shall) assume that every h_i vanishes nowhere on the G-orbit $G \cdot x$ of x. Enlarging the set $\{h_1, \ldots, h_s\}$ by including in it $g \cdot h_i$ for every i and $g \in G$, we may (and shall) assume that $\{h_1, \ldots, h_s\}$ is G-stable. Then $h := h_1 \cdots h_s \in k[X]^G$. Hence the affine open set $X_h := \{z \in X \mid h(z) \neq 0\}$ is G-stable and contains $G \cdot x$. Since $k[X_h] = k[h_1, \ldots, h_s, 1/h]$ we have $h_i \in k[X_h]^*$ for every i. Hence, X_h is toral by (a). This proves (b).

(c) Consider the action of H on X determined by φ. Let $H \cdot x$ be the H-orbit of a point $x \in X$. Since $\mathrm{char}\, k = 0$, $H \cdot x$ is isomorphic to $\mathbf{A}^d$ for some d, see [26, Corollary of Theorem 2]. Since H is unipotent and X is affine, $H \cdot x$ is closed in X, cf. [3, 4.10]. Hence $H \cdot x$ is toral. Since $k[\mathbf{A}^d]^* = k^*$, from (a) we then infer that $d = 0$, i.e., x is a fixed point. This proves (c_1). In turn, (c_1) implies (c_2) by (1.1). □

Corollary 1.15. *If* $\mathrm{ML}(X_1) = k$ *and* X_2 *is toral, then* $\mathrm{ML}(X_1 \times X_2) = k[X_2]$.

Utilizing the above statements one gets many interesting varieties with trivial Makar-Limanov invariant. The following construction is typical (but not the only possible, see Examples 1.21 and 1.22).

Let G be a connected semisimple algebraic group acting on an affine variety X. By Hilbert's theorem, $k[X]^G$ is a finitely generated k-algebra. Let $k[X]^G = k[f_1,\ldots,f_n]$. For every $\alpha_1,\ldots,\alpha_n \in k$, denote by $X(\alpha_1,\ldots,\alpha_n)$ the closed subvariety of X whose underlying topological space is $\{x \in X \mid f_1(x)=\alpha_1,\ldots,f_n(x)=\alpha_n\}$ (warning: in general, the ideal $(f_1-\alpha_1,\ldots,f_n-\alpha_n)$ of $k[X]$ is not radical). Let Y be a G-stable closed suvariety of X. It is well-known that $k[X]^G \to k[Y]^G$, $f \mapsto f|_Y$, is an epimorphism [31, 3.4]. Hence $k[Y]^G = k$ if and only if Y is contained in some $X(\alpha_1,\ldots,\alpha_n)$. From Theorem 1.3 we then infer that the Makar-Limanov invariant of every G-stable closed subvariety of $X(\alpha_1,\ldots,\alpha_n)$ is trivial.

There are many instances where $f_1,\ldots,f_n$ can be explicitly described. E.g., classical invariant theory yields such a description for a number of finite-dimensional modules X of classical linear groups G; for some of them, it is proved that $(f_1-\alpha_1,\ldots,f_n-\alpha_n)$ is radical. If the latter happens, one obtains the instances of affine algebras with trivial Makar-Limanov invariant that are explicitly described by equations.

Below are several illustrating examples.

Example 1.16 (Closures of adjoint orbits). Let f_s be the sum of all principal $s\times s$-minors of the $n\times n$-matrix (x_{ij}) where $x_{11},\ldots,x_{nn}$ are variables considered as the standard coordinate functions on $\mathbf{M}_{n\times n}$. For $\alpha_1,\ldots,\alpha_n \in k$,

$$\mathbf{M}_{n\times n}(\alpha_1,\ldots,\alpha_n) := \{a \in \mathbf{M}_{n\times n} \mid f_1(a)=\alpha_1,\ldots,f_n(a)=\alpha_n\}$$

is the set of all matrices whose characteristic polynomial is $t^n + \sum_{i=1}^{n}(-1)^i\alpha_i t^{n-i}$.

Consider the action of $\mathbf{SL}_n$ on $\mathbf{M}_{n\times n}$ by conjugation. Then $k[\mathbf{M}_{n\times n}]^{\mathbf{SL}_n}$ is freely generated by $f_1,\ldots,f_n$ (cf., e.g., [31, 0.6]). Moreover, $\mathbf{M}_{n\times n}(\alpha_1,\ldots,\alpha_n)$ is irreducible and the ideal $(f_1-\alpha_1,\ldots,f_n-\alpha_n)$ of $k[\mathbf{M}_{n\times n}]$ is radical (see the next paragraph). Hence, $\mathbf{M}_{n\times n}(\alpha_1,\ldots,\alpha_n)$ is a closed subvariety of $\mathbf{M}_{n\times n}$ such that

$$\mathrm{ML}\big(\mathbf{M}_{n\times n}(\alpha_1,\ldots,\alpha_n)\big) = k$$

and $k[\ldots,x_{ij},\ldots]/(f_1-\alpha_1,\ldots,f_n-\alpha_n)$ is the k-domain with trivial Makar-Limanov invariant.

This admits the following generalization. Let G be a connected reductive algebraic group and let $\mathrm{Lie}(G)$ be the Lie algebra of G endowed with the adjoint action of G. By [18] the graded k-algebra $k[\mathrm{Lie}(G)]^G$ is free and, for every minimal system of its homogeneous generators $f_1,\ldots,f_r$ and constants $\alpha_1,\ldots,\alpha_r \in k$,

(i) $\mathrm{Lie}(G)(\alpha_1,\ldots,\alpha_r) := \{a \in \mathrm{Lie}(G) \mid f_1(a)=\alpha_1,\ldots,f_r(a)=\alpha_r\}$ is the closure of a G-orbit;

(ii) the ideal $(f_1-\alpha_1,\ldots,f_r-\alpha_r)$ of $k[\mathrm{Lie}(G)]$ is radical.

Since the center Z of G acts trivially on $\mathrm{Lie}(G)$ and G/Z is semisimple, this yields

$$\mathrm{ML}\big(\mathrm{Lie}(G)(\alpha_1,\ldots,\alpha_r)\big) = k$$

and $k[\mathrm{Lie}(G)]/(f_1-\alpha_1,\ldots,f_r-\alpha_r)$ is the k-domain with trivial Makar-Limanov invariant.

For $G=\mathbf{GL}_n$, we have $\mathrm{Lie}(G)(\alpha_1,\ldots,\alpha_r) = \mathbf{M}_{n\times n}(\alpha_1,\ldots,\alpha_n)$.

Example 1.17 (Determinantal varieties). Given positive integers $n \geqslant m > r$, let $\{x_{ij} \mid i=1,\ldots,n, j=1,\ldots,m\}$ be the set of variables considered as the standard coordinates functions on $\mathbf{M}_{n\times m}$. Let $I_{n,m,r}$ be the ideal of $k[\mathbf{M}_{n\times m}] = k[\ldots,x_{ij},\ldots]$ generated by all $(r+1)\times(r+1)$-minors of the matrix (x_{ij}). Then $I_{n,m,r}$ is radical, cf., e.g., [32]. The (affine) determinantal variety $\mathrm{D}_{n,m,r}$ is the subvariety of $\mathbf{M}_{n\times m}$ defined by $I_{n,m,r}$. Its underlying set is that of $n \times m$-matrices of rank $\leqslant r$. It is stable with respect to the action of $\mathbf{SL}_n \times \mathbf{SL}_m$ on $\mathbf{M}_{n\times m}$ by $(g,h)\cdot a := gah^{-1}$ and contains a dense orbit. Whence

$$\mathrm{ML}(\mathrm{D}_{n,m,r}) = k$$

and $k[\mathbf{M}_{n\times m}]/I_{n,m,r}$ is the k-domain with trivial Makar-Limanov invariant.

Example 1.18 (S-varieties in the sense of [41]). Denote by $\mathsf{S}^d\, k^n$ the dth symmetric power of the coordinate vector space (of columns) k^n. The natural $\mathbf{SL}_n$-action on k^n induces that on $\mathsf{S}^d\, k^n$. The (affine) Veronese morphism

$$\nu_n^d\colon k^n \to \mathsf{S}^d\, k^n, \quad v \mapsto v^d,$$

is $\mathbf{SL}_n$-equivariant. Its image $\nu_n^d(k^n)$ is closed and contains a dense $\mathbf{SL}_n$-orbit. The ideal of $\nu_n^d(k^n)$ is generated by all 2×2-minors of a certain symmetric matrix whose entries are the coordinates on $\mathsf{S}^d\, k^n$, cf. [15]. Thus,

$$\mathrm{ML}\big(\nu_n^d(k^n)\big) = k$$

and the coordinate algebra of $\nu_n^d(k^n)$ is the explicitly described k-domain with trivial Makar-Limanov invariant.

More generally, the following combination of the Veronese and Segre morphisms

$$\nu_{n_1,\ldots,n_s}^{d_1,\ldots,d_s}: k^{n_1} \times \cdots \times k^{n_s} \to \mathsf{S}^{d_1}\, k^{n_1} \otimes \cdots \otimes \mathsf{S}^{d_s}\, k^{n_s},$$
$$(v_1,\ldots,v_s) \mapsto v^{d_1} \otimes \cdots \otimes v^{d_s},$$

is equivariant with respect to the natural $\mathbf{SL}_{n_1} \times \cdots \times \mathbf{SL}_{n_s}$-actions, its image is closed and contains a dense orbit. Whence,

$$\mathrm{ML}\left(\nu_{n_1,\ldots,n_s}^{d_1,\ldots,d_s}(k^{n_1} \times \cdots \times k^{n_s})\right) = k.$$

In turn, this construction admits a further generalization. Namely, any matrix

$$A = \begin{pmatrix} a_{11} & \ldots & a_{1s} \\ \ldots & \ldots & \ldots \\ a_{r1} & \ldots & a_{rs} \end{pmatrix}$$

with the entries in $\mathbf{Z}_{>0}$ defines the diagonal morphism

$$\nu_{n_1,\ldots,n_s}^{A} := \nu_{n_1,\ldots,n_s}^{a_{11},\ldots,a_{1s}} \times \cdots \times \nu_{n_1,\ldots,n_s}^{a_{r1},\ldots,a_{rs}}.$$

This morphism is $\mathbf{SL}_{n_1} \times \cdots \times \mathbf{SL}_{n_s}$-equivariant and its image

$$\mathbf{H}_{n_1,\ldots,n_s}^{A} := \nu_{n_1,\ldots,n_s}^{A}(k^{n_1} \times \cdots \times k^{n_s})$$

is closed and contains a dense orbit. Thus,

$$\mathrm{ML}(\mathbf{H}_{n_1,\ldots,n_s}^{A}) = k.$$

In fact, $\mathbf{H}_{n_1,\ldots,n_s}^{A}$'s are special examples of varieties with trivial Makar-Limanov invariant obtained by the following general construction [41].

Let G be a connected semisimple algebraic group and let $E(\lambda)$ be a simple G-module with the highest weight λ (with respect to a fixed Borel subgroup and its maximal torus). Let v_{λ_i} be a highest vector of $E(\lambda_i)$. For $x = v_{\lambda_1} + \cdots + v_{\lambda_s} \in E(\lambda_1) \oplus \cdots \oplus E(\lambda_s)$, let $X(\lambda_1,\ldots,\lambda_s)$ be the closure of the G-orbit of x. Up to G-isomorphism, $X(\lambda_1,\ldots,\lambda_s)$ depends only on $\lambda_1,\ldots,\lambda_s$. By Corollary 1.5

$$\mathrm{ML}\big(X(\lambda_1,\ldots,\lambda_s)\big) = k.$$

The ideal $I(\lambda)$ of $X(\lambda)$ in $k[E(\lambda)]$ is generated by quadratic forms that can be explicitly described. Namely, $k[E(\lambda)]$ is the symmetric algebra of the dual G-module $E(\lambda)^* = E(\lambda^*)$. The submodule $\mathsf{S}^2 E(\lambda^*)$ of the G-module $k[E(\lambda)]$ contains a unique simple submodule with the highest weight $2\lambda^*$, the Cartan component of $\mathsf{S}^2 E(\lambda^*)$. Hence $\mathsf{S}^2 E(\lambda^*)$ contains a unique submodule L complement to the Cartan component. This L generates $I(\lambda)$, see [4, Theorem 4.1].

For $G = \mathbf{SL}_{n_1} \times \cdots \times \mathbf{SL}_{n_s}$ and $\lambda_i = a_{i1}\varpi_1^{(1)} + \cdots + a_{is}\varpi_1^{(s)}$ where $\varpi_1^{(j)}$ is the highest weight of the natural $\mathbf{SL}_{n_j}$-module k^{n_j}, we have

$$X(\lambda_1, \ldots, \lambda_s) = \mathbf{H}^A_{n_1,\ldots,n_s}.$$

Example 1.19 (Irreducible affine surfaces quasihomogeneous with respect to an algebraic group in the sense of [14]**).** By [26], up to isomorphism, such surfaces are exhausted by the following list (we maintain the notation of Example 1.18):

(i) smooth surfaces:

$$\mathbf{A}^2, \quad \mathbf{A}^1 \times \mathbf{A}^1_*, \quad \mathbf{A}^1_* \times \mathbf{A}^1_*, \quad (\mathbf{P}^1 \times \mathbf{P}^1) \setminus \Delta, \quad \mathbf{P}^2 \setminus C, \tag{1.10}$$

where Δ is the diagonal in $\mathbf{P}^1 \times \mathbf{P}^1$, and C is a nondegenerate conic in $\mathbf{P}^2$;

(ii) singular surfaces:

$$\mathbf{V}(n_1, \ldots, n_r) := \mathbf{H}^A_2 \quad \text{for } A = (n_1, \ldots, n_r)^{\mathsf{T}},\ n_1, \ldots, n_r \geqslant 2.$$

Each of these surfaces but $\mathbf{A}^1 \times \mathbf{A}^1_*$ and $\mathbf{A}^1_* \times \mathbf{A}^1_*$ admits an $\mathbf{SL}_2$-action with a dense orbit. Namely, $(\mathbf{P}^1 \times \mathbf{P}^1) \setminus \Delta = \mathbf{SL}_2 / T$ and $\mathbf{P}^2 \setminus C = \mathbf{SL}_2 / \mathrm{N}(T)$, where T is a maximal torus of $\mathbf{SL}_2$ and $\mathrm{N}(T)$ its normalizer, see [26, Lemma 2]. The surface $\mathbf{V}(n_1, \ldots, n_r)$ is the closure of the $\mathbf{SL}_2$-orbit of $v_1 + \cdots + v_r \in \mathbf{R}_{n_1} \oplus \cdots \oplus \mathbf{R}_{n_r}$, where v_i is a highest vector of the simple $\mathbf{SL}_2$-module $\mathbf{R}_{n_i}$ of dimension n_i+1 (such a module is unique up to isomorphism), see [26, Section 2]. By Corollary 1.5 this yields

$$\mathrm{ML}((\mathbf{P}^1 \times \mathbf{P}^1) \setminus \Delta) = \mathrm{ML}(\mathbf{P}^2 \setminus C) = \mathrm{ML}\big(\mathbf{V}(n_1, \ldots, n_r)\big) = k, \tag{1.11}$$

As $\mathbf{A}^m_* := (\mathbf{A}^1_*)^m$ is toral and $\mathrm{ML}(\mathbf{A}^n) = k$, Corollary 1.15 implies that

$$\mathrm{ML}(\mathbf{A}^n \times \mathbf{A}^m_*) = k[\mathbf{A}^m_*]. \tag{1.12}$$

From (1.12) we get the Makar-Limanov invariants of the remaining three surfaces in (1.10).

Example 1.20 (Irreducible affine threefolds quasihomogeneous with respect to an algebraic group in the sense of [14]**).** We maintain the notation of Examples 1.18 and 1.19. Identify $\mathrm{Pic}\big((\mathbf{P}^1 \times \mathbf{P}^1) \setminus \Delta\big)$ with $\mathbf{Z}$ by a fixed isomorphism φ. Let $\mathbf{X}_n$ be the total space of the one-dimensional vector bundle over $(\mathbf{P}^1 \times \mathbf{P}^1) \setminus \Delta$ corresponding to $n \in \mathbf{Z}$, and let $\mathbf{X}^*_n$ be the complement of the zero section in $\mathbf{X}_n$. In fact, $\mathbf{X}_n$ is isomorphic to $\mathbf{X}_{-n}$ and $\mathbf{X}^*_n$ to $\mathbf{X}^*_{-n}$, so $\mathbf{X}_n$ and $\mathbf{X}^*_n$ do not depend on the choice of φ, see [27].

The group $\mathrm{Pic}\big(\mathbf{P}^2 \setminus C\big)$ has order 2. Let $\mathbf{Y}_0$ and $\mathbf{Y}_1$ be the total spaces of, resp., trivial and nontrivial one-dimensional vector bundles over $\mathbf{P}^2 \setminus C$. Let $\mathbf{Y}^*_n$ be the complement of the zero section in $\mathbf{Y}_n$.

Let $\widetilde{\mathrm{T}}, \widetilde{\mathrm{O}}, \widetilde{\mathrm{I}}$, and $\widetilde{\mathrm{D}}_n$ be, resp., the binary tetrahedral, octahedral, icosahedral, and dihedral subgroup of order $4n$ in $\mathbf{SL}_2$. Put $\mathbf{S}_3 = \mathbf{SL}_2 / \widetilde{\mathrm{T}}$, $\mathbf{S}_4 = \mathbf{SL}_2 / \widetilde{\mathrm{O}}$, $\mathbf{S}_5 = \mathbf{SL}_2 / \widetilde{\mathrm{I}}$, and $\mathbf{W}_n = \mathbf{SL}_2 / \widetilde{\mathrm{D}}_n$.

By [26] up to isomorphism irreducible affine threefolds quasihomogeneous with respect to an algebraic group in the sense of [14] are exhausted by the following list:

(i) smooth threefolds:

$$\begin{gathered}\mathbf{X}_n,\quad \mathbf{X}_n^*,\quad \mathbf{W}_n,\quad \mathbf{Y}_0,\quad \mathbf{Y}_0^*,\quad \mathbf{Y}_1^*,\quad \mathbf{S}_3,\quad \mathbf{S}_4,\quad \mathbf{S}_5,\\ \mathbf{A}^3,\quad \mathbf{A}^2\times\mathbf{A}^1_*,\quad \mathbf{A}^1\times\mathbf{A}^2_*,\quad \mathbf{A}^3_*;\end{gathered}\tag{1.13}$$

(ii) singular threefolds:

$$\begin{aligned}\mathbf{P}(A) &:= \mathbf{H}_3^A \quad \text{where all entries of } A \text{ are} \geqslant 1,\\ \mathbf{Q}(B) &:= \mathbf{H}_{2,2}^B \quad \text{where } \operatorname{rk} B = 1.\end{aligned}$$

By construction, $\mathbf{S}_3, \mathbf{S}_4, \mathbf{S}_5, \mathbf{W}_n$ are homogeneous with respect to $\mathbf{SL}_2$ while $\mathbf{P}(A)$ and $\mathbf{Q}(B)$ admit an action of resp. $\mathbf{SL}_2\times\mathbf{SL}_2$ and $\mathbf{SL}_3$ with a dense orbit. In fact, $\mathbf{X}_n^*$ for $n\neq 0$ is homogeneous with respect to $\mathbf{SL}_2$ as well (it is the quotient of $\mathbf{SL}_2$ modulo a cyclic subgroup of order $|n|$). By [27, Theorem 9] every $\mathbf{X}_n$ is homogeneous with respect to the nonreductive linear algebraic group $\mathbf{SL}_{2,|n|} := \mathbf{SL}_2 \ltimes \mathbf{R}_{|n|}$ (see Example 1.19); the radical of $\mathbf{SL}_{2,|n|}$ is unipotent. By [27, Proposition 18], $\mathbf{Y}_0$ is homogeneous with respect to $\mathbf{SL}_{2,d}$ for every even $d>0$. From Theorem 1.7 we then deduce that

$$\begin{aligned}&\mathrm{ML}(\mathbf{S}_3)=\mathrm{ML}(\mathbf{S}_4)=\mathrm{ML}(\mathbf{S}_5)=\mathrm{ML}(\mathbf{Y}_0)=k,\\ &\mathrm{ML}(\mathbf{X}_n)=\mathrm{ML}(\mathbf{W}_n)=\mathrm{ML}\big(\mathbf{P}(A)\big)=\mathrm{ML}\big(\mathbf{Q}(B)\big)=k,\\ &\mathrm{ML}(\mathbf{X}_n^*)=k \quad \text{for } n\neq 0.\end{aligned}$$

As $\mathbf{X}_0^* = \big((\mathbf{P}^1\times\mathbf{P}^1)\setminus\Delta\big)\times\mathbf{A}^1_*$ and $\mathbf{Y}_0^*=(\mathbf{P}^2\setminus C)\times\mathbf{A}^1_*$, we deduce from (1.11) and Corollary 1.15 that

$$\mathrm{ML}(\mathbf{X}_0^*)=\mathrm{ML}(\mathbf{Y}_0^*)=k[\mathbf{A}^1_*].$$

By [27, Proposition 16]], $\mathbf{Y}_1^*$ is homogeneous with respect to $\mathbf{SL}_2\times\mathbf{G}_m$. Since the latter is a reductive group with one-dimensional center, Theorem 1.7 implies that $\operatorname{tr}\deg_k \mathrm{ML}(\mathbf{Y}_1^*)\leqslant 1$. On the other hand, by [27, Proposition 19], $k[\mathbf{Y}_1^*]/k^*$ is a free abelian group of rank 1. Since $k[X]^*\subseteq \mathrm{ML}(X)$ for every X, this yields

$$\operatorname{tr}\deg_k \mathrm{ML}(\mathbf{Y}_1^*)=1.$$

Finally, (1.12) yields the Makar-Limanov invariants of the last four threefolds in (1.13).

Example 1.21 (Schubert varieties). Let G be a connected semisimple algebraic group and let $\mathbf{P}E$ be the projective space of 1-dimensional linear subspaces in a nonzero simple G-module E. There is a unique closed G-orbit $\mathcal{O}$ in $\mathbf{P}E$. Let U be a maximal unipotent subgroup of G. There are only finitely many U-orbits in $\mathcal{O}$; their closures are called Schubert varieties, cf., e.g., [39, 8.3–8.5]. Let $X\subseteq\mathcal{O}$ be a Schubert variety and let $\widehat{X}$ be the affine cone over X in E. As U is unipotent, Corollary 1.8 yields

$$\operatorname{tr}\deg_k \mathrm{ML}(\widehat{X})\leqslant 1.$$

The ideal of $\widehat{X}$ in $k[E]$ is generated by certain forms of degree $\leqslant 2$ that admit an explicit description, see, e.g., [2, 2.10].

Example 1.22 (Not stably rational smooth affine varieties with trivial Makar-Limanov invariant). In [20] a construction of nonrational singular affine threefolds with trivial Makar-Limanov invariant is exhibited. Our approach yields, for every integer d, examples of not stably rational (hence a fortiori nonrational) smooth affine varieties of dimension $\geqslant d$ with trivial Makar-Limanov invariant. Here is the construction.

Let F be a linear algebraic group. By [29, Theorem 1.5.5] the following properties are equivalent:

(a) for every locally free finite-dimensional algebraic kF-module V, the field $k(V)^F$ is stably rational over k;

(b) there is an embedding $F \subseteq H$, where H is a special group in the sense of Serre, such that the variety H/F is stably rational.

Here "locally free" means that F-stabilizers of points in general position in V are trivial (see [29, 1.2.2]); for finite F, this is equivalent to triviality of the kernel of action. About special groups see, e.g., [29, 1.4; 31, 2.6].

Now let F be a finite group whose Schur multiplier $H^2(F, \mathbf{Q}/\mathbf{Z})$ contains a nonzero element α such that $\alpha|_A = 0$ for every abelian subgroup A of F. It is known that such groups exist and, for every locally free finite-dimensional algebraic kF-module V, the field $k(V)^F$ is not stably rational over k (see, e.g., [38]). By [39, 2.3.7] we can (and shall) embed F in $\mathbf{SL}_n$ for some n. As $\mathbf{SL}_n$ is special (cf. [29, 1.4; 31, 2.6]), the aforesaid yields that the smooth variety $X := \mathbf{SL}_n/F$ is not stably rational. As F is reductive and $\mathbf{SL}_n$ is a connected algebraic group that has no nontrivial characters, Corollary 1.6 implies that X is a not stably rational smooth affine variety with trivial Makar-Limanov invariant.

1.2. The Derksen invariant. Let X be a variety. Recall that the *Derksen invariant* $\mathrm{D}(X)$ of X is the k-subalgebra of $k[X]$ generated by all $k[X]^H$'s where H runs over all subgroups of $\mathrm{Aut}(X)$ isomorphic to $\mathbf{G}_a$. If there are no such subgroups, we put $\mathrm{D}(X) = \varnothing$.

Example 1.23. If X is toral, then $\mathrm{D}(X) = \varnothing$ by Lemma 1.14(c_1).

In this section we deduce some information on $\mathrm{D}(X)$ in case when $\mathrm{Aut}(X)$ contains a connected noncommutative reductive algebraic subgroup.

Recall that if an algebraic group G acts linearly on a (not necessarily finite-dimensional) k-vector space V, then the G-module V is called *algebraic* if every element of V is contained in an algebraic finite-dimensional G-submodule of V, cf., e.g., [31, 3.4].

The starting point is

Lemma 1.24. *Let G be a connected noncommutative reductive algebraic group. Then every algebraic G-module V is a k-linear span of the set*

$$\bigcup_{H \subset G} V^H, \tag{1.14}$$

where H in (1.14) runs over all one-parameter unipotent subgroups of G.

PROOF. The assumptions that G is reductive, $\mathrm{char}\, k = 0$, and V is algebraic imply that V is a sum of simple G-submodules. Hence we may (and shall) assume that V is a nonzero simple G-module. Since G is a connected noncommutative reductive algebraic group, it contains a one-dimensional unipotent subgroup U

(indeed, since (G,G) is a nontrivial semisimple group, a root subgroup of (G,G) with respect to a maximal torus may be taken as U). By the Lie – Kolchin theorem $V^U \neq \{0\}$. Let v be a nonzero vector of V^U. As $g \cdot v \in V^{gUg^{-1}}$ for every element $g \in G$, the G-orbit $G \cdot v$ of v is contained in set (1.14). Hence the k-linear span of $G \cdot v$ is contained in the k-linear span of this set. But the k-linear span of $G \cdot v$ is G-stable and therefore coincides with V since V is simple. This completes the proof. □

Theorem 1.25. *Let X be a variety. If* $\mathrm{Aut}(X)$ *contains a connected noncommutative reductive algebraic subgroup, then*

$$\mathrm{D}(X) = k[X]. \tag{1.15}$$

PROOF. Let G be a connected noncommutative reductive algebraic subgroup of $\mathrm{Aut}(X)$. Since the G-module $k[X]$ is algebraic (see [31, Lemma 1.4]), the claim follows from Lemma 1.24 and the definition on $\mathrm{D}(X)$. □

Remark 1.26. The following are equivalent:

(i) $\mathrm{Aut}(X)$ contains a connected noncommutative reductive algebraic subgroup;

(ii) $\mathrm{Aut}(X)$ contains SL_2 or PSL_2.

Indeed, SL_2 and PSL_2 are connected noncommutative reductive algebraic groups and every connected noncommutative reductive algebraic group contains SL_2 or PSL_2, cf. [3, Theorem 13.18(4); 39, 7.2.4].

The following example shows that the assumption of noncommutativity in Theorem 1.25 cannot be dropped.

Example 1.27. By [8], for the Koras – Russell cubic threefold X, the following inequality distinguishing X from $\mathbf{A}^3$ holds:

$$\mathrm{D}(X) \neq k[X]. \tag{1.16}$$

On the other hand, since X is defined in $\mathbf{A}^4$ by $x_1 + x_1^2 x_2 + x_3^2 + x_4^3 = 0$ where $x_1, \ldots, x_4$ are the standard coordinate functions on $\mathbf{A}^4$, it is stable with respect to the action of $\mathbf{G}_m$ on $\mathbf{A}^4$ defined by $t \cdot (a_1, a_2, a_3, a_4) = (t^6 a_1, t^{-6} a_2, t^3 a_3, t^2 a_4)$. Hence $\mathrm{Aut}(X)$ contains a one-dimensional connected commutative reductive subgroup, cf. [11, Section 3].

One can apply Theorem 1.25 to proving that, for some varieties X, there are no connected noncommutative reductive algebraic subgroups in $\mathrm{Aut}(X)$.

Example 1.28. For the Koras – Russell cubic threefold X, Theorem 1.25 and (1.16) imply that $\mathrm{Aut}(X)$ contains no connected noncommutative reductive algebraic subgroups.

Since $\mathrm{Aut}(\mathbf{A}^n)$ for $n \geqslant 2$ contains a connected noncommutative reductive algebraic subgroup (for instance, $\mathbf{GL}_n$), the next corollary generalizes the well-known fact that $\mathrm{D}(X \times \mathbf{A}^n) = k[X \times \mathbf{A}^n]$ for $n \geqslant 2$ (see, e.g., [5]).

Corollary 1.29. *Let Z be a variety such that* $\mathrm{Aut}(Z)$ *contains a connected noncommutative reductive algebraic subgroup. Then, for every variety X,*

$$\mathrm{D}(X \times Z) = k[X \times Z]. \tag{1.17}$$

Proof. Consider the natural action of $\mathrm{Aut}(Z)$ on Z and its trivial action on X. Then the diagonal action of $\mathrm{Aut}(Z)$ on $X\times Z$ identifies $\mathrm{Aut}(Z)$ with a subgroup of $\mathrm{Aut}(X\times Z)$. Whence the claim by Theorem 1.25. □

The following example shows that the assumption of noncommutativity in Corollary 1.29 cannot be dropped.

Example 1.30. Let x_1, x_2 be the standard coordinate functions on $\mathbf{A}^2$. The principal open set Y in $\mathbf{A}^2$ defined by $x_1 \neq 0$ is isomorphic to $\mathbf{A}^1_* \times \mathbf{A}^1$ and

$$k[Y] = k[t, t^{-1}, s], \quad \text{where } t := x_1|_Y,\ s := x_2|_Y. \tag{1.18}$$

Since t is the unit of $k[Y]$, for every action of $\mathbf{G}_a$ on Y we have

$$t, t^{-1} \in k[Y]^{\mathbf{G}_a}. \tag{1.19}$$

As, clearly, $\mathrm{Aut}(Y)$ contains a one-dimensional unipotent subgroup, (1.19) and the definition of $\mathrm{D}(Y)$ yield $k[t,t^{-1}] \subseteq \mathrm{D}(Y)$. We also deduce from (1.19) that, for every point $y \in Y$, the $\mathbf{G}_a$-orbit of y lies in the line defined by the equation $t = t(y)$. But this orbit is closed in Y since Y is affine and $\mathbf{G}_a$ is unipotent, cf. [3, 4.10]. Hence, if y is not a fixed point, this orbit coincides with the aforementioned line. Therefore, if $\mathbf{G}_a$ acts on Y nontrivially, t separates orbits in general position. Since $\mathrm{char}\, k = 0$, by [31, Lemma 2.1] this means that $k(Y)^{\mathbf{G}_a} = k(t)$. Whence by (1.18) we have $k[Y]^{\mathbf{G}_a} = k[t,t^{-1}]$. From this, (1.12) and (1.18) we then infer that

$$k[t,t^{-1}] = \mathrm{ML}(\mathbf{A}^1_* \times \mathbf{A}^1) = \mathrm{D}(\mathbf{A}^1_* \times \mathbf{A}^1) \subsetneq k[\mathbf{A}^1_* \times \mathbf{A}^1] = k[t,t^{-1},s].$$

Thus, (1.17) does not hold for $X = \mathbf{A}^1_*$, $Z = \mathbf{A}^1$ while both $\mathrm{Aut}(\mathbf{A}^1_*)$ and $\mathrm{Aut}(\mathbf{A}^1)$ contain a one-dimensional connected commutative reductive algebraic subgroup.

Theorem 1.31. *If X_i is a variety such that $\mathrm{ML}(X_i) \neq k[X_i]$, $i = 1, 2$, then*

$$\mathrm{D}(X_1 \times X_2) = k[X_1 \times X_2].$$

Proof. As $\mathrm{ML}(X_1) \neq k[X_1]$, there is a nontrivial $\mathbf{G}_a$-action α on X_1. The diagonal $\mathbf{G}_a$-action on $X_1 \times X_2$ determined by α and trivial action on X_2 is a nontrivial $\mathbf{G}_a$-action for which $k[X_2]$ lies in the algebra of invariants. Hence, $k[X_2] \subseteq \mathrm{D}(X_1 \times X_2)$. Similarly, $k[X_1] \subseteq \mathrm{D}(X_1 \times X_2)$. As $k[X_1 \times X_2]$ is generated by $k[X_1]$ and $k[X_2]$, the claim follows. □

Example 1.32. If X is the Koras–Russell cubic threefold X, then $\mathrm{D}(X) \neq k[X]$ by [8]. But for the square of X we have $\mathrm{D}(X \times X) = k[X \times X]$—since $\mathrm{ML}(X) = k[x_1|_X] \neq k[X]$ (cf., e.g., [13, Chapter 9]), this follows from Theorem 1.31.

1.3. Generalizations. The Makar-Limanov and Derksen invariants can be naturally generalized.

Namely, let X be a variety and let F be an algebraic group.

Definition 1.33. The *F-kernel* of X is the following k-subalgebra of $k[X]$:

$$\mathrm{Ker}_F(X) := \bigcap_H k[X]^H, \tag{1.20}$$

where H in (1.20) runs over the images of all algebraic homomorphisms $F \to \mathrm{Aut}(X)$.

Definition 1.34. The *F-envelope* of X is the k-subalgebra

$$\mathrm{Env}_F(X)$$

of $k[X]$ generated by all $k[X]^H$'s where H runs over all subgroups of $\mathrm{Aut}(X)$ isomorphic to F. If there are no such subgroups, we put $\mathrm{Env}_F(X) = \varnothing$.

Example 1.35. The definitions imply that

$$\mathrm{Ker}_{\mathbf{G}_a}(X) = \mathrm{ML}(X), \quad \mathrm{Env}_{\mathbf{G}_a}(X) = \mathrm{D}(X).$$

Definition 1.36. We say that an algebraic group G is *F-generated* if it is generated by the images of all homomorphisms $F \to G$.

Examples 1.37. (1) By Lemma 1.1 a connected linear algebraic group G is $\mathbf{G}_a$-generated if and only if G has no nontrivial characters that, in turn, is equivalent to the condition $\mathrm{Rad}\, G = \mathrm{Rad}_u\, G$.

(2) Every connected reductive algebraic group G is $\mathbf{G}_m$-generated. This is clear if G is a torus. The general case follows from the case of torus because of the following two facts: (a) the subgroup of G generated by connected algebraic subgroups is closed (see, e.g., [39, 2.2.7]); and (b) the union of maximal tori of G contains a dense open subset of G ([39, 6.4.5(iii), 7.6.4(ii)]).

(3) Clearly, the subgroup generated by the images of all homomorphisms $F \to G$ is normal. Hence, if G is simple as abstract group and there exists a nontrivial homomorphism $F \to G$, then G is F-generated.

The following are the generalizations of the above statements on $\mathrm{ML}(X)$ and $\mathrm{D}(X)$.

Theorem 1.38. *If a variety X is endowed with an action of an F-generated algebraic group G, then* $\mathrm{Ker}_F(X) \subseteq k[X]^G$.

PROOF. This follows from Definitions 1.33 and 1.36. □

Corollary 1.39. *If a variety X is endowed with an action of an F-generated algebraic group G and X contains a dense G-orbit, then* $\mathrm{Ker}_F(X) = k$.

Corollary 1.40. *If H is a reductive subgroup of an F-generated linear algebraic group G, then G/H is an affine variety with* $\mathrm{Ker}_F(G/H) = k$.

Corollary 1.41. *Let X be an irreducible variety. If there is an action of $\mathbf{G}_m$ on X with a fixed point and without other closed orbits, then*

$$\mathrm{Ker}_{\mathbf{G}_m}(X) = k. \tag{1.21}$$

PROOF. The assumptions imply that the fixed point is unique and lies in the closure of every $\mathbf{G}_m$-orbit; whence $k[X]^{\mathbf{G}_m} = k$. In turn, this and (1.20) yield (1.21). □

Remark 1.42. If X in Corollary 1.41 is normal, then by [30] it is affine.

Corollary 1.43. *Let X be a closed subset of $\mathbf{P}^n$ and let $\widehat{X} \subseteq k^{n+1}$ be the affine cone over X. Then* $\mathrm{Ker}_{\mathbf{G}_m}(\widehat{X}) = k$.

Example 1.44. Consider the case $F = \mathbf{G}_m$. If G is a connected reductive subgroup of $\mathrm{Aut}(X)$ and X contains a dense G-orbit, then Corollary 1.39 and

Example 1.37(2) imply that (1.21) holds. In particular, this is so for every toric variety X; for instance,

$$\mathrm{Ker}_{\mathbf{G}_m}(\mathbf{A}^n \times \mathbf{A}^m_*) = k.$$

(compare with (1.12)). Applying this to the varieties considered in Examples 1.16 – 1.20, we see that (1.21) holds for every X from the following list:

$\mathrm{Lie}(G)(\alpha_1, \dots, \alpha_n)$	(see Example 1.16);
$\mathrm{D}_{n,m,r}$	(see Example 1.17);
$X(\lambda_1, \dots, \lambda_s)$	(see Example 1.18);
$(\mathbf{P}^1 \times \mathbf{P}^1) \setminus \Delta$, $\mathbf{P}^2 \setminus C$, $\mathbf{V}(n_1, \dots, n_r)$ where $n_1, \dots, n_r \geqslant 2$	(see Example 1.19);
$\mathbf{S}_3$, $\mathbf{S}_4$, $\mathbf{S}_5$, $\mathbf{W}_n$, $\mathbf{P}(A)$, $\mathbf{Q}(B)$, $\mathbf{X}^*_n$ where $n \neq 0$, $\mathbf{Y}^*_1$	(see Example 1.20).

The threefold $\mathbf{X}_n$ from Example 1.20 is homogeneous with respect to $\mathbf{SL}_{2,|n|}$. One can prove that $\mathbf{SL}_{2,|n|}$ is $\mathbf{G}_m$-generated; whence $\mathrm{Ker}(\mathbf{X}_n) = k$.

The remaining threefolds $\mathbf{X}^*_0$, $\mathbf{Y}_0$, and $\mathbf{Y}^*_0$ from Example 1.20 are considered in Example 1.47 below.

The same proof as that of Lemma 1.9 yields

Lemma 1.45. *For any varieties X_1 and X_2,*

$$\mathrm{Ker}_F(X_1 \times X_2) \subseteq \mathrm{Ker}_F(X_1) \otimes_k \mathrm{Ker}_F(X_2).$$

Corollary 1.46. *For any varieties X_1 and X_2, the following are equivalent:*

(i) $\mathrm{Ker}_F(X_1)$ *and* $\mathrm{Ker}_F(X_2)$ *lie in* $\mathrm{Ker}_F(X_1 \times X_2)$;

(ii) $\mathrm{Ker}_F(X_1 \times X_2) = \mathrm{Ker}_F(X_1) \otimes_k \mathrm{Ker}_F(X_2)$.

Example 1.47. Since $\mathbf{X}^*_0 = \big((\mathbf{P}^1 \times \mathbf{P}^1) \setminus \Delta\big) \times \mathbf{A}^1_*, \mathbf{Y}_0 = (\mathbf{P}^2 \setminus C) \times \mathbf{A}^1$, and $\mathbf{Y}^*_0 = (\mathbf{P}^2 \setminus C) \times \mathbf{A}^1_*$ (see Example 1.20), we deduce from Lemma 1.45 and Example 1.44 that $\mathrm{Ker}_{\mathbf{G}_m}(\mathbf{X}^*_0) = \mathrm{Ker}_{\mathbf{G}_m}(\mathbf{Y}_0) = \mathrm{Ker}_{\mathbf{G}_m}(\mathbf{Y}^*_0) = k$.

Lemma 1.48. *For any connected algebraic group F that has no nontrivial characters,*

$$k[X]^* \subseteq \mathrm{Ker}_F(X). \tag{1.22}$$

PROOF. Let H be the image of an algebraic homomorphism $F \to \mathrm{Aut}(X)$. We claim that $k[X]^* \subseteq k[X]^H$; by virtue of Definition 1.33 this inclusion implies (1.22). Since H is connected, every irreducible component of X is H-stable, so proving the claim we may (and shall) assume that X is irreducible. In this case every element of $k[X]^*$ is H-semiinvariant by [31, Theorem 3.1], hence lies in $k[X]^H$ since H has no nontrivial characters. This completes the proof. □

Corollary 1.49. *Let F be a connected algebraic group that has no nontrivial characters. Let X_1 and X_2 be varieties such that $\mathrm{Ker}_F(X_1)$ and $\mathrm{Ker}_F(X_2)$ are generated by units. Then $\mathrm{Ker}_F(X_1 \times X_2) = \mathrm{Ker}_F(X_1) \otimes_k \mathrm{Ker}_F(X_2)$.*

Lemma 1.50. *Let G be a connected reductive algebraic group of rank $\geqslant 2$. Then every algebraic G-module V is a k-linear span of the set*

$$\bigcup_{H \subseteq G} V^H, \tag{1.23}$$

where H in (1.23) runs over all one-dimensional tori of G.

PROOF. Like in the proof of Lemma 1.24 we may (and shall) assume that V is a nonzero simple G-module. Let T be a maximal torus of G and let $v \in V$, $v \neq 0$, be a weight vector of T. Since $\dim T \geqslant 2$, the T-stabilizer T_v of v is a diagonalizable group of dimension $\geqslant 1$. Hence T_v contains a one-dimensional torus S. Thus, $v \in V^S$. Like in the proof of Lemma 1.24 we then conclude that the orbit $G \cdot v$ is contained in set (1.23). Since V is simple, the k-linear span of $G \cdot v$ coincides with V; whence the claim. □

Theorem 1.51. *Let X be a variety such that* $\mathrm{Aut}(X)$ *contains a connected reductive algebraic group G of rank $\geqslant 2$. Then*

$$\mathrm{Env}_{\mathbf{G}_m}(X) = k[X].$$

PROOF. Since the G-module $k[X]$ is algebraic, the claim follows from Lemma 1.50 and Definition 1.34. □

Remark 1.52. Clearly, $\mathrm{Env}_{\mathbf{G}_m}(\mathbf{A}^1) = k$. This shows that in Lemma 1.50 and Theorem 1.51 the condition "$\geqslant 2$" can not be replaced by "$\geqslant 1$."

2. Finite automorphism groups of algebraic varieties

2.1. Jordan groups. The following definition is inspired by the classical Jordan theorem (Theorem 2.3 below).

Definition 2.1. A group G is called a *Jordan group* if there exists a positive integer J_G, depending only on G, such that every finite subgroup K of G contains a normal abelian subgroup whose index in K is at most J_G.

Remark 2.2. Dropping the assumption of normality in Definition 2.1 we do not obtain a more general notion. Indeed, as is known (see, e.g., [19, Exercise 12 to Chapter I]), if a group P contains a subgroup Q of finite index, then there is a normal subgroup N of P such that $[P{:}N] \leqslant [P{:}Q]!$ and $N \subseteq Q$.

Jordan's theorem (see [6, Theorem 36.13]) can be then reformulated as follows:

Theorem 2.3. *Every* $\mathbf{GL}_n(k)$ *is Jordan.*

Remark 2.4. For $G = \mathbf{GL}_n(k)$, the explicit upper bounds J_G are known, see [6, Section 36].

Since subgroups of Jordan groups are Jordan and every linear algebraic group is isomorphic to a subgroup of some $\mathbf{GL}_n(k)$ (see [39, 2.3.7]), Theorem 2.3 yields the following more general

Theorem 2.5. *Every linear algebraic group is Jordan.*

Lemma 2.6. *Let H be a finite normal subgroup of a group G. If G is Jordan, then G/H is Jordan.*

PROOF. Let $\pi\colon G \to G/H$ be the natural projection and let K be a finite subgroup of G/H. Since H is finite, $\pi^{-1}(K)$ is a finite subgroup of G. As G is Jordan, $\pi^{-1}(K)$ contains a normal abelian subgroup A whose index is at most J_G. Hence $\pi(A)$ is a normal abelian subgroup of K whose index is at most J_G. □

Lemma 2.7. *Let H be a normal torsion-free subgroup of a group G. If G/H is Jordan, then G is Jordan with $J_G = J_{G/H}$.*

Proof. Let $\pi\colon G \to G/H$ be the natural projection and let K be a finite subgroup of G. Since H is torsion free, $K \cap H = \{1\}$. Therefore, $\pi|_K\colon K \to \pi(K)$ is an isomorphism. As G/H is Jordan, this implies that K contains a normal abelian subgroup whose index in K is at most $J_{G/H}$. □

Lemma 2.8. *If the groups G_1 and G_2 are Jordan, then $G_1 \times G_2$ is Jordan.*

Proof. Let $\pi_i\colon G := G_1 \times G_2 \to G_i$ be the natural projection and let K be a finite subgroup of G. Since G_i is Jordan, $K_i := \pi_i(K)$ contains an abelian normal subgroup A_i such that

$$[K_i{:}A_i] \leqslant J_{K_i}. \tag{2.1}$$

The subgroup $\tilde{A}_i := \pi_i^{-1}(A_i) \cap K$ is normal in K and $K/\tilde{A}_i$ is isomorphic to K_i/A_i. From (2.1) we then conclude that

$$[K{:}\tilde{A}_i] \leqslant J_{K_i}. \tag{2.2}$$

Since $A := \tilde{A}_1 \cap \tilde{A}_2$ is the kernel of the diagonal homomorphism

$$K \to \prod_{i=1}^{2} K/\tilde{A}_i$$

determined by the canonical projections $K \to K/\tilde{A}_i$, we infer from (2.2) that

$$[K{:}A] = |K/A| \leqslant \left|\prod_{i=1}^{2} K/\tilde{A}_i\right| \leqslant J_{K_1} J_{K_2} \tag{2.3}$$

By construction, $A \subseteq A_1 \times A_2$. Since A_i is abelian, this implies that A is abelian. As A is normal in K, this and (2.3) complete the proof. □

The following definition distinguishes a special class of Jordan groups.

Definition 2.9. A group G is called *bounded* if there is a positive integer b_G, depending only on G, such that the order of every finite subgroup of G is at most b_G.

Examples 2.10. (1) Finite groups and torsion free groups are bounded.

(2) Every finite subgroup of $\mathbf{GL}_n(\mathbf{Q})$ is conjugate to a subgroup of $\mathbf{GL}_n(\mathbf{Z})$ (see, e.g., [6, Theorem 73.5]). On the other hand, by Minkowski's theorem (see, e.g., [16, Theorem 39.4]) $\mathbf{GL}_n(\mathbf{Z})$ is bounded. Hence $\mathbf{GL}_n(\mathbf{Q})$ is bounded. Note that H. Minkowski and I. Schur obtained explicit upper bounds of the orders of finite subgroups in $\mathbf{GL}_n(\mathbf{Z})$, see [16, Section 39].

(3) It is immediate from the definition that every extension of a bounded group by bounded is bounded as well.

Lemma 2.11. *Let H be a normal subgroup of a group G such that G/H is bounded. Then G is Jordan if and only if H is Jordan.*

Proof. A proof is needed only for the sufficiency. So assume that H is Jordan; we have to prove that G is Jordan. Let K be a finite subgroup of G. By Definition 2.1

$$L := K \cap H \tag{2.4}$$

contains an abelian normal subgroup A such that

$$[L{:}A] \leqslant J_H. \tag{2.5}$$

Let g be an element of K. Since L is a normal subgroup of K, we infer that gAg^{-1} is a normal abelian subgroup of L and

$$[L{:}A] = [L{:}gAg^{-1}]. \tag{2.6}$$

Consider now the group

$$M := \bigcap_{g\in K} gAg^{-1}. \tag{2.7}$$

It is a normal abelian subgroup of K. We claim that $[K{:}M]$ is upper bounded by a constant not depending on K. To prove this, fix the representatives $g_1, \dots, g_{|K/L|}$ of all cosets of L in K. Then (2.7) and normality of A in L imply that

$$M = \bigcap_{i=1}^{|K/L|} g_iAg_i^{-1}. \tag{2.8}$$

From (2.8) we deduce that M is the kernel of the diagonal homomorphism

$$L \to \prod_{i=1}^{|K/L|} L/g_iAg_i^{-1}$$

determined by the canonical projections $L \to L/g_iAg_i^{-1}$. This, (2.6), and (2.5) yield

$$[L{:}M] \leqslant [L{:}A]^{|K/L|} \leqslant J_H^{|K/L|}. \tag{2.9}$$

Let $\pi\colon G \to G/H$ be the canonical projection. By (2.4) the finite subgroup $\pi(K)$ of G/H is isomorphic to K/L. Since G/H is bounded, this yields $|K/L| \leqslant b_{G/H}$. We then deduce from (2.9) and $[K{:}M] = [K{:}L][L{:}M]$ that

$$[K{:}M] \leqslant b_{G/H} J_H^{b_{G/H}}.$$

This completes the proof. □

Remark 2.12. There are non-Jordan extensions (even semidirect products) of Jordan groups by Jordan ones, see in Section 2.2 below the discussion of $\mathrm{Bir}(\mathbf{P}^1 \times B)$ where B is an elliptic curve. Therefore, “bounded” in Lemma 2.11 cannot be replaced by “Jordan.”

Corollary 2.13. *Let H be a finite normal subgroup of a group G such that the center of H is trivial. If G/H is Jordan, then G is Jordan.*

Proof. The conjugating action of G on H determines a homomorphism $\varphi\colon G \to \mathrm{Aut}(H)$. The definition of φ and triviality of the center of H implies that

$$H \cap \ker\varphi = \{1\}. \tag{2.10}$$

In turn, (2.10) yields that the restriction of the natural projection $G \to G/H$ to $\ker\varphi$ is an embedding $\ker\varphi \hookrightarrow G/H$. Hence $\ker\varphi$ is Jordan since G/H is Jordan. But $G/\ker\varphi$ is finite since it is isomorphic to a subgroup of $\mathrm{Aut}(H)$ for the finite group H. Whence G is Jordan by Lemma 2.11. This completes the proof. □

We shall now discuss the notion of Jordan group in the frame of automorphism groups of algebraic varieties.

Example 2.15, Theorems 2.16, 2.19 and their corollaries below give, for some varieties X, the affirmative answer to the following

Question 2.14. Let X be an irreducible affine variety. Is it true that $\mathrm{Aut}(X)$ is Jordan?

Example 2.15. $\mathrm{Aut}(\mathbf{A}^n)$ is Jordan for $n \leqslant 2$. For $n = 1$ this is clear, for $n = 2$ follows from Theorem 2.3 and the well-known fact that every finite subgroup of $\mathrm{Aut}(\mathbf{A}^2)$ is linearizable, i.e., conjugate to a subgroup of $\mathbf{GL}_2(k)$ (see also Section 2.2 below).

Theorem 2.16. *The automorphism group of every irreducible toral variety* (*see Definition* 1.13) *is Jordan.*

PROOF. By [34], for any irreducible variety X, the abelian group

$$\Gamma := k[X]^*/k^*$$

is free and of finite rank. Let X be toral and let H be the kernel of the natural action of $\mathrm{Aut}(X)$ on Γ. We claim that H is abelian. Indeed, for every element $f \in k[X]^*$, the line spanned by f in $k[X]$ is H-stable. Since $\mathbf{GL}_1$ is abelian, this yields that

$$h_1h_2 \cdot f = h_2h_1 \cdot f \quad \text{for any elements} \quad h_1, h_2 \in H. \tag{2.11}$$

As X is toral, $k[X]^*$ generates the k-algebra $k[X]$ by Lemma 1.14. Hence (2.11) holds for every $f \in k[X]$. Since X is affine, the automorphisms of X coincide if and only if they induce the same automorphisms of $k[X]$. Whence H is abelian, as claimed.

Let n be the rank of Γ. Then $\mathrm{Aut}(\Gamma)$ is isomorphic to $\mathbf{GL}_n(\mathbf{Z})$. By the definition of H, the natural action of $\mathrm{Aut}(X)$ on Γ induces an embedding of $\mathrm{Aut}(X)/H$ into $\mathrm{Aut}(\Gamma)$. Hence $\mathrm{Aut}(X)/H$ is isomorphic to a subgroup of $\mathbf{GL}_n(\mathbf{Z})$. Example 2.10(2) then implies that $\mathrm{Aut}(X)/H$ is bounded. Thus, $\mathrm{Aut}(X)$ is an extension of a bounded group by an abelian group, hence Jordan by Lemma 2.11. This completes the proof. □

Remark 2.17. Maintain the notation of the proof of Theorem 2.16. Let $f_1, \ldots f_n$ be a basis of Γ. There are the homomorphisms $\lambda_i \colon H \to k^*$, $i = 1, \ldots, n$, such that $g \cdot f_i = \lambda(g) f_i$ for every $g \in H$ and i. Since $k[X]^*$ generates $k[X]$, the diagonal map $H \to (k^*)^n$, $h \mapsto \big(\lambda_1(g), \ldots, \lambda_n(g)\big)$, is injective. This and the proof of Theorem 2.16 show that the automorphism group of X is an extension of a subgroup of $\mathbf{GL}_n(\mathbf{Z})$ by a subgroup of the torus $(k^*)^n$.

The following lemma is well-known (see, e.g., [12, Lemma 2.7(b)]).

Lemma 2.18. *Let X be a variety and let G be a reductive algebraic subgroup of* $\mathrm{Aut}(X)$. *Let $x \in X$ be a fixed point of G. Then the kernel of the induced action of G on $\mathrm{T}_{x,X}$ is trivial.*

Theorem 2.19. *Let $\sim$ be the equivalence relation on the set of points of a variety X defined by*

$$x \sim y \iff \textit{the local rings of } X \textit{ at } x \textit{ and } y \textit{ are } k\textit{-isomorphic.}$$

If there is a finite equivalence class of $\sim$, then $\mathrm{Aut}(X)$ *is Jordan.*

PROOF. Every equivalence class of $\sim$ is $\mathrm{Aut}(X)$-stable. Let C be a finite equivalence class of $\sim$ and let G be the kernel of the action of $\mathrm{Aut}(X)$ on C. Then G is a normal subgroup of finite index in $\mathrm{Aut}(X)$. By Lemma 2.11 it suffices to prove that G is Jordan.

Let K be a finite subgroup of G and let x be a point of C. As x is fixed by K, the action of K on X induces an action of K on $\mathrm{T}_{x,X}$. The latter is linear and hence determined by a homomorphism $\tau\colon K \to \mathbf{GL}(\mathrm{T}_{x,X})$. Being finite, K is reductive. Hence τ is injective by Lemma 2.18. Theorem 2.3 then yields that K contains an abelian normal subgroup A such that $[K{:}A] \leqslant J_{\mathbf{GL}_n(k)}$, $n := \dim \mathrm{T}_{x,X}$. This completes the proof. □

Given a variety X, we say that its point x is a *vertex* of X if

$$\dim \mathrm{T}_{x,X} \geqslant \dim \mathrm{T}_{y,X} \quad \text{for every point } y \in X.$$

Clearly, an irreducible X is smooth if and only if every its point is a vertex.

Corollary 2.20. *The automorphism group of every variety with only finitely many vertices is Jordan.*

Corollary 2.21. *Let $\approx$ be the equivalence relation on the set of points of a variety X defined by*

$$x \approx y \iff \textit{the tangent cones of } X \textit{ at } x \textit{ and } y \textit{ are isomorphic.}$$

If there is a finite equivalence class of $\approx$, then $\mathrm{Aut}(X)$ *is Jordan.*

Corollary 2.22. *The automorphism group of every nonsmooth variety with only finitely many singular points is Jordan.*

Corollary 2.23. *Let $\widehat{X} \subset k^{n+1}$ be the affine cone of a smooth closed proper subvariety X in $\mathbf{P}^n = \mathbf{P}(k^{n+1})$ that does not lie in any hyperplane. Then* $\mathrm{Aut}(\widehat{X})$ *is Jordan.*

PROOF. The assumptions imply that the singular locus of $\widehat{X}$ consists of a single point, the origin; whence the claim by Corollary 2.22. □

Remark 2.24. Smoothness in Corollary 2.23 may be replaced by the assumption that X is not a cone. Indeed, in this case the origin constitutes a single equivalence class of $\approx$ for points of $\widehat{X}$; whence the claim by Corollary 2.21.

Theorem 2.25. *For every variety X, every finite subgroup G of* $\mathrm{Aut}(X)$ *such that $X^G \neq \varnothing$ contains an abelian normal subgroup whose index in G is at most $J_{\mathbf{GL}_d(k)}$ where $d = \max_x \dim \mathrm{T}_{x,X}$.*

PROOF. Like in the above proof of Theorem 2.19, this follows from Lemma 2.18 and Theorem 2.3. □

Corollary 2.26. *Let p be a prime number. Then every finite p-subgroup G of* $\mathrm{Aut}(\mathbf{A}^n)$ *contains an abelian normal subgroup whose index in G is at most $J_{\mathbf{GL}_n(k)}$.*

PROOF. This follows from Theorem 2.25 since in this case $(\mathbf{A}^n)^G \neq \varnothing$, see [37, Theorem 1.2]. □

Remark 2.27. To date, it is not known whether or not $(\mathbf{A}^n)^G \neq \varnothing$ for every finite subgroup G of $\mathrm{Aut}(\mathbf{A}^n)$. By Theorem 2.25 the affirmative answer would imply that $\mathrm{Aut}(\mathbf{A}^n)$ is Jordan.

Remark 2.28. The statement of Corollary 2.26 remains true if $\mathbf{A}^n$ is replaced by any p-acyclic variety X and n in $J_{\mathbf{GL}_n(k)}$ by $\max_x \dim \mathrm{T}_{x,X}$. This is because in this case $X^G \neq \varnothing$ for every finite p-subgroup G of $\mathrm{Aut}(X)$, see [37, Sections 7–8].

Theorem 2.29. *For every variety X, there is an integer m_X such that any finite subgroup G of any connected linear algebraic subgroup L of* $\mathrm{Aut}(X)$ *contains an abelian normal subgroup whose index in G is at most m_X.*

PROOF. Being reductive, G is contained in a maximal reductive subgroup R of L. Then R is a Levi subgroup, i.e., L is a semidirect product of R and $\mathrm{Rad}_u L$, cf., e.g., [25, Chaptre 6]. As L is connected, R is connected as well. Since the kernel of the action of R on X is trivial, $\mathrm{rk}\, R \leqslant \dim X$, see [27, Section 3]. The claim then follows from Theorem 2.5 as there are only finitely many connected reductive groups of rank at most $\dim X$. □

2.2. Generalizations. One may ask whether "affine" in Question 2.14 can be dropped:

Question 2.30. Is there an irreducible variety X such that $\mathrm{Aut}(X)$ is not Jordan?

The negative answer to Question 2.30 would follow from that to

Question 2.31. Is there an irreducible variety X such that $\mathrm{Bir}(X)$ is not Jordan?

In Theorem 2.32 below we answer Question 2.31 for curves and surfaces.

Curves. If X is a curve, then the answer to Question 2.31 is negative.

Proving this we may assume that X is smooth and projective. Then $\mathrm{Bir}(X) = \mathrm{Aut}(X)$.

If $g(X)$, the genus of X, is 0, then $X = \mathbf{P}^1$, hence $\mathrm{Bir}(X) = \mathbf{PGL}_2(k)$, so $\mathrm{Bir}(X)$ is Jordan by Theorem 2.5.

If $g(X) = 1$, then X is an elliptic curve; whence $\mathrm{Bir}(X)$ is the extension of a finite group by the abelian algebraic group X, hence Jordan by Lemma 2.11.

If $g(X) \geqslant 2$, then $\mathrm{Bir}(X)$ is finite, hence Jordan.

Note that all curves (not necessarily smooth and projective) with infinite automorphism group are classified in [28].

Surfaces. Answering Question 2.31 for surfaces X, we may assume that X is a smooth projective minimal model.

If X is of general type, then by Matsumura's theorem $\mathrm{Bir}(X)$ is finite, hence Jordan.

If X is rational, then $\mathrm{Bir}(X)$ is the planar Cremona group over k, hence Jordan by [35, Theorem 5.3; 36, Théorème 3.1].

If X is a nonrational ruled surface, it is birationally isomorphic to $\mathbf{P}^1 \times B$ where B is a smooth projective curve such that $g(B) > 0$; we may then take $X = \mathbf{P}^1 \times B$. As $g(B) > 0$, there are no dominant rational maps $\mathbf{P}^1 \dashrightarrow B$, hence the elements of $\mathrm{Bir}(X)$ permute the fibers of the natural projection $\mathbf{P}^1 \times B \to B$. The set of elements inducing trivial permutation is a normal subgroup $\mathrm{Bir}_B(X)$ of $\mathrm{Bir}(X)$. The definition implies that $\mathrm{Bir}_B(X) = \mathbf{PGL}_2(k(B))$, hence Jordan by Theorem 2.5. Naturally identifying $\mathrm{Aut}(B)$ with the subgroup of $\mathrm{Bir}(X)$, we get the decomposition $\mathrm{Bir}(X) = \mathrm{Bir}_B(X) \rtimes \mathrm{Aut}(B)$. Note that $\mathrm{Aut}(X) = \mathbf{PGL}_2(k) \times \mathrm{Aut}(B) \neq \mathrm{Bir}(X)$ (see [21, pp. 98–99]), so $\mathrm{Aut}(X)$ is Jordan by Lemma 2.8. Let $g(B) \geqslant 2$. Then $\mathrm{Aut}(B)$ is finite, hence $[\mathrm{Bir}(X)\colon \mathrm{Bir}_B(X)] < \infty$. Lemma 2.11 then implies that $\mathrm{Bir}(X)$ is Jordan. For $g(B) = 1$, this argument does not work as B is an elliptic curve and hence $\mathrm{Aut}(B)$ is infinite. In fact, by [42], if B is an elliptic curve, then $\mathrm{Bir}(X)$ is *not* Jordan (this dispelled a hope expressed in the earlier preprint of the present paper `arXiv:1001.1311v2`).

The canonical class of all other surfaces X is numerically effective, so, for them, $\mathrm{Bir}(X) = \mathrm{Aut}(X)$, cf. [17, Section 7.1, Theorem 1 and Section 7.3, Theorem 2].

Let X be such a surface. The group $\mathrm{Aut}(X)$ has the structure of a locally algebraic group with finite or countably many components, i.e., there is a normal subgroup $\mathrm{Aut}(X)^0$ in $\mathrm{Aut}(X)$ such that

(i) $\mathrm{Aut}(X)^0$ is a connected algebraic group; and
(ii) $\mathrm{Aut}(X)/\mathrm{Aut}(X)^0$ is either finite or countable group,

see [23]. By (i) and the structure theorem on algebraic groups [1, 33] there is a normal connected linear algebraic subgroup L of $\mathrm{Aut}(X)^0$ such that $\mathrm{Aut}(X)^0/L$ is an abelian variety. By [22, Corollary 1] nontriviality of L would imply that X is ruled. As we assumed that X is not ruled, this means that L is trivial, i.e., $\mathrm{Aut}(X)^0$ is an abelian variety. Hence, $\mathrm{Aut}(X)^0$ is an abelian and, therefore, a Jordan group.

By (i) the group $\mathrm{Aut}(X)^0$ is contained in the kernel of the natural action of $\mathrm{Aut}(X)$ on $H^2(X, \mathbf{Q})$ (we may assume that $k = \mathbf{C}$). Therefore, this action defines a homomorphism $\mathrm{Aut}(X)/\mathrm{Aut}(X)^0 \to \mathbf{GL}\big(H^2(X, \mathbf{Q})\big)$. The kernel of this homomorphism is finite by [9, Proposition 1], and the image is a bounded by Example 2.10(2). By Example 2.10(1), (3) this yields that $\mathrm{Aut}(X)/\mathrm{Aut}(X)^0$ is bounded. In turn, as $\mathrm{Aut}(X)^0$ is Jordan, by Lemma 2.11 this implies that $\mathrm{Aut}(X)$ is Jordan.

This completes the proof of the following

Theorem 2.32. *Let X be an irreducible variety of dimension $\leqslant 2$. Then the following properties are equivalent*:

(a) *the group* $\mathrm{Bir}(X)$ *is Jordan*;
(b) *the variety X is not birationally isomorphic to $\mathbf{P}^1 \times B$, where B is an elliptic curve.*

References

1. I. Barsotti, *Structure theorems for group-varieties*, Ann. Mat. Pura Appl. (4) **38** (1955), 77–119.
2. S. Billey and V. Lakshmibai, *Singular loci of Schubert varieties*, Progr. Math., vol. 182, Birkhäuser, Boston, MA, 2000.
3. A. Borel, *Linear algebraic groups*, 2nd ed., Grad. Texts in Math., vol. 126, Springer, New York, 1991.
4. M. Brion, *Représentations exceptionnelles des groupes semi-simples*, Ann. Sci. École Norm. Sup. (4) **18** (1985), no. 2, 345–387.
5. A. Crachiola and S. Maubach, *The Derksen invariant vs. the Makar-Limanov invariant*, Proc. Amer. Math. Soc. **131** (2003), no. 11, 3365–3369 (electronic).
6. C. W. Curtis and I. Reiner, *Representation theory of finite groups and associative algebras*, Pure Appl. Math., vol. 11, Wiley, New York – London, 1962.
7. D. Daigle, *Affine surfaces with trivial Makar-Limanov invariant*, J. Algebra **319** (2008), no. 8, 3100–3111.
8. H. Derksen, *Constructive invariant theory and the linearization problem*, Ph.D. Thesis, Univ. Basel, 1997.
9. I. Dolgachev, *Infinite Coxeter groups and automorphisms of algebraic surfaces*, The Lefschetz Centennial Conference. Part I (Mexico City, 1984) (D. Sundararaman, ed.), Contemp. Math., vol. 58, Amer. Math. Soc., Providence, RI, 1986, pp. 91–106.
10. A. Dubouloz, *Completions of normal affine surfaces with a trivial Makar-Limanov invariant*, Michigan Math. J. **52** (2004), no. 2, 289–308.
11. A. Dubouloz, L. Moser-Jauslin, and P.-M. Poloni, *Inequivalent embeddings of the Koras – Russell cubic 3-fold*, Michigan Math. J. **59** (2010), no. 3, 679–694.

12. H. Flenner and M. Zaidenberg, *Locally nilpotent derivations on affine surfaces with a $\mathbb{C}^*$-action*, Osaka J. Math. **42** (2005), no. 4, 931–974. MR2196000 (2007a:14070)
13. G. Freudenburg, *Algebraic theory of locally nilpotent derivations*, Encyclopaedia Math. Sci., vol. 136, Springer, Berlin, 2006.
14. M. H. Gizatullin, *Affine surfaces that are quasihomogeneous with respect to an algebraic group*, Izv. Akad. Nauk SSSR Ser. Mat. **35** (1971), 738–753 (Russian); English transl., Math. USSR-Izv. **5** (1971), no. 4, 754–769.
15. J. Harris, *Algebraic geometry*, Grad. Texts in Math., vol. 133, Springer, New York, 1992.
16. B. Huppert, *Character theory of finite groups*, de Gruyter Exp. Math., vol. 25, de Gruyter, Berlin, 1998.
17. V. A. Iskovskikh and I. R. Shafarevich, *Algebraic surfaces*, Algebraic Geometry. II, Encyclopaedia Math. Sci., vol. 35, Springer, Berlin, 1996, pp. 127–262.
18. B. Kostant, *Lie group representations on polynomial rings*, Amer. J. Math. **85** (1963), 327–404.
19. S. Lang, *Algebra*, Addison-Wesley, Reading, MA, 1965.
20. A. Liendo, *Affine $\mathbb{T}$-varieties of complexity one and locally nilpotent derivations*, Transform. Groups **15** (2010), no. 2, 389–425.
21. M. Maruyama, *On automorphism groups of ruled surfaces*, J. Math. Kyoto Univ. **11** (1971), 89–112.
22. H. Matsumura, *On algebraic groups of birational transformations*, Atti Accad. Naz. Lincei Rend. Cl. Sci. Fis. Mat. Natur. (8) **34** (1963), 151–155.
23. T. Matsusaka, *Polarized varieties, fields of moduli and generalized Kummer varieties of polarized abelian varieties*, Amer. J. Math. **80** (1958), 45–82.
24. S. Mukai and Y. Namikawa, *Automorphisms of Enriques surfaces which act trivially on the cohomology groups*, Invent. Math. **77** (1984), no. 3, 383–397.
25. A. L. Onishchik and E. B. Vinberg, *Lie groups and algebraic groups*, Springer Ser. Soviet Math., Springer, Berlin, 1990.
26. V. L. Popov, *Classification of affine algebraic surfaces that are quasihomogeneous with respect to an algebraic group*, Izv. Akad. Nauk SSSR Ser. Mat. **37** (1973), 1038–1055 (Russian); English transl., Math. USSR-Izv. **7** (1973), no. 5, 1039–1055.
27. ———, *Classification of three-dimensional affine algebraic varieties that are quasihomogeneous with respect to an algebraic group*, Izv. Akad. Nauk SSSR Ser. Mat. **39** (1975), no. 3, 566–609 (Russian); English transl., Math. USSR-Izv. **9** (1975), no. 3, 535–576.
28. ———, *Algebraic curves with an infinite automorphism group*, Mat. Zametki **23** (1978), no. 2, 183–195 (Russian); English transl., Math. Notes **23** (1978), no. 1-2, 102–108.
29. V. Popov, *Sections in invariant theory*, The Sophus Lie Memorial Conference (Oslo, 1992) (O. A. Laudal and B. Jahren, eds.), Scand. Univ. Press, Oslo, 1994, pp. 315–361.
30. V. L. Popov, *Algebraic cones*, Mat. Zametki **86** (2009), no. 6, 947–949 (Russian); English transl., Math. Notes **86** (2009), no. 6, 892–894.
31. V. L. Popov and E. B. Vinberg, *Invariant theory*, Algebraic Geometry. IV (A. N. Parshin and I. R. Shafarevich, eds.), Encyclopaedia Math. Sci., vol. 55, Springer, Berlin, 1994, pp. 123–284.
32. C. Procesi, *Lie groups*, Universitext, Springer, New York, 2007.
33. M. Rosenlicht, *Some basic theorems on algebraic groups*, Amer. J. Math. **78** (1956), 401–443.
34. ———, *Some rationality questions on algebraic groups*, Ann. Mat. Pura Appl. (4) **43** (1957), 25–50.
35. J.-P. Serre, *A Minkowski-style bound for the orders of the finite subgroups of the Cremona group of rank 2 over an arbitrary field*, Mosc. Math. J. **9** (2009), no. 1, 193–208.
36. ———, *Le groupe de Cremona et ses sous-groupes finis*, Séminaire Bourbaki. Volume 2008/2009. Exposés 997–1011, Astérisque, vol. 332, Exp. No. 1000, 2010, pp. 75–100.
37. ———, *How to use finite fields for problems concerning infinite fields*, Arithmetic, Geometry, Cryptography and Coding Theory (Marseilles, 2007) (G. Lachaud, C. Ritzenthaler, and M. A. Tsfasman, eds.), Contemp. Math., vol. 487, Amer. Math. Soc., Providence, RI, 2009, pp. 183–193.
38. I. R. Shafarevich, *The Lüroth problem*, Trudy Mat. Inst. Steklov. **183** (1990), 199–204, 229 (Russian); English transl., Proc. Steklov Inst. Math. **1991** (1991), no. 4, 241–246.
39. T. A. Springer, *Linear algebraic groups*, 2nd ed., Progr. Math., vol. 9, Birkhäuser, Boston, MA, 1998.

40. J. H. Sampson and G. Washnitzer, *A Künneth formula for coherent algebraic sheaves*, Illinois J. Math. **3** (1959), 389–402.
41. E. B. Vinberg and V. L. Popov, *A certain class of quasihomogeneous affine varieties*, Izv. Akad. Nauk SSSR Ser. Mat. **36** (1972), 749–764 (Russian); English transl., Math. USSR-Izv. **6** (1972), 743–758.
42. Yu. G. Zarhin, *Theta groups and products of abelian and rational varieties*, available at `arXiv:1006.1112`.

Steklov Mathematical Institute, Russian Academy of Sciences, Gubkina 8, Moscow 119991, Russia

E-mail address: `popovvl@mi.ras.ru`

Centre de Recherches Mathématiques
CRM Proceedings and Lecture Notes
Volume **54**, 2011

Embeddings of Hyperbolas

Avinash Sathaye

1. Introduction

Let k be an algebraically closed field of characteristic zero. By an abstract hyperbola, we mean a plane curve defined by $f(X,Y)=0$ biregular to the usual hyperbola. This means that its coordinate ring $k[X,Y]/(f)$ is isomorphic to $k[X,Y]/(XY-1) \approx k[t,t^{-1}]$.

A more common term for an abstract hyperbola is $\mathbb{C}^*$ since when $k=\mathbb{C}$ the field of complex numbers, a hyperbola corresponds to an affine line $\mathbb{C}$ minus one point.

A lot of work has been done on determining the isomorphism types of the equations of abstract hyperbolas. A paper by Borodzik and Zoladek [5] promises a complete classification of all abstract hyperbolas subject to final verification of some plausible but unproved "local estimates." Their paper also includes families of embeddings of $\mathbb{C}^*$ into the plane which may not be biregular (and hence possibly singular).

Cassou-Noguès, Koras and Russell are in the process of verifying the various families using powerful results in the surface theory and affine algebraic geometry [6].

The known different families have been discovered by various mathematicians including Kaliman [7], Russell, Cassou-Noguès and so on with the latest comprehensive list by Borodzik and Zoladek.

All these families of hyperbolas have two interesting features, which we now describe.

Recall that an asymptote of an affine plane curve is defined as a line in the affine plane which is tangent to the curve at a point at infinity. We know that the "usual" hyperbola $XY-1$ has two distinct asymptotes, namely the X and the Y axes. It is also easy to find plane curves without any asymptotes (any curve with only one place at infinity, for example) and by a non linear change of coordinates, any curve can be transformed into one without any asymptotes.[1]

2010 *Mathematics Subject Classification.* Primary 14H50, 14R05; Secondary 14H45.
This is the final form of the paper.

[1]To see this, suppose that a curve $f(X,Y)$ has been prepared to be a monic polynomial in X as well as Y with its highest degree form as Y^n where n is the total degree of f.

Then clearly, the only possible asymptote is a line of the form $Y+c$. However, since f is also monic in X of some degree $m<n$, we see that all such lines have the same total intersection

The two features of hyperbolas are as follows.

• All the known abstract hyperbolas can be arranged to have at least one asymptote after a suitable possibly nonlinear change of coordinates.

There are examples which have, unlike the usual hyperbolas, only one asymptote.

For example the simplest version of the Kaliman examples discussed at the end:

$$f(X,Y) = (XY-1)^2 = Y^5 \quad \text{or parametrically: } X = t^3 + t^{-2},\ Y = t^2.$$

has only $Y = 0$ as an asymptote.

• For convenience, let us call an asymptote to be "very good " if it does not meet the curve at finite distance (as in the case of an ordinary hyperbola), just "good," if it meets just at one point with intersection multiplicity 1 and "bad" otherwise.

If we believe the current known classification to be complete, then there are only finitely many examples together with one infinite family of hyperbolas without a good asymptote and even these seem *not to meet the affine part of the curve more than twice.*

The aim of this paper is to present some relatively simpler calculations in support of the proposed classification of abstract hyperbolas.

In Section 2, we discuss some generalities about abstract hyperbolas.

In Section 3, we prove a simple but useful criterion for an irreducible plane curve to be an abstract hyperbola. Theorem 3.1. Then we use the criterion to give a different verification for the known families.

In Section 4, we review necessary parts of the Abhyankar – Moh theory about places at infinity. We discuss the notion of Abhyankar – Moh data near any given CD data and prove its existence in Theorem 4.1 (Section 4.4).

In Section 5, we develop a useful intersection formula for the discriminant of a curve with two places at infinity.

In Section 6, we discuss uses of our criterion of a hyperbola with an asymptote.

Finally, we prove the structure of the families with a very good asymptote (one of the families described by Kaliman) Theorem 6.1 (Section 6.3).

In Section 7, we present some thoughts on a possible proof of the full classification based on application of the criterion.

2. Rational curves with two places at infinity

We will systematically use the notation θ (*Abhyankar's nonzero*) to denote any nonzero element in our ground field k. This notation allows us to use the same symbol multiply without forcing the different values to be equal.

Under our assumption of an algebraically closed ground field k of characteristic zero, we can assume that a rational irreducible curve $f(X,Y)$ with two places at

---

multiplicity m with f at finite distance. If any of them were to be tangent at infinity, its finite intersection has to be smaller than the general intersection m. This shows that none of them are tangents to the curve at infinity.

We can arrange any f to have such preparation by first changing variables to make it monic in X of degree m and then a further change of the form $Y' = Y$, $X' = X + Y^d$ for a sufficiently large $d > 0$ will clearly make its highest degree form as Y^n with $n = dm$.

In case of a curve with one place of infinity, it is well known that the curve is automatically prepared after a possible linear change of variables to insure that X and Y have distinct orders at the unique place at infinity.

infinity may be assumed to have a parametrization by two meromorphic polynomials as:

$$X = p = p(t), Y = q = q(t), \quad \text{with } p, q \in k[t, t^{-1}] \text{ such that } f(p, q) = 0$$

and moreover, $k(p, q) = k(t)$.

Let $A = k[t, t^{-1}]$.

Moreover, the condition that the curve has two places at infinity means that p, q cannot be both contained in $k[t]$ or $k[t^{-1}]$.

Note that this curve f is an abstract hyperbola, if further it satisfies the condition $k[p, q] = A$.

The two places at infinity for the curve are simply the t-adic and the t^{-1}-adic valuations of $k(t)$ and we shall respectively call these v_0 and v_∞.

Thus, we note that

$$v_0\big(h(t)\big) = \operatorname{ord}_t\big(h(t)\big) \quad \text{and} \quad v_\infty\big(h(t)\big) = -\deg_t\big(h(t)\big) \qquad \text{for } 0 \neq h(t) \in A.$$

Extended notation. Having fixed a parametrization (p, q) of f, for a polynomial $g = g(X, Y)$ we will write $v_0(g)$ for $v_0\big(g(p, q)\big)$ and similarly $v_\infty(g)$ for $v_\infty\big(g(p, q)\big)$ provided $g(p, q) \neq 0$.

2.1. Asymptote. Now we analyze the concept of an asymptote for such a curve $f(X, Y) = 0$. Let a line $g(X, Y) = aX + bY + c$ be an asymptote to it for the t-adic branch.

By a linear change of variables, if necessary, we *may arrange that* $v_0(1), v_0(X)$, $v_0(Y)$ are all distinct.

Then we see that

$$v_0(g) = \max\{v_0(1), v_0(X), v_0(Y)\},$$

i.e., it has a maximum v_0-order among all lines. Since $\{p, q\}$ $//subsetk[t^{-1}]$, the maximum is clearly positive. Hence $v_0(g) > 0$ and hence $g(p, q) \in k[t] \setminus k$.

Similarly, if g is an asymptote for the t^{-1}-adic branch, then we deduce that $g(p, q) \in k[t^{-1}]$.

Thus, we have the conclusion that the line Y is an asymptote to f if either $q \in k[t] \setminus k$ or $q \in k[t^{-1}] \setminus k$.

3. Hyperbola criterion

Theorem 3.1. *Assume that* $f(X, Y)$ *is* irreducible *and* monic in **Y** of degree **N**. *Let* $\operatorname{Disc}_{f,Y}(X)$ *be its discriminant with respect to* Y, *i.e., the* Y*-resultant of* $f(X, Y)$ *and* $f_Y(X, Y)$.

Then f *is an abstract hyperbola if and only if* $\deg_X(\operatorname{Disc}_{f,Y})(X) = N$.

Comment. It is enough to assume $f(X, Y)$ to be submonic, i.e., the highest degree coefficient of Y is a nonzero constant, since we can then modify $f(X, Y)$ by a nonzero constant multiple to make it monic.

PROOF. We recall the rank of a reduced plane curve $f(X, Y)$ given by :

$$\rho(f) = 2P_g(f) + (\nu_\infty(f) - 1) + \sum_{P \text{ on } f} (\nu_P(f) - 1) \qquad \text{(Rank formula 1)}$$

where $P_g(f)$ is the usual geometric genus of the reduced curve f, $\nu_\infty(f)$ is the number of places at infinity for f and $\nu_P(f)$ gives the number of branches of f at various (affine) points P on f.

It is obvious that $\rho(f) = 1$ if f is an abstract hyperbola since its geometric genus is 0, it has two places at infinity and has single branches at various finite points, since they are all simple points.

Now we invoke an equivalent formula to compute this rank with the assumption of monicness in Y:

$$\rho(f) = 1 - N + \langle f, f_Y \rangle - \langle f_X, f_Y \rangle_f \qquad \text{(Rank formula 2)}$$

A detailed analysis of the two versions of this formulas appears in [3, Section 9 p. 92]. Please note that in that paper, our ρ is denoted as ρ_a for technical reasons and is shown to be equivalent to it in (9.1), (9.2).

Here, the quantity $\langle f, f_Y \rangle$ measures the total intersection multiplicity of f, f_Y at finite distance and due to the monicness it is easily seen to be equal to $\deg_X(\mathrm{Disc}_{f,Y}(X))$.

The last term $\langle f_X, f_Y \rangle_f$ denotes the total intersection multiplicity of f_X, f_Y at points of f. Clearly this is zero for nonsingular curves by the Jacobian criterion of nonsingularity.

Thus, if f is an abstract hyperbola, then this equation reduces to

$$1 = 1 - N + \deg_X(\mathrm{Disc}_{f,Y}(X)) - \langle f_X, f_Y \rangle_f = 1 - N + \deg_X(\mathrm{Disc}_{f,Y}(X)).$$

The last term is dropped since our hyperbola is nonsingular.

This proves that

$$\deg_X(\mathrm{Disc}_{f,Y}(X)) = N$$

as required.

Now we prove the converse. Thus, we assume that $\deg_X(\mathrm{Disc}_{f,Y}(X)) = N$ for a monic irreducible curve f and prove that it is an abstract hyperbola.

The rank formula 2 for our curve now reduces to:

$$\rho(f) = 1 - N + N - \langle f_X, f_Y \rangle_f = 1 - \langle f_X, f_Y \rangle_f.$$

Since the curve is irreducible, the rank formula 1 clearly says that the rank is nonnegative and hence we must have $\langle f_X, f_Y \rangle_f \leq 1$.

We consider the two possible cases:

- If $\langle f_X, f_Y \rangle_f = 0$, then our curve is non singular and has rank 1. From the rank formula 1, it follows that there are no contributions from finite points and hence

$$\rho(f) = 1 = 2P_g(f) + (\nu_\infty(f) - 1)$$

and since both terms on the right-hand side are nonnegative, clearly the only choice is

$$P_g(f) = 0 \quad \text{and} \quad \nu_\infty(f) = 2.$$

Thus, we have an abstract hyperbola as claimed.

- If $\langle f_X, f_Y \rangle_f = 1$, then we have a singular curve f with $\rho(f) = 0$. Fortunately, we have the famous Lin – Zaidenberg theorem [9] which classifies all such curves as rational curves with one place at infinity and equivalent to a curve $X^m - Y^n = 1$ by an automorphism. All these curves have a single singular point with a single branch (a cusp) and for any cusp, its contribution to $\langle f_X, f_Y \rangle_f$ is at least 2.

Hence we can rule this case out!

This proves our criterion. □

Comments. (1) The criterion is quite useful, especially if a computer is used to test a curve to be an abstract hyperbola.

(2) The irreducibility hypothesis is essential. To see this, consider the example $(Y^2 + X^2)^2 - 1$. Its Y-discriminant has X-degree 4 equal to its Y-degree, but the curve is a pair of circles not meeting at finite distance.

4. Extended Abhyankar – Moh theory

We recall necessary parts of the Abhyankar – Moh theory of places at infinity and present some generalizations to two places.

4.1. Branches and valuations.

1. By a branch of a plane curve defined by $f(X, Y) = 0$ at infinity, we mean a non constant fractional power series $\psi^*(X)$ in X^{-1} such that $f\big(X, \psi^*(X)\big) = 0$. Moreover, the fractional power series is assumed to have a bounded denominator, meaning, it can be assumed to be a power series with integral powers of $X^{-1/d}$ for some integer d.

2. If we set $t = X^{-1/d}$, then we can reformulate the description of a branch as $(-d, \psi(t))$ where d is a positive integer and $\psi(t) \in k(\!(t)\!)$ is a power series with integral exponents such that $f\big(t^{-d}, \psi(t)\big) = 0$.

We further *normalize the expression* by requiring that the GCD of d and the support of the meromorphic power series $\psi(t)$ is 1. Two such normalized branches $\big(d, \psi(t)\big)$ and $\big(d', \psi'(t)\big)$ are said to be equivalent if $d = d'$ and $\psi'(t) = \psi(\omega t)$ for some ω such that $\omega^d = 1$.

3. There is no reason to restrict to infinity. The branches of a curve at finite distance can be similarly defined, except the fractional power series is assumed to have non negative exponents and hence the reformulation becomes a pair $\big(d, \psi(t)\big)$ with $f\big(t^d, \psi(t)\big) = 0$ where $d > 0$ and $\psi(t) \in k[\![t]\!]$.

4. Associated with any branch $\big(-d, \psi(t)\big)$ at infinity is a valuation defined thus: Given any rational function $h(X, Y)$ which has no factors in common with $f(X, Y)$, we define the order $v\big(h(X, Y)\big) = \mathrm{ord}_t\left(h\big(t^{-d}, \psi(t)\big)\right)$. It is not hard to see that this valuation is *independent of which normalized expression* is used.

5. A similar valuation is defined for a finite branch except we evaluate $v\big(h(X, Y)\big) = \mathrm{ord}_t\big(h\big(t^d, \psi(t)\big)\big)$ where $\psi(t) \in k[\![t]\!]$.

4.2. Characteristic data.

1. Given a branch $\big(-d, \psi(t)\big)$ at infinity, we define its characteristic data as per the Abhyankar – Moh theory thus:

$$\begin{pmatrix} -d_1 & q_1 & q_2 & \dots & q_h \\ & d_1 & d_2 & \dots & d_h \end{pmatrix}$$

where $q_1 = \mathrm{ord}_t\big(\psi(t)\big)$ and $d_2 = \mathrm{GCD}(q_1, d_1)$. Moreover, if the data is inductively defined through q_i, d_i then we define $d_{i+1} = \mathrm{GCD}(q_i, d_i)$ and $q_1 + \cdots + q_i + q_{i+1}$ equals the first member of the support of $\psi(t)$ which is not divisible by d_{i+1}.

Finally, the process stops at h when $d_{h+1} = \mathrm{GCD}(q_h, d_h)$ divides all members of the support of $\psi(t)$, i.e., q_{h+1} if formally calculated, would be ∞. In the Abhyankar – Moh theory it is customary to add this last pair (∞, d_{h+1}) for the sake of completeness.

Note that our normalization implies $d_{h+1} = 1$ and thus we have no further GCD-drop possible and our data is terminated. We often describe this by saying that the characteristic data has h characteristic pairs $(q_1, d_1), \dots, (q_h, d_h)$.

Formally, this makes $q_{h+1} = \infty$ and signals the end. We have simply chosen not to include this termination pair $(\infty, 1)$.

For technical convenience, it is useful to allow non normalized data in calculations which simply correspond to positive integral multiples of a normalized data.

2. Notation. We may write the characteristic data in a compact form as a sequence of characteristic pairs

$$(q_1, d_1), (q_2, d_2), \dots, (q_h, d_h)$$

where the initial $-d_1$ is deduced for a branch at infinity.

3. As before, the definitions can be extended to finite branches, but we do not need them here.

4. Several auxiliary expressions and notations are introduced for ease of calculations based on this characteristic data. These are:

$$s_i = \sum_{j=1}^{i}(q_j d_j), \quad r_i = \frac{s_i}{d_i}, \quad \bar{s}_i = \sum_{j=2}^{i} q_j(d_j - 1).$$

We note that $\bar{s}_i$ is not a usual formula, but we shall find it useful for our work below. Its summation starts from $j = 2$ and is thus nontrivial only when $h > 1$.

In addition, we define: $n_i = d_i/d_{i+1}$ for $i = 1, \dots, h$ and then note as a consequence we get a new expression $r_i = q_i + n_{i-1}r_{i-1}$.

5. Note that $n_i > 2$ for all $i > 1$. If $n_1 = 1$, then r_0 divides r_1 and this is the so-called principal case. One typically makes a change of variables to avoid this case, if convenient.

6. The key point established by the Abhyankar – Moh theory is a standard generation of the semigroup generated by $\{r_i\}$, namely the fact that

$$n_i r_i = \sum_{j=0}^{i-1} a_{ij} r_j$$

where the integers a_{ij} satisfy $0 \le a_{ij} < n_j$ for $1 \le j < i$ and $0 \le a_{i0}$.

If we drop the nonnegativity of a_{i0}, then this is an obvious fact from number theory, but the positivity comes from the structure of a one place at infinity curve.

4.3. Abhyankar – Moh theory. The main feature of the Abhyankar – Moh theory is the characterization of the characteristic data for the branch at infinity for a curve with one place at infinity.

Thus, assume that we have the following normalized characteristic data for the branch at infinity of such a curve:

$$\mathrm{CD} = \begin{pmatrix} -d_1 & q_1 & q_2 & \cdots & q_h \\ & d_1 & d_2 & \cdots & d_h \end{pmatrix}$$

or in compact form

$$(q_1, d_1), (q_2, d_2), \dots, (q_h, d_h).$$

Then we have the following properties.

1. There are three important conditions.

(a) We have: $q_1 < 0$ and for $i \ge 2$ all $q_i > 0$.

(b) for $i = 1, \dots, h$, the s_i and hence equivalently r_i are negative.

(c) Finally, $n_i r_i$ is in the semigroup generated by $r_0, r_1, \dots, r_{i-1}$ where, we define $r_0 = -d_1$.

2. The original papers of Abhyankar – Moh established the necessity of these conditions [4]. In [8], I proved that the conditions are sufficient to guarantee the existence of a plane curve with one place (branch) at infinity which has the CD data. There, the semigroup generated by such $\{r_i\}$ was shown to have many interesting properties and was also characterized by these conditions.

3. It is possible to show that the last two conditions simply force an upper bound for values of q_i for $i \geq 2$.

4. Let us call a characteristic data "an Abhyankar – Moh data," if it satisfies the above three conditions.

5. If a certain characteristic data is not an Abhyankar – Moh data, it can still correspond to a branch at infinity for some curve with multiple branches at infinity.

4.4. Nearby Abhyankar – Moh data. We now introduce the concept of an Abhyankar – Moh data near any given CD-data.

For a characteristic data CD and an integer λ, consider a shifted data CD_λ obtained by replacing the q_1 in CD by $q_1 - \lambda d_1$ and leaving other quantities unchanged. This has the effect of replacing the first term of the power series $\psi(t)$ by a much smaller exponent term but it leaves d_2 with the same value and hence the rest of the data is unaffected. Indeed, we are simply replacing the branch $(-d_1, \psi(t))$ by a new branch $(-d_1, t^{-\lambda d_1}\psi(t))$.

Theorem 4.1. *Given any normalized characteristic data* CD *at infinity the shifted data* CD_λ *is Abhyankar – Moh for sufficiently large values of* λ.

PROOF. We assume the above notation for CD.

1. Let the r-sequence for the data CD_λ be denoted by r_i^* for $1 \leq i \leq h$.

We note that $r_0^* = r_0 < 0$ and we have $r_1^* = r_1 + \lambda r_0$. Inductively, we argue that $r_i^* = r_i + \lambda(d_1/d_i)r_0^*$. This is clear for $i = 1$ and for $i > 1$ we note

$$r_i^* = n_{i-1}r_{i-1}^* + q_i = n_{i-1}r_i + q_i + \lambda n_{i-1}\frac{d_1}{d_{i-1}}r_0^* = r_i + \lambda\frac{d_1}{d_i}r_0^*.$$

Thus, it is clear that since $r_0 < 0$, we get the negativity of r_i for large enough λ.

2. Now we show that the third (semigroup) condition also holds for large enough λ. For this, we write the standard relations:

$$n_i r_i = \sum_{j=1}^{i-1} a_{ij} r_j + a_{i0} r_0$$

where we assume $0 \leq a_{ij} < n_j$ are integers and a_{i0} is an integer which may be negative. We wish to transform this relation for the r_i^* and argue that its coefficient for r_0^* is non negative for large enough λ.

We get the relation:

$$n_i r_i^* = \sum_{j=1}^{i-1} a_{ij} r_j^* + a_{i0} r_0 + \lambda r_0^* \left(n_i \frac{d_1}{d_i} - \sum_{j=1}^{i-1} a_{ij} \frac{d_1}{d_j} \right).$$

Consider the coefficient of λr_0^*:

$$d_1 \left(\frac{n_i}{d_i} - \sum_{j=1}^{i-1} \frac{a_{ij}}{d_j} \right)$$

If we show this expression to be positive, then for sufficiently large λ the coefficient for r_0^* in the expression for r_i^* will be positive and the characteristic data CD_λ will satisfy the semigroup condition.

The following lemma helps us prove this.

3. Lemma. *We claim*:

$$\sum_{j=1}^{p-1} \frac{a_{ij}}{d_j} \leq \frac{\theta_p}{d_p}$$

where θ_p is a rational number less than 2.

PROOF. We use induction on p. For $p = 2$, we have

$$\frac{a_{i1}}{d_1} = \frac{a_{i1}}{n_1} \cdot \frac{1}{d_2}$$

and the claim is verified by taking $\theta_2 = a_{i1}/n_1 < 1$.

Assuming the result true for p, we deduce it for $p+1$. Thus, using induction hypothesis we get:

$$\sum_{j=1}^{p} \frac{a_{ij}}{d_j} = \sum_{j=1}^{p-1} \frac{a_{ij}}{d_j} + \frac{a_{ip}}{d_p} < \frac{\theta_p}{d_p} + \frac{a_{ip}}{d_p}.$$

The last term can be rearranged as:

$$\frac{(\theta_p/n_p + a_{ip}/n_p)}{d_{p+1}}.$$

Since p is at least 2, we have $n_p \geq 2$ and hence the numerator is at most $\theta_p/2 + 1$ and our result follows by taking $\theta_{p+1} = \theta_p/2 + 1$. □

4. Using the lemma, we get that

$$\frac{n_i}{d_i} - \sum_{j=1}^{i-1} \frac{a_{ij}}{d_j} > \frac{n_i}{d_i} - \frac{2}{d_i} \geq 0$$

since we only need this when $i \geq 2$ and hence $n_i \geq 2$.

5. If the original CD is obtained from a branch $\bigl(-d, \psi(t)\bigr)$ so that $f\bigl(t^{-d}, \psi(t)\bigr) = 0$, then this may be said to correspond to the substitution $X = t^{-d}$, $Y = \psi(t)$.

We note that the CD_λ then corresponds to a substitution $X = t^{-d}$, $Y = t^{-d\lambda}\psi(t)$ which may be thought of as belonging to the curve $g(X,Y) = 0$, where $g(X,Y)$ is the numerator of $f(X, Y/X^\lambda)$. This can be interpreted as a related curve obtained by a birational transformation.

6. Comment. It would be interesting to study the closest "one place at infinity curve" given by a minimal value of λ and compare its properties with the given curve. □

5. An intersection formula

We now consider the case of a plane curve with two branches at infinity and develop an intersection formula to help us calculate the intersection $\langle f, f_Y\rangle$. This will be used in applying the hyperbola criterion.

1. Let $f(X,Y) = 0$ be an irreducible curve with two places at infinity. We shall assume it to be monic in Y of some degree $N > 1$ and the two branches are given by the two normalized characteristic data

$$\text{CD} = (q_1, d_1), \ldots, (q_h, d_h) \quad \text{and} \quad \text{CD}^* = (q_1^*, d_1^*), \ldots, (q_{h^*}^*, d_{h^*}^*).$$

2. Note that then $d_1 + d_1^* = N$.

3. We are making *a special assumption* that the fractional power series of our branches separate from each other at an exponent present in the support of one and not in the other.

In terms of explicit expansions, this means we assume the existence of fractional expansions for the two branches $\psi(X), \psi^*(X)$ such that $\text{ord}_X\big(\psi(X) - \psi^*(X)\big) = s$ and moreover, the coefficient of X^s is zero in $\psi(X)$ or $\psi^*(X)$.

Note that this assumptions requires us to choose appropriate equivalent expansions for the two branches. It is clear that if we replace the two expansions $\psi(X), \psi^*(X)$ by any equivalent expansions, then the coefficient of X^s will not cancel in the difference $\psi(X) - \psi^*(X)$ and hence the order s is the maximal order of the difference of the two expansions among all equivalent expansions.

If the special assumption does not hold, then we have to carefully calculate the maximal order for the difference of the two expansions among equivalent choices and the formula gets more involved. For our current application, the special assumption will be valid for a trivial reason, hence we make it.

4. Let l be the largest integer so that

$$\frac{q_1}{d_1} = \cdots = \frac{q_{l-1}}{d_{l-1}}$$

and $q_l d_l^* < q_l^* d_l$.

Notation. We shall call l the separation spot for the two branches.

This is guaranteed by the special assumption.

5. Let $\theta = \sum_1^{l-1}(q_i d_i^*) + q_l d_l^*$.

6. Let V and V^* be the respective valuations of the two branches.

7. Then we claim:

$$\text{ord}_V(f_Y) = \theta + \sum_{j=1}^{h} q_j(d_j - 1).$$

Similar formula holds, for the second branch:

$$\text{ord}_{V^*}(f_Y) = \theta + \sum_{j=1}^{h^*} q_j^*(d_j^* - 1).$$

Idea of the proof. The proof is similar to the calculations in the one place case of the Abhyankar – Moh theory. We assume that the two branches are given by fractional power series

$$Y = \eta\big(X^{(-1/d_1)}\big) \quad \text{and} \quad Y = \eta^*\big(X^{(-1/d_1^*)}\big)$$

from which we create the normalized critical data.

In the field of fractional power series with bounded denominators in X^{-1}, we can assume a factorization

$$f(X,Y) = \Bigg(\prod_{\omega^{d_1}=1}\Big(Y - \eta\big(\omega X^{(-1/d_1)}\big)\Big)\Bigg)\Bigg(\prod_{\omega^{*d_2}=1}\Big(Y - \eta^*\big(\omega^* X^{(-1/d_1^*)}\big)\Big)\Bigg).$$

Then the required V-order is obtained by substituting $Y = \eta\big(X^{(-1/d_1)}\big)$ in $f_Y(X,Y)$ and multiplying the X-order by d_1.

It is easy to see that this is the same as

$$\text{substituting } Y = \eta(X^{(-1/d_1)}) \text{ in } \frac{f(X,Y)}{Y - \eta\big(X^{(-1/d_1)}\big)}$$

and multiplying the result by d_1.

The part θ gives the evaluation of the substitution in the second part of $f(X,Y)$ while the remaining expression comes from the substitution in the first part after removal of one linear factor as explained.

The second formula is similarly worked out. □

6. Hyperbolas with asymptotes

We now show how the hyperbola criterion can be used to analyze various families of abstract hyperbolas with asymptotes.

6.1. Notations. First we set up notations.

1. We consider an irreducible monic curve $f(X,Y) = 0$ which is an abstract hyperbola with an asymptote. We assume that this asymptote is $Y = 0$ and we also arrange the equation to be monic in Y so our hyperbola criterion is applicable.

2. From our analysis of an asymptote, this means that we may assume the following parametric form for our curve:

- The parametrization is $X = p(t)$, $Y = q(t)$, with $k[p(t), q(t)] = k[t, t^{-1}]$.
- We have $q(t) = t^m + \cdots + \theta t^n \in k[t]$ So that $m \geq n > 0$.

We may allow $m = n$, but then we assume $q(t) = t^m$.

- $p(t) = t^\alpha + \cdots + \theta t^{-\beta}$, where $\alpha > 0 > -\beta$. We note that $0 > -\beta$ is necessary since otherwise we don't have two places at infinity.

Note that the monicness of $f(X,Y)$ in Y gives an integral equation $f\big(p(t), q(t)\big) = 0$ satisfied by $q(t)$ over $k[p(t)]$. If α were to be non positive, then $p(t) \in k[t^{-1}]$ and then $q(t) \in k[t] \setminus k$ cannot be integral over $k[p(t)]$. Hence our monicness assumption implies $\alpha > 0$.[2]

We note that for $p(t)$ having given description, the ring $k[t, t^{-1}]$ is integral over $k[p(t)]$ since $p(t)$ has negative values at both the valuations at infinity for the ring $k[t, t^{-1}]$. Thus, assumption of monicness for $f(X,Y)$ can be easily satisfied.

3. We now assume the two characteristic data CD for the t^{-1}-adic and CD^* t-adic valuations to have the following characteristic pairs:

$$\text{CD has compact form } (-m, \alpha) = (q_1, d_1), \ldots, (q_h, d_h)$$

and

$$\text{CD}^* \text{ has compact form } (n, \beta) = (q_1^*, d_1^*), \ldots, (q_{h^*}^*, d_{h^*}^*).$$

4. Since we have

$$\frac{q_1}{d_1} = \frac{-m}{\alpha} < \frac{n}{\beta} = \frac{q_1^*}{d_1^*}$$

we conclude that the separation spot l is equal to 1 and $\theta = q_1 d_1^* = -m\beta$.[3]

---

[2]In case $\alpha \leq 0$ to start with, we would make a replacement of X by $X+Y$. This would make $\alpha > 0$ and achieve submonicness as well. Note that in this case we get that $m = \alpha$.

[3]This shows why our special assumption easily holds in our situation and hence we use it.

6.2. Basic identity.

1. Note that we can evaluate the quantity $\langle f, f_Y\rangle$ as the negative of the sum of orders of f_Y at the two branches at infinity. In other words:

$$\langle f, f_Y\rangle = -\operatorname{ord}_V(f_Y) - \operatorname{ord}_{V^*}(f_Y).$$

Using $\theta = -m\beta$, the intersection formula established in Section 5 gives:

$$\begin{aligned}
-\langle f, f_Y\rangle &= \left(-m\beta - m(\alpha-1) + \sum_2^h q_j(d_j-1)\right) \\
&\qquad + \left(-m\beta + n(\beta-1) + \sum_2^{h^*} q_j^*(d_j^*-1)\right) \\
&= -2m\beta - m(\alpha-1) + n(\beta-1) + \bar{s}_h + \bar{s}_{h^*}^*.
\end{aligned}$$

Noticing that $N = \alpha+\beta$, and using our criterion for a hyperbola, we get:

$$-N = -(\alpha+\beta) = -2m\beta - m(\alpha-1) + n(\beta-1) + \bar{s}_h + \bar{s}_{h^*}^*.$$

Rearranging terms, we deduce:

$$2m\beta + m\alpha - m - n\beta + n - \alpha - \beta = \bar{s}_h + \bar{s}_{h^*}^*$$

or

$$(m-1)(\alpha+\beta) + (\beta-1)(m-n) = \bar{s}_h + \bar{s}_{h^*}^*. \qquad \text{(Basic identity)}$$

We shall use this basic identity to analyze the various families of abstract hyperbolas with an asymptote.

6.3. The Kaliman family. Now we apply the above calculations and the basic identity in Section 6.2 to the analysis of the family of hyperbolas discovered by Kaliman. Actually, this is the simpler of the two families in his paper [7] having very good asymptotes.

1. This family has a very good asymptote, so we assume a parametrization

$$X = p(t) = t^\alpha + \cdots + \theta t^{-\beta}, \quad Y = q(t) = t^m.$$

2. Here, $m = n$, so the basic identity reduces to:

$$(\alpha+\beta)(m-1) = \bar{s}_h + \bar{s}_{h^*}^*.$$

3. The characteristic pairs (q_i, d_i) can be read off from a power series expansion obtained by writing

$$p(t) = \tau^{-\alpha}, \quad q(t) = \eta(\tau) = \tau^{-m} + \cdots$$

and then reading off the characteristic terms in the expansion of $\eta(\tau)$.

4. We already have $q(t) = (t^{-1})^{-m}$ and $p(t)$ being a polynomial in t can be considered as a meromorphic power series in t^{-1}. As such its characteristic terms can be read off from the support of $p(t)$.

The inversion formula formula of Abhyankar states that for $i \geq 2$, the characteristic pairs (q_i, d_i) for $\eta(\tau)$ as a meromorphic series in τ and those for $p(t)$ viewed as a meromorphic series in t^{-1} are identical [1].[4]

[4]Abhyankar has a recent sequel *Inversion and invariance of characteristic pairs.* I, in which he generalizes the concepts of characteristic terms and also allows fields of positive characteristic in some calculations. Part II is also in the works.

From this observation, we note that all the characteristic terms appear in the span of $\alpha+\beta$ monomials possible in the expansion of $p(t)$. Hence

$$\sum_2^h q_i \leq \alpha+\beta.$$

5. By a similar analysis for the t-adic expansion, we get that

$$\sum_2^{h^*} q_i^* \leq \alpha+\beta.$$

6. Moreover, since each d_j and d_j^* is a factor of m, we get two conclusions:

- If $j>2$ then $d_j-1<(m-1)/2$ and similarly $d_j^*-1<(m-1)/2$.
- Either $d_2=m$ and hence $d_2-1=m-1$ or $d_2-1<(m-1)/2$.
- Similarly either $d_2^*=m$ and hence $d_2^*-1=m-1$ or $d_2^*-1<(m-1)/2$.

7. We now use the above observations to deduce the following

Theorem 6.1. *For the Kaliman family of hyperbolas, we have exactly one of the following three cases.*

(a) $h^*=1, h=2$ *and* $q_2=\alpha+\beta$ *with* $d_2=m$.
(b) $h=1,\ h^*=2$ *and* $q_2^*=\alpha+\beta$ *with* $d_2^*=m$.
(c) $h=h^*=2$ *and* $q_2+q_2^*=\alpha+\beta$ *with* $d_2=d_2^*=m$.

Proof. We consider various possible values of d_2, d_2^*.

Situation 1. $d_2=m$ and $d_2^*<m$. Since

$$p(t)=t^\alpha+\cdots+\theta t^{(\alpha-q_2)}+\cdots+\theta t^{-\beta}$$

and the support of $p(t)$ between α and $\alpha-q_2$ is divisible by m, we see that the last characteristic term for the t-adic expansion must end at or before $\theta t^{(\alpha-q_2)}$.

Thus, $\sum_2^{h^*} q_j^* \leq (\alpha+\beta-q_2)$.

This coupled with $d_2^*<m$ shows that

$$\bar{s}_{h^*}^* = \sum_2^{h^*} q_j^*(d^*j-1) \leq (\alpha+\beta-q_2)\frac{m-1}{2}.$$

Note that here the equality is possible if and only if $(\alpha+\beta-q_2)=0$ and thus $h^*=1$.

Moreover, we see that

$$\bar{s}_h = q_2(m-1)+\sum_3^h q_j(d_j-1) \leq q_2(m-1)+\left(\sum_3^h (q_j)\right)\frac{m-1}{2}$$
$$= q_2(m-1)+(\alpha+\beta-q_2)\frac{m-1}{2}.$$

Note that here the equality is possible if and only is $h=2$.

Combining the above two calculations, we see that

$$\bar{s}_h+\bar{s}_{h^*}^* \leq q_2(m-1)+(\alpha+\beta-q_2)\frac{m-1}{2}+(\alpha+\beta-q_2)\frac{m-1}{2}$$
$$\leq (\alpha+\beta)(m-1).$$

We conclude that the only possibility for equality is when $q_2=\alpha+\beta$, $h=2$ and $h^*=1$.

Thus this is the case (a) of the statement of the theorem.

SITUATION 2. $d_2 < m$ and $d_2^* = m$. This situation is similar and *we conclude that* $q_2^* = \alpha + \beta$, $h^* = 2$ and $h = 1$.

Thus this is the case (b) of the statement of the theorem.

SITUATION 3. $d_2 = d_2^* = m$. By calculations similar to the above situations, we get that $h = h^* = 2$ and there is only one characteristic exponent $\alpha - q_2 = -\beta + q_2^*$, i.e., $q_2 + q_2^* = \alpha + \beta$.

Thus this is the case (c) of the statement of the theorem.

SITUATION 4. $d_2 < m$ and $d_2^* < m$. We note that

$$\bar{s}_h = \sum_2^h q_j(d_j - 1) \leq \left(\sum_2^h q_j\right)(d_2 - 1) < (\alpha + \beta)\frac{m-1}{2}$$

and similarly

$$\bar{s}^*_{h^*} = \sum_2^{h^*} q_j^*(d_j^* - 1) \leq \left(\sum_2^{h^*} q_j^*\right)(d_2^* - 1) < (\alpha + \beta)\frac{m-1}{2}.$$

So, the basic identity fails and this situation does not arise. □

8. Corollary. *The above theorem shows that the Kaliman family of abstract hyperbolas can be identified by a parametrization* $Y = t^m$, $X = p(t)$ *where* $p(t) = \theta t^d + p_1(t)$ *where* $p_1(t) \in k[t^m, t^{-m}]$ *with* $\mathrm{GCD}(m, d) = 1$.

By a Kaliman family, we are simply referring to the family described by in Proposition 2.9 of [7] (where the X, Y are interchanged), or the family named "(a)" in [5].

Here the exponent d is identified with the above cases as follows.

- In case (a) of the statement of the theorem, we identify $d = \alpha$.
- In case (b) of the statement of the theorem, we identify $d = -\beta$.
- In case (c) of the statement of the theorem, we identify d with the unique characteristic exponent between α and $-\beta$ so that $d = \alpha - q_2 = q_2^* - \beta$.

7. Conclusion

The analysis of the Kaliman family was relatively easy because we had a very precise estimate on $\bar{s}_h$ and $\bar{s}^*_{h^*}$.

An abstract hyperbola with a "good " asymptote is given by a parametrization with $n = m - 1$, i.e.:

$$X = p(t) = t^\alpha + \cdots + \theta t^{-\beta}, \quad Y = t^m + \theta t^{m-1}.$$

Our basic identity now reduces to:

$$(\alpha + \beta)(m - 1) + (1)(\beta - 1) = \bar{s}_h + \bar{s}^*_{h^*}.$$

Unlike the Kaliman family, we can no longer read off the characteristic terms from the exponents in $p(t)$, but must make power series transformations rewriting $q(t) = \tau^{-m}$ or τ^{m-1} to get the characteristic data from the transformed $p(t)$.

While we no longer have $\sum_2^h q_h \leq \alpha + \beta$ we can still estimate it as

$$\sum_2^h q_h \leq \alpha + \beta + 1 + \frac{\alpha + \beta}{m}.$$

This can be improved to $\sum_2^h q_h \leq \alpha + \beta + 1 + \beta/m$ provided we also use restrictions implied by the basic identity.

It is possible to verify all the known families from the Borodzik – Zoladek paper [5] using the basic identity and rule out various degree configurations as impossible using the basic identity. In general, the problem of determining the possible characteristic data subject to the basic identity is under investigation.

It would be interesting to complete a proof of the classification of abstract hyperbolas based on these techniques.

References

1. S. S. Abhyankar, *Inversion and invariance of characteristic pairs*, Amer. J. Math. **89** (1967), 363–372.
2. ———, *Inversion and invariance of characteristic terms*. I, The Legacy of Alladi Ramakrishnan in the Mathematical Sciences (K. Alladi, J. R. Klauder, and C. R. Rao, eds.), Springer, New York, 2010, pp. 93–168.
3. S. S. Abhyankar, W. J. Heinzer, and A. Sathaye, *Translates of polynomials*, A Tribute to C. S. Seshadri (Chennai, 2002) (V. Lakshmibai, V. Balaji, V. B. Mehta, K. R. Nagarajan, K. Paranjape, P. Sankaran, and R. Sridharan, eds.), Trends Math., Birkhäuser, Basel, 2003, pp. 51–124.
4. S. S. Abhyankar and T. T. Moh, *Newton – Puiseux expansion and generalized Tschirnhausen transformation*. I, J. Reine Angew. Math. **260** (1973), 47–83; II, **261** (1973), 29–54.
5. M. Borodzik and H. Zoladek, *Complex algebraic curves. Annuli*, available at `arXiv:0708.1661`.
6. P. Cassou-Noguès, M. Koras, and P. Russell, *Closed embeddings of* $\mathbb{C}^*$ *in* $\mathbb{C}^2$. I, J. Algebra **322** (2009), no. 9, 2950–3002.
7. Sh. Kaliman, *Rational polynomials with a* $\mathbf{C}^*$*-fiber*, Pacific J. Math. **174** (1996), no. 1, 141–194.
8. A. Sathaye and J. Stenerson, *Plane polynomial curves*, Algebraic Geometry and Its Applications (West Lafayette, IN, 1990) (C. L. Bajaj, ed.), Springer, New York, 1994, pp. 121–142.
9. M. G. Zaidenberg and V. Ya. Lin, *An irreducible, simply connected algebraic curve in* $\mathbf{C}^2$ *is equivalent to a quasihomogeneous curve*, Dokl. Akad. Nauk SSSR **271** (1983), no. 5, 1048–1052 (Russian); English transl., Soviet Math. Dokl. **28** (1983), 200–204.

Department of Mathematics, University of Kentucky, Lexington, KY 40506-0027, USA

E-mail address: `sathaye@uky.edu`

Centre de Recherches Mathématiques
CRM Proceedings and Lecture Notes
Volume **54**, 2011

Groups of Russell Type and Tango Structures

Yoshifumi Takeda

Abstract. The group of Russell type is a form of the additive group and the Tango structure is a certain invertible sheaf of locally exact differentials on a curve in positive characteristic. By using the notion of Tango structure, we can construct a group of Russell type over a projective curve, whose completion induces some pathological phenomena in positive characteristic. We consider the locus of curves which have such Tango structures, in the moduli space of curves.

1. Groups of Russell type

Let k be a field of positive characteristic which is not perfect. Take an element a in k such that $a \notin k^p$. Consider a scheme $X_k = \operatorname{Spec} k[x, y]$ subjected to $y^p = x + ax^p$. We know that X_k is a k-group scheme with group structures

$$\begin{aligned} k[x, y] &\to k[x, y] \otimes_k k[x, y] \\ x &\mapsto x \otimes 1 + 1 \otimes x \\ y &\mapsto y \otimes 1 + 1 \otimes y \end{aligned}$$

and so on.

Take an extension field $k' = k(a^{1/p})$ and consider the base extension

$$X_{k'} = X_k \otimes_k k'.$$

We then obtain that $X_{k'}$ is isomorphic to the additive group $\mathbb{G}_{ak'}$ as a k'-group scheme. Indeed, we can rewrite

$$y^p = x + ax^p \quad \text{as } (y - a^{1/p}x)^p = x,$$

so we have $k'[x, y] = k'[t]$, where $t = y - a^{1/p}x$. Namely, we have a diagram

$$\begin{array}{ccc} \mathbb{G}_{ak'} \cong X_{k'} & \longrightarrow & X_k \\ \downarrow & \square & \downarrow \\ k' & \longrightarrow & k, \end{array}$$

which is compatible with the group structures.

2010 *Mathematics Subject Classification.* Primary 14L15; Secondary 14D22, 14H10, 14H51, 14J29, 14J50.

This is the final form of the paper.

The following celebrated theorem was given by Russell in 1970:

Theorem 1.1 ([8]). *Let X be a k-group scheme such that $X \otimes_k K \cong \mathbb{G}_{aK}$ for some field $K \supset k$. We then have that X is isomorphic to a subgroup scheme $\operatorname{Spec} k[x,y]/I$ of $\mathbb{G}_{ak}^2 = \operatorname{Spec} k[x,y]$, where $I = (y^{p^n} - a_0 x - a_1 x^p - \cdots - a_m x^{p^m})$ with $a_0 \neq 0$ and $m, n \in \mathbb{N}$.*

We call such a group scheme a *group of Russell type.*

2. Tango structures

Let C be a nonsingular projective curve of genus $g > 1$ over an algebraically closed field k. For the sake of simplicity, we assume that the characteristic of k is greater than 2, from now on.

Consider the relative Frobenius morphism $F\colon \widetilde{C} \to C$ over k and let $\mathcal{B}$ be the quotient $F_*\mathcal{O}_{\widetilde{C}}/\mathcal{O}_C$. In other words, $\mathcal{B}$ is the direct image of the first coboundary of the de Rham complex $F_*\mathrm{B}^1(\Omega^\bullet_{\widetilde{C}})$. This is a locally free $\mathcal{O}_C$-module of rank $p-1$.

Definition 2.1. We call an invertible subsheaf $\mathcal{L} \subset \mathcal{B}$ a *Tango structure* if $\mathcal{L}^p \cong \Omega_C$.

Now we suppose that we have a Tango structure $\mathcal{L}$ on C. Suppose, furthermore, that there exists an invertible sheaf $\mathcal{N}$ such that $\mathcal{N}^{p-1} \cong \mathcal{L}$. We then have an extension

$$\begin{array}{ccccccccc} 0 & \longrightarrow & \mathcal{O}_C & \longrightarrow & \mathcal{E} & \longrightarrow & \mathcal{N}^{p-1} & \longrightarrow & 0 \\ & & \| & & \cap & & \cap & & \\ 0 & \longrightarrow & \mathcal{O}_C & \longrightarrow & F_*\mathcal{O}_{\widetilde{C}} & \longrightarrow & \mathcal{B} & \longrightarrow & 0. \end{array}$$

Take an affine open covering $\{U_i\}_{i\in I}$, local generators 1 and q_i of $\mathcal{E}$ subjected to

$$q_i = b_{ij}\,1 + d_{ij}^{p-1} q_j,$$

where $\{d_{ij}\}_{i,j\in I}$ are transition functions of $\mathcal{N}$ and where

$$\left\{\begin{pmatrix} 1 & b_{ij} \\ 0 & d_{ij}^{p-1} \end{pmatrix}\right\}_{i,j\in I}$$

are transition matrices of $\mathcal{E}$.

Set $a_i = q_i^p$. We know that $a_i = b_{ij}^p + d_{ij}^{p(p-1)} a_j$ and a_i is a local section of $\mathcal{O}_C$. Consider a group of Russell type

$$X_i = \operatorname{Spec} \mathcal{O}_C(U_i)[x_i, y_i]/(y_i^p - x_i - a_i x_i^p)$$

over each U_i.

Let us glue the X_i's together under the relations

$$x_i = d_{ij}^{-p} x_j, \quad y_i = b_{ij} d_{ij}^{-p} x_j + d_{ij}^{-1} y_j.$$

Note that we have

$$\begin{aligned} y_i^p - x_i - a_i x_i^p &= (b_{ij} d_{ij}^{-p} x_j + d_{ij}^{-1} y_j)^p - d_{ij}^{-p} x_j - (b_{ij}^p + d_{ij}^{p(p-1)} a_j) d_{ij}^{-p^2} x_j^p \\ &= b_{ij}^p d_{ij}^{-p^2} x_j^p + d_{ij}^{-p} y_j^p - d_{ij}^{-p} x_j - b_{ij}^p d_{ij}^{-p^2} x_j^p - d_{ij}^{-p} a_j x_j^p \\ &= d_{ij}^{-p}(y_j^p - x_j - a_j x_j^p), \end{aligned}$$

which is compatible with the group structures. Thus we obtain a *group of Russell type X over C.*

Indeed, over $\widetilde{C}$, we can rewrite

$$y_i^p = x_i + a_i x_i^p \quad \text{as } (y_i - q_i x_i)^p = x_i.$$

Set $t_i = y_i - q_i x_i$. We then obtain $t_i = d_{ij}^{-1} t_j$ and

$$X \times_C \widetilde{C}|_{U_i} \cong \operatorname{Spec} \mathcal{O}_{\widetilde{C}}(U_i)[t_i].$$

Hence, by considering the vector bundle

$$\mathbb{S}\mathrm{pec}\, \mathcal{S}ymm(F^*\mathcal{N}^{-1}) \to \widetilde{C}$$

without scalar-multiplication, we have

$$\begin{array}{ccccc}
\mathbb{S}\mathrm{pec}\, \mathcal{S}ymm(F^*\mathcal{N}^{-1}) \cong & X \times_C \widetilde{C} & \longrightarrow & X \\
\searrow & \downarrow & \square & \downarrow \\
& \widetilde{C} & \xrightarrow[F]{} & C.
\end{array}$$

For more generalized discussions on groups of Russell type over a curve, see [13].

3. Completion of X

Consider the vector bundle

$$\mathbb{S}\mathrm{pec}\, \mathcal{S}ymm(\mathcal{N}^{-p}) \to C.$$

We have a diagram

$$\begin{array}{ccccc}
\mathbb{S}\mathrm{pec}\, \mathcal{S}ymm(F^*\mathcal{N}^{-1}) & \longrightarrow & X & \longrightarrow & \mathbb{S}\mathrm{pec}\, \mathcal{S}ymm(\mathcal{N}^{-p}) \\
\downarrow & \square & \downarrow & \swarrow & \\
\widetilde{C} & \longrightarrow & C. & &
\end{array}$$

The composite of the two top morphisms is the relative Frobenius morphism over k. Consider, furthermore, the $\mathbb{P}^1$-bundles

$$\mathbb{P}\mathrm{roj}\, \mathcal{S}ymm(\mathcal{N}^{-p} \oplus \mathcal{O}_C) \to C,$$
$$\mathbb{P}\mathrm{roj}\, \mathcal{S}ymm(F^*\mathcal{N}^{-1} \oplus \mathcal{O}_{\widetilde{C}}) \to \widetilde{C},$$

which are completions of $\mathbb{S}\mathrm{pec}\, \mathcal{S}ymm(\mathcal{N}^{-p}) \to C$ and of $\mathbb{S}\mathrm{pec}\, \mathcal{S}ymm(F^*\mathcal{N}^{-1}) \to \widetilde{C}$, respectively.

Take the normalization Y of $\mathbb{P}\mathrm{roj}\, \mathcal{S}ymm(\mathcal{N}^{-p} \oplus \mathcal{O}_C)$ in $k(X)$. We then know that Y is a completion of X and obtain a diagram

$$\begin{array}{ccccc}
\mathbb{P}\mathrm{roj}\, \mathcal{S}ymm(F^*\mathcal{N}^{-1} \oplus \mathcal{O}_{\widetilde{C}}) & \longrightarrow & Y & \longrightarrow & \mathbb{P}\mathrm{roj}\, \mathcal{S}ymm(\mathcal{N}^{-p} \oplus \mathcal{O}_C) \\
\cup & & \cup & & \cup \\
\mathbb{S}\mathrm{pec}\, \mathcal{S}ymm(F^*\mathcal{N}^{-1}) & \longrightarrow & X & \longrightarrow & \mathbb{S}\mathrm{pec}\, \mathcal{S}ymm(\mathcal{N}^{-p}) \\
\downarrow & \square & \downarrow & \swarrow & \\
\widetilde{C} & \longrightarrow & C. & &
\end{array}$$

The composite of the two top morphisms is the relative Frobenius morphism over k.

Here the following theorem, originally given by Raynaud [7], holds:

Theorem 3.1. *In the same notation and under the same assumption as above, we have that the completion Y is a minimal nonsingular projective surface having a fibration such that each fibre is a rational curve with one cusp of type $u^p + v^{p-1} = 0$. Moreover,*

(1) *If $p = 3$, then Y is a quasi-elliptic surface of $\kappa = 1$.*
(2) *If $p > 3$, then Y is a surface of general type.*

In any case, Y gives a counter-example to the Kodaira vanishing theorem in positive characteristic.

For the proof and more detailed discussions on the completion Y, see [11, 13].

Needless to say, Y cannot be lifted over the ring of Witt-vectors of length two(see for example [1]). In our case, furthermore, it holds

Theorem 3.2 (Russell, Ganong, Kurke, W. Lang, Mukai, ...). *Retain the same notation and assumption as above, we have*

$$H^0(Y, \Theta_Y) = H^0(C, \mathcal{N}),$$

where Θ_Y is the tangent sheaf of Y.

For the proof, see [2–4, 6, 9]. See also [12].

These theorems imply that, if $H^0(C, \mathcal{N}) \neq 0$ and $p > 3$, then Y is a surface *of general type with nontrivial vector fields.* On the other hand, it is well-known that the automorphism groups of varieties of general type are finite(see for example [5]). Hence we know that the automorphism group scheme of Y is not reduced.

Example 3.3. Consider the plane curve defined by the equation

$$C : y^{p(p-1)} - y = x^{p(p-1)-1}$$

in $p > 3$. We then have a Tango structure

$$\mathcal{L} \cong \mathcal{O}_C((p-1)(p^2 - p - 3)P_\infty)$$

arising from the exact form dx whose divisor

$$(dx) = p(p-1)(p^2 - p - 3)P_\infty,$$

where P_∞ is the point at infinity. Let $\mathcal{N} = \mathcal{O}_C((p^2 - p - 3)P_\infty)$. We then obtain

$$\mathcal{N}^{p-1} \cong \mathcal{L}, \quad H^0(C, \mathcal{N}) \neq 0.$$

Hence we get a nonsingular projective surface of general type with nontrivial vector fields.

Thus we obtain a really pathological surface in positive characteristic. In the circumstances, the following problems seem interesting:

Are such surfaces really exceptional ones?
Do they form discrete points in the moduli space?

It is, however, likely difficult to solve these problem straightforward, because of the difficulty of the moduli space of surfaces. So, for the first step, let us consider the following problems:

Are the curves having such Tango structures really exceptional ones?
Do they form discrete points in the moduli space?

4. Main theorem

In this last section, we shall give a negative answer to the latter problems. Consider the following locus in the moduli space $\mathcal{M}_g$ of curves of genus $g > 1$ such that $p(p-1) \mid 2g-2$ with $p > 3$:

$$\mathcal{T}_g = \{C \in \mathcal{M}_g \mid C \text{ has a Tango structure } \mathcal{N}^{p-1} \text{ with } H^0(C, \mathcal{N}) \neq 0\}.$$

The main theorem is the following:

Theorem 4.1 (cf. [14]). *The locus $\mathcal{T}_g$ contains a variety of* $\dim \geq g-1$ *provided* $p \equiv 3 \pmod 4$.

PROOF. Consider the subvariety

$$\mathcal{H}_g = \{C \in \mathcal{M}_g \mid C \text{ is a hyperelliptic curve}\}.$$

We have a rational mapping

$$\begin{array}{ccc} \mathbb{A}^{2g-1} & \longrightarrow & \mathcal{H}_g \\ \Cup & & \Cup \\ (a_1, \dots, a_{2g-1}) & \longmapsto & y^2 = x(x-1)(x-a_1)\cdots(x-a_{2g-1}). \end{array}$$

Let C be an image, that is a hyperelliptic curve, and let

$$\begin{pmatrix} c_{11} & \cdots & c_{1g} \\ \vdots & & \vdots \\ c_{g1} & \cdots & c_{gg} \end{pmatrix}$$

be the Hasse–Witt matrix of C with respect to

$$\left\{ \frac{dx}{y}, \dots, \frac{x^{g-1}\,dx}{y} \right\}.$$

Note that each entry is a polynomial in $a_1, \dots, a_{2g-1}$ (see for example [17]). Consider the subvariety S of $\mathbb{A}^{2g-1}$ defined by $c_{11} = \cdots = c_{g1} = 0$, i.e., the first column is zero. We then know that $\dim S \geq g-1$ if S is not empty. Let T be the image of S in $\mathcal{H}_g$.

$$\begin{array}{ccc} \mathbb{A}^{2g-1} & \longrightarrow & \mathcal{H}_g \\ \cup & & \cup \\ S & \longrightarrow & T. \end{array}$$

Since the mapping $\mathbb{A}^{2g-1} \to \mathcal{H}_g$ is induced from an action of a certain finite group, we also know that $\dim T \geq g-1$ if T is not empty. Moreover, we have that

$$\mathcal{C}\left(\frac{dx}{y}\right) = 0$$

on every curve in T, where $\mathcal{C}$ is the Cartier operator(see for example [10]). In other words, the form dx/y is exact.

On the other hand, dx/y induces the divisor

$$\left(\frac{dx}{y}\right) = (2g-2)P_\infty$$

and $2g-2 = np(p-1)$ with $n \in \mathbb{N}$. Therefore, from this exact form, we obtain a Tango structure

$$\mathcal{O}_C(nP_\infty)^{p-1} \hookrightarrow \mathcal{B} \quad \text{with } H^0(\mathcal{O}_C(nP_\infty)) \neq 0.$$

So we know that T is contained in the locus $\mathcal{T}_g$.

Now we have only to verify that T is not empty. Consider the hyperelliptic curve defined by

$$y^2 = x^d - x$$

with $d = 2g+1$, $2g-2 = np(p-1)$, $n \in \mathbb{N}$. We can then compute as follows:

$$y\,\mathcal{C}\left(\frac{dx}{y}\right) = \mathcal{C}(y^{p-1}dx) = \mathcal{C}\big((x^d - x)^{(p-1)/2}dx\big).$$

In $(x^d - x)^{(p-1)/2} = x^{(p-1)/2}(x^{d-1}-1)^{(p-1)/2}$, only terms

$$x^{(p-1)/2}, \quad x^{(p-1)/2+d-1}, \quad x^{(p-1)/2+2(d-1)}, \quad \dots, \quad x^{d(p-1)/2}$$

appear. More precisely, since $d = 3 + np(p-1)$, we know that only terms

$$x^{(p-1)/2}, \quad x^{(p-1)/2+2+np(p-1)}, \quad x^{(p-1)/2+2\cdot 2+2np(p-1)}, \quad \dots,$$
$$x^{(p-1)/2+(p-1)/2\cdot 2+(p-1)/2\cdot np(p-1)}$$

appear. Note that

$$\frac{p-1}{2} < p-1, \quad p < \frac{p-1}{2} + \frac{p-1}{2}\cdot 2 < 2p-1.$$

We can verify that $\mathcal{C}(dx/y) \neq 0$ if and only if there exists an integer l such that

$$\frac{p-1}{2} + 2l = p-1 \quad \text{with } 1 \leq l \leq \frac{p-1}{2},$$

i.e., $4l = p-1$. In other words,

$$\mathcal{C}\left(\frac{dx}{y}\right) \neq 0 \text{ if and only if } 4 \mid p-1.$$

Otherwise dx/y is exact and so we know that our curve is lying in T. That is the required assertion. □

Remark 4.2. By applying Tsuda's method ([16]), we get a slightly better estimate. Retain the same notation and assumption as in the previous theorem. Consider a hyperelliptic curve defined by

$$y^2 = x(x-1)(x-a_1)\cdots(x-a_{2g-1})$$

and the differential form dx/y. We can compute as follows:

$$y\,\mathcal{C}\left(\frac{dx}{y}\right) = \mathcal{C}(y^{p-1}dx) = \mathcal{C}\Big(\big(x(x-1)(x-a_1)\cdots(x-a_{2g-1})\big)^{(p-1)/2}dx\Big).$$

Let

$$(x(x-1)(x-a_1)\cdots(x-a_{2g-1}))^{(p-1)/2}$$
$$= b_{(p-1)/2}x^{(p-1)/2} + b_{(p-1)/2+1}\,x^{(p-1)/2+1} + \cdots + b_{d(p-1)/2-1}x^{d(p-1)/2-1} + x^{d(p-1)/2}.$$

Note that each b_i is a polynomial in $a_1, \dots, a_{2g-1}$. Since $d = 2g+1 = np(p-1)+3$, we know

$$d\frac{p-1}{2} = \frac{n(p-1)^2}{2}p + 3\frac{p-1}{2}.$$

It follows that we have $\mathcal{C}(dx/y) = 0$ if the coefficients

$$b_{p-1}, \quad b_{2p-1}, \quad \dots, \quad b_{(n(p-1)^2/2+1)p-1}$$

are zero. Consider the subvariety of $\mathbb{A}^{2g-1}$ defined by

$$b_{p-1} = b_{2p-1} = \cdots = b_{(n(p-1)^2/2+1)p-1} = 0$$

and let U be its image in $\mathcal{H}_g$. We then know that U is contained in $\mathcal{T}_g$. Moreover, the number of the above-mentioned coefficients is $n(p-1)^2/2+1$, which is less than $g = np(p-1)/2+1$.

On the other hand, the hyperelliptic curve defined by $y^2 = x^d - x$, mentioned in the previous proof, is lying in U provided $4 \nmid p-1$. Hence we obtain

$$\dim U \geq 2g - \frac{n(p-1)^2}{2} > g - 1.$$

Acknowledgements. The author expresses his sincere gratitude to the organizers of the conference for their invitation.

References

1. P. Deligne and L. Illusie, *Relèvements modulo p^2 et décomposition du complexe de de Rham*, Invent. Math. **89** (1987), no. 2, 247–270.
2. R. Ganong and P. Russell, *The tangent bundle of a ruled surface*, Math. Ann. **271** (1985), no. 4, 527–548.
3. H. Kurke, *Example of false ruled surfaces*, Proceedings of Symposium on Algebraic Geometry (Kinosaki, 1981), pp. 203–223.
4. W. E. Lang, *Examples of surfaces of general type with vector fields*, Arithmetic and Geometry. Vol. II (M. Artin and J. Tate, eds.), Progr. Math., vol. 36, Birkhäuser, Boston, MA, 1983, pp. 167–173.
5. H. Matsumura, *On algebraic groups of birational transformations*, Atti Accad. Naz. Lincei Rend. Cl. Sci. Fis. Mat. Natur. (8) **34** (1963), 151–155.
6. S. Mukai, *On counter-examples to the Kodaira vanishing theorem and the Yau inequality in positive characteristic*, Proceedings of Symposium on Algebraic Geometry (Kinosaki, 1979), pp. 9–31 (Japanese).
7. M. Raynaud, *Contre-exemple au "vanishing theorem" en caractéristique $p > 0$*, C. P. Ramanujam – A Tribute (K. G. Ramanathan, ed.), Tata Inst. Fund. Res. Studies in Math., vol. 8, Springer, Berlin, 1978, pp. 273–278.
8. P. Russell, *Forms of the affine line and its additive group*, Pacific J. Math. **32** (1970), 527–539.
9. ———, *Factoring the Frobenius morphism of an algebraic surface*, Algebraic Geometry, Bucharest 1982 (Bucharest, 1982) (L. Bădescu and D. Popescu, eds.), Lecture Notes in Math., vol. 1056, Springer, Berlin, 1984, pp. 366–380.
10. J.-P. Serre, *Sur la topologie des variétés algébriques en caractéristique p*, Symposium internacional de topologí a algebraica/International symposium on algebraic topology, Universidad Nacional Autónoma de México and UNESCO, Mexico City, 1958, pp. 24–53.
11. Y. Takeda, *Fibrations with moving cuspidal singularities*, Nagoya Math. J. **122** (1991), 161–179.
12. ———, *Vector fields and differential forms on generalized Raynaud surfaces*, Tohoku Math. J. (2) **44** (1992), no. 3, 359–364.
13. ———, *Groups of Russell type over a curve*, J. Pure Appl. Algebra **128** (1998), no. 1, 93–108. Corrigendum, **148**, (2000), 317–318.
14. Y. Takeda and K. Yokogawa, *Pre-Tango structures on curves*, Tohoku Math. J. (2) **54** (2002), no. 2, 227–237. Errata and addenda, **55**, (2003), 611–614.
15. H. Tango, *On the behavior of extensions of vector bundles under the Frobenius map*, Nagoya Math. J. **48** (1972), 73–89.
16. N. Tsuda, *Pre-Tango structures on hyperelliptic cures*, Master's Thesis, Osaka University, 2003 (Japanese).
17. N. Yui, *On the Jacobian varieties of hyperelliptic curves over fields of characteristic $p > 2$*, J. Algebra **52** (1978), no. 2, 378–410.

Department of Mathematics and Statistics, Wakayama Medical University, Wakayama City 6418509, Japan
E-mail address: ytakeda@wakayama-med.ac.jp